Structural and Resonance Techniques in Biological Research

Physical Techniques in Biology and Medicine

Edited by

DENIS L. ROUSSEAU
AT&T Bell Laboratories
Murray Hill, New Jersey

WILLIAM L. NASTUK
Columbia University
New York, New York

Denis L. Rousseau (ed.), Structural and Resonance Techniques in Biological Research
Denis L. Rousseau (ed.), Optical Techniques in Biological Research

Structural and Resonance Techniques in Biological Research

Edited by

DENIS L. ROUSSEAU

Chemical Physics Research Laboratory
AT&T Bell Laboratories
Murray Hill, New Jersey

1984

ACADEMIC PRESS, INC.
(Harcourt Brace Jovanovich, Publishers)
Orlando San Diego New York London
Toronto Montreal Sydney Tokyo

QH
505
.S74
1984

COPYRIGHT © 1984 BELL TELEPHONE LABORATORIES, INCORPORATED.
ALL RIGHTS RESERVED.
NO PART OF THIS PUBLICATION MAY BE REPRODUCED OR
TRANSMITTED IN ANY FORM OR BY ANY MEANS, ELECTRONIC
OR MECHANICAL, INCLUDING PHOTOCOPY, RECORDING, OR ANY
INFORMATION STORAGE AND RETRIEVAL SYSTEM, WITHOUT
PERMISSION IN WRITING FROM THE PUBLISHER.

ACADEMIC PRESS, INC.
Orlando, Florida 32887

United Kingdom Edition published by
ACADEMIC PRESS, INC. (LONDON) LTD.
24/28 Oval Road, London NW1 7DX

Library of Congress Cataloging in Publication Data

Main entry under title:

Structural and resonance techniques in biological research.

(Physical techniques in biology and medicine series)
Includes bibliographies and index.
Contents: Biophysical measurements using nuclear
magnetic resonance / Truman R. Brown and Kâmil
Uğurbil -- Electron spin resonance / Daniel J. Kosman --
Mössbauer spectroscopy / D.P.E. Dickson and C.E. Johnson
-- [etc.]
 1. Biophysics--Addresses, essays, lectures.
I. Rousseau, Denis L.   II. Series.
QH505.S74 1984   574'.028   84-9372
ISBN 0-12-599320-X (alk. paper)

#1075 2015

PRINTED IN THE UNITED STATES OF AMERICA

84 85 86 87   9 8 7 6 5 4 3 2 1

Contents

List of Contributors   vii
Series Introduction   ix
Preface   xi

1. Nuclear Magnetic Resonance
Truman R. Brown and Kâmil Uğurbil

I.	Introduction	2
II.	Introduction to Nuclear Magnetic Resonance	2
III.	Theory	6
IV.	Experimental Considerations	31
V.	Applications	42
VI.	Concluding Remarks	79
	Appendix	80
	References	84

2. Electron Spin Resonance
Daniel J. Kosman

I.	Introduction	90
II.	Principles of Electron Spin Resonance	90
III.	Establishing and Detecting Electron Spin Resonance	102
IV.	Applications of Direct (Continuous-Wave) ESR	109
V.	Resonance as a Perturbation	183
VI.	An ESR Perspective	229
	Appendix A. ESR Hardware, Glassware, and Supplies	230
	Appendix B. ESR Software	231
	Appendix C. Calibration of ESR Spectra	233
	Bibliography	235
	References	238

3. Mössbauer Spectroscopy
D. P. E. Dickson and C. E. Johnson

I.	Introduction	246
II.	Principles of Mössbauer Spectroscopy	249

III.	Experimental Techniques	257
IV.	Applications to Isolated Biomolecules	265
V.	Applications to Molecular Systems	274
VI.	Applications to Tissue	282
VII.	Applications to Whole or Part Organisms: Uptake and Metabolism	284
VIII.	Measurement of Vibration and Movement	288
IX.	Conclusions	290
	References	291

4. X-Ray Absorption Spectroscopy
Robert A. Scott

I.	Introduction	295
II.	Theory	297
III.	Experimental Considerations	308
IV.	EXAFS as a Structural Technique in Biology	344
V.	Conclusion	358
	References	358

5. Macromolecular Crystallography
Keith Moffat

I.	Introduction	364
II.	Crystallization of Macromolecules	367
III.	X-Ray Diffraction from Crystals	374
IV.	Phase Determination	391
V.	Heavy-Atom Derivatives	408
VI.	The Electron-Density Map	414
VII.	New Directions	423
	Appendix A. The Basic Mathematics of Crystallography	428
	Appendix B. A Specific Strategy for Protein Crystallization	430
	Appendix C. A Crystallographic Paper Primer	431
	Appendix D. Suggestions for Further Reading	432
	References	433

6. Small-Angle X-Ray Scattering and Diffraction
J. Stamatoff

I.	Introduction	437
II.	X-Ray Scattering and Diffraction	438
III.	X-Ray Technology	447
IV.	Biological Applications	451
V.	Summary	467
	References	468

Index 469

List of Contributors

Numbers in parentheses indicate the pages on which the authors' contributions begin.

TRUMAN R. BROWN[1] (1), Molecular Biophysics Research Department, AT&T Bell Laboratories, Murray Hill, New Jersey 07974

D. P. E. DICKSON (245), Department of Physics, University of Liverpool, Liverpool, England

C. E. JOHNSON (245), Department of Physics, University of Liverpool, Liverpool, England

DANIEL J. KOSMAN (89), Department of Biochemistry, School of Medicine, State University of New York at Buffalo, Buffalo, New York 14214

KEITH MOFFAT (363), Section of Biochemistry, Molecular and Cell Biology, Cornell University, Ithaca, New York 14853

ROBERT A. SCOTT (295), School of Chemical Sciences, University of Illinois, Urbana, Illinois 61801

J. STAMATOFF (437), Celanese Research Company, Summit, New Jersey 07901

KÂMIL UĞURBIL (1), Department of Biochemistry and Gray Freshwater Biological Institute, College of Biological Sciences, University of Minnesota, Navarre, Minnesota 55392

[1] PRESENT ADDRESS: Department of Radiology, Fox Chase Cancer Center, Philadelphia, Pennsylvania 19111.

Series Introduction

With the appearance of "Structural and Resonance Techniques in Biological Research" and "Optical Techniques in Biological Research," both edited by one of us (D. L. R.), we mark the introduction of a new series of volumes, Physical Techniques in Biology and Medicine. This series is intended to replace a previous treatise, Physical Techniques in Biological Research, as many of the diverse physical methods used to address biological systems were in their infancy when the original works were published. For those techniques that had attained relative maturity, the earlier contribution has been cited in the present chapters, which provide an updated view and discuss contemporary developments.

The subject matter we hope to include in this series will form volumes of a topical nature. The biophysical sciences are populated by physicists who have become interested in living systems and by biologists who find they have a need for a particular physical technique in which they may have had no formal training. It is this duality of audience we hope to reach, making every attempt to ensure that each chapter is sufficiently methods-oriented and illustrative of a range of applications that the reader will be provided with an adequate entrée to the desired technique. We shall also stress the importance of a comprehensive bibliography to permit ready access to the literature.

As technology continues to advance our instrumentational and computational capabilities, we look forward to a plethora of both novel and renewed topics to be covered by volumes in this series.

Denis L. Rousseau
William L. Nastuk

Preface

The first two books in this series are composed of chapters loosely organized into a volume on structural and resonance techniques and one on optical techniques. Included in this volume are discussions of nuclear magnetic resonance, electron spin resonance, Mössbauer spectroscopy, x-ray absorption spectroscopy, macromolecular crystallography, and small-angle x-ray scattering and diffraction.

The chapters in this volume are aimed at a level such that only a general understanding of chemistry and biology is required. The objective, which I believe has been largely achieved, is to present material in a way that allows the research worker to assess quickly the applicability, utility, and significance of the specific technique to his or her problem or field of interest. With these guidelines the authors have written chapters in which the diversity of the particular technique and its application to different types of problems have been stressed, occasionally at the expense of timeliness and excessive detail. Similarly, these chapters are not intended to be exhaustive reviews of all the literature, but instead include examples which most appropriately illustrate the application of the technique. Extensive derivations of the basic principles underlying the physics of each technique have been kept to a minimum and do not form the central theme of any chapter. However, the underpinning for each technique is outlined, and sufficient references have been provided for the reader who wants a more in-depth understanding.

Finally, I wish to thank all the authors for being willing and able to take time away from their other responsibilities in order to write these pedagogical chapters. I am especially thankful to those who submitted their manuscripts on time. I hope that they are sufficiently pleased with the final product that they are able to excuse the delays that went into the completion of these volumes.

Denis L. Rousseau

Structural and Resonance Techniques in Biological Research

1
Nuclear Magnetic Resonance

TRUMAN R. BROWN

Department of Radiology
Fox Chase Cancer Center
Philadelphia, Pennsylvania

KÂMIL UĞURBIL

Department of Biochemistry and
Gray Freshwater Biological Institute
University of Minnesota
Navarre, Minnesota

I.	Introduction	2
II..	Introduction to Nuclear Magnetic Resonance	2
III.	Theory	6
	A. Zeeman Energy	6
	B. Boltzmann Distribution and Bulk Magnetization	9
	C. Bloch Equations	10
	D. Chemical Shift	15
	E. Spin–Spin Interactions	19
	F. Relaxation Times and the Nuclear Overhauser Effect	23
	G. Chemical Exchange	29
IV.	Experimental Considerations	31
	A. The Nuclear Magnetic Resonance Spectrometer	31
	B. Measurement Techniques	36
V.	Applications	42
	A. Structure	42
	B. Dynamics	53
	C. Studies with Quadrupolar Nuclei	56
	D. Kinetics	57
	E. Cellular and Metabolic Studies	67
VI.	Concluding Remarks	79
	Appendix A	80
	References	84

I. Introduction

The phenomenon of nuclear magnetic resonance (NMR) was discovered in 1946 (Block et al., 1946; Purcell et al., 1946). At first, the discovery was exploited primarily by physicists interested in understanding the structure of the nucleus. However, the discovery in 1950 that the exact resonance frequency depends on the details of the molecular environment (Proctor and Yu, 1950) led chemists to exploit the technique for probing the structure of molecules. As NMR techniques increased in sensitivity and sophistication, biologists and biochemists started to employ this form of spectroscopy as a research tool. The range of biological applications in which NMR has proven to be useful is too wide to cover adequately in a single chapter. Necessarily, we have restricted our discussion and, in doing so, have tried to emphasize the broad areas of application such as structure, dynamics, kinetics, and cellular metabolism. There are obviously numerous other specific applications, some of which are summarized in Section VI.

We would like to emphasize that this chapter is not meant to provide a review of the NMR work conducted in these areas; its aim is to help investigators decide whether it is worth the effort to learn the details of a complicated technique to aid their research. Therefore, the discussion is at an introductory level and is meant to provide only an intuitive picture. See Abragam (1961) or Slichter (1978) for a fuller treatment.

II. Introduction to Nuclear Magnetic Resonance

This section is intended to provide a qualitative understanding of the overall physical processes observed and the type of information provided in NMR spectroscopy. The ideas are presented heuristically and will be repeated later with greater detail and rigor.

The phenomenon of nuclear magnetic resonance arises as a result of interactions between magnetic fields and those atomic nuclei which possess magnetic properties. In the presence of an external field, such nuclei behave like microscopic bar magnets and experience forces which tend to align them parallel to the external field. As a result of these forces, the energy levels of these nuclei change and acquire new values which depend on their orientation. Unlike the bar magnets, however, these nuclear magnets must behave according to the laws of quantum mechanics. Consequently, they can acquire only a limited number of discreet orientations and energies. For example, in the case of "spin-$\frac{1}{2}$" nuclei, such as ^1H, ^{13}C, and ^{31}P, the nuclear energy levels split in two in the presence of the magnetic field; each level corresponds to one of the two possible orientations these nuclei can have relative to the direction of the field. Transitions between these two energy states can be induced by a secondary magnetic field oscillating in time with the appropriate frequency. These transitions are detected and presented as absorption peaks in an NMR spectrum (Fig. 1). In NMR spectroscopy one is generally interested in the values of four parameters which are directly or indirectly obtainable from the NMR peaks. These are the resonance frequency, the area of the NMR peak, and the spin relaxation times T_1 and T_2. These parameters in turn yield a wealth of information about the atoms and the molecules being studied.

The resonance frequency of a magnetic nucleus is determined primarily by three factors; the properties intrinsic to the magnetic nucleus, the electronic environment of the nucleus within the atom or molecule, and the magnitude of the external field. The first is the reason why, at a constant magnetic field, NMR spectroscopy of different nuclei are performed at very different frequencies (Table I). This is a much larger effect than the one induced by the electronic environment (Fig. 2); however, it is the latter which provides chemical information and which renders NMR spectroscopy so useful.

The peak area is simply the integral of the absorption peak. It is proportional to the total number of spins contained in the sample giving rise to the observed peak. Combined with the sensitivity of NMR to chemical environments, measurement of peak areas tells us how much of a certain chemical species our sample contains.

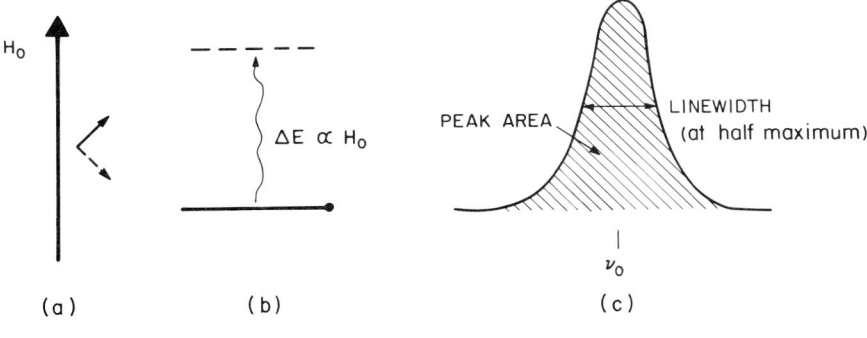

Fig. 1. Schematic representation of spin orientations (a) and energy levels (b) in the presence of a magnetic field, and the resultant absorption peak (c) that would be detected as a result of transitions between these levels. The peak-area linewidth and frequency are the experimental parameters determined by such a measurement.

Table I Magnetic Properties of Some Nuclei

Nucleus	Spin	Natural abundance (%)	Resonance frequency at 70.46 kG (MHz)
^{1}H	1/2	99.99	300.00
^{2}H	1	0.015	46.05
^{13}C	1/2	1.1	75.43
^{14}N	1	99.63	21.67
^{15}N	1/2	0.37	30.40
^{17}O	5/2	0.037	40.67
^{19}F	1/2	1.00	282.23
^{23}Na	3/2	100	79.35
^{25}Mg	5/2	10.13	18.36
^{31}P	1/2	100	121.44
^{35}Cl	3/2	75.53	29.40
^{37}Cl	3/2	24.47	24.47
^{39}K	3/2	93.1	14.00
^{41}K	3/2	6.9	7.68
^{43}Ca	7/2	0.15	20.18
^{111}Cd	1/2	12.75	63.62
^{113}Cd	1/2	12.26	66.55
^{195}Pt	1/2	33.8	64.50
^{199}Hg	1/2	16.84	53.48
^{205}Tl	1/2	70.5	173.12

Spin-relaxation parameters T_1 and T_2 are generally not directly measurable from the NMR peaks and require execution of other experiments, which are discussed in Section IV.B. Qualitatively, the nature of T_1 and techniques for its measurements can be understood if one considers the fact that the NMR sample is not a single nucleus but consists of many such nuclei. These nuclei distribute themselves between the allowed energy states; ultimately, the distribution is determined only by the sample temperature and the energy of the allowed states. When this time-independent distribution is reached, the spins are said to be in thermal equilibrium. The rate constant for the approach to thermal equilibrium is T_1^{-1}.

At thermal equilibrium, there is a net difference between the spin populations of the allowed nuclear-energy states, with more of the spins residing in the lower energy state. This difference is the source of the NMR absorption peaks. The distribution of spins between the energy states can be perturbed by the application of radio frequency (rf) radiation of appropriate power and frequency. For example, it is possible to invert the spin populations (Fig. 3). If an NMR spectrum is taken immediately after such a perturbing pulse, one observes a negative or emission peak as shown in the lower part of Fig.

3. This new population state, however, is unstable. The spins must return to the distribution dictated by the temperature. They do so at a rate determined by the rate constant T_1^{-1}. Thus, one can measure T_1 by perturbing the spin populations and monitoring the relaxation back to the stable state.

T_2 is connected with an analogous but somewhat more complicated relaxation which is explained in Section III.F. In the event the laboratory magnetic field is perfectly homogeneous over the sample, T_2 is proportional to the inverse of the resonance linewidth. If, however, the external field is not sufficiently homogeneous, the observed linewidths contain contributions from this inhomogeneity as well as the relaxation process associated with T_2.

Both T_1 and T_2 are sensitive to molecular motion. This dependence is generally well understood and is used to extract detailed information on molecular dynamics. The most apparent manifestation of this sensitivity to motion is the increase in observed resonance linewidths with increasing molecular weight. Small molecules, such as those generally encountered in organic chemical applications, execute rapid random rotations in solution and yield NMR spectra with very narrow linewidths which are typically less than 1 Hz for spin-½ nuclei. NMR peaks from biological macromolecules,

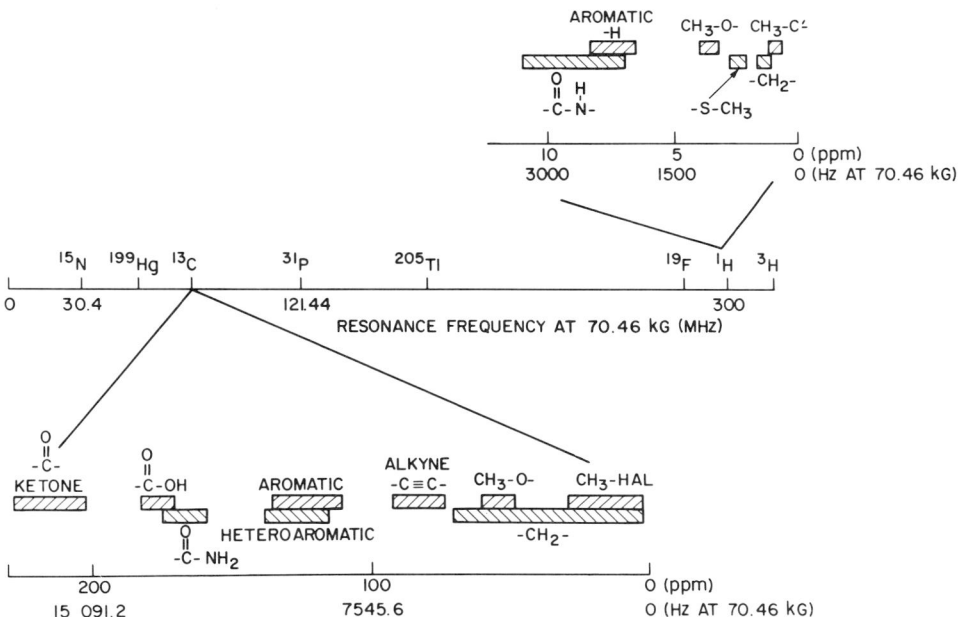

Fig. 2. Nuclear resonance frequencies of several magnetic nuclei at 70.46 kG and the chemical shift range ^{13}C and ^{1}H nuclei.

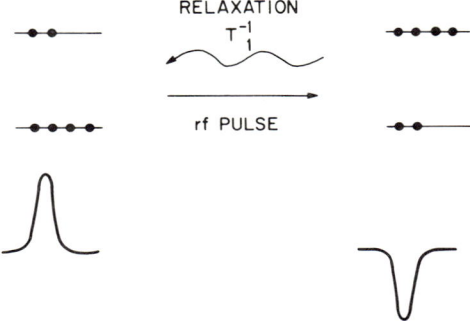

Fig. 3. Spin populations at thermal equilibrium (left) and after they are inverted by a pulse of oscillating magnetic field (right).

on the other hand, are much broader due to the slower rotation of these larger molecules. For example, proton resonances from lysozyme can have linewidths greater than 5 Hz. In cases where the T_1 and the T_2 are dominated by magnetic interaction among the various spins in a molecule, they are a function of the internuclear distances as well as the molecular motion. This distance dependence is the source of most of the structural studies with NMR spectroscopy.

In the following sections, we will discuss in greater detail the physical basis of these parameters and the information that can be extracted from them. As we have done in this section, we will limit our examples, except when stated otherwise, to spin-$\frac{1}{2}$ nuclei, which constitute the simplest and generally most useful set of nuclear spins.

III. Theory

A. Zeeman Energy

In the presence of a magnetic field, magnetic dipoles experience a force which tends to align them along the direction of the field. Consequently, the dipole acquires a potential energy which depends on its orientation relative to the field. This dependence is expressed by the equation

$$E = -\mu \cdot \mathbf{H} = -H\mu \cos \theta, \tag{1}$$

where μ and H represents the magnitudes of the magnetic dipole moment and the magnetic field, respectively, and θ is the angle between the direction of the magnetic field and the direction of μ. By convention, the direction of H is taken as the z axis and $\mu \cos \theta = \mu_z$, the projection of μ along the direction of H. Thus, the energy of the dipole varies between $-\mu H$ and $+\mu H$, and the projections of μ along H which correspond to the minimum and maximum energy are those for which θ is equal to 0 and 180 degrees, respectively. Note that the azimuthal orientation with respect to the magnetic field has no effect on the energy.

A macroscopic magnetic dipole, such as a circular loop of wire carrying a current, can assume all possible orientations relative to a magnetic field. Therefore, its energy can vary continuously between $-\mu H$ and $+\mu H$. For electrons and atomic nuclei, however, quantum mechanics dictates that the angular momentum and the dipole moment can have only a limited number of discreet projections along a particular direction. The angle θ and therefore the energy acquired by a magnetic nucleus in the presence of the magnetic field become discrete as well. These different projections correspond to the different energy levels mentioned in the introduction.

The number of possible different projections for a particular kind of particle is fixed and determined by its spin quantum number I, which characterizes both the spin angular momentum and magnetic moment of the particle. For any particular nucleus, I is an invariant quantity and must be an integer or half integer. For spin-$\frac{1}{2}$ nuclei such as ^{1}H, ^{13}C, and ^{31}P, I is equal to $\frac{1}{2}$. For deuterium, I is 1 (spin-one), and for ^{23}Na, a magnetic isotope of Na, I is $\frac{3}{2}$ (see Table I for other nuclei).

The number of possible projections the dipole moment can have is limited to $2I + 1$; each projection is described by specifying another spin quantum number, m_I, which can vary between I and $-I$ in integer steps. The magnitude of the overall magnetic moment and its z component depend on I and m_I in the following way:

$$\mu = \gamma\hbar \sqrt{I(I+1)}, \tag{2}$$

$$\mu_z = \gamma\hbar m_I, \tag{3}$$

where $\hbar$ is Plank's constant divided by 2π and γ the gyromagnetic ratio, so called because it relates the angular momentum of a particle to its magnetic moment. The energy of the nucleus is given by

$$E = -\mathbf{\mu}\cdot\mathbf{H} = -\mu_z H_0 = -\gamma\hbar m_I H_0, \tag{4}$$

where the last equality follows from Eq. (3).

For spin-$\frac{1}{2}$ nuclei, m_I can be $+\frac{1}{2}$ or $-\frac{1}{2}$, corresponding to a parallel or antiparallel orientation of the spin (and hence the magnetic moment) with respect to the magnetic-field direction. Figure 4 represents these two possibilities for a spin-$\frac{1}{2}$ particle. For higher spin particles, μ can acquire $2I + 1$ allowed orientations.

For spin-$\frac{1}{2}$ particles, the energies of the two orientations are $-\frac{1}{2}\gamma\hbar H_0$ and

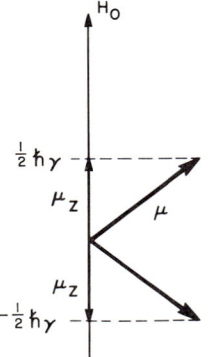

Fig. 4. Allowed orientations for a spin-½ nucleus, where $|\mu| = \hbar\gamma\sqrt{\frac{3}{4}}$.

$+\frac{1}{2}\gamma\hbar H_0$. The energy separation between these states is then $\gamma\hbar H_0$. For higher-spin particles, the separation between adjacent levels will still be $\gamma\hbar H_0$ since μ_z can change only in steps of $\hbar\gamma$. Because of the H_0 factor, the energy separation will increase with increasing magnetic field and, at a fixed field strength, will be higher for nuclei with the higher gyromagnetic ratio. For example, the γ of the ^{1}H nucleus is approximately fourfold higher than that of the ^{13}C nucleus. Thus, at a fixed magnetic field strength, the energy separation between the adjacent spin states of the ^{1}H nucleus is approximately fourfold larger than that of the ^{13}C nucleus.

If a *totally isolated* magnetic dipole is in one of the energy states specified by the quantum number m_I, it will stay in that state forever. For this particle to undergo a transition to another state with a different m_I value, it has to interact with another system and either gain or lose energy to that system. Such an interaction can occur with magnetic fields whose magnitudes are oscillating in time. From time-dependent perturbation theory it can be shown that an oscillating magnetic field will cause transitions between different energy states of a magnetic dipole, provided the oscillatory magnetic field is perpendicular to the main static field and the frequency of oscillation is equal to γH_0, the energy separation between adjacent states divided by $\hbar$. This frequency is the natural resonant frequency of the spin system and is referred to as the Larmor frequency. Time-dependent perturbation theory also dictates that $\Delta m_I = \pm 1$; in other words, transitions can be induced only between adjacent states which differ in the quantum number m_I by 1. The probability of a transition is the same whether the quantum number changes from m_I to $m_I + 1$ or $m_I + 1$ to m_I. Thus, in a sample with an equal number of nuclei in each state, no net absorption of energy would occur. Fortunately, in real samples at finite temperatures the lower energy states are populated in excess, and a net absorption of energy does occur. This is discussed in greater detail in Section III.B.

In reality, the energy levels of the spins in the presence of an external field are never infinitely sharp. Consequently, a spread of frequencies centered about γH_0 can induce transitions. The transition probability is highest when the frequency of oscillation ω is equal to γH_0; it decreases as the difference between ω and γH_0 increases. As a result one observes an NMR peak entered about γH_0 with a finite linewidth which comes from the widths of the energy levels (Fig. 1).

In contemporary NMR spectrometers H_0 is typically 20–120 kG (kilogauss) (10 kG = 1 T). The Larmor or resonance frequencies of several nuclei at 70.46 kG are given in Table I. Note that γH_0 has units of radians per second; however, traditionally, frequencies are expressed as $\gamma H_0/2\pi$ using units of megahertz (MHz).

B. Boltzmann Distribution and Bulk Magnetization

In an NMR measurement, as in many other forms of spectroscopy, one does not deal with a single nucleus; instead, the sample contains an extremely large number of spins. For example, a typical sample for ^{1}H NMR determinations is ~ 0.5 ml, and its solute concentration is $\gtrsim 1$ mM; thus, it contains $\gtrsim 3 \times 10^{17}$ spins.

In the absence of external perturbations which induce transitions, these spins distribute themselves between their different energy states. As in any system at thermal equilibrium, the probability of finding a spin with a certain energy E is proportional to $e^{-E/kT}$, where k is Boltzmann's constant and T the absolute temperature. Therefore, at a given instant, a larger fraction of the spins in the sample will occupy the lower energy state. It is this difference in the spin population which gives rise to an NMR signal.

Let us consider the spin-$\frac{1}{2}$ case. There are two energy states characterized by $m_I = \pm\frac{1}{2}$. As previously discussed, these two energy states correspond to two different orientations of the magnetic dipole with $\mu_z = \gamma \hbar m_I$. Since more spins will occupy the lower energy state, the sum of magnetic moments along the z direction will not cancel out. There will thus be a net bulk magnetic moment along the z axis whose magnitude will be $(\frac{1}{2}\gamma \hbar \, \Delta n)$, where Δn is the population difference between the two states. Calculations for the general case of spin I nuclei yield

$$M_z^0 = [\gamma^2 \hbar^2 I(I+1) H_0 N]/3kT,$$

where M_z^0 is the bulk magnetization along the z axis, H_0 the magnetic field strength, N the total number of spins in the sample; and the term $I(I+1)$ arises from the quantum mechanical nature of the spin states. This equation is known as Curie's law. In the presence of just the external field along the z direction, there is no preferred orientation of the spin magnetic dipoles along the x and y directions. As a result, the components of the bulk magnetization along these two axes is zero.

It is the behavior of this magnetization after it has been disturbed from equilibrium in various ways (discussed in Section III.C) which provides the basic information in NMR. For example, if it could be tipped away from the magnetic field, what would it do? In fact, it behaves like a gyroscope, rotating around the z axis as its characteristic nuclear, or Larmor, frequency. This rotation will gradually decay until thermal equilibrium is reestablished. If there is a coil wrapped around the sample, then there will be a voltage induced in the coil at the Larmor frequency due to the rotating magnetic field of the bulk magnetization.

C. Bloch Equations

1. Relaxation Terms

When a spin system is perturbed away from thermal equilibrium, it will ultimately recover its equilibrium population distribution. As in chemical reactions, there is a rate constant associated with this process; it is designated by the symbol T_1^{-1}, where T_1 is called (for historical reasons) the spin–lattice relaxation time. Thus, if thermal equilibrium is disturbed, as it is during measurement of NMR signals, the z component of magnetization, M_z, recovers its thermal equilibrium value M_z^0 with the time constant T_1. The phenomonological equation describing this recovery is

$$dM_z/dt = -T_1^{-1}(M_z - M_z^0).$$

It is part of one of the three Bloch equations which describe the overall behavior of **M**.

At thermal equilibrium, $M_x = M_y = 0$ because there is no orientational preference along either the x or the y direction. However, by appropriate applications of high-power rf radiation in the form of a short pulse, spin populations can be rearranged so that the system is left with a nonvanishing bulk magnetic moment in the x–y plane. Again, in the absence of the perturbing pulse, the spin system evolves to establish thermal equilibrium. Therefore, the net magnetization established on the x–y plane immediately after the perturbation decays back to zero. The rate constant associated with this decay is T_2^{-1}, where T_2 is called the spin–spin relaxation time. The time course of this decay follows an equation analogous to that for M_z:

$$dM_i/dt = -T_2^{-1}M_i,$$

where $i = x$ or y.

2. Magnetic-Field Terms

In addition to the relaxation process mentioned above, the Bloch equations also predict how the magnetization **M** will behave in the presence of either a static or time-varying field. Temporarily ignoring the relaxation, we now consider the forces on a magnetic moment **M** in a magnetic field **H**. Classically, there is a torque on **M** given by **M** × **H**. If **M** possessed no angular momentum, then it would start to rotate toward alignment with **H**, overshoot, and oscillate about its minimum energy orientation. Any friction in the system would gradually bring it to rest pointing along **H**. If, on the other hand, **M** does possess angular momentum **J**, then the equation of motion will be

$$d\mathbf{J}/dt = \mathbf{M} \times \mathbf{H}.$$

In the case of a magnetic nucleus (and many others as well), the angular momentum and the magnetic moment are related by $\mathbf{M} = \gamma \mathbf{J}$, where γ is the previously mentioned gyromagnetic ratio. This then gives an equation

$$d\mathbf{J}/dt = \gamma^{-1} d\mathbf{M}/dt = \mathbf{M} \times \mathbf{H}$$

or

$$d\mathbf{M}/dt = \gamma \mathbf{M} \times \mathbf{H}, \tag{5}$$

which is analogous to the equation describing the motion of a gyroscope.

To analyze the behavior of **M** it is very helpful to introduce a frame of coordinates rotating about the z axis with some angular frequency ω. Such a system is called a "rotating frame" and, as we will see below, aids considerably in the understanding of this problem. It can be shown that the time derivative of a vector **V** in the laboratory frame, $d\mathbf{V}/dt$, is related to its time derivative in the rotating frame, $\delta \mathbf{V}/\delta t$, by the following equation (Slichter, 1978):

$$d\mathbf{V}/dt = \delta r/\delta t + \omega \hat{z} \times \mathbf{V}.$$

If this relation is used in Eq. (5), then we have

$$d\mathbf{M}/dt = \delta \mathbf{M}/\delta t + \omega \hat{z} \times \mathbf{M} = \gamma \mathbf{M} \times \mathbf{H},$$

or

$$\delta \mathbf{M}/\delta t = \mathbf{M} \times (\gamma \mathbf{H} + \omega \hat{z}),$$

using the fact that $\hat{z} \times \mathbf{M} = -\mathbf{M} \times \hat{z}$. Thus, the equation describing the behavior of **M** in the two frames is the same, provided we replace our original field in the laboratory frame $H_0 \hat{z}$ with an effective field in the rotating frame H_e equal to $(H_0 + \omega/\gamma)\hat{z}$.

We can now see an easy way to solve Eq. (5) for a static field, namely, inspect the behavior of **M** in a rotating frame with $\omega = -\gamma H_0$. In this frame $\delta \mathbf{M}/\delta t = 0$, so **M** does not change. Since this frame is rotating around the z axis at $-\gamma H_0$ with respect to the laboratory frame, **M** will also precess in a fixed cone around the z axis in the laboratory frame with frequency γH_0 (Fig. 5).

This is the behavior we referred to at the end of Section III.B except that we have no decay here because of the absence of the relaxation terms. With these terms included, then **M** would rotate around the z axis, gradually losing its components along the x and y axes and reestablishing a net magnetic moment along the z axis.

Putting these components together we have

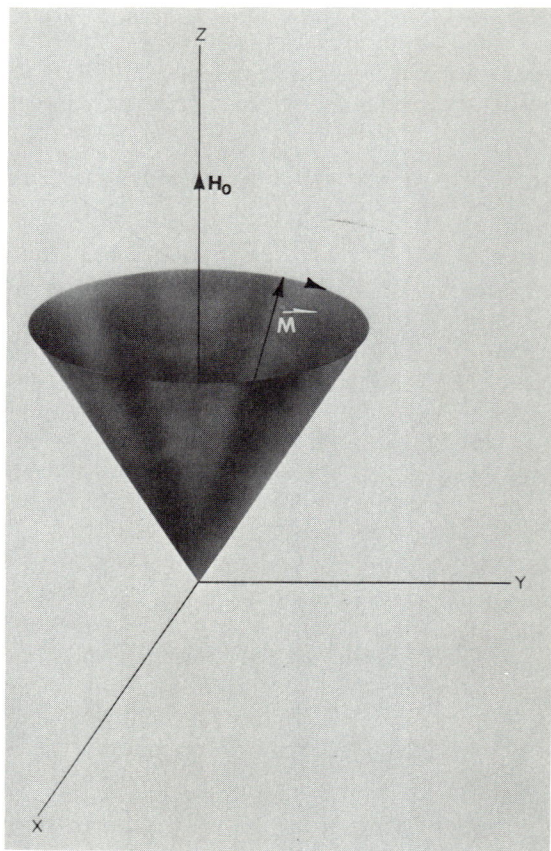

Fig. 5. Precession of **M** around **H₀**.

1. NUCLEAR MAGNETIC RESONANCE

$$dM_z/dt = -(M_z - M_z^0)/T_1 + \gamma(\mathbf{M} \times \mathbf{H})_z, \tag{6}$$

$$dM_x/dt = -M_x/T_2 + \gamma(\mathbf{M} \times \mathbf{H})_x,$$
$$dM_y/dt = -M_y/T_2 + \gamma(\mathbf{M} \times \mathbf{H})_y. \tag{7}$$

This combined group of equations is known as the Bloch equation after Felix Bloch, who first proposed them in 1946 (Bloch, 1946). Although phenomenological in character, they are very useful in a wide variety of cases, particularly in the case of weakly interacting spins.

It should be noted that nothing in the derivation of the Bloch equations limits **H** to a constant. In fact, the equations are valid for a time-dependent **H** as well. This allows us to calculate the effects of irradiating the spin system with radio-frequency fields.

Let us consider the case where $\mathbf{H} = H_0\hat{z} + 2H_1\hat{y}\cos\omega t$, namely, a static field H_0 in the z direction plus a field $2H_1$ in the y direction oscillating at frequency ω. To analyze this we first need to separate the linear alternating field $2H_1$ into two counter-rotating fields of strength H_1, as shown in Fig. 6. When ω is near the Larmor frequency, it is necessary to consider only the effects of the component rotating in the same sense as **M** about H_0. In a frame rotating at frequency ω around the z axis, there will be a static magnetic field of magnitude H_1 in the x–y plane and a static field H_0 along the z axis. In the presence of these stationary fields, the resultant equation for **M** is

$$d\mathbf{M}/dt = \gamma \mathbf{M} \times [(H_0 - |\omega|/\gamma)\hat{z} + H_1\hat{y}], \tag{8}$$

where we have ignored the relaxation terms for the moment and taken H_1 to lie along the y axis. In going to the rotating frame, we have modified the expression for the field along the z direction by $-|\omega|/\gamma$; the minus sign comes from the fact that ω is taken to be negative if it describes a clockwise rotation. The total effective field in the rotating frame, $\mathbf{H}_{\text{eff}}$, is the vector sum of $(H_0 - |\omega|/\gamma)\hat{z}$ and $H_1\hat{y}$. As in the case of Eq. (5) in the laboratory frame, the solution of Eq. (8) in the rotating frame is a precession of **M** around the effective field $\mathbf{H}_{\text{eff}}$ at the angular frequency γH_{eff}, as illustrated in Fig. 7. In the case when $\omega = \gamma H_0$ (the resonance condition), the effective field in the rotating frame lies along the y axis and is equal to H_1. Thus, **M** will precess in the z–x plane, alternately pointing along the z axis, the negative x axis, the

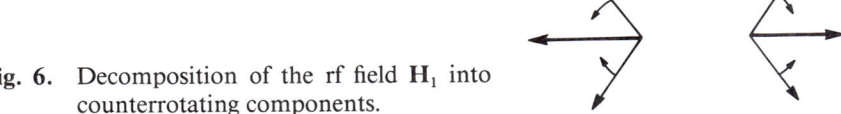

Fig. 6. Decomposition of the rf field $\mathbf{H}_1$ into counterrotating components.

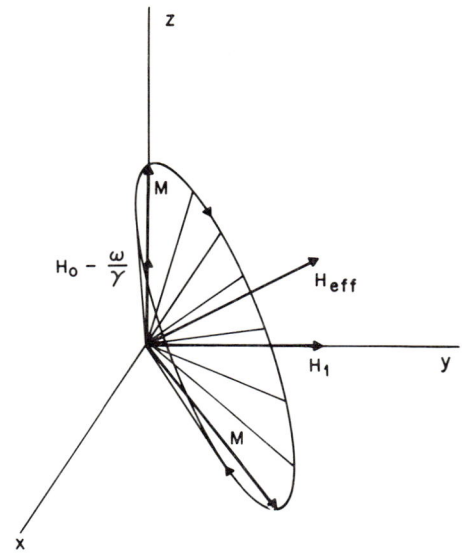

Fig. 7. Precession of **M** and **H**$_{\text{eff}}$ in the rotating frame.

negative z axis, the x axis, and so on. Its frequency of rotation in this plane will be given by γH_1. Since H_1 is typically 10 G, the rotation period for this motion is 50–100 μsec.

We have dealt with this at some length because it is the basis for understanding how modern Fourier transform spectrometers function. The magnetization is tipped into the x–y plane by turning on an rf field at the resonant frequency of the spins, leaving it on for a quarter of a rotation time and then turning it off. This leaves the magnetization in the x–y plane. Such a procedure is referred to as a 90° pulse; obviously, a 180° pulse would leave the spins along the negative z axis. The steps in a 90° pulse are shown in Fig. 8, including the behavior of **M** after the rf is turned off. After the rf pulse is turned off, **M** remains stationary in the rotating frame; in the laboratory frame, however, it precesses about the z axis at the Larmor frequency.

If we include the relaxation terms involving T_1 and T_2, it is clear that they will cause the magnetization in the x–y plane, M_{xy}, to decay exponentially with the time constant T_2, whereas M_z will recover exponentially toward M_z^0 with the time constant T_1. Therefore, in the laboratory frame, M_{xy} will precess about the z axis at the Larmor frequency and decay to zero with time. If there is a pickup coil of appropriate geometry around the sample, an oscillating voltage will be generated in this coil by the magnetization in the x–y plane as it rotates about the z axis in the laboratory frame.

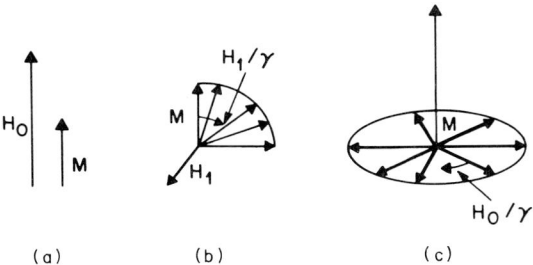

Fig. 8. Sequence of events in a 90° pulse, leaving **M** rotating at the Larmor frequency in the $x-y$ plane: (a) before rf; (b) rf on, in rotating frame at resonance; and (c) rf off.

This voltage will vanish exponentially as the magnetization in the $x-y$ plane decays to zero. The signal induced in the coil as a result of the rotating and decaying magnetization is known as the free induction decay (FID). The Fourier transform (Appendix A) of this signal is an absorption peak centered around the precession frequency. Note that if there were nuclei with slightly different frequencies in the sample (as discussed in Section III.D), they would all be tipped onto the $x-y$ plane, provided the H_1 field is strong enough ($H_1/\gamma > \Delta f$). After the rf pulse is turned off, a magnetization vector arising from each type of spin will precess about the z axis at its characteristic Larmor frequency. Consequently, the resultant FID will contain a multiplicity of frequencies which, upon Fourier transformation, will lead to a multiplicity of peaks in the absorption spectrum. It is for this reason that modern Fourier transform (FT) spectrometers have superseded the process of continuous-wave NMR in which the H_1 field is swept through the nuclear frequencies and the absorption of energy by the sample is monitored to obtain the NMR spectrum. With FT spectrometers, all of the frequencies are sampled simultaneously, resulting in a considerable improvement in sensitivity if there are many lines in the spectrum, as is usually the case.

D. Chemical Shift

Previously, we discussed a collection of isolated nuclei in a magnetic field. In a molecule or an atom, however, the magnetic field actually experienced by the nucleus is different from the external magnetic field. This is because, in the presence of the external field, the electrons generate a local field, H_{loc}, the magnitude of which is proportional to the strength of the external field. The proportionality constant is called the "shielding constant" and is designated by $-\sigma$ since, in general, the local field opposes the external field. The source of this internal field is the current induced in the atomic or molecular electrons by the external field. In a molecule, the actual field seen by a nucleus becomes the sum of the external magnetic field H_0 and the local field generated by the electrons, $-\sigma H_0$. As a result, the Larmor frequencies of the magnetic dipoles become $\omega = \gamma H_0(1 - \sigma)$.

To generate a local field parallel to the external field, the electrons must be able to move in the plane perpendicular to the external field. As an example, let us consider a benzene molecule in two configurations: when the external field is perpendicular to the molecular plane and when it is parallel to the molecular plane. Since the electrons are constrained to move in molecular orbitals, they obviously cannot execute the same motion in the plane perpendicular to the external field in these two orientations (Fig. 9). Thus the actual magnetic field at the nuclei will be different, and the frequencies of absorption will be different as well. Because of this dependence on orientation, shielding must be described by a tensor, which can considerably broaden the NMR lines in solids and slowly moving molecules. In a liquid, however, a small molecule changes its orientation so rapidly that the shielding tensor is averaged into a single number.

This effective shielding constant depends on the electronic environment of the nucleus. Consider benzene again: each carbon atom is identical with the others, and the distribution of electrons about each carbon nuclei must be the same. Consequently, all of these carbon nuclei have the same σ. In toluene, however, the situation is different. Introduction of the methyl group breaks the symmetry of the benzene molecule and results in five different types of carbon atoms, hence up to five different σ's. Therefore,

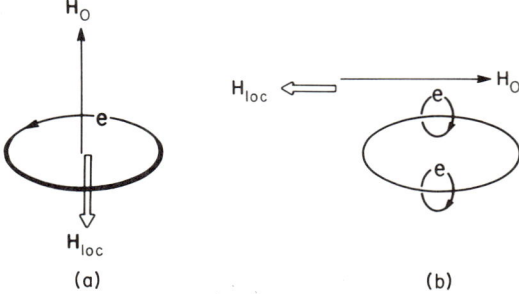

Fig. 9. Differing response of the π electrons in a benzene molecule at two orientations of the molecule relative to H_0: when the external field is (a) perpendicular to the molecular plane and (b) parallel to the molecular field.

five distinct peaks are detected in the ^{13}C spectrum of toluene as opposed to the one from benzene (Fig. 10).

In typical applications, the chemical shifts of a compound are measured and reported relative to a standard sample. The resonances of two nuclei of the same type (i.e., with the same gyromagnetic ratio) but with different

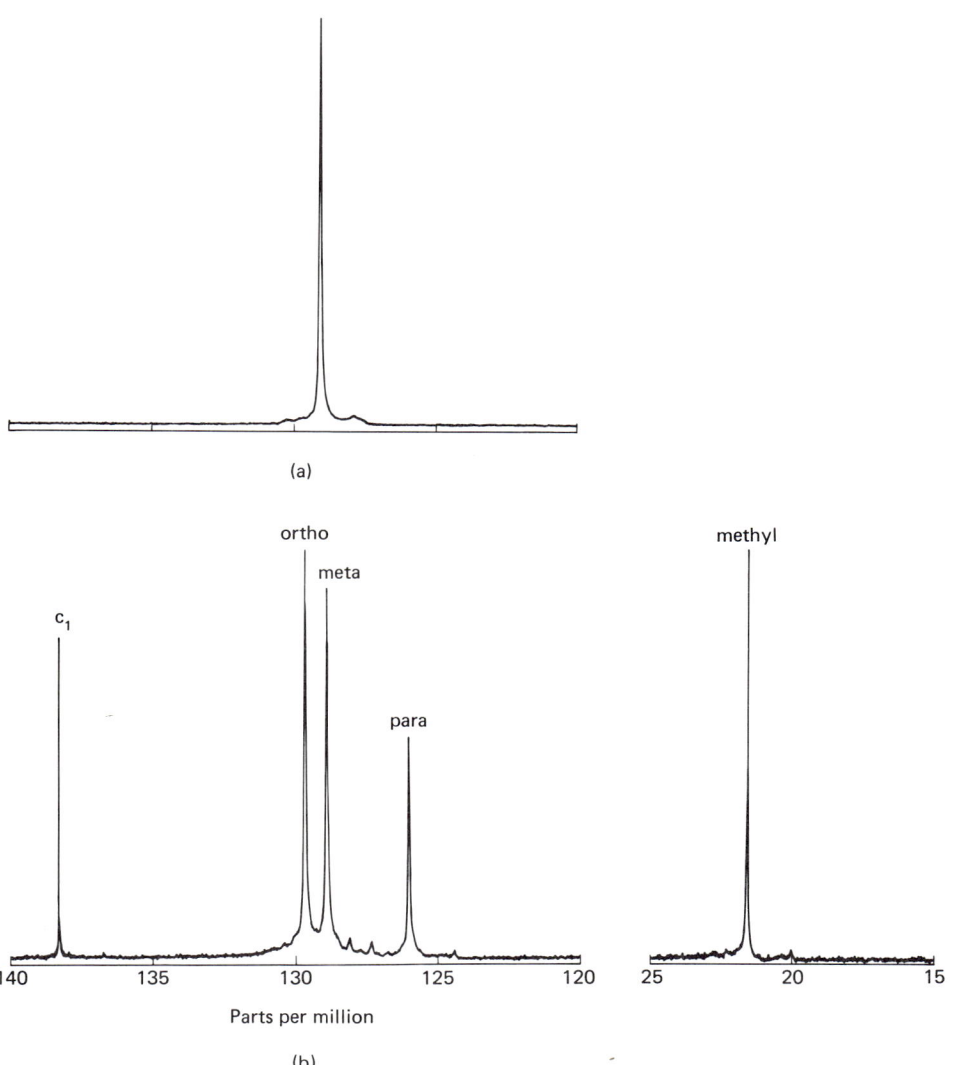

Fig. 10. ^{13}C NMR spectra of (a) benzene and (b) toluene.

shielding constants σ_1 and σ_2, will be separated by $(\sigma_1 - \sigma_2)\gamma H_0$. This quantity, divided by the operating frequency of the spectrometer for that nucleus, γH_0, is dimensionless and expresses the differences between the shielding constants. It is called the chemical shift and is reported in parts per million (ppm). The utility of this procedure is that a reported chemical shift reflects the influene of the electronic environment and does not depend on the field strength of the spectrometer used.

For historical reasons having to do with the fact that the first NMR spectrometers were operated at constant frequency and variable magnetic field, the relative direction between two chemical shifts is given by the terms "upfield" and "downfield." Conventionally, spectra are plotted with the upfield direction to the right. Note this means that frequency increases to the left and, if the ppm axis is given in frequency terms, as it usually is on modern, fixed-field spectrometers, the axis will be labeled by increasingly negative numbers going to the right.

A special type of local field is encountered in aromatic molecules. Here, the electrons of the Π orbitals flow freely through the conjugated Π bond system and generate a local field which can cause significant shifts in resonances. This local field affects not only the nuclei of that particular aromatic moiety but other nuclei which may be located in close proximity. For example, in nucleotides, H1' of the ribose moiety experiences a shift from the local field of the purine or pyrimidine group. Similarly, in proteins the methyl group of an alanine residue would experience shifts if it were located near an aromatic amino acid; the shift can be in either direction, depending on the position of the methyl group relative to the aromatic ring. In proteins, such shifts are of considerabe help in distinguishing between resonances from the same type of amino acids located at different positions in the protein.

The range of chemical shifts observed for a given type of nuclei in different molecular environments varies dramatically with different nuclei. For proton resonances, the range of known chemical shifts (except in those molecules which contain an unpaired electron spin) covers a 10-ppm range (Fig. 2). For ^{13}C and ^{31}P nuclei, the chemical-shift range covers ~ 250 and 600 ppm, respectively (Fig. 2). The spread of resonance frequencies caused by all possible different chemical environments, however, is much smaller than the variations in the value of the resonant frequencies among nuclei. For example, in a spectrometer with a 70.5-kG H_0 field, protons resonate at ~ 300 MHz. In this spectrometer, a 10-ppm chemical-shift range corresponds to a frequency spread of 3000 Hz; thus, the resonance frequency of two ^1H nuclei in different chemical environments will differ from each other by, at most, ~ 3000 Hz. In the same spectrometer, ^{13}C nuclei will resonate at ~ 75 MHz; the difference in the resonance frequencies of different carbons will be less than ~ 20 kHz.

E. Spin–Spin Interactions

Each spin generates a magnetic field due to its magnetic dipole moment. This field is experienced by all other spins and causes two major spin–spin interactions. These are the scalar and the dipole–dipole couplings. In NMR spectra of liquids the most common manifestation of scalar interactions is the multiplet structure observed from coupled nuclei. On the other hand, at least in liquids, dipole–dipole coupling is usually the major source of relaxation of the spins.

1. Scalar or J Coupling

The scalar coupling is an interaction between two magnetic nuclei in the same molecule and is mediated by electrons in the chemical bonds. Consider a simple case with two magnetic nuclei, A and B, with two electrons in the bonding orbital between them. The electron spins are paired so that the net electron spin vanishes. Therefore, in first order, there is no interaction between the electron spins and the nuclear spins. But since A has a magnetic nucleus, the electron closer to A will tend to orient itself in the magnetic field of the nuclear dipole; this in turn would induce an orientational preference for the electron near B and, consequently, for the nuclear spin of B. Thus, the net effect is a coupling between A and B nuclei which depends on the relative orientation of the nuclear spins alone and not on the orientation of the molecule as a whole with respect to the external field.

The scaler interaction energy takes on the form $\alpha \mu_A \cdot \mu_B$, where μ_A and μ_B are the nuclear magnetic dipoles of atoms A and B, respectively, and α a measure of the strength of the interaction. This interaction is usually expressed in units of hertz using the symbol J, where $J = \alpha \gamma^2 \hbar / 2\pi$.

This interaction causes the multiplet structure observed in liquids by making the differences in energy of the spin states at one nucleus dependent on the spin state of the other nucleus. Although the details are too complicated to discuss here, there are two limiting types of multiplet structure, one occurring when $J \ll \Delta\nu$ and the other when $J \gg \Delta\nu$, where $\Delta\nu$ is the frequency difference which would exist between A and B in the absence of the scalar interaction. These limits are shown in Fig. 11 in the case of spin-$\frac{1}{2}$ nuclei. Note that $\Delta\nu$ depends on the magnetic field strength; hence, a spectrum in the $J \gg \Delta\nu$ limit on a low-field spectrometer may shift into the $J \ll \Delta\nu$ on a high-field spectrometer. In the case of coupling between different nuclei, e.g., $^{13}C-^{1}H$ or $^{31}P-^{1}H$, we are always in the $J \ll \Delta\nu$ limit.

In NMR spectra, higher-order multiplet structures are observed when a nucleus is coupled to several magnetically identical spin-$\frac{1}{2}$ particles. For example, in a freely rotating methyl group all protons are magnetically identical. Consequently, they behave like a particle with $I = \frac{3}{2}$ with allowed

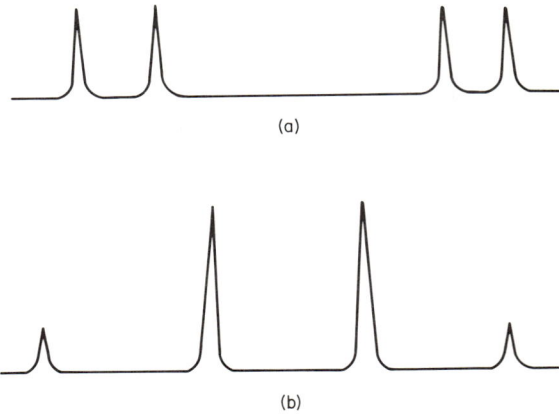

Fig. 11. Absorption spectra for a pair of scalar-coupled spin-$\frac{1}{2}$ nuclei in the extreme limits of (a) $J \ll \Delta\nu$ and (b) $J \gg \Delta\nu$.

m_I values of $\frac{3}{2}$, $\frac{1}{2}$, $-\frac{1}{2}$, and $-\frac{3}{2}$. Any nucleus which experiences a scalar coupling to these methyl protons will be split into a quartet. For example, the ^{13}C resonance of a methyl carbon splits into a quartet. The magnetically identical protons, however, do not affect each other.

Multiplet structures resulting from nearby magnetic nuclei with spin quantum numbers greater than $\frac{1}{2}$, such as deuterium, are also observed. For some quadrupolar nuclei, however, transitions between the energy states of the magnetic dipoles may be very rapid. In this case, nearby spins experience the average of all possible magnetic fields the quadrupolar nucleus can generate and, consequently, are not split into a multiplet.

In molecules, the scalar coupling can occur between nuclei separated by several bonds. For example, protons bonded to adjacent carbons are coupled to each other by scalar interactions. This is the most common source of multiplet structures in ^{1}H NMR spectra. A weaker coupling also exists between protons separated by a larger number of bonds.

2. Dipole–Dipole Interactions

The magnitude and the direction of a dipolar magnetic field at a point in space depends on the inverse cube of the distance from the magnetic dipole and the relative orientation of the point with respect to the magnetic dipole vector. This magnetic field is given by the equation

$$\mathbf{H}_d = [3(\mu \cdot \hat{r})\hat{r} - \mu]/r^3,$$

where μ is the magnetic dipole moment generating the field, r the distance from the dipole to the point at which H_d is being measured, and $\hat{r}$ a unit vector along the line joining the dipole and the point in space where H_d is being measured.

If there is more than one magnetic dipole present, each will generate a field of its own and each will experience the field generated by all the others. In the simple case of two nuclear spins A and B, there will be dipole fields $\mathbf{H}_{d,A}$ and $\mathbf{H}_{d,B}$ coming from nuclei A and B, respectively. A magnetic nucleus acquires an energy in the presence of any magnetic field according to Eq. (1). Consequently, the energy of nuclei A will change by $-\mu_A \cdot \mathbf{H}_{d,B}$, and the energy of nuclei B will change by $-\mu_B \cdot \mathbf{H}_{d,A}$.

Unlike the scalar interaction, the dipole–dipole interaction depends on orientational parameters and internuclear distance as well as on the nuclear spin states. Consequently, it is a more complicated interaction to analyze. For rapidly tumbling molecules, we have to average the dipolar interaction over all possible orientations. Due to this averaging, the first-order effects vanish and there is no net change in resonance frequencies. However, there are second-order terms which are the primary source of spin relaxation in liquids. In solids, however, this averaging does not occur, and the dipolar energies make a significant contribution to the total energy of a spin and hence its resonant frequency.

In the event molecules contain unpaired electron spin(s), dipolar interaction also exists between magnetic nuclei and the electrons. This is a much stronger interaction, because the gyromagnetic ratio of an electron is approximately 10^3 times larger than that of a proton. Again, for molecules which can rotate rapidly, no net changes in the spin energies of either the electron or the magnetic nuclei are induced. However, for the nuclei in these molecules this interaction provides a very effective mechanism of spin–spin and spin–lattice relaxation.

3. Contact and Pseudocontact Shifts

These interactions occur when a molecule contains a net electron spin. They manifest themselves as changes in the resonance frequency of the nuclear spins in the paramagnetic molecule.

Contact shifts require the existence of net electron-spin polarization over the magnetic nucleus. The interaction alters the energy of the nuclear spin according to the equation $\delta E_{cs} = \alpha \mu_n \cdot \mu_e$, where μ_n and μ_e represent magnetic dipole moments of the nucleus and the electron, respectively. Note that this equation is identical in form to the scalar coupling term. However, it does not induce splitting of the nuclear resonances. The reason for this is that the electron jumps between its allowed energy levels so rapidly that on the nuclear-spin time scale the nucleus sees an average value for the electron dipole moment. The interaction becomes a net change in energy, and therefore resonance frequency, rather than the appearance of multiplet structures.

Electron magnetic moments are difficult to deal with due to the presence of orbital momentum terms, highly anisotropic interactions with the static H_0 field, and the multiplicity of spin energy levels when more than one unpaired electron exists. However, it is instructive to consider the simple case of a single unpaired electron which interacts isotropically with H_0. In this case one can expand the scalar product as follows:

$$\delta E_{cs} = \alpha \mu_{nz} \mu_{ez}^0 + \alpha[\{\mu_{nz}(\mu_{ez} - \mu_{ez}^0) + \mu_{nx}\mu_{ex} + \mu_{ny}\mu_{ey}\}],$$

where μ_{ez}^0 is the average magnetic moment of the electron, which is determined by the Boltzmann distribution. In fact, if the sample contains N unpaired electrons, $N\mu_{ez}^0$ is the bulk electron magnetic moment and would be given by Curie's law (Section III.B). All except the first term in this equation would average out to zero due to rapid electron-spin flips. The terms which vanish contribute to spin relaxation but not to the net energy of the nuclear spin. From the non vanishing term, one can calculate the change in energy levels to be

$$\delta E_{cs} = \alpha \hbar^2 \gamma_e^2 [S(S+1)/3kT] \gamma_N \hbar m_I,$$

where γ_e and γ_N are the electron and nuclear gyromagnetic ratios and S the spin quantum number of the electrons; for a single electron, $S = \frac{1}{2}$.

The shift in frequency in radians per second is

$$\Delta\omega = \alpha \gamma_N [\gamma_e^2 \hbar^2 S(S+1)/3kT]. \tag{9}$$

This expression was first derived by Bloembergen (1957). Generally, it is presented in terms of the hyperfine coupling constant A, which is equal to $\alpha \gamma_N \gamma_e \hbar$ in our notation.

In biological samples a well-known example of contact shifts is the heme-containing proteins such as cytochrome c. In such a molecule the resonances from the heme experience large shifts in the paramagnetic protein due to the unpaired electrons of the heme iron. In the diamagnetic state, all the electrons are paired, and the shifts disappear. The magnitudes of the contact shifts are temperature dependent, as predicted by Eq. (9). In fact, the temperature dependence is generally used to identify the nature of these shifts.

Pseudocontact shifts arise when the electron–nuclear dipolar coupling is not averaged to zero. This happens when the electron environment and, consequently, electron interactions with the static field are highly anisotropic. Unlike the scalar or the contact interaction, it is a "through-space" effect and does not require net electron-spin density over the nucleus. It is a considerably more complicated phenomenon than the isotropic contact shift; therefore we will not attempt a derivation here. A simplification occurs when the molecule has axial symmetry about the electron. In this case the pseudocontact shift is given by

$$\Delta\omega/\omega = D[(3\cos^2\theta - 1/r^3)],$$

where θ is the polar angle defining the position of the nucleus relative to the axis of symmetry. This axis is fixed on the molecule, and θ does not change if the molecule rotates in the laboratory frame. Note that pseudocontact shifts depend on structural parameters; therefore, they can and have been used to obtain structural information.

F. Relaxation Times and the Nuclear Overhauser Effect

1. Relaxation

In this section we will discuss the underlying mechanisms which determine T_1 and T_2. Since both T_1 and T_2 are phenomonological constants which describe the approach to thermal equilibrium, they depend on interactions which can induce transitions between the various magnetic levels. As discussed in Section III.A, such transitions are caused by oscillating magnetic fields with frequencies which match the energy separation between the adjacent states. In fact, any time-dependent magnetic field can induce such transitions. This can be understood from the fact that any time-dependent field can be regarded as a sum of oscillating fields. Several interactions among spins lead to time-dependent magnetic fields. Dipole–dipole interactions have already been mentioned as an example.

In the dipole–dipole interaction, the dipole field depends on the distance between the spins and the relative orientation of the two spins in space. Thus, the magnitude of the dipolar field encountered by each spin will be a function of r, θ, and ϕ (Fig. 12). If the two spins are on the same molecule, r is generally fixed; if they are not, r will be randomly varying as a result of diffusion. Regardless of whether the spins are on the same molecule or not, rotational motion of the molecules will cause random fluctuations in θ- and ϕ-dependent terms. Because of these random fluctuations, each spin will be subject to a time-dependent magnetic field with components in all directions. Those components perpendicular to the direction of the static H_0 field will induce transitions. The fluctuating component along the axis of $\mathbf{H}_0$ will not contribute to T_1, because it cannot induce transitions between the spin states. However, because it can perturb the magnetization perpen-

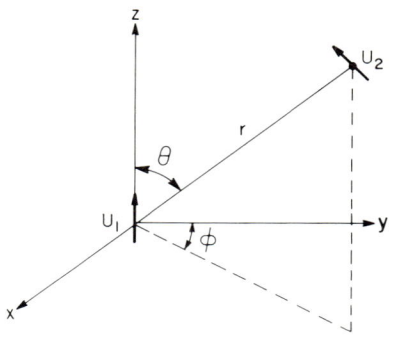

Fig. 12. Parameters determining the dipole–dipole interaction energy of two magnetic moments $\mathbf{U}_1$ and $\mathbf{U}_2$.

dicular to $\mathbf{H}_0$, this component will contribute to T_2. This effect causes T_2 to be always less than or equal to T_1.

In the case of chemical shielding, the anisotropic nature of the shielding leads to a fluctuating local field at the nucleus as the molecules execute random rotational motion. This provides a mechanism for transitions and relaxation similar to the dipole–dipole one. This mechanism is referred to as the chemical-shift anisotropy (CSA) and is generally significant only at high magnetic fields and for heavier nuclei, such as ^{19}F and ^{31}P.

We conclude from this discussion that the T_1 and T_2 for each mechanism will depend on the following parameters: distance (for dipole–dipole), a parameter(s) describing the nature and rapidity of random molecular motions (e.g., translational or rotational diffusion constants), the magnitude of the static magnetic field H_0 (for CSA), and the Larmor frequency. For CSA, H_0 is particularly important because it is H_0 which induces the electron motions which subsequently generate the shielding local field. The dependence of T_1 and T_2 on these parameters for the dipole–dipole interactions between a pair of like spins of spin $\frac{1}{2}$ (e.g., protons interacting with other protons), and in the case of isotropic rotational motion, are given by the following equations:

$$\frac{1}{T_1} = \frac{3}{10} \frac{\hbar^2 \gamma^4}{r^6} \left(\frac{\tau_c}{1 + \omega^2 \tau_c^2} + \frac{4\tau_c}{1 + 4\omega^2 \tau_c^2} \right), \tag{10}$$

$$\frac{1}{T_2} = \frac{3}{20} \frac{\hbar^2 \gamma^4}{r^6} \left(3\tau_c + \frac{5\tau_c}{1 + \omega^2 \tau_c^2} + \frac{2\tau_c}{1 + 4\omega^2 \tau_c^2} \right), \tag{11}$$

where τ_c is the rotational correlation time, r the distance between the two spins, and ω the resonance or Larmor frequency of the spins in radians per second. With nonidentical nuclei, the formulae for T_1 and T_2 are slightly different:

$$\frac{1}{T_{1A}} = \frac{1}{10} \frac{\hbar^2 \gamma_A^2 \gamma_B^2}{r^6} \left(\frac{\tau_c}{1 + (\omega_A - \omega_B)^2 \tau_c^2} \right.$$
$$\left. + \frac{3\tau_c}{1 + \omega_A^2 \tau_c^2} + \frac{6\tau_c}{1 + (\omega_A + \omega_B)^2 \tau_c^2} \right), \quad (12)$$

$$\frac{1}{T_{2A}} = \frac{1}{20} \frac{\hbar^2 \gamma_A^2 \gamma_B^2}{r^6} \left(4\tau_c + \frac{\tau_c}{1 + (\omega_A - \omega_B)^2 \tau_c^2} + \frac{3\tau^2}{1 + \omega_A^2 \tau_c^2} \right.$$
$$\left. + \frac{6\tau_c}{1 + \omega_B^2 \tau_c^2} + \frac{6\tau_c}{1 + (\omega_A + \omega_B)^2 \tau_c^2} \right), \quad (13)$$

where T_{1A} and T_{2A} are the spin–lattice and spin–spin relaxation times for nucleus A and ω_A and ω_B the Larmor frequencies for spins A and B, respectively. Equations (12) and (13) are applicable to electron–nuclear interactions as well; therefore they are applicable to the case of paramagnetic ions interacting with nuclear spins in isotropically rotating molecules.

In the limit of rapid motion, defined as $\omega \tau_c \ll 1$, the dipole–dipole interactions for like spins simplify to the following:

$$T_1^{-1} = T_2^{-1} = \tfrac{3}{2}(\hbar^2 \gamma^4 / r^6) \tau_c.$$

Except for protons at fields higher than 80 kG ($\omega > 3 \times 10^9/\text{sec}^{-1}$), a τ_c of less than 10^{-10} sec is in this limit. To indicate the order of magnitude of this interaction, its contribution to a proton T_1 and T_2, with $\tau_c = 10^{-10}$ sec and $r = 2$ Å, is 1.35 sec. A shorter correlation time would, of course, yield a longer T_1.

In the limit of very slow motion, or $\omega \tau_c \gg 1$, the behavior of T_1 and T_2 is quite different. T_1 becomes very long whereas T_2 becomes very short. Physically, this is because it is hard to flip a single spin in this limit but easy to flip two if they are oppositely oriented. Note that many NMR studies of macromolecules are in this limit, making the lines very broad and difficult to untangle.

The contributions to T_1 and T_2 from CSA are similar:

$$T_1^{-1} = \frac{6}{20} \gamma^2 H_0^2 \bar{\sigma}^2 \left(\frac{\tau_c}{1 + \omega^2 \tau_c^2} \right),$$

$$T_2^{-1} = \frac{1}{20} \gamma^2 H_0^2 \bar{\sigma}^2 \left(\frac{3\tau_c}{1 + \omega^2 \tau_c^2} + 2\tau_c \right),$$

where $\bar{\sigma}$ is the chemical-shift anisotropy.

In cases where the spin relaxation is dominated by one mechanism, from T_1 and T_2 it is possible to extract information on molecular motion (dynamics) and distances. For example, for ^{13}C nuclei which are directly bonded to protons, the primary relaxation mechanism is dipole–dipole coupling with these protons. In this case, r is known and fixed. Since ω is also known, it is possible to extract a τ_c which describes the molecular motion. Similarly, if τ_c is known, distance information can be extracted from the relaxation times.

For a hard sphere executing isotropic rotation, τ is given by the Stokes–Einstein equation,

$$\tau_c = M\overline{V}\eta/RT,$$

where M is the molecular weight of the sphere, $\overline{V}$ the partial specific volume, η the viscosity of the medium in which the sphere is rotating, and T the absolute temperature. This equation can be used as an approximation for spherical proteins in a dilute solution. Generally, however, it is inadequate for macromolecules which have nonspherical shapes. Double-stranded polynucleotides and lipids are specific examples. In these cases, the long, rod-shaped molecules clearly cannot execute isotropic rotation. At least two different rotational correlation times are required to describe the motion of a rigid rod. Even in the case of near-spherical macromolecules, the Stokes–Einstein equation is inadequate because of the existence of local motion; for example, amino-acid side chains located on the surface of proteins usually are able to execute rapid rotation about C—C single bonds. Therefore, somewhat more sophisticated treatment of the motion is required to deduce dynamic information from the T_1 and T_2 data; such treatments are feasible and are employed for several different types of macromolecules. In all these procedures it must be kept in mind that they are model dependent and often do not allow for averaging of possible molecular conformations.

In systems where an unpaired electron exists, such as those that contain paramagnetic metal ions, nearby nuclei can relax via scalar interactions as well as the dipole–dipole mechanism. This interaction requires the unpaired electron to have a nonvanishing density over the nucleus in consideration. Consequently, it is operative only for the nuclei of the ligands of the metal ion.

All other relaxation mechanisms essentially operate on the same principle; a local magnetic field fluctuates because of molecular motion and consequently induces transitions. For nuclei with spin 1 or greater, quadrupolar interactions are the dominant relaxation mechanism because of the strength of the interaction between the local electric-field gradients and the nuclear quadrupole moment. This interaction is used extensively to obtain dynamic information; a specific example is the deuterium NMR studies of lipid motions (Seelig and Seelig, 1980; Smith et al., 1978).

2. The Nuclear Overhauser Effect

It is appropriate to discuss another phenomenon, the nuclear Overhauser effect (NOE), in this section. The NOE exists as a consequence of the relaxation of dipole-coupled spins. Previously, we have discussed how the spin system returns to equilibrium by exchanging energy with the random motions of its molecules through the dipolar coupling. With the NOE, we impose a steady-state, nonequilibrium condition on some of the spins and ask how the others respond. A comprehensive reference is the monograph by Noggle and Schirmer (1978).

For a pair of dipole–dipole-coupled spins, if the population distribution of one spin is disturbed away from thermal equilibrium by irradiating it selectively, then the distributions of the other spin will also be affected. The fractional deviation from the thermal-equilibrium population induced in the *nonirradiated* spin is defined as the nuclear Overhauser enhancement factor (NOEF), and the NOE is defined as $1 + $ NOEF. Because it is mediated by the dipole–dipole coupling, the NOE has an r^{-6} dependence, where r is the distance between the two spins; it is also a function of τ_c, the correlation time of the fluctuations of the dipole–dipole interaction, and the resonance frequency of the spins. Homonuclear NOE between protons in a molecule is extensively used for structural studies, as discussed in greater detail in Section V.A. Heteronuclear NOEs between protons and nuclei with low gyromagnetic ratios such as ^{31}P, ^{13}C, and ^{15}N are a source of signal enhancement for these low-sensitivity nuclei; in this case, since r is generally fixed by the bond lengths, the NOE provides a measure of τ_c.

In typical NOE measurements between a coupled pair of spins, the procedure is to irradiate the sample at the exact resonance frequency of one of the spins with sufficient power so that the difference in the population between the low- and high-energy spin states is annihilated. This usually is referred to as saturation (Fig. 13). If an NMR spectrum is recorded immediately after the irradiation, a signal from this spin is not observed. Under these conditions, the signal intensity of the coupled spin which is not directly irradiated is also affected as a result of the NOE. The fractional change in the integrated intensity of the resonance signal, the NOEF, is usually reported as a measure of the NOE. It has the form $\eta = (I - I_0)/I_0$, where I and I_0 are the integrated intensities with and without irradiation, respectively.

For homonuclear NOEs (i.e., NOE between like spins), η can vary from 0.5 to -1 depending on $\omega\tau_c$ (Fig. 14). The latter limit corresponds to total saturation of both resonances when only one is irradiated. When $\omega\tau_c > 1$ the coupling between the two spins is such that they exchange energy via the dipole–dipole coupling by one spin flipping up and the other simulta-

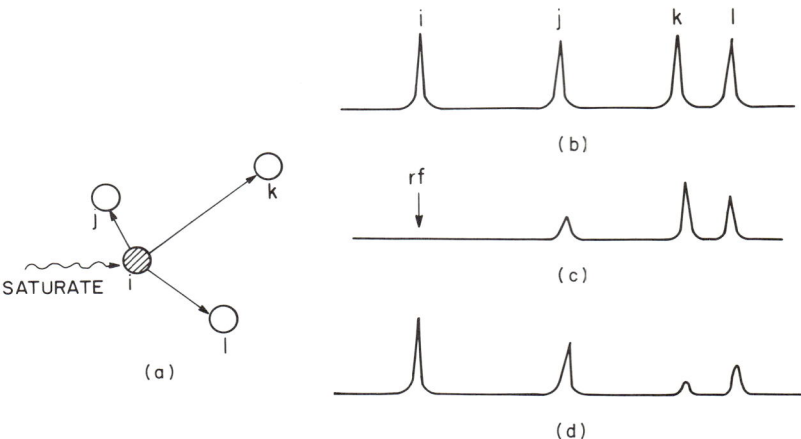

Fig. 13. Effects of saturating an individual resonance in a coupled group: (a) coupling strengths from the saturated spin (hatched) to the others in the group, (b) spectrum without saturation, (c) spectrum with saturation of spin i, (d) the difference, or NOE, spectrum.

neously flipping down. This process distributes the energy pumped into the irradiated spin between the coupled pair. Consequently, saturation of one spin leads to a decrease in the intensity of the other ($\eta < 0$). As $\omega\tau_c$ gets much larger than one, this mode of exchange becomes dominant over any other mechanism of energy dissipation, and irradiation of one spin leads to the total saturation of both ($\eta = -1$). In the limit $\omega\tau_c < 1$, however, the dominant mode of energy dissipation is simultaneous spin–flips from a state where both spins are down ($m = -\frac{1}{2}$ for both) to a state where both spins are up ($m = +\frac{1}{2}$ for both). For reasons discussed in the appendix, this results in an increase in the intensity of the nonirradiated peak ($\eta > 0$).

In the heteronuclear case, with the exception $^{19}F-^{1}H$ coupling, the NOE cannot cover a positive and a negative range. For a given pair of spins, it is either negative or positive regardless of the value of $\omega\tau_c$; the sign is determined by the signs of the gyromagnetic ratios. If both gyromagnetic ratios are of the same sign (e.g., ^{13}C and ^{1}H, where $\gamma > 0$ for both), $\eta \geq 0$. If on the other hand the gyromagnetic ratios are of opposite signs (e.g., ^{15}N and ^{1}H, where $\gamma < 0$ for ^{15}N), $\eta \leq 0$. The distinction between hetero and homonuclear NOEs stems from the fact that the process by which spins exchange energy by one spin flipping up and the other down is never an efficient process for unlike spins, whereas it is efficient for like spins. For unlike spins, there is always a large discrepancy between the energies involved in

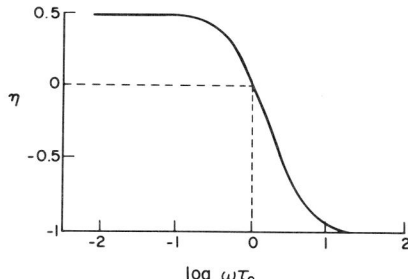

Fig. 14. Nuclear Overhauser enhancement factor as a function of $\omega\tau_c$.

spin flips for the different nuclei. This discrepancy, however, is very small for like spins, and the total change in energy of the pair of spins is very small when one spin flips up and the other down.

There are two major complications to interpretations of NOE data for structural or dynamic information. One is the possible presence of other relaxation mechanisms, such as CSA and scalar coupling. The other is the presence of more than a single pair of spins, all of which are coupled to each other via dipole–dipole interactions. In homonuclear NOE studies with protons, only the latter complication is encountered. This property, however, can be productively used for sequential assignments of resonances, as discussed in Section V.A. When this type of multiple interaction exists, one uses NOEs obtained from short irradiation periods for structural information. Generally, saturation by external rf radiation of a spin is much faster than the transfer of energy between spins; so, instead of irradiating long enough to establish steady-state NOEs, one irradiates for much shorter periods (typically 100–200 msec) and measures the effect on other spins. In this limit, the NOE is directly proportional to the irradiation time and inversely proportional to the sixth power of the internuclear distance.

G. Chemical Exchange

In many biological and chemical applications, molecules may be involved in chemical reactions while NMR measurements are being conducted. Consider a sample containing two molecules A and B, each of which possesses a nuclear spin of the same type but with different chemical shifts. If there is no interconversion between these molecules, two separate signals will be detected with resonance frequencies ν_A and ν_B Hz. They may also possess different T_1's, T_2's, and linewidths. If chemical exchange occurs, however, all of these parameters can be affected by the exchange process.

Two extreme limits in this type of situation are those of "fast" and "slow" exchange. The criteria for each limit depend on the lifetime of the chemical species and the differences in the NMR parameter being considered. In the "fast" limit, rapid exchange effectively averages the parameters of interest. Let us consider the averaging of chemical-shift differences. To detect a resonance from a set of spins, the bulk magnetization vector for those spins must precess on the x–y plane several times; only by measuring the rate of precession can we determine the resonance frequency of the spins. If chemical conversion occurs during the precession, the chemical shift and, therefore, the resonance frequency of the spins will be altered. An extreme case is when this interconversion happens several times before each set of spins completes one full revolution about the z axis at its characteristic frequency; in this case, it is not possible to detect two separate precession frequencies, and one observes a single NMR peak. This condition actually corresponds to an unnecessarily restrictive limit for fast exchange. Strictly, the criterion for fast exchange in chemical-shift averaging is $|\nu_A - \nu_B|\tau \ll 1$, where τ is the lifetime of the chemical species (we are assuming equal lifetime for both species). Since the bulk magnetization vectors of spins A and B will be precessing at ν_A and ν_B, $|\nu_A - \nu_B|$ represents the rate at which the two vectors will be separating from each other; if they cannot separate by very much (i.e., the angle between them is much less than 2π) before they are interconverted, one cannot distinguish between them (Fig. 15).

Slow exchange corresponds to $|\nu_A - \nu_B|\tau \gg 1$. In this case two separate signals are detected even though chemical exchange is occurring. In the event $|\nu_A - \nu_B|\tau \simeq 1$, exchange is in the intermediate limit, and one detects a very broad peak spread over the chemical-shift range of the two peaks (Fig. 15).

Since the exchange criteria is based on differences in resonance frequency, they are dependent on the strength of the H_0 field being used for the measurements. Thus, intermediate exchange on one instrument may in fact shift into the slow exchange limit on a higher-field instrument.

For the spin-relaxation parameters, exchange limits are defined in a similar way. Both T_1^{-1} and T_2^{-1} define the rate at which spin relaxation occurs. The chemical interconversion rate is compared with the differences in these relaxation constants. Thus

$$|T_{Ai}^{-1} - T_{Bi}^{-1}|\tau \gg 1 \quad \text{(slow)},$$

$$|T_{Ai}^{-1} - T_{Bi}^{-1}|\tau \ll 1 \quad \text{(fast)},$$

$$|T_{Ai}^{-1} - T_{Bi}^{-1}|\tau \simeq 1 \quad \text{(intermediate)},$$

where i is 1 or 2 defining T_1 or T_2.

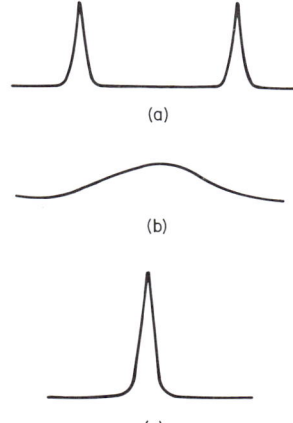

Fig. 15. (a) Slow ($|v_A - v_B|\tau \gg 1$), (b) intermediate ($|v_A - v_B|\tau \approx 1$), and (c) fast ($|v_A - v_B|\tau \ll 1$) exchange spectra for two exchanging species.

IV. Experimental Considerations

A. The Nuclear Magnetic Resonance Spectrometer

An NMR spectrometer consists of five more or less distinct components: the magnet, to polarize the spins; a radio-frequency coil to couple electrically to the spins; an rf source to provide the oscillating magnetic fields which drive the spins out of equilibrium and allow their subsequent detection; a sensitive amplifier and demodulation electronics to detect and amplify the voltages induced in the pickup coil by the spins and to reduce the actual frequencies from tens or hundreds of megahertz down to tens of kilohertz; and a computer to average, manipulate, and store the resultant data. A block diagram indicating the interrelations of these components for a high-field spectrometer is shown in Fig. 16. An excellent review of an NMR receiver and electronics is the article by Hoult and Richards (Hoult and Richards, 1975).

1. Magnet

In most modern spectrometers, the need for high fields (> 20 kG) means that a superconducting magnet of solenoidal geometry is employed. The magnetic field is carefully adjusted by auxiliary shim coils to be homogeneous to 1 in 10^8 or better for the highest-resolution work. To achieve this homogeneity, it is necessary to adjust the shim coils for each sample individually, since the diamagnetic susceptibility of the sample itself is usually 50–100 parts in 10^7 in liquid samples. Fortunately, most spectrometers have standard shim settings for various samples which can be expected to reproduce a

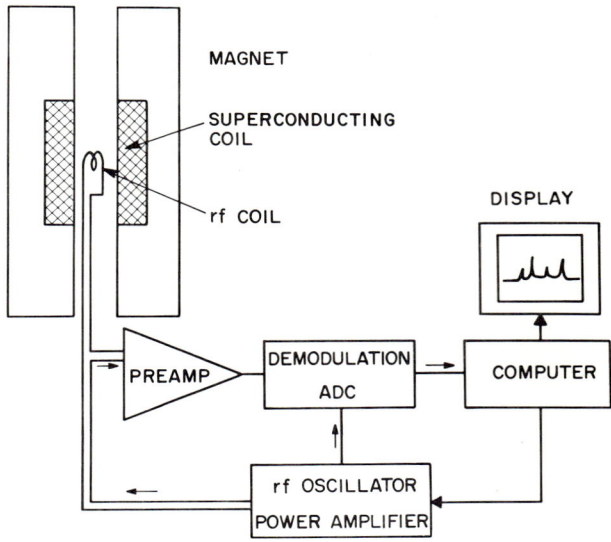

Fig. 16. Block diagram of an NMR spectrometer.

homogeneity of 5–10 parts in 10^9, which is often adequate when dealing with proteins or other large molecules. Further improvement is usually accomplished by spinning the sample around an axis parallel to the field, which averages out any inhomogeneities in the azimuthal direction. With this technique an experienced operator can obtain resonance lines 0.1 Hz wide at a primary frequency of 400 MHz, or 2.5 parts in 10^{10}.

2. Probe and rf Coil

Each different magnetic nuclei has a unique gyromagnetic ratio γ. Thus different nuclei in the same field will resonate at different frequencies (see Table I). This should not be confused with different molecules containing the same nucleus resonating at different frequencies (chemical shifts); in this latter case, the actual magnetic field at the nuclear position is different in different molecules, but the γ is the same. Further, the quantitative variations in the various γ's among the different nuclei are very large, while the chemical shifts are, at the largest, 10^3 ppm, and in most cases of interest to us here ~ 100 ppm (see Fig. 2). It is this difference in scale that allows us to distinguish the different types of nuclei from each other in spite of the fact that they are all in the magnetic field simultaneously. This is accomplished by tuning the rf coil to the gross frequency of a particular nucleus. Such rf circuits generally have resonances as sharp as 1 part in 10^3, so there is no chance of detecting the wrong nucleus. In addition, of course, the rf driving frequency will be set to this frequency as well.

For very high-resolution work on small molecules, the standard probes supplied by the spectrometer manufacturer are used for the particular nucleus of interest. These are carefully designed, high-performance components but generally are only able to accommodate standard tubes. For work on isolated organs and other cellular studies, it is often better to modify an old probe (or make a new one) specifically designed for the job at hand. This is possible because the resolution required in these studies is 10–100 times less than that needed for solution work. The resultant probe and rf coil do not have to be capable of spinning the sample, for example. Neither is it necessary to maintain the careful tolerances and symmetry of construction of the commercial probes.

Figure 17 shows such a probe, designed for studying isolated perfused hearts. The rf coil is a single horizontal turn of copper sheet into which fits a glass tube containing the isolated heart. Perfusion lines and other connections are made through stoppers fitted into the glass. For temperature control, the perfusate is brought through a double-jacketed section in the lower part of the probe. The single-turn coil is tuned to the correct frequency by fixed and adjustable capacitors. Without going deeply into the details, two of the adjustable capacitors are to tune the frequency and the other is to insure proper coupling between the resonant circuit of the coil and the coaxial line which connects it to the preamp and power transmitter. These can be adjusted by minimizing the reflected rf power at the correct nuclear frequency, using an rf sweep generator and reflection bridge.

As pointed out, to perturb the spin populations an rf field with a magnetic component perpendicular to the main field must be applied. For reasons of signal intensity, the best way to generate this is to use a solenoidal geometry with the sample inside rather than two separate coils with the sample in between them. The signal is roughly 2.5 times larger in the former case compared with the latter. In the heart probe, which was designed for ^{31}P studies, a single turn was used so that the proton NMR signals from water could be observed to shim the magnet. A single turn is desired because at the much higher proton frequency (360 MHz in this application) the rf currents flowing in a multiturn coil will not be in phase with each other and, therefore, not result in the same rf field distribution across the sample. An effect of this sort makes it difficult to shim the magnet properly, because one is observing different regions of the sample with the protons than with the ^{31}P nuclei and can adjust the shims so the proton signal is homogeneous but the ^{31}P signal is not.

We have discussed this in some detail because, for cellular and organ studies, optimal results are obtained with a dedicated probe, and researchers

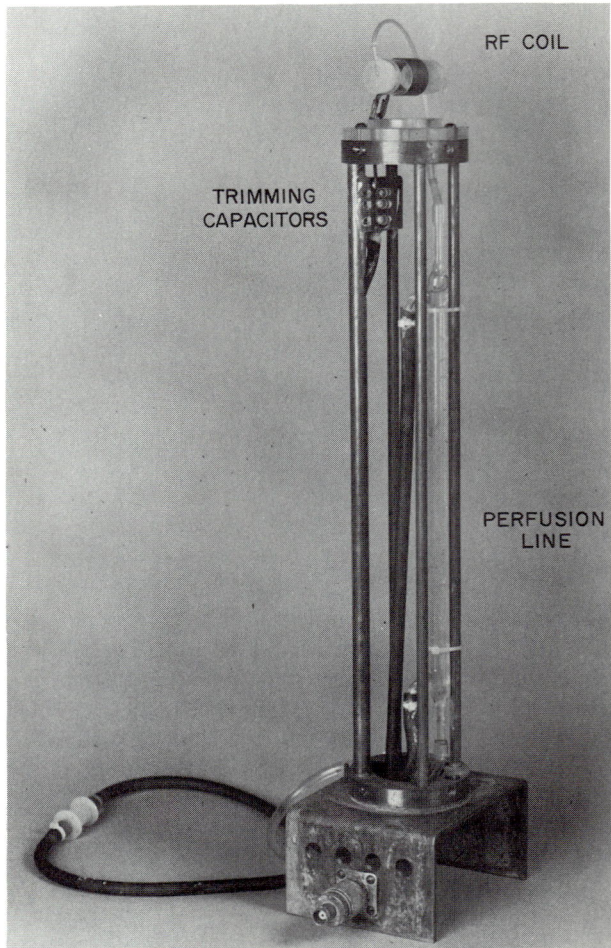

Fig. 17. An NMR probe for isolated heart studies.

are often intimidated by the thought of making their own. In fact, because of the relaxed resolution requirements, it is not difficult to build such a probe. The main knowledge required is that of rf resonant circuits and how to match them to a coaxial line. One wants the Q, or sharpness in frequency, of the tuned circuit to be as high as possible. This is achieved by keeping the circuit components clean and any tubes containing saline solutions away from them insofar as is possible. A discussion of the factors affecting sensitivity can be found in the articles by Hoult and Richards (1976).

A final type of probe geometry which should be mentioned is a surface coil. Originally proposed by Ackermann (Ackermann *et al.*, 1980), it has been used extensively to study intact animals and humans. From an rf point of view, it is nothing more than a flat coil with appropriate capacitors added to make a resonant circuit. The rf field generated by such a coil has a spatial profile which roughly resembles an ice-cream scoop of the same radius. Thus the nuclear spins detected by it lie mostly within this volume, and it is able to provide a certain degree of spatial localization.

3. rf Transmitter and Power Amplifiers

The rf source is a high-quality adjustable sine-wave generator, usually under computer control. It determines the frequency being excited and must be set to the correct nuclear frequency. Its output is then gated, amplified, and sent to the rf coil to generate the rf magnetic field which excites the spins. A typical power level at the output stage is 100 W, which might correspond to a oscillating field of 5 G depending on the details of the probe. One calculates the oscillating field strength by measuring the length of a 90° pulse, τ, and using the formula $H_1 = 2\pi/4\gamma\tau$, where γ is the gyromagnetic ratio for the observed nuclei.

4. Preamplifier and Demodulation Electronics

The purpose of the preamplifier is to amplify the signal to levels where it is insensitive to interference. The demodulation circuits are used to reduce the frequency of the signal into a range where the analog-to-digital converter can function. This is possible because the actual spread in frequency of interest is only a few 100 ppm due to the chemical shifts. Thus, if the nuclear frequency is 100 MHz, the spread of interesting frequencies is only ~ 10 kHz. The reduction in frequency is accomplished by standard demodulation techniques. The only point which deserves mention is that the final stage demodulates both phases of the signal (the sine and the cosine parts), and both are detected. This quadrature detection scheme results in frequencies near the original excitation frequency falling near the center of the final Fourier transformed spectrum.

Because of the low sensitivity of NMR, signal-to-noise (S/N) considerations are very important. The two critical areas in an NMR spectrometer, as far as sensitivity is concerned, are the Q of the rf coil and the quality of the preamplifier. This is simply because those are where the signal is at a low level and hence most susceptible to interference. The performance of the preamp in this regard is determined by its noise figure, which is a measure of how much noise the preamp adds while amplifying the signal. The simplest way to measure this very important parameter is to replace the probe with a

50-Ω resistor, and compare the thermal noise which is detected by the spectrometer at ambient temperature with that detected when the resistor is immersed in liquid nitrogen. The size of the latter should be at least 15–25% smaller than the former for a good preamp. If no change can be seen, then substantial improvement in the present instrumental S/N is possible.

5. Computer

The computer, as has been mentioned already, controls the various frequencies, tuning, and detection circuits and can generate quite complex patterns of pulses and frequencies if desired. It also Fourier transforms the signals, processes, stores, and displays them. Details vary from one spectrometer to another, but the overall structure of the operating software is usually similar.

B. Measurement Techniques

This section is intended as a brief overview of how to measure simple experimental parameters. In addition, various pitfalls will be mentioned. Obviously, it is not possible to present an operational manual for a spectrometer; what we hope to do is make the initial interaction with a complex instrument somewhat less foreboding. For a "hands on" description of operating a spectrometer see Fukusima and Roeder (1981). To save space a number of mathematical results are presented without derivations.

The basis of obtaining an NMR spectrum on contemporary FT instruments have alrady been discussed (Section III.C). A relatively strong rf pulse with a magnetic component on the x–y plane is applied to tilt the bulk magnetic moment **M** away from the z axis. The pulse duration is adjusted to get the desired tilt angle θ, which is always 90° or less in standard measurements. Immediately after the pulse, the component of **M** on the x–y plane (**M** sin θ) precesses about the z axis at its Larmor frequency and decays exponentially with a time constant of T_2. This precession and decay in the x–y plane is recorded as the FID (Fig. 18). Even though the precession is at the Larmor frequency, for instrumental reasons already discussed, one always records only the difference between the Larmor frequency and the operating frequency (i.e., the frequency of the rf pulse). To improve the signal-to-noise ratio, the several rf pulses are applied with a suitable delay time between them, and the FIDs resulting from each pulse are summed up. Summation of n FIDs yield an S/N improvement of $\sqrt{n}$ in the final spectrum. The number of FIDs summed, n, is determined by the desired S/N ratio and how dilute the sample is. The dependence of the signal intensity on the pulse angle and the time between the pulses is given by Eq. (15).

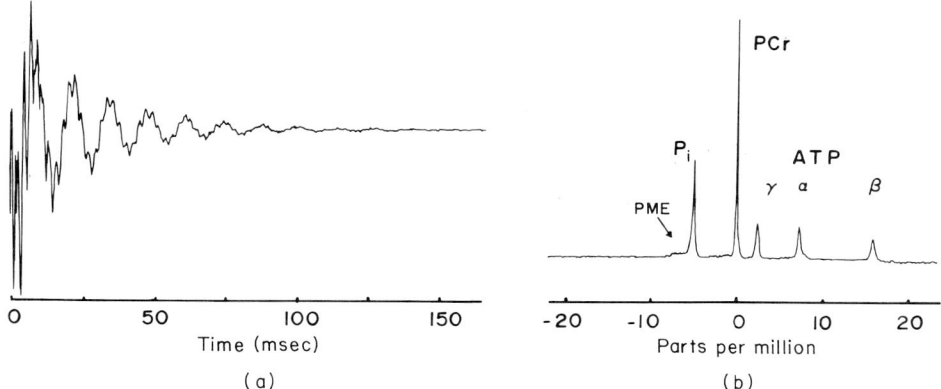

Fig. 18. Free induction decay (FID) signal (a) and its Fourier transform (b) from an isolated cat biceps muscle. Twenty FIDs were summed to increase S/N. The absorption peaks are identified as the three phosphates of ATP (α, β, and γ), phosphocreatine (PCr), inorganic phosphate (P_i), and a small amount of phosphomonoesters (PME).

The FID represents an NMR spectrum in the time domain. For a sample with a single resonance it will be of the form

$$F(t) = \cos[(\omega_L - \omega_R)t] \exp(-t/T_2^*)$$

for one of the two quadrature detection channels. In this expression, ω_L and ω_R are the Larmor frequency and the rf pulse frequency, respectively; T_2^* the apparent T_2, which contains contributions from the true T_2 and the inhomogeneities in the H_0 field; T_2^* equals $(\pi \Delta \nu)^{-1}$, where $\Delta \nu$ is the linewidth observed in the NMR spectrum. The second quadrature channel will record a similar function, which will be 90° out of phase with the first; in other words, $\cos(\omega_L - \omega_R)t$ will be replaced by $\sin(\omega_L - \omega_R)t$. When the sample contains several different signals owing to different chemical shifts, the FID will be the sum of signals coming from each type of spin precessing at its characteristic frequency and decaying with its own T_2^*. Thus we have

$$F(t) = \sum_i \cos[(\omega_{Li} - \omega_R)t] \exp(-t/T_{2i}^*).$$

The Fourier transform of $F(t)$ is the conventional NMR spectrum (Fig. 18); the Fourier transformation simply converts the time domain spectrum (i.e., the FID) into the frequency domain spectrum. (See the appendix for a discussion of Fourier transformation).

The primary experimental parameters obtainable from an NMR spectrum are the frequency of a resonance, its width, and its area. Generally, one is not interested in the absolute frequency of a resonance but rather its separation from some standard resonance. For reasons discussed in the section on chemical shifts, this separation is measured in parts per million (ppm) so as to be independent of the field strength of the spectrometer. Thus the reported frequency of a resonance is so-and-so many ppm relative to such and such as the standard. If one is studying the shift in frequency of some resonance, say, upon titration, obviously the standard used must be independent of the titrating compound. An apparently simple way to guarantee this, namely, to put the standard in a different compartment within the same pickup coil, should be used with caution, because under these circumstances the magnetic field in the standard is related to the magnetic field in the actual sample by the magnetic susceptibility of both solutions. This can cause systematic errors in the measured frequency difference between the lines. The best answer to this problem is to find a reference compound which is unaffected by the titration and which can also be added to the solution.

To measure the linewidth of a resonance line is easy (provided resonance overlaps, or steeply sloped baselines are not present). Difficulties arise, however, when trying to assign the causes of the width. Generally, several components need to be untangled. The simplest is that caused by the variation of the magnetic field over the sample. This can be minimized by adjusting the shims, as discussed previously. In homogeneous solutions, this contribution can usually be reduced to negligible proportions; in cellular suspensions, organs, and whole animals, it is often the dominant contribution to the linewidth. There are techniques for measuring the "true" or the intrinsic linewidth a resonance; the intrinsic width is equal to $1/\pi T_2$. T_2 can be measured by the method of spin echoes described farther on. In these measurements, repeated 180° rf pulses cancel the variations of the external magnetic field and allow the true decay of the resonance to be observed. The method was first described by Carr and Purcell (1954) and was later refined by Mieboom and Gill (1958).

The area of a resonance line is always a difficult measurement to make, even though it is simply the integral of the peak. The difficulties arise from the noise in the base line and uncertainty as to where to stop integrating. Theoretically, if the line shape and position are *known,* a condition not always fulfilled, then the optimal estimate of the area is the convolution of the theoretical shape with the experimental points appropriately normalized. Effectively, one is weighting the different portions of the line in proportion to their height.

In addition to the difficulties of determining the area of an observed resonance, there is a further problem in NMR which is not encountered in most other spectroscopic techniques; namely, the possibility that the resonance is saturated, so that the area is less than it should be for the number of nuclei present in the sample. Mathematically, the Bloch equation for the z component of magnetization shows that it will recover exponentially with time constant T_1. Thus, following a 90° pulse which renders $M_z = 0$, we have

$$M_z(t) = M^0[1 - \exp(-t/T_1)]. \tag{14}$$

Clearly, full recovery of the M_z to the thermal equilibrium value of M^0 requires a delay which is several T_1's. If another 90° pulse is applied before full relaxation occurs, the observed magnetization will be smaller than M^0. In general, the steady-state magnetization with rf pulses of angle θ repeated every T_0 seconds is

$$M_{ss} = M_0 \frac{1 - \exp(-T_0/T_1)}{1 - \cos\theta \exp(-T_0/T_1)}. \tag{15}$$

Thus, as the repetition rate increases the steady-state magnetization decreases. The signal observed under these conditions is $M_{ss} \sin\theta$, since only the portion of the magnetization in the x-y plane will rotate around the H_0 field in the laboratory frame and produce a signal in the pickup coil. We see, then, if the signal areas are to be accurate representations of the number of nuclei present, we have to pulse either very slowly or very lightly (i.e., small θ). Of course, if the T_1's of the resonances are known, then any saturation effects can be calculated and corrected.

Unfortunately, the conditions of no saturation are not the optimal ones for data accumulation, because considerable time is spent letting the signal recover all the way back to its equilibrium value. A more optimal approach, from a signal-to-noise point of view, is to pulse more rapidly, suffering some saturation but accumulating more pulses in a given time. In fact, using Eq. (15) it is possible to show that an optimal S/N for a resonance with a particular T_1 is obtained when the pulse angle θ has the following relationship to the repetition time T_0:

$$\cos\theta = \exp(-T_0/T_1). \tag{16}$$

This relation is graphed in Fig. 19. Note that at a repetition time equal to the T_1 the optimum pulse angle is 68°. In general, for maximum sensitivity one should pulse at least as fast as T_1, although, if θ is chosen according to Fig. 19, even at $T_0 = 3T_1$ the sensitivity is down by only 25%. Fortunately, the S/N is only a slow function of θ around the optimum, and it is not necessary to set the rf pulse length more accurately than 10% for adequate results.

Note that Eq. 15 also provides a basis for determining T_1's by measuring the magnitude of the steady-state magnetization (i.e., the peak intensities) as

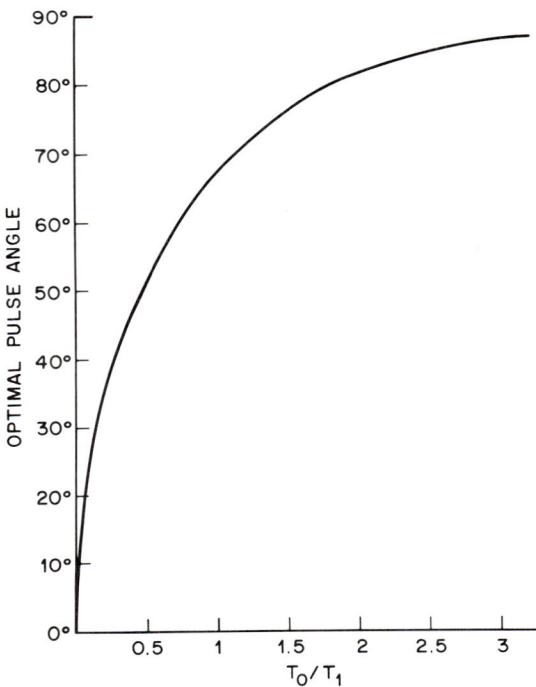

Fig. 19. Optimal pulse angle as a function of the ratio of pulse interval T_0 to spin–lattice relaxation time T_1.

a function of different repetition times, using a fixed pulse angle. This method of T_1 measurement is called saturation recovery (Freeman and Hill, 1971). Generally, pulse angles are set to 90° so that the θ dependence is eliminated and M_{ss} becomes a simple function of the repetition time alone. This method of T_1 measurement is particularly useful in the limit $T_1 \gg T_2$, such as in the case of macromolecules and intact cells. This is because, before a 90° pulse is applied, the magnetization on the x–y plane from the previous pulse must have vanished; since this magnetization decays exponentially with a time constant T_2, $T_0 \gtrsim 3T_2$ must be the case.

Another commonly used method of T_1 measurement is inversion recovery. In this method, the spin populations are first inverted by a 180° pulse, leaving the macroscopic magnetization **M** pointing in the $-z$ direction. Relaxation is allowed to occur for a time τ; subsequently, a 90° pulse is applied to measure the magnitude of $\mathbf{M}_z$. This sequence is repeated for n scans and the FIDs for a fixed τ are summed up for S/N improvement. Then, the whole procedure is repeated for different delay times τ. The pulse sequence therefore can be outlined as $(180°-\tau-90°-T_0)_n$, generating N spectra for N different τ values. The measured peak intensities depend on τ according to

$$M_z(\tau) = M_0\{1 - [2 - \exp(-T_0/T_1)]\exp(-\tau/T_1)\}.$$

Generally, T_0 is set equal to or greater than $5T_1$, and the equation is simplified to

$$M_z(\tau) = M_0[1 - 2\exp(-\tau/T_1)].$$

This derivation assumes a perfect 180° pulse. In practice this is difficult to obtain. Hence, the data analysis should be (but often is not) performed using a modified equation:

$$M_z(\tau) = M_0[1 - 2\alpha \exp(-\tau/T_1)],$$

where $(1 - 2\alpha)M_0$ is the intensity of the magnetization vector along the $-z$ direction immediately after the imperfect 180° pulse.

Measuring T_2 is, in general, more difficult than measuring T_1 because any variation of the magnetic field over the sample volume spreads out the resonance line, making the apparent T_2 derived from the linewidth appear shorter. Overcoming this problem requires the use of more complex pulse sequences than those we have discussed so far. The basis for all the sequences used for this problem is the idea of a spin echo in which the spins in different parts of the sample initially dephase in the $x-y$ plane following a 90° pulse, then, following a 180° pulse, rephase until they are coherent again. We can illustrate this process by considering just two spins in different parts of the sample at slightly different frequencies, ω_1 and $\omega_1 + \Delta\omega$, due to field inhomogeneity. As shown in Fig. 20a, immediately following a 90° pulse the magnetization from the two points in the sample will be parallel, pointing along the y axis. We will consider the evolution of this problem from the frame rotating with frequency ω_1. In this frame one vector will remain stationary along the y axis and the other will rotate with frequency $\Delta\omega$, so that after a time τ it will have opened up an angle of $\Delta\omega\tau$, as indicated in Fig. 20b. If we now apply a 180° pulse around the x axis, as shown in Fig. 20c, the angle between the magnetization vectors will remain constant, but now the M_2 vector will be moving to close the angle instead of opening it. After a time τ the two vectors will be in phase again, and an echo will be seen (Fig. 20d). Obviously, this argument will apply to a spread of frequencies as well, so that the entire distribution of frequencies caused by the field variations will be refocused causing echoes. The echo amplitude will then reflect the true spin-spin relaxation time T_2.

In actual experiments one uses a series of multiple 180° pulses developed by Carr and Purcell (1954) to avoid difficulties due to diffusion of the spins

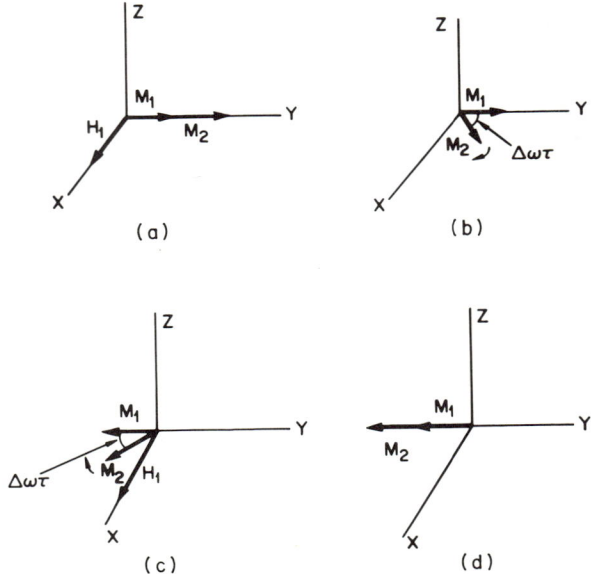

Fig. 20. Events in the formation of an echo following a 90°–τ–180° pulse sequence. See text for details.

from one field value to another. For more details and possible pulse sequences, we refer the interested reader to the book on Fourier transform NMR by Farrar and Becker (1971).

V. Applications

In this section we will present examples of the use of NMR in the areas of structural determination, dynamics, kinetics, cellular studies of metabolism, and a variety of other applications.

A. Structure

Nuclear magnetic resonance (NMR) spectroscopy has been extensively used to obtain structural information of varying complexity and sophistication. The routine use of NMR for identification of organic compounds stems from the sensitivity of NMR parameters to structure. The resonance frequencies of a methyl carbon or methyl protons give rise to peaks which are well separated from those of a methylene carbon or protons, respectively.

Therefore, just the knowledge of the chemical shifts provides considerable constraint on the possible structures available to the molecule under examination. On identification of the observed resonances with particular nuclei, traditionally through chemical modifications, more detailed structural information can be obtained using the measured scalar coupling constants T_1, T_2 and the NOE to construct a model. Of course, it must be kept in mind that what is being observed is the average-solution structure.

When molecules with an aromatic moiety are being studied, structural information is also contained in the chemical shifts of the observed resonances. The magnitude and the direction of the field at a given point in space depends on the distance and the position relative to the ring. In fact, at distances larger than the aromatic ring, the magnetic field of the ring current appears as that of a magnetic dipole. When a nearby nucleus encounters this field, its resonance frequency and apparent chemical shifts change. Such ring-current shifts are seen when aromatic rings are stacked, as in DNA or polynucleotides. Also in proteins, due to the constraints imposed by the three-dimensional structure of the molecule, many side chains are positioned sufficiently close to the aromatic amino acids to experience a ring-current shift.

We will illustrate the use of some of these techniques with the ATP molecule. Subsequently, more sophisticated applications to peptide and protein structure will be discussed. Several structural questions about the ATP molecule have been examined by NMR. These include the geometry of meta-ion binding to ATP, the solution conformation of the ATP molecule (*anti* versus *syn*), and the structure of molecular complexes formed between ATP and other aromatic molecules, such as the biogenic amines.

Very frequently, reactions which utilize ATP have an absolute requirement for a divalent cation such as Mg^{2+} or Mn^{2+}; the cation may be tightly bound to the enzyme, or it may be bound to the ATP, and the enzyme recognizes and uses the cation–ATP complex. In cells most of the ATP exists as Mg^{2+}–ATP. Naturally, there have been efforts made to delineate the metal binding site(s) and the overall structure of the ATP–cation complex. When the cation used is the paramagnetic Mn^{2+} ion, distance information is yielded by dipole–dipole interactions between the electron spin localized on the metal and the various spins (^1H, ^{31}P, and ^{13}C) of the ATP molecule. Due to the very large gyromagnetic ratio of an electron, the magnitude of its dipole field is very large; consequently, in the presence of such paramagnetic ions, the dominant relaxation of the nuclei is through dipole–dipole coupling to the unpaired electron(s). This causes the T_1's and the T_2's to become very short and increases the linewidths. The coupling, however, diminishes very rapidly as the separation from the electron(s) increases, so that not all resonances are affected to the same extent.

Addition of divalent cations to a solution of ATP strongly affects the ^{31}P resonances of this molecule. If the divalent cations are diamagnetic, such as Mg^{2+}, Zn^{2+}, or Ca^{2+}, the primary effect is a shift of the γ- and β-phosphate resonances. If the cations are paramagnetic, ^{31}P resonances become broader (Fig. 21), in particular, the β and γ resonances. These ATP–cation interactions were first demonstrated by Cohn and Hughes (1962), using ^{31}P and ^{1}H NMR, and subsequently were studied extensively by others [e.g., Sternlicht et al. (1968), Brown et al. (1973), and Tanswell et al. (1975)]. It was concluded from these studies that the nature of the binding of these ions to the ATP molecule is such that they bridge the β and γ phosphates.

ATP and other nucleotides have two possible configurations about the glycosidic bond, *syn* and *anti* (Fig. 22). In many macromolecules where aromatic rings are stacked, the *anti* conformation is preferred. In dilute solution, however, they exist primarily in the *syn* form, as has been deduced

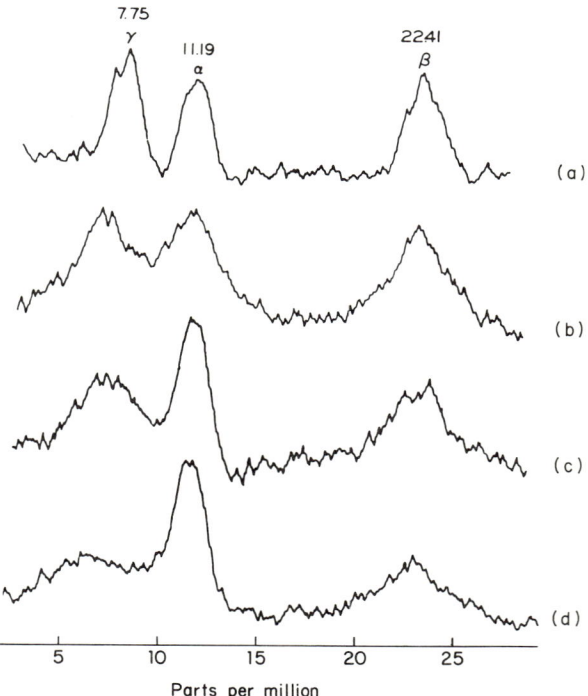

Fig. 21. Effects of paramagnetic divalent cations on the ^{31}P spectrum of ATP. The sample was 0.1 M ATP at pH 7.2 with no additions (a), with 8×10^{-5} M $MnCl_2$ (b), with 1.65×10^{-5} M $CuCL_2$ (c), with 3.15×10^{-5} M $CuCL_2$ (d). (From Cohn and Hughes, 1962.)

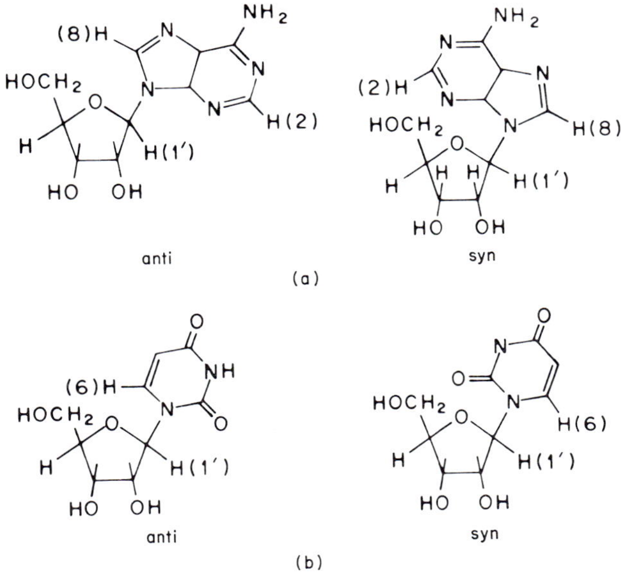

Fig. 22. Syn and anti configurations about the glycosidic bond of nucleosides in solution: (a) adenosine and (b) uridine.

from the NOE measurements using ^{1}H NMR (Son et al., 1972; Bothner-by, 1979).

In the absence of paramagnetic ions, the main relaxation of an ^{1}H spin is dipole–dipole coupling to other ^{1}H spins. On the basis of this, one can predict that the H8 proton of the adenine moiety would have a slower relaxation rate (i.e., a longer T_1) in the *anti* conformation than in the *syn*. This is because of the close proximity of the H8 proton to the ribose protons in the *syn* conformation. Similarly, the adenine H2 proton is closer to the H1′ of the ribose in the *syn* conformation and further away in the *anti* (Fig. 22). The extent of dipole–dipole coupling between the H1′ and H2 or H1′ and H8 can be examined by the NOE effect. Irradiation and saturation of the H1′ resonance should lead to an NOE effect, the magnitude of which will depend on the conformation. In fact, NOEs were observed on the H2 but not the H8. Given the fact that the H8–H1′ distance is comparable to the H2–H1′ distance in the *anti* but much larger in the *syn*, it was concluded that *syn* conformation dominates in dilute solution (Son et al., 1972). Interestingly, this NOE disappeared at higher nucleotide concentrations, presumably due to some degree of stacking and, consequently, a preference for the *anti* conformation.

When aromatic molecules form molecular complexes, all the nuclei of each molecule experience shifts due to the ring currents of the other. The magnitude of the shifts are determined by the proximity and relative orientation of the nuclei to the aromatic rings. From such shifts, one can deduce the geometry of the complex. An example of this type of work with small molecules is the NMR studies of catecholamine–ATP interactions (Granot and Fiat, 1977; Granot, 1978).

Catecholamines are sequestered together with ATP at high concentrations within subcellular vesicles of the adrenal medulla and neurons. Since both the catecholamines and ATP contain aromatic rings, formation of complexes between the two with stacking of the aromatic rings was anticipated.

In aqueous mixtures of ATP and catecholamines, the protons of both molecules experience upfield shifts. If the binding is not strong enough, a solution of these molecules will contain a mixture of uncomplexed ATP, uncomplexed amine, as well as the ATP–amine complexes. Each molecule will be exchanging between the free and the bound forms. With ATP and catecholamines, the exchange is "fast" (see Section III) compared to the chemical-shift difference between the free and bound forms. This results in the observation of a single resonance for each nuclei despite the presence of multiple solution forms. The chemical shift of this resonance is an average of the intrinsic chemical shifts of the ATP–amine complex and the free forms. To obtain the geometry of the complex, a knowledge of the intrinsic chemical shifts within the complex is required. Typically, one keeps the concentration of one of the chemical species constant and varies the concentration of the other. A plot of the dopamine (DA) protons is shown as a function of increasing ATP concentration (Fig. 23) (Granot and Fiat, 1977). These curves can be analyzed to extract the formation constants and the chemical shifts of these resonances in the bound state.

To determine the geometry of the associating molecules from the chemical shifts in the complex, calculations of magnetic fields resulting from the ring currents of aromatic rings are needed. Such a calculation was initially performed for benzene (Johnson and Bovey, 1958) and subsequently extended to several other aromatic molecules such as amino acids, purines, and pyrimidines (Giesner–Prettre and Pullman, 1970, 1971; Giesner–Prettre et al., 1976). In the ATP–catecholamine binding study (Granot, 1978) isoshielding contours for the ATP and the catechol rings were calculated. These contours and the geometries consistent with the calculated intermolecular shielding effects and the chemical shifts of the bound complex are illustrated in Fig. 24. It should be mentioned that the stoichiometries of the association affects the analysis. For the ATP–DA association, it was concluded that 1:1 and 1:2 complexes are the dominant species.

1. NUCLEAR MAGNETIC RESONANCE   47

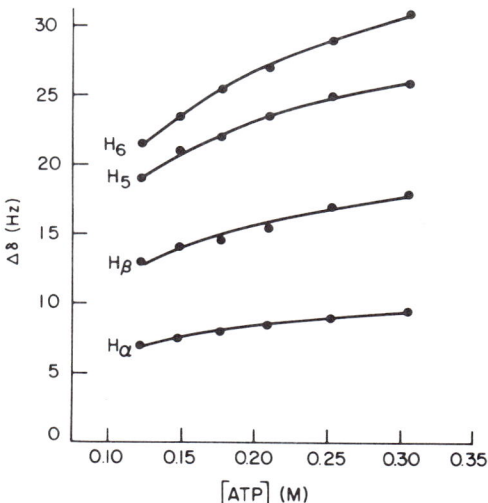

Fig. 23. Shifts induced in dopamine protons by increasing concentrations of ATP in aqueous mixtures of dopamine and ATP. (From Granot and Fiat, 1977. © 1977 American Chemical Society.)

1. Peptides, Proteins, and Polynucleotides

Early studies on peptide conformation primarily used scalar coupling constants. The most commonly used scalar coupling in structural studies is the vicinal coupling between protons bonded to adjacent atoms. This coupling is mediated through three bonds; its magnitude depends on the torsional angle θ, which defines the rotation about the bond between the atoms to which the hydrogens are attached. For the specific case of hydrogens bonded to two adjacent sp^3 carbon atoms, the following angle dependence of the coupling constant was derived (Karplus, 1959):

$$J = 8.5 \cos^2 \theta - 0.28, \quad 0 \leq \theta \leq 90°,$$
$$J = 9.5 \cos^2 \theta - 0.28, \quad 90° \leq \theta \leq 180°.$$

The minimum and maximum values of J occur when θ equals 90° and 180°, respectively. For adjacent sp^3 carbons, there are three values of θ which correspond to stable, minimum-energy configurations (Fig. 25). These are 60°, 180°, and 300°; the first and the last are equivalent. From the given equations, we see that J is ~ 1.8 Hz for the *gauche* and ~ 9.2 Hz for the *trans* form. Caution must be exercised in using these equations, because

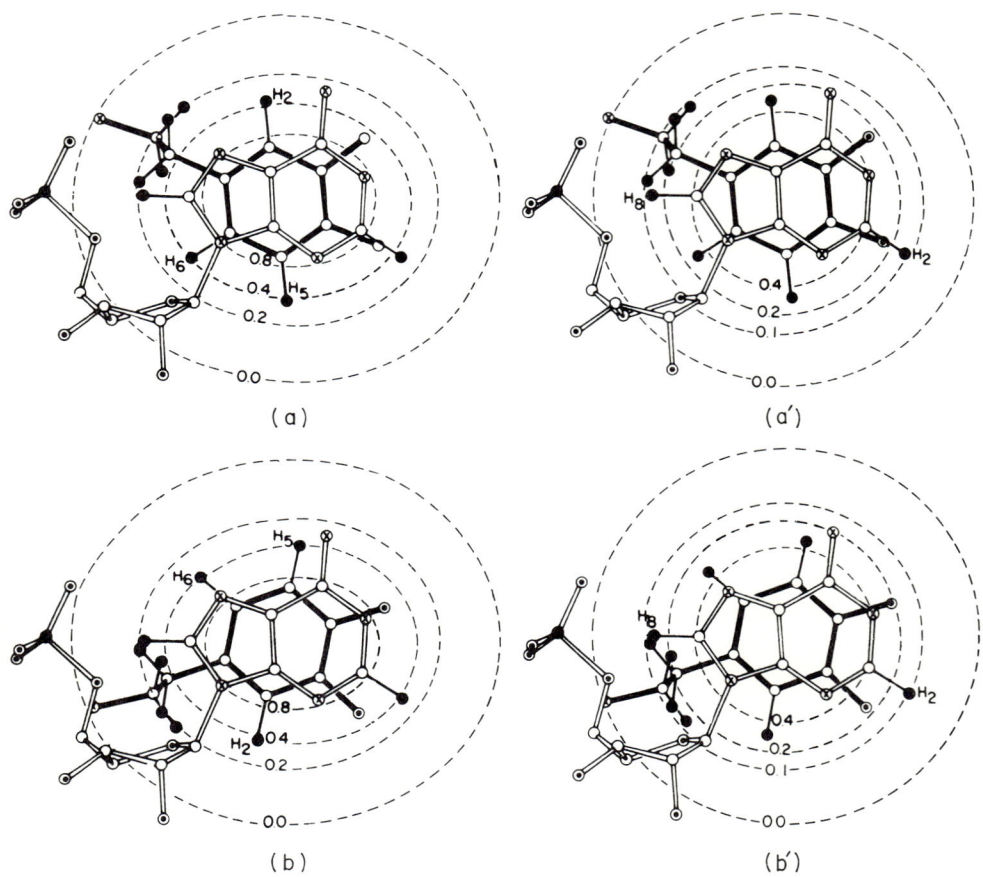

Fig. 24. Ring-current isoshielding contours and the two (a and b) preferred intermolecular geometries for a 1:1 ATP–dopamine complex. Isoshielding contours are drawn for the adenine (a and a') and the catechol rings (b and b'). The aromatic ring protons experiencing the ring-current shifts are labeled H_2, H_5, and H_6 for Da, and H_2 and H_8 for ATP. (From Granot, 1978). © 1978 American Chemical Society.)

parameters such as the nature of the other atoms or groups bonded to the carbons also affect the magnitude of J.

In biological systems, the vicinal coupling has been used extensively to determine the torsion angle ϕ in peptides (Fig. 26). If it is assumed that the peptide bond is planar as a consequence of its partial double-bond character, then the polypeptide backbone conformation can be uniquely specified with a set of torsion angles ϕ_i and ψ_i (Fig. 26). The coupling constant between protons attached to the α-carbon and the nitrogen of the same amino acid residue depends on ϕ. This angular dependence is generally expressed as

$$J = A \cos^2 \theta + B \cos \theta + C \sin^2 \theta,$$

where $\theta = |\phi - 60°|$ defines the relationship between θ and the standard torsion angle ϕ used with peptides. A, B, and C are coefficients which are determined empirically. Typical values proposed and used for these constants vary from 8.6 to 9.8 for A, -3.5 to -0.4 for B, and 0.1 to 1.5 for C (Bystrov et al., 1969; Thong et al., 1969; Ramachandran et al., 1971). These are determined from studies on model compounds with fixed geometries. The ambiguities in these empirically determined constants arise from the uncertainties in the configuration of the model compounds, from the small but nonnegligible effects of the ψ angle on this coupling constant, and from the effects of the groups attached to the α carbon. The presence of motion in peptide chains dictates that there cannot be a single, fixed value of ϕ and ψ; instead an average over the allowed values for these angles must be considered. This, however, generates an additional ambiguity because how the different values of ϕ should be weighed in the average is not straightforward. The simplest procedure is to consider all values of ϕ equally possible.

This type of analysis is ultimately limited by the ability to separately detect and identify all $C_\alpha H$ and NH resonances. Even in the very high fields available in contemporary spectrometers, such resolution can be observed only with very small peptides ($\lesssim 20$ amino acids). Other practical factors, such as the fact that protons of some of the NH groups may exchange with solvent protons or deuterons (in deuterated solvents) and that solvents also affect the vicinal coupling constants, introduce further complexities into this type of analysis (Bystrov et al., 1973).

As in the example with ATP, interactions with paramagnetic ions have

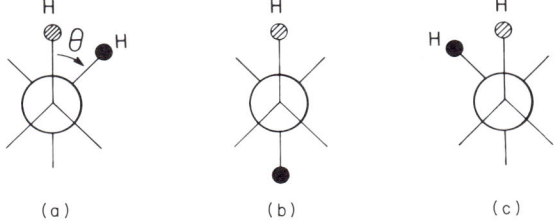

Fig. 25. Stable rotational states about the carbon–carbon bond for two sp^3 carbon atoms: (a) $\theta = 60°$, gauche; (b) $\theta = 180°$, trans; (c) $\theta = 300°$ ($=60°$), gauche.

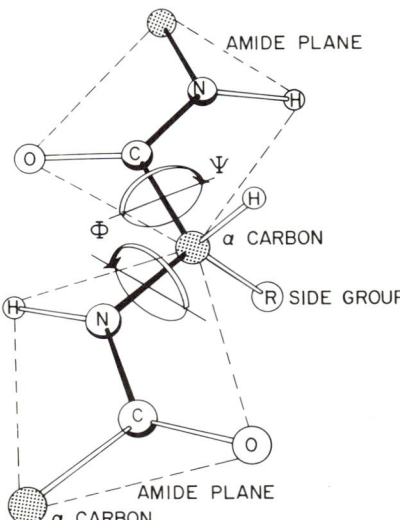

Fig. 26. Torsional angles Ψ and Φ in a peptide bond.

been used for obtaining structural information with proteins. Many of these applications utilized the fact that some proteins contain a tightly bound paramagnetic ion. An example of this is the Cu^{2+} containing protein azurin; this is a small-molecular-weight protein from *Pseudomonas,* which contains a single copper ion per protein molecule. The copper is easily converted between its two redox states using electron donors and acceptors with the appropriate redox potentials. Cu^+ is diamagnetic, whereas Cu^{2+} is a paramagnetic ion which has a single unpaired electron. The presence of the tightly bound copper provides a center from which distances to other nuclei can be measured based on dipolar coupling between the electron spin and the magnetic nuclei. In this type of analysis a knowledge of τ_c is required. However, if there exist three or more nuclei whose relaxation parameters are affected by the Cu^{2+} and whose positions relative to each other are known, both the correlation time and distances to the copper can be derived. In the ^{13}C spectra of azurin, only two such resonances exist, the C^γ and C^{δ_2} carbons of trp-48; these two nuclei can only be separated by their bond length. This information was used to get an upper limit for the distances from the Cu to these carbon atoms, as well as an upper limit on the dipolar correlation time (Uğurbil and Bersohn, 1977; Uğurbil *et al.,* 1977). Subsequently, using ^1H as well as ^{13}C NMR data on azurin and this upper limit on the correlation time, several predictions were made on the location of the various residues. These in general were found to be consistent with the x-ray structure of azurin and an analogous protein, plastocyanin.

Generally, to calculate accurate distances from this type of data, one needs to measure changes in the relaxation parameters, preferably for both T_1 and T_2, at several different field strengths. With such data, a prior knowledge of the correlation time is not needed, since the problem is overdetermined. An extensive discussion of this type of analysis is found in Chapter 10 of the monograph by Dwek (1973).

A final point to note in applications of this type is that often resonances from the paramagnetic protein or the molecular complex where the paramagnetic ion is bound cannot be observed because of very large linewidths. In such cases, one examines mixtures of the paramagnetic and diamagnetic species. If *fast exchange* occurs (Section III), then the observed linewidths, T_2's and T_1's, are averages of the values for both species. By varying the relative amounts and measuring the NMR parameters in the mixtures, one can deduce what the T_2's and T_1's of the nuclei are in the purely paramagnetic species, in spite of the fact that these resonances are not directly observable.

More recent efforts in the field of structural studies with biological molecules utilize the homonuclear NOE among ^1H nuclei. As previously discussed, NOE between a pair of spins is inversely proportional to the sixth power of the distance separating them. To extract exact distances from the NOE data, the correlation time τ_c for the fluctuations of the dipole–dipole interactions is needed. This is usually available from other experiments; for example, relaxation of ^{13}C nuclei is generally dominated by the dipole–dipole coupling to the directly bonded protons. Since the bond lengths are fixed, knowledge of T_1, T_2, and heteronuclear NOE between ^{13}C and ^1H nuclei can be used to calculate τ_c (Section III).

The NOE has been utilized for distance calculations with several biological molecules. NOEs between a $C_\alpha H$ proton and the NH proton of the next amino acid residue were used to calculate distances between these nuclei to determine backbone and side-chain conformation in several small peptides (Wüthrich *et al.*, 1981). This distance was shown to be a function of the torsional angle ψ (Fig. 26), assuming that the peptide bond is planar. Such measurements have been performed on numerous other peptides (review: Bothner-by, 1979).

In another example with larger biological molecules NOEs between T_{54} methyl and ψ_{55} amino protons of tRNA were used for distance calculations (Tropp and Redfield, 1981); based on this data, the geometry of the stacked T_{54} and ψ_{55} rings were deduced. In this study, for example, τ_c obtained from T_1's of the ribose carbons and fluorescence depolarization measurements were used in the calculations. In experiments with lysozyme (Poulsen *et al.*, 1980), it was shown that the magnitude of the NOE between several pairs of nuclei with varying separations agreed well with the distances known from the x-ray structure of the molecule; this work indicates that, even with macromolecules of this size and even in the presence of multiple dipolar couplings among ^{1}H nuclei, reliable distance information can be extracted.

A recent approach for protein structure determinations proceeds on the premise that knowledge of a large number of relatively inaccurate distances between pairs of atoms provides sufficient constraint on the possible conformations and can therefore be used to arrive at the correct structure. The very existence of NOEs between pairs of spins is adequate for establishing such inaccurate distances. This type of analysis has been successfully used on BPTI and lipid-bound peptide glucagon (Billeter et al., 1981; Wager and Wüthrich, 1981; Wüthrich et al., 1981; Wider et al., 1981).

In many instances, just the knowledge of the presence or absence of NOEs between pairs of spins provides highly desirable structural information even without any distance calculations. An example of this is the studies with polynucleotides (Patel et al., 1982). It was shown that a large NOE is observed between H1′ and H8 protons of guanine when the polynucleotide is in the Z-DNA conformation. This is because in Z-DNA the guanyl residues are in the *syn* conformation, which brings the H1′ and H8 protons within ~2.6 Å of each other. In the *anti* conformation, which is adopted in the B-DNA form, the separation is much larger, and an NOE is not observed.

The primary restriction in all these studies utilizing different methodologies is the ability to resolve and assign resonances to single protons or groups of protons attached to a single carbon. The ability to resolve ^{1}H resonances is improved with higher magnetic fields. The resolution problem gets worse as the molecular weight and, consequently, the number of ^{1}H nuclei which the molecule contains increase. In addition, as the molecular weight increases, the rotational motion slows down, and the interactions which contribute to relaxation are not averaged out as completely. The result is shorter T_2's and therefore larger linewidths, which compound the lack of resolution. So far, the largest molecule which has been studied with any degree of success is lysozyme. The strategies used for assigning the observed ^{1}H resonances in BPTI and in lysozyme primarily involve chemical modification and studies of proteins from different species, with minor alterations and a systematic study of NOE between peptide NH and C_αH of adjacent residues, as well as the two-dimensional (2D) FT techniques (see Freeman and Morris, 1979 for a review).

The 2D FT techniques are a major advance in assigning resonances, comparable to 2D gels in biochemistry. This is so for much the same reasons, namely, another axis along which to spread things out makes overlapping peaks less likely. In the case of 2D NMR spectra, the two dimensions are the ordinary chemical-shift axis and either an axis sensitive to only the J coupling between the spins or an axis which is just the chemical-shift axis but created in such a way that resonances which are coupled (by NOE or chemical exchange or J coupling) have cross peaks. This is accomplished by a sequence of pulses which excite the spins, allow them to evolve for a time t_1, then perturb them in some way (depending on the desired second axis), and finally excite them again and collect the FID. This process is repeated for different values of t_1 until an entire series of FIDs has been collected, each corresponding to a different t_1. This series is then Fourier transformed with respect to t_1 as well as the normal time variable of the FID to provide the 2D frequency spectrum.

B. Dynamics

The potential of NMR spectroscopy for studying molecular motions was indicated by the discussion on spin relaxation (Section III). Relaxation parameters T_1 and T_2, the NOE phenomenon, and resonance linewidths are all dependent on molecular motion; in those cases where the relaxation mechanism can be identified and where the nonmotional parameters entering into the relaxation are known, it is possible to extract dynamical information. Two such cases are the relaxation of protonated ^{13}C nuclei and deuterium nuclei. The former is dominated by the dipolar coupling between the ^{13}C and the directly bonded protons and the latter by the quadrupolar interactions. In both cases, the spatial parameters of the coupling are determined by the carbon–hydrogen or carbon–deuterium bonds, and therefore are known. The majority of the studies aimed at obtaining information on the motions of biological molecules in fact utilize these two nuclei. To a lesser extent ^{19}F, ^{15}N, and ^{31}P NMR have also been employed.

NMR studies do not provide a detailed time course of the motion giving rise to the spin relaxation. Instead, one tries to obtain a measure of the correlation time τ for the motion in question. In most cases, as we shall discuss, the relaxation data is in fact insufficient to extract this information in a unique way. Hence, the approach becomes very model dependent; based on physical constraints and other physical considerations, possible motions are considered, and a model is developed. Subsequently, the ability of the model to account for the NMR relaxation data is tested. Even though this approach in some cases cannot identify a single specific model, it still can provide important insights into the dynamical behavior of molecules or sections of molecules.

Let us consider different types of motions that can contribute to nuclear relaxation. First of all, there are constraints on both the amplitude and the rate of motion. Very slow processes, such as unfolding and refolding of the peptide chains or motions of very small amplitude, such as a methyl group executing a small-amplitude oscillation in a deep potential well, do not contribute to spin relaxation. In solution studies, especially with macromolecules, interactions between spins of a different molecule also are insignificant. Hence, translational motions become unimportant. As a result, one generally is concerned with overall rotation and the relatively rapid motions executed by the flexible parts of the molecule.

The simplest case is when the molecule is approximately spherical and rigid. Then the motion in question is isotropic rotational diffusion of a hard sphere, which is characterized by a single correlation time. For a protonated ^{13}C nucleus in such a molecule, the relaxation parameters are given by Eqs. (12) and (13), and the correlation time is unambiguously determined from measurements of T_1, T_2, and NOE. The rotation of many globular proteins is approximately isotropic. In small globular proteins, this is the only motion which appears to contribute to the relaxation of the backbone α-carbons.

For nonspherical molecules, the rotation is anisotropic, requiring three rotational-diffusion constants to describe its motion. For an ellipsoidal molecule, relaxation rates for anisotropic rotational diffusion have been derived (Shimuzu, 1962; Woesner, 1962a,b). In the new expressions for T_1, T_2, and the NOE, the spectral density terms which are of the form

$$J(\Omega) \propto \tau/(1 + \Omega^2\tau^2)$$

and which appear in the case of isotropic motion [Eq. (10)–(14), where $|\Omega|^2$ takes on values of ω^2, $4\omega^2$, $(\omega_A - \omega_B)^2$, $(\omega_A + \omega_B)^2$, etc.] are each replaced by a sum over five similar terms, each with its own correlation time:

$$J(\Omega) = \sum_{n=1}^{5} \frac{a_n\tau_n}{1 + \Omega^2\tau_n^2}.$$

The situation is simplified considerably if the molecule is axially symmetric (i.e., approximates a rigid rod). Then each $J(\Omega)$ is reduced to a sum of two terms with only two distinct correlation times.

Superimposed on the overall rotation of the molecule is the motion executed by its segments within the constraints of the molecule. For example, in a protein, amino acid side chains can execute rotations about C—C bonds. A surface methyl group of an alanine residue may be able to rotate freely about the C_α—C_β bond, thereby affecting the relaxation of the $^{13}C_\beta$ nucleus. A lysine side chain can execute rotations about its various C—C bonds; if the segments of the side chains are not hindered by the presence of a nearby atom, the motions will be cumulative, and each carbon of the side chain from C_α to C_ϵ will undergo increased motion in addition to the overall tumbling of the molecule. Thus, qualitatively, one can predict that in the absence of specific interactions and molecular packing that restrict the motion, the relaxation of carbons should become slower (i.e., longer T_1's), and

NOEs should increase as one goes from the α carbon to the terminal carbon of the sidechain. This is illustrated for a poly-L-lysine sample (Table II) (Wittebort et al., 1980). To extract quantitative information about the motion of the side chains, however, requires consideration of specific models. Initial studies primarily considered free internal rotation and utilized the model developed by Woesner (Woesner, 1962b) for a single internal rotation (e.g., that of a methyl group). For side chains with more than one carbon, a model was used in which each successive carbon was assumed to be undergoing independent axial diffusion about the C—C bonds (Wallach, 1967). However, in biological macromolecules most side chains cannot execute such free rotations; exceptions appear to be the terminal groups and some of the surface residues with side chains projecting into the solvent. Therefore, models were developed in which the amplitude of the motions were restricted (London and Avitabile, 1978; Wittebort and Szabo, 1978). The poly-L-lysine data reproduced in Table II (Wittebort et al., 1980) was analysed according to the restricted-amplitude-motion mode (Wittebort and Szabo, 1978) with good agreement between calculated and experimentally observed data; a measure of the overall tumbling motion of the poly-L-lysine polymer was obtained from C_α, T_1, and NOE. The calculated values for the allowed rotation span as $\pm 60°$, $\pm 50°$, and $\pm 120°$ for C_β, C_γ, and C_δ, suggesting highly restricted motion about C_α—C_β and C_β—C_γ bonds and somewhat freer motion about the C_γ—C_δ bond.

In a protein, the motion of the lysine side chains is probably more restricted due to the nearby side chains of other amino acids, unless they are protruding into the solvent away from the protein surface. Similar studies concerned with the motion of aliphatic side chains of lysines, isoleucines and

Table II Measured NT_1 and NOE Values for Lysine Carbons in Poly-L-Lysine[a] at 67.9 MHz

Carbon	NT_1[b]	NOE
α	170 ± 8	1.6 ± 0.2
β	202 ± 8	2.0 ± 0.2
γ	264 ± 12	2.4 ± 0.2
δ	430 ± 20	2.3 ± 0.2
ϵ	734 ± 50	2.7 ± 0.2

[a] From Wittebort et al. (1980) NOE defined as $(1 + \eta)$.

[b] N is the number of protons directly bonded to the carbon atom.

methionine residues, and aromatic side chains have been conducted with a large variety of peptides and proteins. The studies with the aromatic side chains are especially noteworthy because NMR data indicate that the aromatic rings execute 180° flips (Campbell and Dobson, 1979), even though theoretical studies (Gelin and Karplus, 1975) of the crystal structure of BPTI suggest the existence of very large barriers to this motion. For detailed reviews of these dynamical studies, readers are referred to articles by London (1980), Gurd et al. (1982), and the recent book by Jardetsky and Roberts (1981). The general conclusion from such studies is the existence of considerable amount of motion of side chains, libations of the α-carbon backbone, superimposed on the overall tumbling of the molecule. Given the tight packing of atoms within a protein interior, many of these motions, such as the flips of aromatic side chains must be accompanied by complementary motions in the neighboring segments of the molecule. Thus a highly correlated picture of the protein dynamics emerges with numerous implications for the function of these macromolecules.

More recent studies have emphasized development of methodologies which do not require specific models at the outset (King et al., 1978; Riberio et al., 1980; Lipari and Szabo, 1982a,b). In these analyses, one still does not extract a detailed time course of the motion, and a physical description of the motion still requires interpretation of the data within the context of a model.

As in all NMR studies, before any information about the dynamics of a particular group can be obtained, individual resonance(s) from that group must be detected and assigned. In ^{13}C studies, this requirement is helped by the fairly large dispersion of ^{13}C chemical shifts. However, many resonances in a natural abundance ^{13}C spectra of macromolecules are not resolved, especially if the overall motion is not fast and the resonance linewidths are broad. In some cases the specific labeling is used to overcome the problem. In deuterium NMR studies, which we have not discussed here [but see Jelinski et al. (1980) and Seelig and Seelig (1980)], specific labeling is an absolute requirement. The result is a simple spectrum with resonances coming from the few types of deuterium atoms.

C. Studies with Quadrupolar Nuclei

As previously mentioned, we have dealt primarily with spin-$\frac{1}{2}$ nuclei in this chapter. Although research using these nuclei predominate the NMR field, the use of quadrupolar nuclei in various different types of studies has seen a rapid rise. In this section we will give a very cursory and definitely incomplete summary of the type of biological applications which utilize quadrupolar nuclei. Interested readers are referred to various review articles and references therein.

In the case of higher-spin nuclei ($I > \frac{1}{2}$), there is a multiplicity of spin states and hence transitions. In the absence of other interactions, these transitions are degenerate (i.e., have the same frequency) so that a single resonance line is observed. The presence of an electric-field gradient at the nucleus, due to an asymmetric change distribution, will cause transitions starting at different m_I levels to have different frequencies, leading to a multiline spectrum. The details of the interactions can be quite complex, particularly for $I > 1$, and we will not discuss them. The same sensitivity to correlation times is exhibited by these resonances as in the $I = \frac{1}{2}$ case, although the averaging is more complex and it is possible to have multiple T_1's and T_2's for different transitions (i.e., $-\frac{1}{2}$ to $\frac{1}{2}$ would be different from $\frac{1}{2}$ to $\frac{3}{2}$ for an $I = \frac{3}{2}$ nucleus). Because of these complexities, any conclusions regarding different binding sites or multiple conformations based on NMR observations of quadrupolar nuclei need to be made with great caution.

One of the most extensive biological uses of quadrupolar nuclei is the investigation of lipid membranes using ^{2}H NMR. ^{2}H is a spin-1 nucleus with three Zeeman levels, $m_I = 1, 0, -1$. In the presence of an electric gradient, these levels are shifted unequally so that two lines are observed in the NMR spectrum of an oriented sample. In powders or samples undergoing incomplete averaging, the situation is more complicated. For membranes these problems are usually subsumed under the concept of an "order parameter," which is a measure of the average axial order in the membrane. Studies on a wide variety of membranes have indicated that the hydrocarbon chains of the lipids making up the bilayer are well ordered up to about four carbons from the end. Here, the deuterium results indicate, the chains start to fray until, by the last atom, there is almost no ordering left. We refer the reader to the review by Seelig and Seelig for further details on this and other aspects of ^{2}H NMR (Seelig, 1977; Seelig and Seelig, 1980).

Another interesting biological problem studied by quadrupolar nuclei is the nature of Na and K in the intracellular environment. The initial studies on samples of muscle, brain, and kidney observed only a single resonance measuring 40% of the intensity found after ashing and redissolving the samples. From this it was concluded that two populations existed, one bound (60%) and one free (40%) (Cope, 1965, 1967). This conclusion was reexamined by Shporer and Civan (Shporer and Civan, 1972), who pointed out that the data could equally well be explained by ordering of the Na^+ around the charged macromolecules in the tissue. Further experiments have confirmed this view [see Civan and Shporer 1978) for a review].

D. Kinetics

A variety of NMR techniques can be used to obtain kinetic rate information. These include linewidth analysis, magnetization transfer, and isotopic exchange. Each of these is useful under different circumstances.

1. Linewidth Analysis

The physical basis for linewidth analysis is the relationship between the species lifetime and its resonance linewidth. If we ask how accurately can a particular resonant frequency be measured in the absence of other perturbations such as magnetic-field inhomogeneity, the answer depends on how long the spins have to precess, i.e., the longer they precess the more accurately can the frequency be measured and hence the narrower the resonance line due to those spins. Thus, any chemical exchange or other perturbation which shortens the lifetime of the observed species will contribute to the linewidth.

Generally, the regime of applicability of linewidth analysis in determining rate constants is between 1 and 100 sec^{-1}. These limits are set at the lower end by other sources of width such as natural relaxation, field inhomogeneity, etc., and at the upper end by the frequency difference between the exchanging species; as the exchange rate becomes faster than the frequency difference, the fast-exchange limit is reached, and only a single line is observed. Of course, if it is known which two resonances are being averaged, it is still possible to extract lifetimes (see Section III for a discussion of the fast and slow exchange limits).

An example of linewidth analysis is the work of Midelfort *et al.* (1976) in which they determined the anomerization rate of fructose-bis-phosphate (Fru-P$_2$) among its open chain and cyclic α and β forms. Interestingly, the α anomer, although approximately $\frac{1}{6}$ the concentration of the β anomer, has a longer lifetime. This apparently paradoxical result is explained by the existence of the open chain form (the keto form), present at only a few percent, through which the other two anomers must pass in order to exchange. Upon analysis of this model, Midelfort *et al.* derived the kinetic scheme

$$\alpha \rightleftarrows O_1 \rightleftarrows \beta$$

in which the β and the keto are in much faster exchange than the α and the keto.

2. Magnetization Transfer Techniques

These techniques depend upon the fact that a low-power rf field directly affects only those nuclear spins with resonant frequencies very near the frequency of the rf field. This means that it is possible to selectively perturb the spin populations at a particular resonant frequency while leaving other spin populations essentially untouched. In other words, nuclei in a particular molecule can be labeled, at least for times on the order of their T_1's. In the presence of chemical exchange this "labeling" can be transferred to other spin populations. Consequently, a change in the signal intensity is observed and used to calculate the exchange rates.

To illustrate the effect in a simple system, consider the Bloch equations, modified to include chemical exchange (McConnell, 1958), for the z components of the magnetization of two exchanging species, A and B:

$$dM_z^A/dt = -(M_z^A - M_0^A)/T_1^A - k_1 M_z^A + k_2 M_z^B$$

and

$$dM_z^B/dt = -(M_z^B - M_0^B)/T_1^B + k_1 M_z^A - k_2 M_z^B,$$

where k_1 and k_2 are the rate constants for A going to B and vice versa, respectively, and T_1^A and T_1^B are the individual spin–lattice relaxation times of A and B in the *absence* of exchange. At equilibrium, $k_1 M_0^A = k_2 M_0^B$. Now, if a low-power rf field is continuously applied at the resonant frequency of B, ω_B (which is assumed to be distinct from ω_A), with just sufficient power to saturate the B spins, the equation for M_z^A will become

$$dM_z^A/dt \ -(M_z^A - M_0^A)/T_1^A - k_1 M_z^A.$$

If the system is being observed with 90° pulses every T seconds, then the observed signal at ω_A will be the amount that M_z^A has recovered from the last 90° pulse by the time the next one arrives. This is simply

$$M_z^A(T) = [M_0^A/(1 + k_1 T_1^A)] [1 - \exp(-T/T_1^{\text{sat}})],$$

where $T_1^{\text{sat}} = T_1^A/(1 + k_1 T_1^A)$. Thus, provided that $T \gg T_1^{\text{sat}}$, the M_z^A observed under these conditions will be reduced by a factor of $(1 + k_1 T_1^A)$ from that observed when the B spins are not being saturated. The relative change in magnetization, $\Delta M/M_0^A$, between these two conditions is then $k_1 T_1^A/(1 + k_1 T_1^A) = k_1 T_1^{\text{sat}}$. A measurement of the apparent "T_1" during saturation gives T_1^{sat}, hence k_1 can be obtained simply as

$$k_1 = (\Delta M/M_0^A)(1/T_1^{\text{sat}}).$$

Clearly, a similar procedure involving the saturation of A would yield k_2.

Several points should be mentioned: If T is too short, very little effect due to exchange is observed during saturation since, if $T \ll T_1^{\text{sat}}$,

$$M_z^A(T) = M_0^A T/T_1^A$$

and is therefore independent of k_1. Thus, for reliable estimates of $\Delta M/M_0$ and hence of k_1 and k_2, the magnetization must have sufficient time to recover to its steady state. If $k_1 T_1^A$ is too small, there will be very little change in M_z^A, and signal-to-noise considerations will usually prevent the measurement of k_1. If $k_1 T_1^A$ is too large, on the other hand, the measurement of T_1^{sat}

becomes very difficult, and again a reliable measurement of k_1 is usually not possible. In this instance however, one can use inversion-transfer techniques (Dahlquist et al., 1975; Alger and Prestegard, 1977; Brown and Ogawa, 1977; Campbell et al., 1978) to measure these rate constants. Finally, the equations for more complicated reaction schemes are considerably more complex, and thus the application of these straightforward formulas to the data obtained from complex systems should be carried out with great care. For example, in multiple reactions, what is essentially measured by $\Delta M/M_0$ is the ratio of the lifetime of the species to its free T_1. In spite of this caveat, however, this technique has the advantage that it is one of the few capable of providing unidirectional rates in living systems.

The first application of these techniques to biological molecules was by Redfield and Gupta, who studied the exchange between oxidized and reduced cytochrome c (Redfield and Gupta, 1971). By saturating a proton resonance from the oxidized form in a mixture of both forms they observed a reduction at one of the resonances from the reduced form, deducing an exchange-rate constant of 10^4 M^{-1} sec^{-1} under their conditions.

Another example is the work of Otvos et al. (1979), who studied the enzyme alkaline phosphatase with ^{31}P NMR. Figure 27 shows an example of saturation transfer between inorganic phosphate (P_i), an enzyme phosphate complex (E·P), and a phospho–enzyme intermediate (E—P). Knowledge of the T_1's enabled Otvos et al. to calculate exchange rates of 0.19 and 0.23 sec^{-1} for the forward and reverse directions between E·P and E—P. The rates from E·P to P_i were not calculable because of the complete saturation of E·P upon saturating P_i (Fig. 27).

Although we have not discussed inversion transfer in detail, it is very similar to saturation transfer except that the perturbation is a selective inversion rather than a saturation. It is generally used when the rate constants are much faster than the relaxation times. Operationally, the selective inversion is followed by a variable waiting time to allow the exchange to take place. The state of the spin system is then sampled with an ordinary nonselective 90° pulse. Adenylate kinase, an enzyme which catalyzes the equilibrium between ATP, ADP, and AMP, was studied (Brown and Ogawa, 1977) using inversion transfer. In this study part of the pathway of catalysis was probed by simultaneously measuring the kinetic rates of several of the intermediate steps. An inversion-transfer experiment is shown in Fig. 28. As can be seen in the figure, when the time between the selective 180° pulse and the nonselective 90° pulse was increased, there was a diminution of ATP$_\gamma$ resonance, indicating an exchange between the two molecular species. By studying this exchange as a function of substrate concentration, the rates for the different steps in the reaction were determined.

These techniques can be extended to the direct measurement of exchange

1. NUCLEAR MAGNETIC RESONANCE

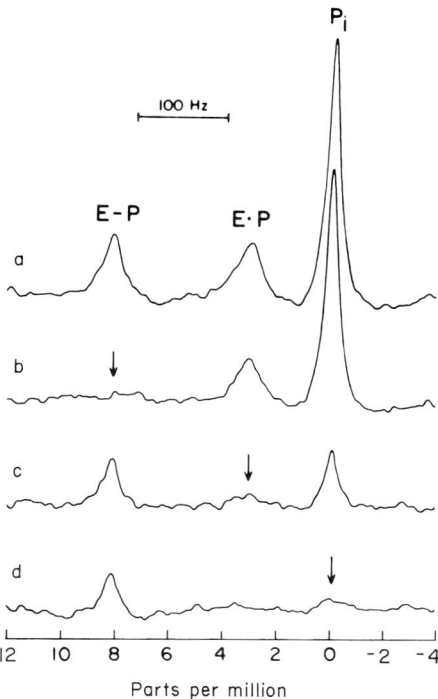

Fig. 27. ^{31}P NMR saturation transfer in a solution of $Zn_4^{2+}Mg_2^{2+}$ of alkaline phosphatase (2.4 mM) in the presence excess P_i (9.6 mM). Spectra a–d are identical except for the frequency of the saturating rf power indicated by the arrows. (From Otuos *et al.*, 1979.)

rates in cells and other intact biological systems. This is a unique application of NMR, not duplicated by any other method, allowing measurement of the biochemical rates under true intracellular conditions. Since in all applications to date ^{31}P nuclei have been used to perform these measurements, the relevant T_1's to consider are those of ^{31}P in the phosphorylated metabolites (PCr, ATP, P_i, etc.) found in large concentrations in cells and organs. These times are typically 1–3 sec at fields of 4–8 T, where the experiments have been carried out, implying that exchange rates on the order of 0.1 sec^{-1} and faster can be measured. These experiments are discussed in Section 5.

3. Isotopic Exchange

A type of chemical-exchange measurement commonly performed with ^{1}H NMR spectroscopy is the exchange of NH hydrogens in proteins and poly-

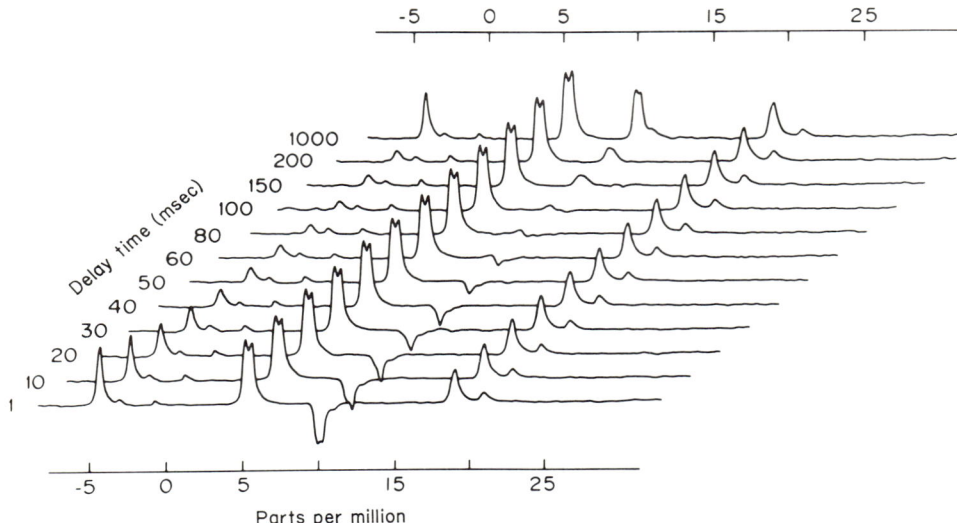

Fig. 28. Inversion transfer observation of adenylate kinase catalyzed exchange between AMP and ATP, ADP. The selective 180° was applied to the overlapping α-phosphate resonances of ATP and ADP. The delay between this pulse and the nonselective 90° pulse used to detect the spectrum is indicated for each spectrum on the left. AMP resonance shows the inverson transfer effect. ATP_β and $ATP_\gamma + ADP_\beta$ resonances are unaffected. (From Brown and Ogawa, 1977.)

nucleotides with solvent hydrogens or deuterons. These measurements employ some of the NMR techniques which are used in other chemical exchange studies. The directly obtained information is a rate constant; the inferences, however, generally concern the structure and dynamics of the macromolecule under consideration.

In aqueous solution, the NH hydrogens of amino acids and nucleotides exchange with solvent hydrogens in the subsecond time domain. Therefore, if one dissolves them in D_2O, the NH hydrogens are rapidly replaced with deuterium atoms and can no longer be detected in ^1H NMR spectra. Even in H_2O, proton resonances from these NH moieties are normally not observed, because the exchange with the H_2O hydrogens is not in the slow-exchange domain (Section III.G). In native proteins or polynucleotides, however, the exchange rate may be considerably reduced owing to shielding of the NH group from the solvent or to participation of the NH hydrogens in intermolecular hydrogen bonds. If the reduction in the exchange rate is sufficiently large, resonances from the NH protons become detectable.

Glickson et al. (1969) were the first to report the detection of such resonances from the NH protons of the tryptophan residues in lysozyme. Subsequently, NH resonances have been observed from histidine, tryptophan, and backbone amide moieties of a variety of proteins; examples include BPTI (Dubs et al., 1979; Wüthrich et al., 1980; Woodward and Hilton, 1980), ribonuclease (Patel et al., 1975), lysozyme (Campbell et al., 1975; Wedin et al., 1982), azurin (Uğurbil and Bersohn, 1977), superoxide dismutase (Lippard et al., 1977; Stoesz et al., 1979), serine proteinases (Robillard and Shulman, 1974a and b), and cytochrome c (Stellwagen and Shulman 1973). With nucleic acids, Kearns et al., (1971) were able to observe hydrogen-bonded ring NH protons of uridine and guanine bases in transfer RNAs. This original observation has been followed by extensive studies of polynucleotides, tRNAs, and DNA fragments using the hydrogen-bonded NH proton resonances as reporters of structure and dynamics of these molecules (reviews: Hilbers, 1979; Krugh and Nuss, 1979; Patel, 1979; Robillard and Reid, 1979; Schimmel and Redfield, 1980).

In both proteins and polynucleotides, the detectable NH resonances generally appear downfield of the resonances of protons bonded to carbon atoms, between approximately 8 and 16 ppm from the commonly used chemical-shift reference 2,2-dimethyl-4-silapentane-1-sulfonate (DSS); the CH protons of aromatic rings are the only other protons which partially overlap the NH chemical shift range. The NH resonances are also well separated from the H_2O peak, which generally resonates at about 5 ppm. This large separation is helpful in the detection of these resonances in H_2O solutions; because the H_2O protons exist at 10^4-10^5 times higher concentrations than the protein or nuclei acids in solution, technically it becomes difficult to detect signals from both H_2O and solute protons in straightforward FT NMR measurements. Hence, one must use methods which selectively excite and observe signals from the protons of interest. These methods include continuous-wave NMR, correlation spectroscopy, or pulsed techniques (Redfield et al., 1975; Plateau and Gueron, 1983). In general, the successful application of such methods improve dramatically as the resonances of interest are well separated from the H_2O peak.

The exchange rates of the NH protons with the solvent protons are measured using a variety of techniques, the choice of which is generally dictated by the range in which the exchange rate falls. If the exchange occurs in several minutes or longer (i.e., with apparent first-order rate constants of 10^{-2} sec^{-1} or smaller), the appropriate method is to rapidly replace the H_2O with D_2O and to start collecting consecutive NMR spectra in which the intensity of the exchangeable resonances are followed as a function of time. An example of such a "real-time" solvent exchange is shown for the yeast tRNA^{phe} (Fig. 29); it is seen that most of the NH resonances observed in the

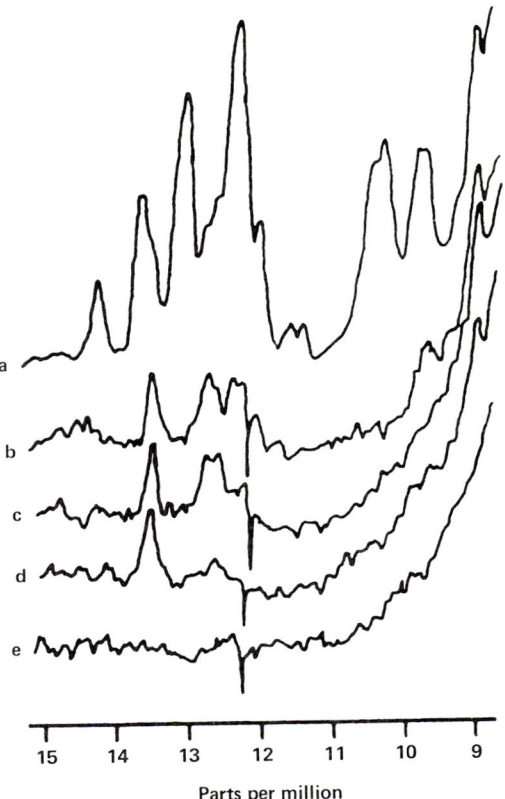

Fig. 29. Proton NMR spectra of yeast tRNA[phe] as a function of time following the replacement of H_2O with D_2O. Spectrum a was recorded in H_2O, where the exchangeable NH resonances are all detected; spectrum b, at 6 min; spectrum c, at 11 min; spectrum d, at 136 min; spectrum e, 24 h. (From Schimmel and Redfield, 1980.)

spectrum when the sample is in H_2O are gone in the first spectrum obtained after replacing the solvent with D_2O. When the exchange is characterized with apparent first-order rate constant of the 5 sec^{-1} or greater, the most recently used technique is saturation recovery (Johnston and Redfield, 1977). The Bloch equation for the z component of the bulk magnetization of an exchanging NH would be

$$dM_z/dt = -T_1^{-1}(M_z - M_z^0) - k_1 M_z + k_{-1} M_{z,H_2O},$$

where k_1 and k_{-1} are the apparent first-order rate constants for the forward and reverse directions of the exchange reaction and M_{z,H_2O} the water magnetization. In the limit $k_1 \gg T_1^{-1}$, if the H$_2$O resonance is saturated as in saturation-transfer experiments discussed earlier, one would observe virtually complete saturation of the NH peak as well. This was demonstrated to be the case for the hydrogen-bonded NH protons in yeast tRNA^{phe} (Campbell *et al.*, 1977; Johnston and Redfield, 1977). In this case, measurement of the effective spin–lattice relaxation without saturating or inverting the H$_2$O resonance can be used to obtain the rate constant. This measurement is commonly performed by the saturation–recovery technique. Such an experiment is performed by first saturating the resonance(s) of interest selectively without perturbing H$_2$O magnetization. Subsequently, the saturated spin(s) is allowed to relax for a period of time τ, and then is detected. The procedure is repeated for different τ values to obtain the relaxation rate. In such an experiment M_{z,H_2O} remains at all times practically equal to M^0_{z,H_2O}, the thermal equilibrium magnetization of the H$_2$O peak; transfer of saturation from the saturated NH peak to the H$_2$O peak is minimal due to the fact that $[-NH]/[H_2O]$ is $10^{-4}-10^{-5}$. In this case the relaxation of the NH proton resonance obeys the equation

$$dM_z/dt = -(T_1^{-1} + k_1)(M_z - M_z^0) + \text{constant}.$$

The effective relaxation-rate constant is $T_1^{-1} + k_1$ or k_1 if $k_1 \gg T_1^{-1}$ is the case. A saturation–recovery sequence is shown for the imino protons of a 12-base, double-helix polynucleotide (Fig. 30).

Exchange of the NH protons with the solvent is thought to occur in

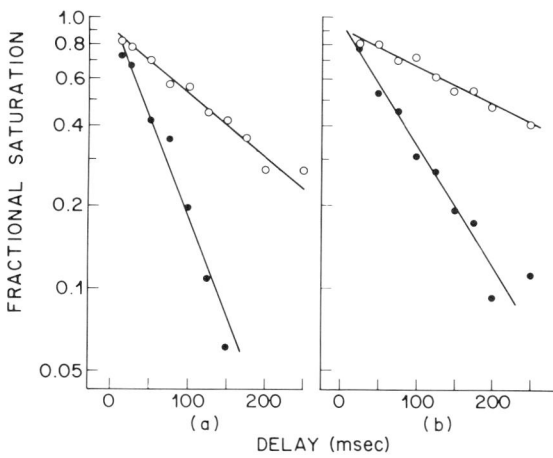

Fig. 30. Saturation recovery measurement of T_1 of the imino protons in a 12-base double-helix polynucleotide; (a) AT 5, (b) AT 6; ○, AATT 12-mer; ●, TATA 12-mer. (From Patel *et al.*, 1983).

two-step processes; the NH in the macromolecule is first exposed to the solvent, followed by the actual exchange. Thus for a hydrogen-bonded proton, the first step is the breaking of the hydrogen bond, and the second step is the exchange:

$$>\text{N}-\text{H}|||||\text{N}< \underset{k_{-1}}{\overset{k_1}{\rightleftarrows}} >\text{NH} + \text{N}<$$

$$>\text{NH} + \text{H*OH} \underset{k_{-2}}{\overset{k_2}{\rightleftarrows}} >\text{NH*} + \text{HOH}$$

The second reaction is acid–base catalyzed, therefore, it is sensitive to the buffer concentration in solution.

With proteins, the first step in the generally accepted two-step exchange-reaction scheme has been thought of either as small fluctuations in structure which allow solvent penetration into the protein matrix, or a local unfolding with consequent breaking of several adjacent hydrogen bonds. In either case, the exchange occurs through a pathway where the overall integrity of the protein structure is more or less conserved. Alternatively, under appropriate conditions, such as in extremes of pH or temperature, a cooperative denaturation which represents a major unfolding of the structure may occur to expose the labile NH group. The two possible pathways are distinguished by their activation energies: the exchange from the folded state occurs by a low-activation energy ($\sim 20-40$ kcal mol^{-1}) process and exchange through a global unfolding is a high-activation energy ($60-120$ kcal mol^{-1}) process. For further details on these rates and their implications in protein dynamics, interested readers are referred to reviews by Woodward *et al.* (1982), Englander and Englander (1978), Gurd and Ruthgeb (1979), and Karplus and McCammon (1981), and references therein.

In the polynucleotide duplexes, the process which exposes the NH hydrogens to the solvent must necessarily involve the disruption of the hydrogen bonds. In analogy with the discussion on proteins, this process may be a transient local opening of the paired bases (i.e., a local fluctuation which does not disrupt the overall structure) or a duplex to single-strand transition, which would be a global unfolding. As in the case of proteins, the two processes are characterized by markedly different activation energies. Measurement of the activation energy for NH exchanges in polynucleotide duplexes indicate that exchange involves only local, single-base-pair openings (Early *et al.*, 1981a,b; Patel *et al.*, 1983).

The two-step exchange reaction can have two extreme limiting cases. One limit is when the exchange occurs every time the base pair opens; in this case the exchange rate determined by NMR is a direct measure of the local duplex openings, and is independent of buffer concentration in the solution. The other limit is when the duplex opens and closes many times before the actual exchange takes place; this limit is characterized by the dependence of rate constants on buffer concentration. In polynucleotide duplexes, the former limit holds for bases in the interior of duplexes (Patel and Hilbers, 1975; Hurd and Reid, 1980; Pardi *et al.*, 1982), and the latter limit is applicable for bases at all ends (Patel and Hilbers, 1975; Pardi *et al.*, 1983), indicating that local duplex openings occur at a much faster rate at the ends of the molecule than in the middle.

E. Cellular and Metabolic Studies

The use of NMR spectroscopy has been extended to studies of intact systems such as cells, organs, and whole animals, including humans [for a review see Gadian *et al.* (1979), Shulman *et al.* (1979), Burt (1981), Roberts and Jardetsky (1981), Radda *et al.* (1982), and the references contained therein]. The motivation for these studies is the ability to probe cellular processes noninvasively. The information obtained includes intracellular concentrations of metabolites, intracellular pH, the physical state of the intracellular constituents, and enzymatic reaction rates. These studies utilized primarily ^{31}P, ^{13}C, and ^{1}H nuclei.

In the metabolic studies, owing to the inherent lack of sensitivity of the NMR method, it is unrealistic to expect information about submillimolar quantities of materials except possibly when observing protons which might be detectable at concentrations as low as a few tenths of a millimolar. This fact immediately excludes the possibility, for example, of monitoring cAMP changes or any other biochemical signal of comparable magnitude. Realistic experimental designs are usually centered around the ability of NMR to probe repeatedly and nondestructively the concentrations and chemical shifts of substrates present in the range of 1 mM or higher no matter what nucleus is being used. This generally restricts NMR applications to the study of the energy producing and utilizing pathways. Most synthetic pathways cannot be followed, since the intermediate and end products do not usually accumulate to these concentrations. Obviously there are synthetic pathways which meet these restrictions, e.g., amino acid synthesis, which shares the TCA cycle with energy production, and gluconeogenesis, where the end product accumulates to quite high levels in some tissues.

1. ^{31}P Studies

A typical example is the observation of the adenosine nucleotides P$_i$ and phosphocreatine in muscle under a variety of stimuli. Here, in addit measuring the levels of these metabolites, one obtains information intracellular pH and free Mg^{2+} from the chemical shifts of P_i and ATP, respectively. The basis for the former determination is fairly important since it is used so widely; therefore, we will consider it at some length.

When a spin is undergoing exchange between two (or more) environments rapidly enough, its apparent chemical shift is the weighted average of the frequencies in each different environment (see Section III.G). Thus, the ^{31}P spin, as it exchanges between HPO_4^{2-} and $H_2PO_4^-$ forms, has an apparent chemical shift which is the weighted average of the chemical shifts of the two separate species. The observed chemical shift is

$$\delta(\text{obs}) = \frac{\delta_{HPO_4^{2-}}[HPO_4^{2-}] + \delta_{H_2PO_4^-}[H_2PO_4^-]}{[HPO_4^{2-}] + [H_2PO_4^-]}.$$

The resulting expression for chemical shift versus pH is shown in Fig. 31 for the pH range 5–8. Similar titrations would occur at the other two pK's of phosphate. Clearly, within the range of pH 6–8, chemical shift can be converted to pH accurately and reliably. One point deserves discussion; the need to know the pK of the phosphate in the particlar medium being measured. This is particularly important when making intracellular determinations. Fortunately, the pK of phosphate depends almost entirely on the ionic strength of the medium and not on the details of its composition (Ogawa et al., 1981). In the range of salt concentrations found intracellularly (~0.15–0.3M) the variation in the pK of P_i is no more than 0.1 pH units. Thus errors of this size are possible in the absolute determination of the pH in situations of uncertain ionic strength. It should be noted, how-

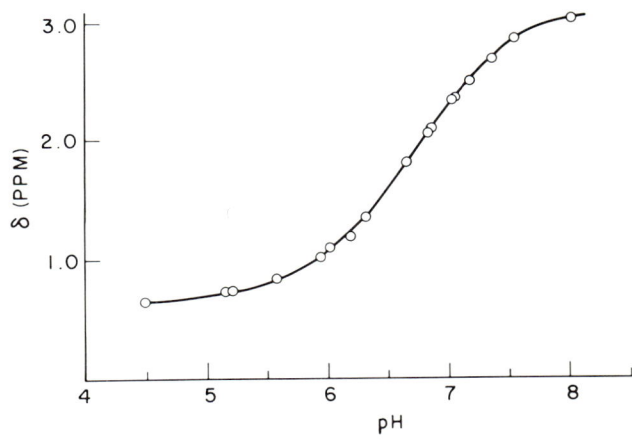

Fig. 31. Chemical shift of inorganic phosphate as a function of pH in 120 mM salt solution.

ever, that any relative changes can be detemined more accurately, assuming that the unknown ionic strength does not vary over the course of the experiment.

Following the initial experiments of Moon and Richards (1973), numerous other workers have used this technique to measure intracellular pH. These studies include the time course of intracellular pH changes in isolated hearts and muscles during ischemia and other stimuli (Garlick *et al.,* 1979; Gadian *et al.,* 1982; Jacobus *et al.,* 1982), correlation of transmembrane pH gradients and nucleotide triphosphate levels in *E. coli* (Uğurbil *et al.,* 1978a,b, 1982), observations of mitochondrial pH in intact hepatocytes (Cohen *et al.,* 1979), and in perfused rat hearts has been reported as well (Garlick *et al.,* 1983), following the intracellular pH changes over the cell cycle in yeast (Gillies *et al.,* 1981). See Roos and Boron (1981) for a general review of intracellular pH. The direct observation of the mitochondrial matrix pH in intact hepatocytes (Cohen *et al.,* 1979) is a good example of the ability of NMR to probe intracellular compartments. The two pools of P_i, cytosolic and mitochondrial, were observed to have different chemical shifts. At the time of the original report it was assumed the ionic strength in the matrix was similar to that in the cytosol, leading to a ΔpH of 0.4 units between the two compartments when the external pH was 7.1. Varying the external pH caused the cytosolic pH to follow the extracellular pH with the mitochondrial pH remaining constant at 7.5. Later work (Ogawa *et al.,* 1978) using FCCP to collapse the ΔpH in isolated liver mitochondria showed the assumption of similar ionic strengths to be correct.

As a final example of pH determination we site the work of Gillies *et al.* (1981) in determining the variation of pH during the cell cycle in yeast. In this work, synchronized yeast cells were allowed to go through an entire cell cycle while their pH was followed at 10-min intervals. Figure 32 shows their results, which demonstrate a transient alkanization of intracellular pH at the start of the cycle regardless of the initial pH of the cells. Observations on unsynchronized cells show only a permanent shift to pH 7.1 upon feeding glucose. It was concluded from these and other data that the pH transient is related to DNA synthesis and not to mitosis, which begins about 1 h after the transient is over (see also Gillies, 1982).

Another example of chemical-shift dependence on the averaging of molecular environments is provided by binding of Mg^{2+} to ATP. The ^{31}P resonances of all three phosphates of ATP are shifted by this interaction. Therefore, their chemical shifts provide a way to measure the free Mg^{2+} concentration, which is analogous to determining the pH through the phosphate chemical shift. The situation is more complicated here, however, because the binding is affected by the pH as well as by the ionic strength. In most cells the free Mg^{2+} concentration (1–3 mM) is much higher than the

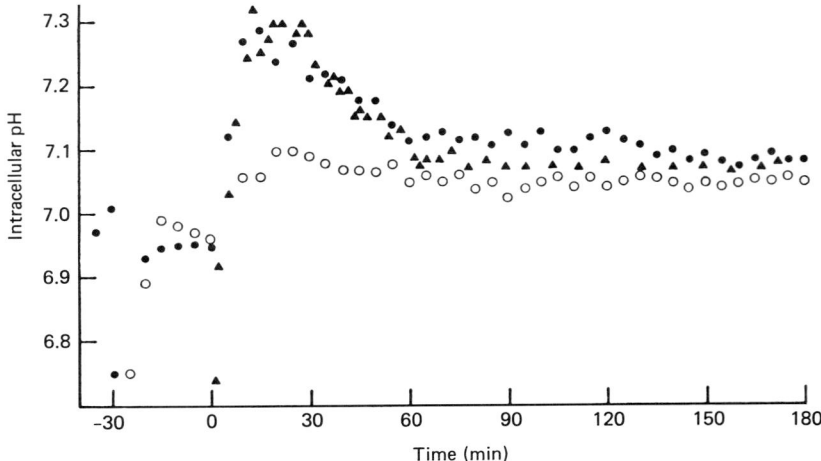

Fig. 32. Intracellular pH as a function of time after refeeding (at time 0) in synchronous (● ▲) and asynchronous (○) cultures of yeast (Saccharomyces cerevisiae). Prior to refeeding, cells were in a medium devoid of glucose but were oxygenated except in case of ▲, which was maintained anaerobic. (From Gillies et al., 1981.)

binding constant (~30 μM), so that the errors in the determination are substantial (Wu et al., 1981). In red blood cells, however, free Mg^{2+} is in the 0.5-mM range; therefore a significant shift of the resonances is observed. This has been used by Gupta et al. (1978) to measure the free Mg^{2+} level in these cells. They determined a value of 0.4 mM, which agrees fairly well with the 0.65 mM measured by the Mg^{2+} ionophore A23187 (Flatman and Lew, 1977). Similar results have been reported for smooth muscle (Dillon et al., 1982).

A unique set of measurements which have been performed in intact cells are the saturation and inversion-transfer determinations of enzyme kinetics. These have provided for the first time direct measurements of enzyme-catalyzed exchange rates under intracellular conditions. The first use of saturation transfer on an *in vivo* system was the study of the ATPase reaction in *E. coli* (Brown et al., 1977). The results from this experiment are shown in Fig. 33, in which the upper trace (a) shows the saturation in the control position as indicated by the arrow, and the middle trace (b) shows the saturation irradiating the ATP_γ. The lower trace (c) shows the difference between these two and clearly indicates a reduction in the intracellular P_i caused by the saturation of the ATP_γ. The peaks are labeled NTP, rather than ATP, because the intracellular nucleotide pool in *E. coli* contains 25%

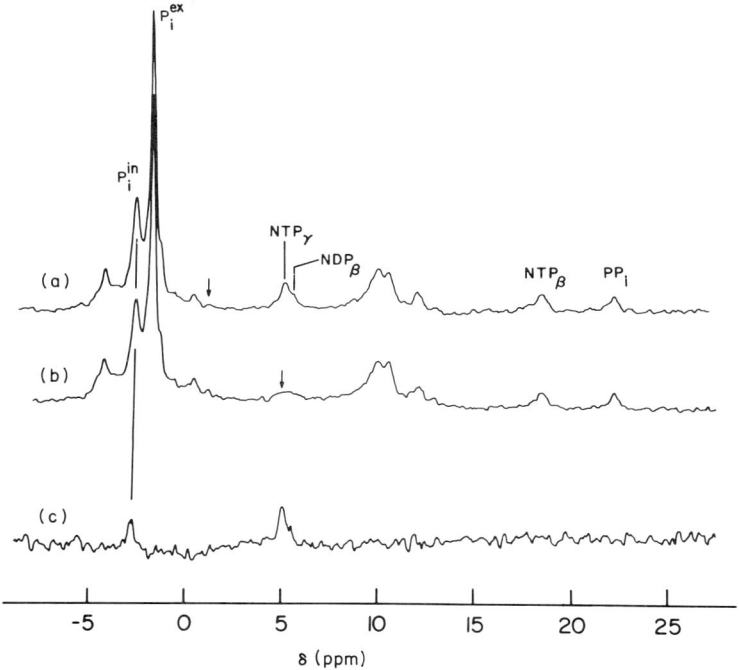

Fig. 33. ^{31}P NMR spectrum of aerobic *E. coli* at 25°C in the absence (a) and during saturation (b) of the NTP$_\gamma$ resonance. The arrows indicate the positions of the low-power rf radiation used for saturation of NTP$_\gamma$ and in the control spectrum. Trace c is a difference spectrum, a − b. (From Brown *et al.*, 1977.)

other nucleotides. Since the others presumably do not take part in the ATPase reaction, we have only discussed ATP here. By measuring the T_1 of the P$_i$ the exchange rate from P$_i$ to ATP$_\gamma$ was calculated to be 0.8 sec^{-1}. Signal-to-noise considerations made it impossible to measure the reverse exchange rate, corresponding to the rate from ATP$_\gamma$ to P$_i$.

Let us consider for a moment the implications of this fairly rapid exchange rate. If the O$_2$ supply is cut off, then the ATP levels in the cells decay away in about 5 min (Uğurbil *et al.*, 1982), suggesting that the background usage of ATP is 0.003 sec^{-1}, considerably slower than the directly measured exchange time of about 1 sec^{-1}. Assuming that the ATPase is not allosterically turned off in the absence of O$_2$, the rate of this reaction compared to the background ATP hydrolysis rate (0.003 sec^{-1}) implies that the ATPase is catalyzing a reaction which is virtually at equilibrium, rather than any sort of one-way step. Similar effects have been observed in yeast (Alger *et al.*, 1982).

Saturation transfer has been used to study the exchange rates between ATP_γ and phosphocreatine (PCr) catalyzed by the enzyme creatine phosphokinase (CPK), widely found in muscle and brain (Brown et al., 1978; Nunnally and Hollis, 1979; Gadian et al., 1981). The fluxes calculated from such data on resting and stimulated frog gastrocnemius are shown in Fig. 34. This result is quite interesting, since a naive prediction of how the fluxes of a reaction catalyzed by CPK during contraction would change is exactly the opposite due to the increase in the ADP concentration during contraction. The conclusion drawn from this is that the enzymatic rate constants during contraction are reduced by allosteric effects.

The sensitivity of relaxation rates and linewidths to molecular motion has already been discussed. With intact cells, this capability was used in studies of nucleotide and amine storage in the dense granules of blood platelets (Costa et al., 1979, 1980; Uğurbil et al., 1979, 1984a,b). These subcellular organelles sequester ATP, ADP, a divalent cation, and biogenic amines. The effective solute concentration within these vesicles exceeds ~ 3 M. Since biological membranes are incapable of tolerating osmotic imbalances which would be generated by such an intragranular solute concentration, it was anticipated that some form of a molecular complex exists within these organelles. The evidence for this was provided by ^{31}P NMR studies; it was observed that ^{31}P resonances from nucleotides contained in human dense granules are undetectably broad because of immobilization, and those from nucleotides in porcine dense granules yield highly temperature-dependent linewidths. This and other observations, including ^{19}F (Costa et al., 1979) and ^{1}H (Uğurbil et al., 1983) NMR data, were used to reach numerous conclusions on the physical state of the dense-granule constituents and interactions responsible for this state.

^{31}P NMR measurements, localized to various organs, have also been performed on whole animals and humans. These experiments relied on two techniques of localization. One is to create a small region over which the external magnetic field is homogeneous and then observe only the narrow lines which come from the whole organism, assuming that any signal from

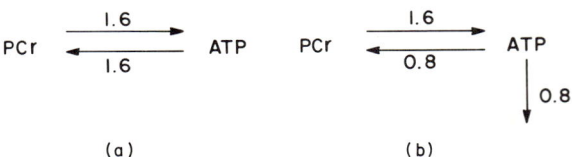

Fig. 34. Fluxes in a frog muscle through creatine kinase during a tetanus measured by ^{31}P saturation transfer: (a) resting muscle and (b) stimulated muscle. Fluxes are in units of $\mu mol/g \cdot sec$. (Adapted from Gadian et al., 1981.)

outside the homogeneous region is broadened out and can be rejected (Gordon et al., 1980). The other is to use an overall homogeneous field as usual but to localize by means of the rf excitation and pickup coil (Ackerman et al., 1980). This technique relies on the fact that the magnetic field from a current loop extends only about a radius away from its center. Thus, the spins in the sample farther away than this are only partially excited by the rf pulses and contribute only marginally to the signal. Because of their obvious surface character, these coils are called "surface coils." Both of these techniques have problems accurately defining the region under observation. However, large amounts of information, particularly using ^{31}P, have been obtained with them on brain, kidney, heart, and muscle in animals (Radda et al., 1982). In humans the observations were limited to arms and legs because of the unavailability of large-enough high-field (2 T) magnets. This restriction is expected to be removed soon.

After approximately one year of observations on human arms in a 20-kG magnet, a number of interesting physiological observations emerged. The muscle response to exercise was followed with both normal volunteers and patients with various muscle disorders. The general pattern was a reduction in PCr levels and intracellular pH with an increase in P_i, although the specific details varied from individual to individual. The time dependence of recovery was quite variable and had some peculiarities such as multiexponential recoveries, different recovery times for P_i than for PCr, etc. Figure 35 shows the PCr, ATP, and P_i levels of a patient with a genetic disorder in her oxidative metabolism (Ross et al., 1981). Owing to her disorder she is unable to make ATP except glycolytically and thus undergoes more acidification than normal when she uses her muscles.

2. ^{13}C Studies

Another obvious area of utility of NMR experimental studies of metabolism is with substrates enriched with ^{13}C. Although the information is similar to that which can be obtained from radiotracer studies using ^{14}C, the ability to sample the isotopic distribution rapidly and sequentially makes the experiments much less time consuming. Further, the position of the label in a molecule is determined automatically, allowing measurement of isotopic scrambling. A final advantage is the ability of NMR to detect a label in a neighboring carbon position due to spin–spin coupling. This is particularly useful in pathways such as gluconeogenesis, where unlabeled and labeled substrates can mix. By studying the amount of labeled versus unlabeled pairs, not only can the different fluxes be measured but information can also be obtained about metabolite channeling among the various pathways.

The first such studies were conducted with *E. coli* and yeast cells (Uğurbil

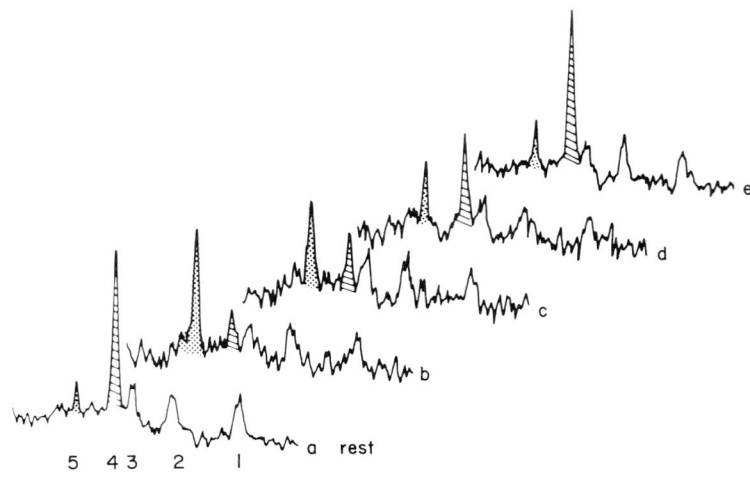

Fig. 35. ^{31}P NMR spectra taken at rest, during exercise, and recovery. The signals are assigned as follows: 1, 2, and 3, the β, α, and γ phosphates of ATP, respectively; 4, phosphocreatine; 5, inorganic phosphate. The inorganic phosphate and phosphocreatine signals are shaded for clarity. All spectra were accumulated at 32.5 MHz dusing a train of rf pulses applied at intervals of 2 sec; the number of pulses was 128 for spectrum a, 32 for spectra b–d, and 64 for spectrum e. Spectrum b was recorded during the last minute of aerobic exercise, and spectra c–e were recorded during the recovery period at 5, 9.5, and 37 min, respectively, after the end of the exercise. (From Radda et al., 1982. Reprinted by permission from Nature **295**, 6809. © 1982 MacMillan Journals Limited.)

et al., 1978a,b; den Hollander et al., 1979), where details of the glucose metabolism were followed after the introduction of ^{13}C-enriched glucose. Studies of the biosynthesis of chlorphyl using enriched acetate have been quite successful because the final location of the label in the molecule could be determined from its NMR spectrum (Scott and Baxter, 1981). Perhaps the best example of the extent of information available from this type of study is the work of Cohen and her collaborators on hepatocytes and perfused livers. They have performed detailed studies on gluconeogenesis using labeled glycerol, pyruvate, and other three carbon precursers (Cohen et al., 1979, 1981). Typical ^{13}C spectra of isolated hepatocytes are shown in Fig. 36. Inspection of the spectrum demonstrates immediately the vast amount of information which can be obtained in a very short time. Detailed consideration of the metabolic pathways involved is not possible here, but Cohen et al. have been able to map out many of the biosynthetic pathways with differing precursors and hormonal states. An example of spin–spin

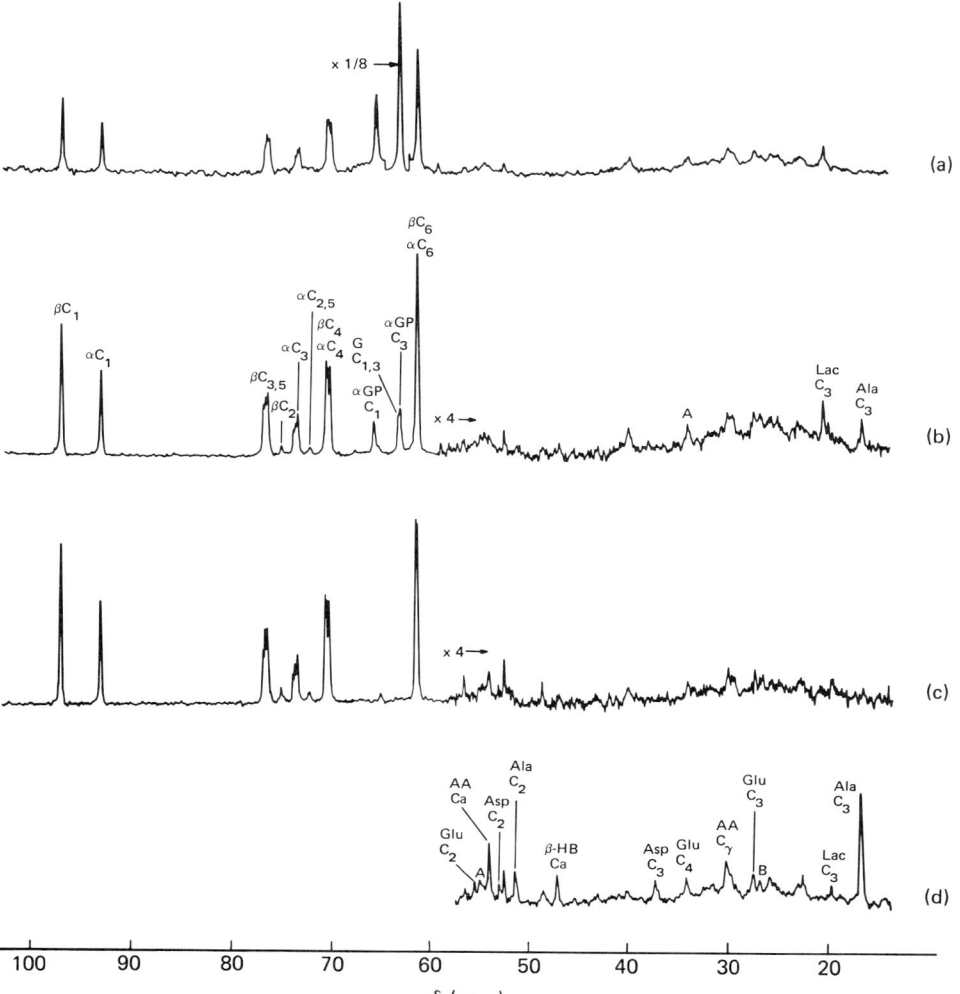

Fig. 36. Part of a sequence of ^{13}C NMR spectra at 25°C taken after a suspension of liver cells isolated from a T3-treated rat was made 22 mM in [1,3-^{13}C]glycerol: (a) accumulated during the period 0–17 min after the addition of substrate; (b) 35–51 min; (c) 85–115 min. The pulse repetition rate was 0.334 sec for spectra a and b and 2 sec for spectrum c. Spectrum d: upfield region of a similar hepatocyte sample made 16 mM in NH_4^+; recorded with increased vertical gain. The abbreviations used include: G C1,3, glycerol C1 and C3; αGP C1, α-glycerophosphate C1; Glu C2, glutamate C2; Asp C2, aspartate C2; AA Cα, acetoacetate CH_2; β-HB Cα, β-hydroxybutyrate CH_2; AA Cγ, acetoacetate CH_3; and Lae C3, lactate C3. (From Cohen *et al.*, 1979.)

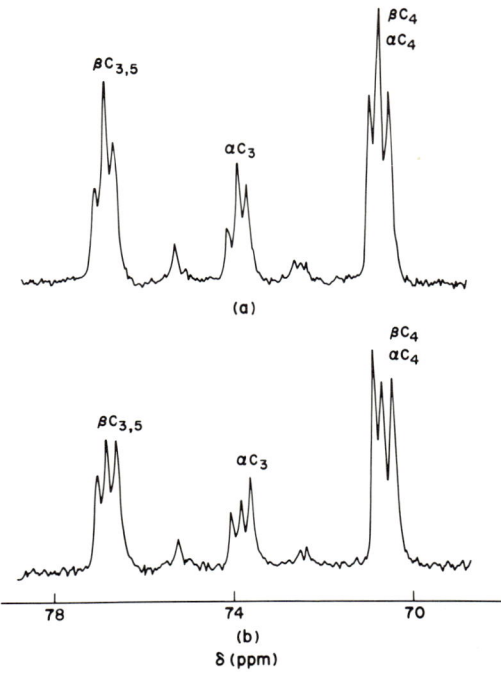

Fig. 37. (a) Expansion of the glucose C3 and C4 region of ^{13}C NMR spectrum of a suspension of liver cells from a rat treated with T3. The spectrum was accumulated 70–100 min after the addition of 22-mM [1,3-^{13}C]glycerol and 8-mM unlabeled fructose. (b) Expansion of the glucose C3 and C4 region of the spectrum of a similar suspension without fructose. (From Cohen et al., 1979.)

coupling may be seen in Fig. 37, which shows an expanded region of spectrum at the C-3 carbon of glucose. The single resonance in the center corresponds to those ^{13}C nuclei without another ^{13}C at C-4, whereas the doublet corresponds to doubly labeled glucose at C-3 and C-4. Quantification of these ratios imply that 22% of the gluconeogenetic flux came from the external unlabeled fructose in the suspension rather than from the labeled glycerol.

3. ^1H Studies

In addition to ^{31}P and ^{13}C, proton NMR has also been utilized for studies with intact cells. It has the advantage of high sensitivity as well as the possibility of isotopic measurements using deuterium as a replacement for protons. In a series of experiments on red blood cells (Brindle et al., 1979,

1980), exchange reactions were observed involving lactate dehydrogenase and other enzymes in the red cell. The difficulty in observing protons is to reduce the strength of the water resonance (110 M in protons) so that the peaks from metabolites present at 1–10 mM can be detected. With cell suspensions, this has been done either by replacing the H_2O with D_2O or by spin-echo techniques, which have the added advantage of removing the broad signals from the protons of macromolecules present in cells; these

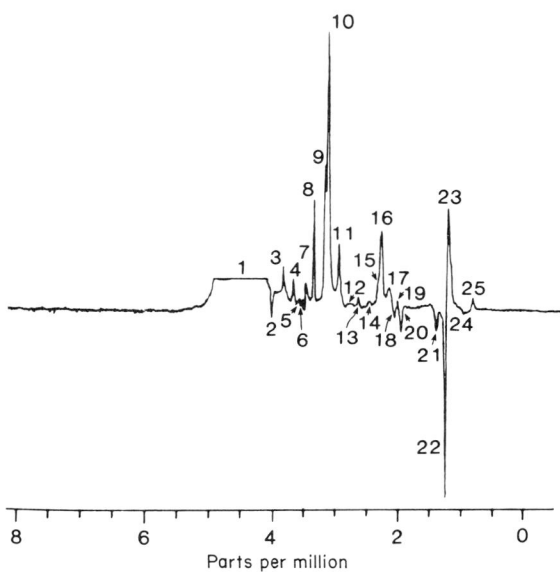

Fig. 38. Proton NMR spectrum of intact friend leukemia cells harvested 60 h after induction of erythroid differentiation. A spin-echo pulse sequence with $\tau = 60$ msec was used to collect the data at 470 MHz and 37°C. Accumulation time was 2.5 min. The most prominent signals are numbered; the signal from H_2O (signal 1) was truncated during plotting. The chemical-shift scale is referred to external sodium 3-trimethylsilylpropionate (TSP) in a capillary. Signal assignments are as follows: 1, H_2O; 2, lactate; 3, phosphocreatine; 4, amino acid Cα; 5 and 6, glycerol and sugars; 7, glycerol; 8, unassigned; 9, phosphorylcholine (glycerophosphorylcholine); 10, choline; 11, phosphocreatine; 12, aspartate (also between signals 13 and 14); 13, citrate; 14, citrate; 15, pyruvate and succinate; 16, glutamate and proline; 17, unassigned; 18, glutamate; 19, proline; 20, isoleucine and proline; 21, alanine; 22, lactate; 23, triglycerides; 24, valine, isoleucine, and leucine; and 25, leucine and isoleucine. (From Agris and Campbell, 1982. © 1982 by the American Association for the Advancement of Science.)

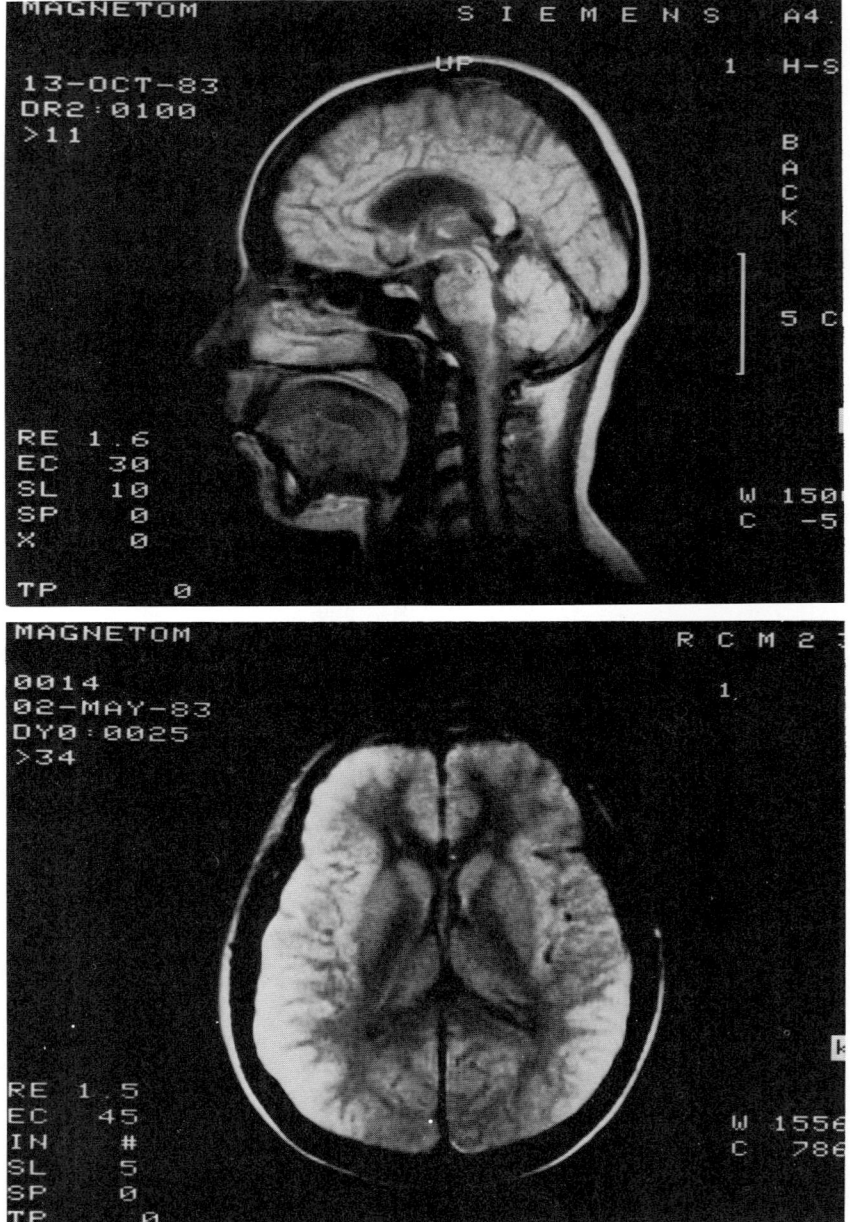

Fig. 39. ¹H NMR image of a human head acquired with a technicare NMR scanner.

1. NUCLEAR MAGNETIC RESONANCE

$$\int f(t) \exp(-i\omega't)\, dt = 2\pi \int F(\omega)\, \delta(\omega - \omega')\, d\omega = 2\pi F(\omega'),$$

using the properties of the delta function. Thus we have

$$F(\omega) = \frac{1}{2\pi} \int f(t)\, e^{-i\omega t}\, dt. \tag{18}$$

As mentioned in Section IV.B, the FID from a single-resonance line at frequency ω_0 and apparent T_2 (T_2^*) has the form $\cos \omega_0 t \exp(-t/T_2^*)$ in one of the detection channels and $\sin \omega_0 t \exp(-t/T_2^*)$ in the other. These two signals are taken as the real and imaginary parts of a single complex signal which is $f(t)$. Thus

$$\begin{aligned}f(t) &= (\cos \omega_0 t + i \sin \omega_0 t) \exp(-t/T_2^*) \\ &= \exp(i\omega_0 t - t/T_2^*).\end{aligned} \tag{19}$$

Applying Eq. (18) to Eq. (19) we find $\mathrm{Re}[F(\omega)]$, which is the NMR spectrum for this FID:

$$F(\omega) = \frac{1}{2\pi} \int_0^\infty \exp(i\omega_0 t - t/T_2^*)\, e^{-i\omega t}\, dt.$$

The lower limit on the integral is zero because $f(t) = 0$ for $t \leq 0$. Doing this integral we find

$$F(\omega) = (2\pi)^{-1}[i(\omega_0 - \omega) - 1/T_2^*]^{-1}.$$

The real part of this is then the NMR absorption spectrum. This is simply

$$\frac{1/T_2^*}{(\omega_0 - \omega)^2 + (1/T_2^*)^2},$$

or

$$\frac{T_2^*}{(\omega_0 - \omega)^2 T_2^{*2} + 1}.$$

This is the expression for a resonance line located at ω_0 of half-width $1/T_2^*$, as expected. This form of a resonance is known as a Lorentzian.

B. Dipolar Relaxation and NOE for a Pair of Spins

The energy diagram for a pair of $s = \frac{1}{2}$ spins i and j is shown in Fig. 40. There are four possible spin states which can be specified with (m_i, m_j). These are $(\frac{1}{2}, \frac{1}{2})$, $(\frac{1}{2}, -\frac{1}{2})$, $(-\frac{1}{2}, \frac{1}{2})$, $(-\frac{1}{2}, -\frac{1}{2})$. The dominant transitions owing to dipole-coupling between the two spins which give rise to the NOE are between $(-\frac{1}{2},$

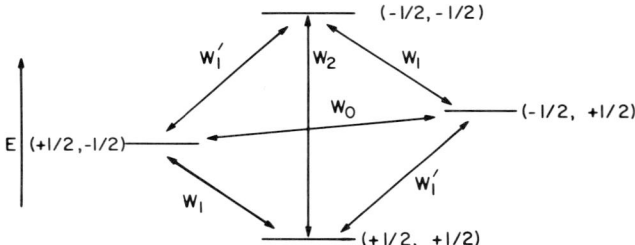

Fig. 40. Energy diagram and the possible transitions for a pair of spins coupled through dipole–dipole interactions.

$+\frac{1}{2}$) and $(+\frac{1}{2}, -\frac{1}{2})$ states when $\omega\tau_c > 1$ and between $(-\frac{1}{2}, -\frac{1}{2})$ and $(+\frac{1}{2}, +\frac{1}{2})$ states when $\omega\tau_c < 1$. The former case is encountered with biological macromolecules. The latter is the case for most small-molecular-weight compounds such as the ones dealt with in most organic-chemical applications.

When two spins are coupled so that any perturbation of the spin populations of one affects the spin population of the other, the Bloch equations for each spin need to be modified. For the z component of the macroscopic magnetization, we now need to write

$$\gamma_i^{-1} \, dM_i/dt = -\rho_i \gamma_i^{-1}(M_i - M_i^0) - \sigma \gamma_j^{-1}(M_j - M_j^0),$$

where M_i^0 and M_j^0 denote the thermal equilibrium value of M_i and M_j, ρ_i and ρ_j equal T_1^{-1} for spins i and j, respectively, and σ the cross-relaxation term that expresses the coupling between the spin populations. An analogous equation exists for the spin j.

If spin j is saturated by external rf irradiation, the $M_j = 0$. When steady state is reached so that spin-populations are no longer changing with time, $dM_i/dt = 0$ and

$$\eta_i = \frac{\overline{M}_i - M_i^0}{M_i^0} = \frac{\sigma}{\rho_i} \frac{\gamma_i}{\gamma_j} \frac{M_j^0}{M_i^0},$$

where $\overline{M}_i$ is the new steady-state magnetization. A similar equation for M_j is also obtained when spin i is saturated.

Values for M_i^0 and M_j^0 are given by Curie's Law (Section III.B). Therefore, they depend on the gyromagnetic ratio, the magnetic-field strength, and the spin quantum number I. For like spins, $M_i^0 = M_j^0$ and $\gamma_i = \gamma_j$. Since integrated intensities of signals are directly proportional to their steady-state magnetization, $\eta_j = \sigma/\rho_j$ and $\eta_i = \sigma/\rho_i$. In case of unlike spins, $M_j^0/M_i^0 = (\gamma_j/\gamma_i)^2$ and therefore $\eta_i = (\sigma/\rho_i)(\gamma_j/\gamma_i)$.

From the energy diagram in Fig. 40, we can see that σ depends on W_0 and W_2; only in these transitions, spin states of both spins i and j simultaneously change. W_1 and W'_1 change the state of only one of the two spins. Thus they contribute to ρ_i and ρ_j but not to σ.

Without any further calculations, from the diagram in Fig. 40 we can write

$$\sigma = W_2 - W_0,$$
$$\rho_i = W_2 + W_0 + 2W'_1,$$

and

$$\rho_j = W_2 + W_0 + 2W_1.$$

The minus sign for W_0 in σ comes from the fact that this transition involves flips of opposite sign for the two spins; when one flips up, the other flips down. From the details of the dipole–dipole coupling, all of the $|W|$'s can be calculated. For random isotropic rotational motion,

$$W_2 = \frac{\gamma_j^2 \gamma_i^2 \hbar^2}{10 r^6} \left[\frac{6\tau_c}{1 + (\omega_i + \omega_j)^2 \tau_c^2} \right],$$

$$W_0 = \frac{\gamma_j^2 \gamma_i^2 \hbar^2}{10 r^6} \left[\frac{\tau_c}{1 + (\omega_j - \omega_i)^2 \tau_c^2} \right],$$

$$W_1 = \frac{\gamma_j^2 \gamma_i^2 \hbar^2}{20 r^6} \left(\frac{3\tau_c}{1 + \omega_j^2 \tau_c^2} \right),$$

where ω_i and ω_j are the resonance frequencies of spins i and j at the magnetic field being used. W'_1 is identical to W_1 except that ω_j is replaced with ω_i. For the homonuclear case $\omega_i \approx \omega_j$. Therefore, $(\omega_i + \omega_j) \approx 2\omega$, $(\omega_j - \omega_i) \approx 0$, and

$$\sigma = (\gamma^4 \hbar^2 / 10 r^6) [6\tau_c / (1 + 4\omega^2 \tau_c^2) - \tau_c].$$

In the limit $\omega \tau_c \gg 1$, W_2 becomes very small and $\sigma \approx -(\gamma^4 \hbar^2 / 10 r^6)\tau_c$. In the other limit when $\omega \tau_c \ll 1$, W_2 is proportional to $6\tau_c$ whereas W_1 remains proportional to τ_c. Consequently, W_2 becomes the dominant term and $\sigma = (\gamma^4 \hbar^2 / 2 r^6)\tau_c$. Thus, σ can be either positive or negative depending on the magnitude of $\omega \tau_c$.

In the common applications of *heteronuclear* NOE, when the Larmor frequencies of the two nuclei are significantly different, W_0 is never larger than W_2; $|\omega_j - \omega_i|$ is always a nonnegligible number. In the limit of slow-motion and long-correlation times, both W_2 and W_0, and hence the heteronuclear cross-relaxation term, vanish and a change in the signal intensity of one spin while the other is saturated is not observed. However, in the limit of rapid motion which is defined as $(\omega_j + \omega_i)\tau_c \ll 1$ in the heteronuclear case, σ is independent of ω_j or ω_i and has the same magnitude as in the homonuclear case. In this limit, it is easy to calculate from the given equations that $\eta_i = \frac{1}{2}\gamma_j/\gamma_i$. Thus, for a ^{13}C nucleus undergoing purely dipolar relaxation caused by coupling to protons, η has a theoretical maximum of 1.989, and the maximal NOE (defined as $1 + \eta$) is ~3.

References

Abragam, A. (1961). "The Principles of Nuclear Magnetism." Oxford University Press, Amen House, London.
Ackermann, J. J. H., Grove, T. M., Wong, G. G., Gadian, D. G., and Radda, G. K. (1980). *Nature* **283**, 167.
Agris, P. F., and Campell, J. D. (1982). *Science* **216**, 1325.
Alger, J. R., and Prestegard, J. M. (1977). *J. Magn. Reson.* **27**, 137.
Alger, J. R., den Hollander, J. A., and Shulman, R. G. (1982). *Biochemistry* **21**, 2957–2963.
Armitage, I. M., and Otvos, J. D. (1982). Principles and applications of ^{113}Cd NMR to biological systems. *In* "Biological Magnetic Resonance" (L. J. Berliner and J. Reuben, eds.), Vol. 4. Plenum, New York.
Bachouchin, W. W., and Roberts, J. D. (1978). *J. Am. Chem. Soc.* **100**, 8041.
Billeter, M., Braun, W., and Wüthrich, K. (1981). *J. Mol. Biol.* **155**, 321.
Bloch, F. (1946). *Phys. Rev.* **70**, 460.
Bloch, F., Hausen, W. W., and Packard, M. (1946). *Phys. Rev.* **69**, 127.
Bloembergen, N. (1957). *J. Chem. Phys.* **27**, 595.
Bothner-by, A. A. (1979). Nuclear Overhauser effects on protons, and their use in the investigation of structures of biomolecules. *In* "Biological Applications of Magnetic Resonance" R. G. Shulman, ed. Academica, New York.
Brindle, K. M., Brown, F. F., Campbell, I. D., Grathwohl, C., and Kuchel, P. W. (1979). *Biochem. J.* **180**, 37.
Brindle, K. M., Brown, F. F., Campbell, I. D., Foxall, L., and Simpson, R. J. (1980). *Biochem. Soc. Trans.* **8**, 645.
Brown, F. F., Campbell, I. D., Henson, R., Hirst, C. W. J., and Richards, R. E. (1973). *Eur. J. Biochem.* **38**, 54.
Brown, T. R., and Ogawa, S. (1977). *Proc. Natl. Acad. Sci.* **74**, 3627.
Brown, T. R., Uğurbil, K., and Shulman, R. G. (1977). *Proc. Natl. Acad. Sci. USA* **74**, 5551.
Brown, T. R., Gadian, D. G., Garlick, P. B., Radda, D. G., Seeley, P. J., and Styles, P. (1978). "Frontiers in Biological Energetics" (J. Leigh, A. Scarpa, P. L. Dutton, eds.), Vol. 2, p. 1341. Academic, New York.
Burt, C. T. (1981). Nuclear magnetic resonance studies of muscle constituents in living tissue. *In* "Cell and Muscle Mobility" (R. M. Dowken and J. W. Shay, eds.), Vol. 1. Plenum, New York.
Bystrov, V. F., Portnova, S. L., Tsetlin, V. I., Ivanov, V. T., and Ovchinnikov, Y. A. (1969). *Tetrahedron* **25**, 493.
Bystrov, V. F., Ivanov, V. T., Portnova, S. L., Balasnova, T. A., and Ovchinnikov, Y. A. (1973). *Tetrahedron*, **29**, 873.
Campbell, I. D., and Dobson, C. M. (1979). *Methods. Biochem. Anal.* **25**, 1.
Campbell, I. D., Dobson, C. M., and Williams, R. J. P. (1975). *Proc. R. Soc. London B* **189**, 485–502.
Campbell, I. D., Dobson, C. M., and Ratcliffe, R. G. (1977). *J. Magn. Reson.* **27**, 455.
Campbell, I. D., Dobson, C. M., Ratcliffe, R. G., and Williams, R. J. P. (1978). *J. Magn. Reson.* **29**, 397.
Carr, H. Y., and Purcell, E. M. (1954). *Phys. Rev.* **94**, 630.

Civan, M. M., and Shporer, M. (1978). "Biological Magnetic Resonance" (L. J. Berhimer and J. Reuben, eds.), Vol. 1. Plenum, New York.
Cohen, S. M., Ogawa S., and Shulman, R. G. (1979). *Proc. Natl. Acad. Sci.* **76,** 1603.
Cohen, S. M., Glynn, P., and Shulman, R. G. (1981). *Proc. Natl. Acad. Sci.* **78,** 60.
Cohn, M. (1982). *Ann. Rev. Biophys. Bioeng.* **11,** 23–42.
Cohn, M., and Hughes, T. J. (1962). *J. Biol. Chem.* **237,** 176.
Cope, F. W. (1965). *Proc. Natl. Acad. Sci.* **54,** 225.
Cope, F. W. (1967). *J. Gen. Physiol* **50,** 1353.
Costa, J. L., Dobson, C. M., Kirk, K. L., Paulsen, F. M., Valerie, C. R., and Becchione, J. J. (1979). *FEBS. Lett.* **99,** 141–146.
Costa, J. L., Dobson, C. M., Kirk, K. L., Paulsen, F. M., Valerie, C. R., and Becchione, J. J. (1980). *Philos. Trans. Roy. Soc. London, Ser. B* **289,** 413–423.
Dahlquist, F. W., Longmuir, K. J., and DuVernet, R. B. (1975). *J. Magn. Reson.* **17,** 406.
den Hollander, J. A., Brown, T. R., Uğurbil, K., and Shulman, R. G. (1979). *Proc. Natl. Acad. Sci.* **76,** 6096.
Dillon, P. F., Meyer, R. A., Kushmerick. M. J. and Brown, T. R. (1982). *Fed. Proc.* **41,** 978.
Dubs, A., Wagner A., and Wüthrich, K. (1979). *Biochim. Biophys. Acta* **577,** 177–194.
Dwek, R. A. (1973). "Nuclear Magnetic Resonance (N.M.R.) in Biochemistry." Clarender, Oxford.
Early, T. A., Kearns, D. R., Hillen, W., and Wells, R. D. (1981a). *Biochemistry,* **20,** 3764–3769.
Early, T. A., Kearns, D. R., Hillen, W., and Wells, R. D. (1981b). *Biochemistry,* **20,** 3756–3764.
Englander, S. W., and Englander, J. J. (1978). *Methods Enzymol.* **49,** 24–39.
Farrar, T. C., and Becker, E. D. (1971). "Pulse and Fourier Transform NMR." Academic Press, New York.
Flatman, P., and Lew, V. L. (1977). *Nature* **267,** 360.
Freeman, R., and Hill, M. D. (1971). *J. Chem. Phys.* **54,** 3367–3377.
Freeman, R., and Morris, C. A. (1979). *Bull. Magn. Reson.* **1,** 5.
Fukusima, E., and Roeder, S. B. (1981). "Experimental Pulse NMR A Nuts and Bolts Approach." Addison-Wesley, Reading, Massachusetts.
Gadian, D. G., Radda, G. K., Richards, R. E., and Seeley, P. J. (1979). ^{31}P NMR in living tissue: the road from a promising to an important tool in biology. *In* "Biological Applications of Magnetic Resonance" (R. G. Shulman, ed.). Academic Press, New York.
Gadian, D. G., Radda, G. K., Brown, T. R., Chance, E. M., Dawson, M. J., and Wilkie, D. R. (1981). *Biochem. J.* **195,** 1.
Gadian, D. G., Radda, G. K., Dawson, M. J., and Wilkie, D. R. (1982). pH measurements of cardiac and skeletal muscle using ^{31}P NMR. *In* "Intracellular pH: Its Measurement, Regulation and Utilization in Cellular Functions" (R. Nuccitelli and D. W. Deamer, eds.). Liss, New York.
Garlick, P. B., Radda, G. K., and Seeley, P. J. (1979). *Biochem. J.* **184,** 547–554.
Garlick, P. B., Brown, T. R., Sullivan, J. H., and Uğurbil, K. (1983). *J. Mol. Cell Cardiol.* **15,** 855.
Gelin, B. R., and Karplus, M. (1975). *Proc. Natl. Acad. Sci. USA* **72,** 2002.
Gerig, J. T. (1978). Fluorine magnetic resonance in biochemistry. *In* "Biological Magnetic Resonance," vol. 1 (L. J. Berliner and J. Reuben, eds). Plenum, New York.
Giesner-Prettre, C., and Pullman, B. (1970). *J. Theor. Biol.* **27,** 87.
Giesner-Prettre, C., and Pullman, B. (1971). *J. Theor. Biol.* **31,** 287.
Giesner-Prettre, C., Pullman, B., Borer, P. M., Kan, L., and Tso, P. O. P. (1976). *Biopolymers* **15,** 2277.

Gillies R. J. (1982). Intracellular pH and proliferation in yeast, tetrahymena and sea urchin eggs. *In* "Intracellular pH: Its Measurement, Regulation and Utilization in Cellular Functions" (R. Nuccitelli and D. W. Deamer, eds.). Liss, New York.

Gillies, R. J., Uğurbil, K., den Hollander, J. A., and Shulman, R. G. (1981). *Proc. Nat. Acad. Sci. USA* **78**, 2125–2129.

Glickson, J. D., McDonald, C. C., and Phillips, W. D. (1969). *Biochem. Biophys. Res. Commun.* **35**, 492.

Gordon, R. E., Manley, P. E., and Shaw, D. (1980). *Nature* **2387**, 736.

Granot, J. (1978). *J. Am. Chem. Soc.* **100**, 1539.

Granot, J., and Fiat, D. (1977). *J. Am. Chem. Soc.* **99**, 4963.

Gupta, R. K., Benovic, J. L., and Rose, Z. B. (1978). *J. Biol. Chem.* **253**, 6172.

Gurd, F. R. N., and Ruthgeb, M. (1979). *Adv. Protein Chem.* **33**, 73–165.

Gurd, F. R., Wittebort, R. J., Rothgeb, T. M., and Neireiter, G., Jr (1982). "Biochem. Struct. Determ. NMR" (A. A. Bothner-By, J. D. Glickson, and B. D. Sykes, eds.), p. 1–29. Dekker, New York.

Hilbers, C. W. (1979). *In* "Biological Applications of Magnetic Resonance" (R. G. Shulman, ed.), pp. 1–44. Academic Press, New York.

Hoult, D. I., and Richards, R. E. (1976). *J. Magn. Res.* **24**, 71–85.

Hoult, D. I., and Richards, R. E. (1975). *Proc. R. Soc. London A* **344**, 311.

Hurd, R. E., and Reid, B. R. (1980). *J. Mol. Biol.* **142**, 181–193.

Jacobus, W. E., Pores, I. H., Lucas, S. K., Kallman, C. H., Weisfeldt, M. L., and Flaherty, J. T. (1982). The role of intracellular pH in the control of normal and ischemic myocardial contractility: a ^{31}P nuclear magnetic resonance and mass spectrometry study. *In* "Intracellular pH: Its Measurement, Regulation and Utilization in Cellular Functions" (R. Nucutelli and D. W. Deanser, eds.). Liss, New York.

Jardetsky, O., and Roberts, G. C. K. (1981). "NMR in Molecular Biology." Academic Press, New York.

Jelinski, L. W., Sullivan, C. E., and Tordira, D. A. (1980). *Nature London* **284**, 531.

Johnson, C. E., and Bovey, F. A. (1958). *J. Chem. Phys.* **29**, 1012.

Johnston, P. D., and Redfield, A. G. (1977). *Nucleic Acids Res.* **4**, 3599–3615.

Karplus, M. (1959). *J. Chem. Phys.* **30**, 11.

Karplus, M., and McCammon, J. A. (1981). *CRC Crit. Rev. Biochem.* **9**, 293–349.

Kearns, D. R., Patel, D. J., and Shulman, R. G. (1971). *Nature*, **229**, 338–339.

King, R., Maas, R., Gassner, M., Nanda, R. K., Connover, W. W., and Jardestky, O.(1978). *Biophys. J.* **6**, 103.

Krugh, T. R., and Nuss, M. E. (1979). *In* "Biological Applications of Nucleic Resonance" (R. G. Shulman, ed.), pp. 113–176. Academic Press, New York.

Lauterbur, P. C. (1977). *In* "NMR in Biology" (R. A. Dwek and I. A. Campbell, eds.). Academic Press, New York.

Lipari, G., and Szabo,A. (1982a), *J. Am. Chem. Soc.* **140**, 4546.

Lipari, G., and Szabo, A. (1982b). *J. Am. Chem. Soc.* **104**, 4559.

Lippard, S. J., Burger, A. R., Uğurbil, K., Pantaliano, M. W., and Valentine, J. S. (1977). *Biochemistry* **16**, 1136.

London, R. E. (1980). *In* "Magnetic Resonance in Biology" (J. J. Cohen, ed.), Vol. 1, p. 1. Wiley, New York.

London, R. E., and Avitabile, J. (1978). *J. Am. Chem. Soc.* **100**, 7159.

McConnell, H. M. (1958). *J. Chem. Phys.* **28**, 430.

Mansfield, P., and Morris, P.G. (1982). "NMR Imaging in Biomedicine." Adv. in Mag. Res. Suppl. 2, (J. S. Waughn, ed.). Academic Press, New York.

Meiboom, S., and Gill, D. (1958). *Rev. Sci. Instrum.* **29**, 688.

Midelfort, C.F., Gupta, R. J., and Rose, I. A. (1976). *Biochemistry* **15**, 2178.
Mildvan, A. S., and Gupta, R. K. (1978). *Methods Enzymol.* **49**, 322.
Moon, R. B., and Richards, J. H. (1973). *J. Biol. Chem.* **248**, 7276.
Noggle, J. H., and Schirmer, R. E. (1971). "The Nuclear Overhauser Effect." Academic Press, New York.
Nunnally, R. L., and Hollis, D. P. (1979). *Biochemistry* **18**, 3642.
Ogawa, S., Rottenberg, H., Brown, T. R., Shulman, R. G., Castillo, C. L., and Glynn, P. (1978). *Proc. Natl. Acad. Sci. USA* **75**, 1796.
Ogawa, S., Boeus, C. C., and Lee, T. M. (1981). *Arch. Biochem. Biophys.* **210**, 740–747.
Otvos, J. D., Alger, J. R., Coleman, J. E., and Armitage, I. M. (1979). *J. Biol. Chem.* **254**, 1778.
Pardi, A., Morden, K. M., Patel, D. J., and Tinoco, I., Jr. (1982). *Biochemistry* **21**, 6567.
Pardi, A., Morden, K. M., Patel, D. J., and Tinoco, I., Jr. (1983). *Biochemistry* **22**, 1107–1113.
Patel, D. J. (1979). *Acc. Chem. Res.* **12**, 118–125.
Patel, D. J., and Hilbers, C. W. (1975). *Biochemistry* **14**, 2651–2656.
Patel, D. J., Canuel, L. L., Woodward, C., and Bovey, F. (1975). *Biopolymers* **14**, 1959.
Patel, D. J., Kawlowski, S., Nordheim, A., and Alexander, R. (1982). *Proc. Natl. Acad. Sci.* **79**, 1413–1417.
Patel, D. J., Ikuta, S., Kozlowski, S., and Itakura, K. (1983). *Proc. Natl. Acad. Sci.* **80**, 2184.
Plateau, P., Dumas, C., and Gueron, M. (1983). *J. Magn. Reson.* **54**, 46.
Poulsen, F. M., Hoch, J. C, and Dobson, C. M. (1980). *Biochemistry* **19**, 2597.
Proctor, W. G., and Yu, F. C. (1950). *Phys. Rev.* **77**, 717.
Purcell, E. M., Torrey, H. C., and Pound, R. V. (1946). *Phys. Rev.* **69**, 37.
Radda, G. K., Gadian, D. G., and Ross, B. D. (1982). Energy metabolism and cellular pH in normal and pathological conditions. A new look through ^{31}P nuclear magnetic resonance. *In* "Metabolic Acidosis." Ciba Sympos. 87. Pitman, London.
Ramachandran, G. N., Chandrasekanan, R., and Kopple, K. D. (1971). *Biopolymers* **10**, 2113.
Redfield, A., and Gupta, R. K. (1971). *Gold Spring Harbor Symp. Quant. Biol.* **36**, 405.
Redfield, A. G., Kunz, S. D., and Ralph, E. K. (1975). *J. Magn. Reson.* **19**, 114.
Riberio, A. A., King, R., Reshvo, C., and Jardetsky, O. (1980). *J. Am. Chem. Soc.* **102**, 4040.
Roberts, J. K. M., and Jardetsky, O. (1983). *Biochim. Biophys. Acta* **639**, 53–76.
Robillard, G. T., and Reid, B. R. (1979). *In* "Biological Applications of Magnetic Resonance" (R. G. Shulman, ed.), pp. 45–112. Academic Press, New York.
Roos, A., and Boron, W. F. (1981). *Physiol. Rev.* **61**, 296.
Ross, B .D., Radda, G. K., Gadian, D. G., Rocker, G., Esiri, M., and Falconer-Smith, J. (1981). *N. E. J. Med.* **304**, 1338.
Schimmel, P.R., and Redfield, A. G. (1980). *Annu. Rev. Biophys. Bioeng.* **9**, 181–221.
Scott, A. I., and Baxter, R. L. (1981). *Annu. Rev. Biophys. Bioeng.* **10**, 151.
Seelig, J. (1977). *Q. Rev. Biophys.* **10**, 353–418.
Seelig, J., and Seelig, A. (1980). *Q. Rev. Biophys.* **13**, 19–61.
Shimuzu, H. (1962). *J. Chem. Phys.* **37**, 765.
Shporer, M., and Civan, M. M. (1972). *Biophys. J.* **12**, 114.
Shulman, R. G., Brown, T. R., Uǧurbil, K., Ogawa, S., Cohen, S. M., and den Hollander, J. A. (1979). *Science* **205**, 160.
Slichter, C. P. (1978). "Principles of Magnetic Resonance," 2nd ed. Springer-Verlag, Berlin and New York.
Smith, I. C. P., Tulloch, A. P., Stockton, G.W., Schreier, S., Joyce, A., Butler, K. W., Boulanger, Y., Blackwell, B., and Bennett, L. G (1978). *Ann. N.Y. Acad. Sci.* **308**, 8.
Son, T. D., Guschlbauer, W., and Guéron, M. (1972). *J. Am. Chem. Soc.* **94**, 7903.
Stellwagen, E., and Shulman, R. G. (1973). *J. Mol. Biol.* **75**, 683.

Sternlicht, J., Jones, D. E., and Kustin, K. (1968). *J. Am. Chem. Soc.* **90,** 7110.
Stoesz, J. D., Malinowski, D. P., and Redfield, A. G. (1979). *Biochemistry,* **18,** 4669–4675.
Tanswell, P., Thornton, J. M., Korda, A. V., and Williams, R. J. P. (1975). *Eur. J. Biochem.* **57,** 135.
Thong, C. M., Canet, D., Granger, P., Marraud, M., and Néel, J. (1969). *C. R. Acad. Sci. Ser. C.* **269,** 580.
Tropp, J., and Redfield, A. G. (1981). *Biochemistry* **20,** 2133–2140.
Uğurbil, K., and Bersohn, R. (1977). *Biochemistry* **16,** 3016.
Uğurbil, K., Norton, R. S., Allerhand, A., and Bersohn (1977). *Biochemistry* **16,** 886.
Uğurbil, K., Brown, T. R., den Hollander, J. A., Glynn, P., and Shulman, R. G. (1978a). *Proc. Natl. Acad. Sci.,* **75,** 3742.
Uğurbil, K., Rottenberg, H., Glynn, P., and Shulman, R. G. (1978b). *Proc. Natl. Acad. Sci. USA,* **75,** 2244–2248.
Uğurbil, K., Fukami, M. H., and Holmsen, H. (1984a). *Biochemistry* **23,** 409–416.
Uğurbil, K., Fukami, M. H., and Holmsen, H. (1984b). *Biochemistry* **23,** 416–428.
Uğurbil, K., Rottenberg, H., Glynn, P., and Shulman, R. G. (1982). *Biochemistry* **21,** 1068.
Uğurbil, K., Holmsen, H., and Shulman, R. G. (1979). *Proc. Natl. Acad. Sci.* **76** (5), 2227–2231.
Wagner ,G., and Wüthrich, K. (1981). *J. Mol. Biol.* **155,**347.
Wallach, D. (1967). *J. Chem. Phys.* **47,** 5258.
Webb, M. R., and Eccleston, J. F. (1981). *J. Biol. Chem.* **256,** 7734.
Wedin, R. E., Delepierre, M., Dobson, C. M., and Poulsen, F. M. (1982). *Biochemistry* **5,** 1098.
Wider, G., Lee, K. H., and Wüthrich, K. (1981). *J. Mol. Biol.* **155,** 367.
Wittebort, R. J., and Szabo, A. (1978). *J. Chem. Phys.* **69,** 1722.
Wittebort, R. J., Szabo, A., and Gurd, F. R. N. (1980). *J. Am. Chem. Soc.* **102,** 5723.
Woesner, D. E. (1962a). *J. Chem. Phys.* **36,** 1.
Woesner, D. E. (1962b). *J. Chem. Phys.* **37,** 647.
Woodward, C. K., and Hilton, B. D. (1980). *Biophys. J.* **32,** 561–575.
Woodward, C., Simon, I., and Tucksen, E. (1982). *Mol. Cell. Biochem.* **48,** 135–160.
Wu, S. T., Pieper, G. M., Salhany, J. M., and Eliot, R. S. (1981). *Biochemistry* **20,** 7399.
Wüthrich, K., Wagner G., Richarz, R., and Braun, W. (1980). *Biophys. J.* **32,** 549–560.
Wüthrich, K., Wider, G., Wagner, G., and Braun, W. (1981). *J. Mol. Biol.* **155,** 311.

Sternlicht, J., Jones, D. E., and Kustin, K. (1968). *J. Am. Chem. Soc.* **90**, 7110.
Stoesz, J. D., Malinowski, D. P., and Redfield, A. G. (1979). *Biochemistry,* **18**, 4669–4675.
Tanswell, P., Thornton, J. M., Korda, A. V., and Williams, R. J. P. (1975). *Eur. J. Biochem.* **57**, 135.
Thong, C. M., Canet, D., Granger, P., Marraud, M., and Néel, J. (1969). *C. R. Acad. Sci. Ser. C.* **269**, 580.
Tropp, J., and Redfield, A. G. (1981). *Biochemistry* **20**, 2133–2140.
Uğurbil, K., and Bersohn, R. (1977). *Biochemistry* **16**, 3016.
Uğurbil, K., Norton, R. S., Allerhand, A., and Bersohn (1977). *Biochemistry* **16**, 886.
Uğurbil, K., Brown, T. R., den Hollander, J. A., Glynn, P., and Shulman, R. G. (1978a). *Proc. Natl. Acad. Sci.,* **75**, 3742.
Uğurbil, K., Rottenberg, H., Glynn, P., and Shulman, R. G. (1978b). *Proc. Natl. Acad. Sci. USA,* **75**, 2244–2248.
Uğurbil, K., Fukami, M. H., and Holmsen, H. (1984a). *Biochemistry* **23**, 409–416.
Uğurbil, K., Fukami, M. H., and Holmsen, H. (1984b). *Biochemistry* **23**, 416–428.
Uğurbil, K., Rottenberg, H., Glynn, P., and Shulman, R. G. (1982). *Biochemistry* **21**, 1068.
Uğurbil, K., Holmsen, H., and Shulman, R. G. (1979). *Proc. Natl. Acad. Sci.* **76** (5), 2227–2231.
Wagner, G., and Wüthrich, K. (1981). *J. Mol. Biol.* **155**, 347.
Wallach, D. (1967). *J. Chem. Phys.* **47**, 5258.
Webb, M. R., and Eccleston, J. F. (1981). *J. Biol. Chem.* **256**, 7734.
Wedin, R. E., Delepierre, M., Dobson, C. M., and Poulsen, F. M. (1982). *Biochemistry* **5**, 1098.
Wider, G., Lee, K. H., and Wüthrich, K. (1981). *J. Mol. Biol.* **155**, 367.
Wittebort, R. J., and Szabo, A. (1978). *J. Chem. Phys.* **69**, 1722.
Wittebort, R. J., Szabo, A., and Gurd, F. R. N. (1980). *J. Am. Chem. Soc.* **102**, 5723.
Woesner, D. E. (1962a). *J. Chem. Phys.* **36**, 1.
Woesner, D. E. (1962b). *J. Chem. Phys.* **37**, 647.
Woodward, C. K., and Hilton, B. D. (1980). *Biophys. J.* **32**, 561–575.
Woodward, C., Simon, I., and Tucksen, E. (1982). *Mol. Cell. Biochem.* **48**, 135–160.
Wu, S. T., Pieper, G. M., Salhany, J. M., and Eliot, R. S. (1981). *Biochemistry* **20**, 7399.
Wüthrich, K., Wagner G., Richarz, R., and Braun, W. (1980). *Biophys. J.* **32**, 549–560.
Wüthrich, K., Wider, G., Wagner, G., and Braun, W. (1981). *J. Mol. Biol.* **155**, 311.

2

Electron Spin Resonance

DANIEL J. KOSMAN

Department of Biochemistry
School of Medicine
State University of New York at Buffalo
Buffalo, New York

I.	Introduction	90
II.	Principles of Electron Spin Resonance	90
	A. The Resonance Condition	90
	B. The g Value	93
	C. Hyperfine and Superhyperfine Splitting	97
III.	Establishing and Detecting Electron Spin Resonance	102
	A. Field Versus Frequency Sweep and the Choice of Resonance Energy	102
	B. Modulation and the Second-Derivative Mode	104
	C. Temperature, Power, and Saturation	106
IV.	Applications of Direct (Continuous-Wave) ESR	109
	A. Detection and Identification of Endogenous Paramagnetic Metal Centers and Flavin-Centered Free Radicals	109
	B. Other Endogenous Paramagnetic Centers	134
	C. Spin Labels	142
	D. Single-Crystal ESR	177
V.	Resonance as a Perturbation	183
	A. Resolution of Inhomogeneously Broadened Lines	183
	B. Electron–Nuclear Double Resonance	184
	C. Electron-Spin Echo	199
	D. Saturation Transfer	222
VI.	An ESR Perspective	229
	Appendix A: ESR Hardware, Glassware, and Supplies	230
	Appendix B: ESR Software	231
	Appendix C: Calibration of ESR Spectra	233
	Bibliography	235
	References	238

I. Introduction

This chapter is written with two goals in mind: to introduce the nonuser of electron spin resonance (ESR) to its fundamentals and its application to problems in biology and to acquaint or reacquaint a current user with new (or old) developments in the use of the technique. The time is past when a chapter of this length could treat the subject in depth and focus on theory as well as practice. Only a broad view can be presented to show how ESR is currently being used in the study of biological systems. What has been sacrificed to a large extent is the theory at the level of the mathematical description of the various physical phenomena discussed.

To help the reader compensate for the brevity of this chapter, the bibliography is organized topically as well as in the standard citation form. Thus, for most sections a list of bibliographic resource texts and articles is provided; these lists are by no means complete. A brief notation for each reference is given indicating this author's sense of its specific attributes. Another feature of the chapter are the appendixes, to which the new user, in particular, is directed. Some practical aspects of ESR are presented there (such as who sells ESR spectrometers) which should help reduce the uncertainties in developing an experimental expertise in its use.

II. Principles of Electron Spin Resonance†

A. The Resonance Condition

Electron spin (paramagnetic) resonance, or ESR (EPR), relies on the behavior of the electron in a magnetic field. This behavior results from the magnetic moment and angular momentum of the electron. The angular momentum of an electron is represented by a spin vector **S**, which can have the values of $\pm \hbar/2$; this is also called the spin of the electron and in units of $\hbar$ can take only the values $\pm\frac{1}{2}$ in any direction specified by the experiment (these are the spin quantum numbers, since $S_z = m_s\hbar$) The magnetic-moment vector of the electron μ is related to the angular momentum or spin by Eq. (1):

$$\mu = -g\beta S/\hbar. \qquad (1)$$

Since S is commonly expressed in units of $\hbar$, Eq. (1) can be rewritten as

$$\mu = \pm\tfrac{1}{2}g\beta. \qquad (2)$$

---

† Please note that general references to chapter sections are collected in the bibliography at the end of this chapter.

In Eqs. (1) and (2), β is the Bohr magneton (μ_0 is sometimes used) and has the value 0.92732×10^{-20} erg/G. The value g is a dimensionless factor which is the proportionality constant between the angular momentum (the spin) and the magnetic moment of an electron. For a "free" electron, $g = 2.002319$. However, the g value of an electron in an atomic, or more commonly, molecular orbital will be different. That is, the characteristic angular momentum or spin of an electron, S, will give rise to quantitatively different magnetic moments depending on the atomic or molecular *orbital* to which the electron is constrained. Although this aspect will be discussed more fully in Section II.B, the reader should recognize that the g value can provide useful information about the electronic system being investigated and about the environment of the system.

The minus sign in Eqs. (1) and (2) shows that the magnetic moment and angular momentum vectors point in opposite directions relative to an external coordinate system. Thus, if the electron is placed in a magnetic field, the electron magnetic moment will align itself so that the angular momentum (spin) will be *opposed* to the field. The energy of an electron moment in such a field is

$$\mathcal{H} = -\mu \cdot \mathbf{H}. \tag{3}$$

Since the direction of the magnetic field is usually defined as H_z, we have

$$E = -\mu H_z \quad \text{or} \quad \mathcal{H} = \frac{g\beta}{\hbar} S H_0. \tag{4}$$

The *Zeeman* energy of the electron can take the values $-\tfrac{1}{2}g\beta H_0$ ($S = -\tfrac{1}{2}\hbar$) and $\tfrac{1}{2}g\beta H_0$ ($S = +\tfrac{1}{2}\hbar$). As noted, the minimum-energy state is that in which the spin angular-momentum vector ($-\tfrac{1}{2}\hbar$) is aligned opposite to the magnetic-moment vector ($\tfrac{1}{2}g\beta$). The ground spin state of an electron in a magnetic field is $-\tfrac{1}{2}$; the "excited" spin state is $+\tfrac{1}{2}$. The ESR experiment involves the excitation of an electron from the gound to the excited state, in essence, the realignment of both the angular-momentum and magnetic-moment vectors in the applied magnetic field H_0. The difference between these states is

$$\Delta E = h\nu = g\beta H_0. \tag{5}$$

Equation (5) defines the "resonance" condition: that combination of electromagnetic radiation ($h\nu$), magnetic field (H_z defined as H_0), and orbital behavior of the electron (the g value) which satisfies the resonance condition, Eq. (5). In the experiment, *only g is invariant*; it is a fundamental property of a particular unpaired electron (β and h are constants for all such electron spins). Thus, resonance can be obtained at a variety of values of ν and H_0, the only requirement being that $h\nu/H_0 = g\beta$. That is, the energy difference

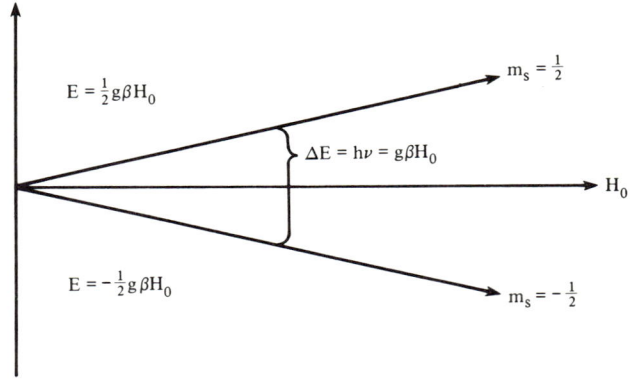

Fig. 1. Splitting diagram indicating the relationship between the energy of the two electron-spin states and the magnitude of H_0.

between the spin states *depends* on the value of H_0. This behavior is commonly represented by a splitting diagram (Fig. 1). One unpaired electron is distinguished from another by the *slopes* of the energy dependencies illustrated in Fig. 1, which are equal to $\pm \frac{1}{2}g\beta$. The larger the g value, the larger the dependence of ΔE on H_0.

This type of behavior distinguishes magnetic resonance, including nuclear magnetic resonance (NMR), from other spectroscopic techniques. The experimentalist can "select" the energy difference between the ground and excited state(s). Since ESR spectrometers are fixed-frequency instruments ($hv \simeq$ constant), resonance is established by varying the external magnetic field H_0. This is illustrated in Fig. 2; unpaired electrons exhibiting larger g values come into resonance at lower values of H_0. Thus, the spectrum generated by going from a low to a high (weak to strong) magnetic field, is, in part, a g-value spectrum, wherein systems with larger g values are at resonance in weaker fields than those with smaller g values.

This aspect relates directly to the resolving power of an ESR spectrometer. For example, assume that v, as indicated in Fig. 2, is ~9 GHz. The $g = 2$ electron comes into resonance at about 3 kG, whereas the $g = 4$ electron is in resonance at half that value, 1.5 kG. There is a 1.5-kG separation. Compare this separation with that obtained using a Q-band instrument, one which operates at a microwave frequency of ~35 GHz. The two g values will be separated by ~6 kG, with the resonances appearing at 6 ($g = 4$) and 12 kG ($g = 2$). Thus, by employing a higher frequency (energy) of exciting electromagnetic radiation, one spreads out the g-value spectrum in the varying magnetic field, improving the potential spectral resolution. The usefulness of employing varying values of v (and H_0) to spread out the g-value spectrum will be noted later in the section on nuclear hyperfine

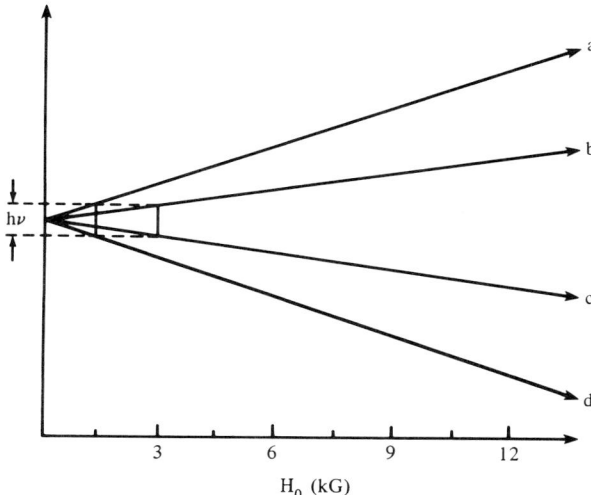

Fig. 2. Splitting diagram for two spins which have g values of 2 (curves b and c) and 4 (curves a and d). The energy of the incident electromagnetic (microwave) radiation is indicated as $h\nu$ chosen to cause resonance (excitation) at 3000 G for $g = 2$. A $g = 4$ electron is in resonance at lower field (1.5 kG in the figure).

couplings (Section II.C); these latter interactions are field independent and are thus distinguishable from g-value differences.

In summary, electrons possess spin and thus angular momentum. Since they are also charged, this angular momentum produces a magnetic moment. The magnetic moment and the spin of an unpaired electron in a metal ion or organic radical will be oriented in an applied magnetic field H_0. Since the spin is quantized ($m_s = \pm\frac{1}{2}$), two energy states will be available, $\pm\frac{1}{2}g\beta H_0$. Excitation of an electron from the ground spin state ($m_s = -\frac{1}{2}$, $E = -\frac{1}{2}g\beta H_0$) to the excited spin state ($m_s = +\frac{1}{2}$, $E = +\frac{1}{2}g\beta H_0$) requires the presence of incident electromagnetic (microwave) radiation, ν, such that $h\nu = g\beta H_0$, the energy difference between the two spin states. In practice, ν is fixed and H_0 is varied; "resonance" occurs when $H_0 = h\nu/g\beta$, i.e., the sample absorbs energy. Note that the g value is a fundamental property, a "fingerprint," of the electron.

B. The g Value

The magnitude of the g value(s) associated with a particular system can, and often does, exhibit directionality or anisotropy. In fact, the anisotropy of the interactions between the various magnetic moments in a molecular system containing an unpaired electron and an external magnetic field is a dominant characteristic of biological ESR. This is so because, in most cases, the molecular system is fixed (as in a frozen solution) or is rotating more slowly than the frequency of the exciting microwave radiation. Therefore, when an unpaired electron is confined to a molecular orbital whose rotation correlates with that of a biological macromolecule, it will itself exhibit a rate of orbital reorientation relative to the external magnetic field so slow as to appear fixed. In most cases, the nature of the ESR spectra derived from paramagnetic biological systems is very dependent on the relative orientations of the field and orbital motion of the electron.

The g value was introduced as the proportionality constant which related the spin (angular momentum) of an electron to the magnetic moment. It is a factor required to compensate for the failure of classical rotational motion to correctly predict the magnetic moment generated by an electron which possesses spin angular momentum, a quantum mechanical property. However, the spin of an electron is not the only motion which an electron exhibits. The electron has orbital motion, also, and thus can generate a magnetic moment associated with this behavior. Whether or not it does so depends on the angular momentum associated with the orbital. An s orbital, for example, has no angular momentum associated with it ($L = 0$), whereas p and d orbitals do. Furthermore, s orbitals are isotropic, that is, are spherically symmetric; they have no directionality. Consequently, an unpaired electron confined to an s orbital would have a g value very close to the "free-spin" value of 2.0023. In contrast, an electron in an orbital which possessed a substantial degree of p or d character would exhibit a g value distinct from 2.0023, and furthermore, since such orbitals are *not* spherically symmetric, the g value measured would very likely depend on the relative orientation of the electron's orbital motion (the magnetic moment associated with its orbital angular momentum L) and the external magnetic field. Thus, the *magnitude* of a g value reflects the orbital angular momentum and, thus, the orbit (or wave function) of the electron. In addition, since orbits which possess angular momentum are inherently anisotropic, the g value of an electron in such an orbit should also exhibit anisotropy.

The g value indicates the relative magnetic moment generated by a given spin angular momentum. Larger values of g suggest that for the same amount of spin angular momentum, a "stronger" moment is generated; one can reasonably picture this as resulting from the contribution to the total moment of the electron by the electron's *orbital* angular momentum. Thus, there is an interaction between the spin and orbital angular momenta which is referred to as the spin–orbit coupling. The efficiency (measured as energy in cm^{-1} or, preferably, the kayser, K) of this coupling varies from system to system, thus spin–orbit coupling is an important component of the g value, which makes characterization of the paramagnetic center possible.

2. ELECTRON SPIN RESONANCE

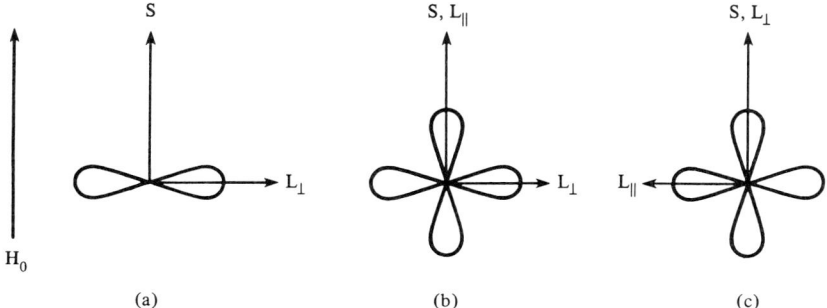

Fig. 3. Relationship between the orientations of the external magnetic field H_0, the spin moment **S**, and the orbital moment **L**. (a) H_0 (and thus **S**) is orthogonal to **L**; there is no coupling between S and L. (b) H_0 and S are again orthogonal to $L_\perp$, but are aligned with (parallel to) $L_\parallel$. Thus **S** and $L_\parallel$ can couple and the orbital motion associated with $L_\parallel$ does contribute to the g value measured ($g_\parallel$) with the system in the orientation indicated. (c) The system in (b) has been rotated 90°; H_0 and $L_\perp$ are now aligned. **S** and $L_\perp$ can couple, and $L_\perp$ contributes to the g value measured ($g_\perp$).

This coupling is also anisotropic. For example, if the spin and orbital magnetic moments are orthogonal (Fig. 3a), there will be no coupling between them. Consequently, the orbital motion associated with the orbital shown makes no contribution to the g value for the system in this orientation. It will be the motion of the electron in orbitals which have significant probability in the same direction as the external magnetic field (and thus have angular momenta which are aligned with the spin moment of the electron) which will couple to the spin. This is illustrated in Fig. 3b and c, in which axes of the system are defined parallel (∥) and perpendicular (⊥) to the external magnetic field.

Figures 3b and c illustrate the extreme orientations of the system; clearly, intermediate orientations are also possible in which H_0 is aligned between $L_\parallel$ and $L_\perp$. Associated with these orientations will be g values to which *both* $L_\parallel$ and $L_\perp$ contribute, as indicated by Eq. (6):

$$g_{meas} = (g_\parallel^2 \cos^2\theta + g_\perp^2 \sin^2\theta)^{1/2}, \qquad (6)$$

where θ is defined as the angle between $L_\parallel$ and H_0.

The preceding discussion and Eq. (6) focused on a system held in a particular orientation relative to H_0, as one might encounter in a study of a single crystal. There are two other situations of interest. One is that associated with a system as in Figs. 3b and c, which has a rotational lifetime similar to or less than a spin-state lifetime; that is, one in which $L_\parallel$ and $L_\perp$ are realigned relative to H_0 during the excitation. This motional averaging occurs if the rotational correlation time τ_c is shorter than $h/\Delta g\beta H$, where Δg is the extreme variation in g. When τ_c satisfies this criterion, the resulting resonance energy ($g\beta H_0$) will be isotropic, that is, it will be independent of the physical orientation of the sample in H_0, and the measured g value will be the average of $g_\parallel$ and $g_\perp$, i.e., $g_{\text{meas}} = \frac{1}{2}(g_\parallel + g_\perp)$. The other situation is that in which this "solution" of randomly oriented, fast-tumbling molecules is frozen. The resulting glass contains all orientations and, for each, a g value as described by Eq. (6). This "powder" spectrum will be a composite of all these values, with the limits defined by $g_\perp$ ($\theta = 90°$) and $g_\parallel$ ($\theta = 0°$). The spectra illustrated for these three cases are presented in Fig. 4a,b,c.

A more complete three-dimensional model is one in which $g_\parallel = g_{zz}$, whereas $g_\perp$ is devolved into two g tensors, g_{xx} and g_{yy}. The term g_{zz} ($g_\parallel$) is designated the most remote (lowest-field) g value, g_{xx} the highest-field one (closest to the free-spin g), and g_{yy} intermediate. If $g_{xx} = g_{yy} = g_{zz}$, the system has octahedral, cubic, or tetrahedral symmetry. In a system with axial symmetry, $g_{xx} = g_{yy} \neq g_{zz}$, a numerical weighting of $g_\perp$ exists, since two principal orientations of g contribute to this value, whereas only one contributes to $g_\parallel$ (g_{zz}). Thus, when H_0 is parallel to $L_\parallel$, only that orientation is contributing to the resonance. When H_0 is perpendicular to $L_\parallel$ (parallel to $L_\perp$) the g_{meas} reflects both g_{xx} and g_{yy} and thus is sampling twice the number of orientations. This is readily apparent in the powder spectrum of an axially symmetric system, as in Fig. 5a. Compare this spectrum to that in Fig. 4c.

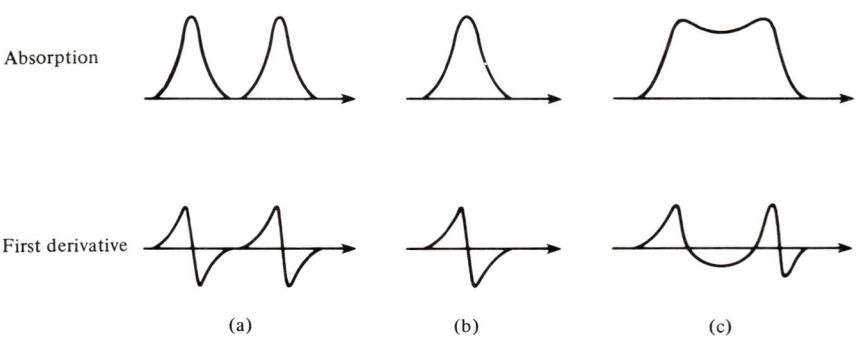

Fig. 4. Absorption and first-derivative curves for the resonance behavior of an electron in an orbital pair, as in Fig. 3. (a) Single crystal (anisotropic); two distinct orientations of the crystal yield the limiting values $g_\parallel$ and $g_\perp$. (b) Rapid tumbling (isotropic) of the molecular framework averages out the orientations and thus the g values. (c) Frozen solution (powder spectrum); the freezing out of all possible orientations generates an envelope of g values. This same envelope can be generated in (a) by rotating $g_\parallel$ into $g_\perp$.

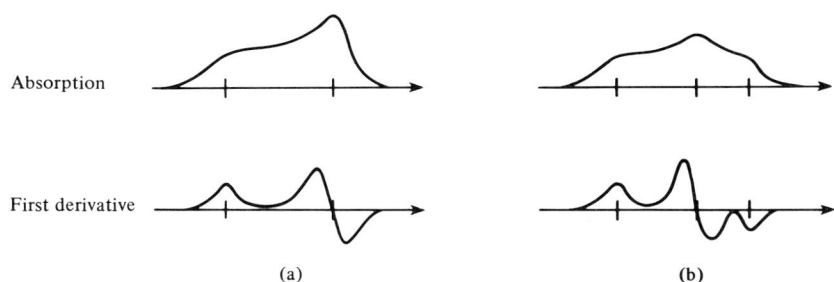

Fig. 5. Powder spectra of the two most common symmetry systems encountered in biological ESR: (a) Axial system: $g_{zz} \neq g_{xx} = g_{yy}$; axially symmetric orbital environment with two characteristic g values; (b) the rhombic system, in which $g_{zz} \neq g_{xx} \neq g_{yy}$ and the resulting three unique principal orientations of the molecular system.

A system with a distortion of the axial symmetry such that $g_{xx} \neq g_{yy} \neq g_{zz}$ exhibits a rhombic spectrum (Fig. 5b). The transitions are spread out in the g-value spectrum. Since the g value is associated directly with specific orbital motion, this anisotropy in g can be used to characterize the electronic structure of the molecular system to which the electron is constrained.

C. Hyperfine and Superhyperfine Splitting

1. Hyperfine Splitting

The field experienced by an unpaired electron, H_0, is modified by the magnetic moment associated with its orbital motion. As noted, this spin–orbit coupling is represented by the deviation of the g value from 2.0023. The field experienced by the electron can be changed by another internal magnetic moment as well, that associated with nearby nuclei which have nonzero nuclear spins. The coupling between the electron and the nuclear moment(s) of the atom(s) on which the electron spin density is centered is known as the hyperfine interaction or splitting (HFS). The weaker interaction(s) this spin has with other nuclei is usually called a superhyperfine splitting (SHFS).

The spin Hamiltonian can be written to include these contributions owing to the magnetic properties of coupled nuclei:

$$\mathcal{H} = \frac{g\beta}{\hbar} \mathbf{S} H_0 - g\beta g_N \beta_N \left[\frac{\mathbf{S} \cdot \mathbf{I}}{r^3} - \frac{3(\mathbf{S} \cdot \mathbf{I})(\mathbf{I'} \cdot \mathbf{r})}{r^5} \right] \\ + \frac{8\pi}{3} g\beta g_N \beta_N |\Psi(0)|^2 \mathbf{S} \cdot \mathbf{I} + \text{nuclear Zeeman term,} \quad (7)$$

where g_N is the magnetogyric ratio or nuclear g factor of the coupled nucleus, β_N the nuclear magneton, r and **r** the distance between and vector joining the electron and nucleus, respectively, and $|\Psi(0)|^2$ the unpaired electron spin density at a given nucleus. The first term is the electron Zeeman interaction energy, Eq. (4). The second describes the through-space, dipolar interaction between the magnetic moments associated with the electron, **S**, and nuclear, **I**, spin angular momenta. The last term, the Fermi-contact term, describes the effect of the field produced at the nucleus when **S** is at the nucleus. These latter two interactions differ in that the dipolar and not the Fermi-contact interaction is distance dependent. Also, the dipolar and not the Fermi-contact interaction is anisotropic, as represented by the radius vector **r**. For this reason, the Fermi-contact term is often called the isotropic hyperfine interaction and can be represented as

$$\mathcal{H}_{\text{iso}} = hA_0 \mathbf{S} \cdot \mathbf{I}, \tag{8}$$

where A_0 is the hyperfine coupling constant and is usually measured in megahertz (MHz). *Only* unpaired s electron density can give rise to a contact interaction, whereas, within the limitations set by the dependence on r^3 and **r**, unpaired spin density in any orbital type can potentially contribute a dipolar interaction.

How these interactions alter the resonance spectra associated with an unpaired electron is a most significant aspect of ESR. These interactions manifest themselves in a splitting of a resonance line (M_s transition) by introducing an additional, local field H_n associated with the nuclear magnetic moment. This local field (magnetic moment) can take the values $g_n \beta_n M_I$, where M_I represents all possible spin states for nucleus of spin = I. Consequently, this local field can either add to or subtract from the laboratory field. The effective field H_{eff} at the electron can be written $H_{\text{eff}} = H_{\text{lab}} + H_{\text{nucl}}$. Since there are $2I + 1$ nuclear spin states, there will be a similar number of H_n fields.

As an example, consider Cu(II). Copper (II) is a d^9 system ($S = \frac{1}{2}$). The nuclear spin of either copper-63 or copper-65 is $\frac{3}{2}$, and associated with this are four nuclear spin states $M_I = +\frac{3}{2}, +\frac{1}{2}, -\frac{1}{2},$ and $-\frac{3}{2}$. Thus, the ground and excited electron-spin states, $-\frac{1}{2}$ and $+\frac{1}{2}$, respectively, are both changed by the presence of the four different nuclear moments associated with the four nuclear-spin states.

The $M_s = -\frac{1}{2}$ electron-spin state is aligned parallel to H_{lab} in the electron-spin ground state. Since μ_e (and not H_{lab}) is the dominant field (magnetic moment) experienced by the M_I states, these states will array themselves with $M_I = +\frac{3}{2}$ as the ground nuclear-spin state. That is, μ_n will be aligned parallel to *both* H_{lab} and μ_e, and will increase H_{eff}. In the $M_s = +\frac{1}{2}$ "excited" state, however, in which the $M_I = -\frac{3}{2}$ is the ground nuclear-spin state (again, because M_I will correlate with μ_e, *not* with H_{lab}), μ_n and H_{lab} will be antiparallel, and H_{eff} will be less than H_{lab}. Since resonance occurs when $h\nu = g\beta H_{eff}$, a lower H_{lab} will be required for the M_s transition associated with the $M_I = +\frac{3}{2}$ state, whereas a higher value will be needed to stimulate the electron-spin transition associated with the $M_I = -\frac{3}{2}$ state. This is shown in Fig. 6. This hyperfine splitting is illustrated in a "stick" diagram (Fig. 7) which also notes the relative orientations of H_{lab}, M_s, and M_I. Note that $\Delta M_I = 0$ is a selection rule. The mixing of the forbidden M_I transitions with those associated with the electron-spin states will be discussed in Section V.

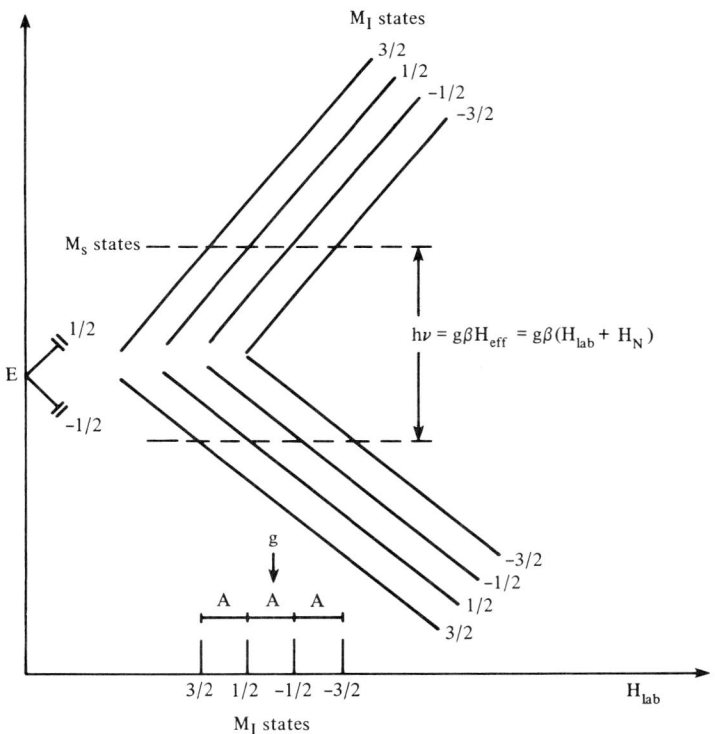

Fig. 6. Idealized splitting diagram for an $S = \frac{1}{2}$, $I = \frac{3}{2}$ electron–nuclear spin system illustrating the relationship between H_{lab} and specific M_s transitions from individual M_s, M_I states. The appropriate position of the g value and the hyperfine *splitting, A* (measured in gauss), is indicated. This is *not* A_0, the coupling constant associated with the Fermi contact (isotropic hyperfine) interaction.

If *isotropic*, the hyperfine splitting constant A in Fig. 6, can be related to A_0 (and to T, the tesla, often used when couplings are expressed as a frequency):

$$A \text{ (in G)} = hA_0/g\beta \quad \text{or} \quad A_0 \text{ (in mK)} = 0.046686gA$$

and

$$T \text{ (in MHz)} = 1.399gA.$$

Generally, the relationships are

$$1 \text{ G} = (g/0.7145) \text{ MHz} = g/(2.142 \times 10^4) \text{ cm}^{-1}$$
$$= 10^{-4}\text{T} = 10^{-1} \text{ mT}.$$

The tesla is the fundamental SI unit for the magnetic induction (magnetic flux density), B. H, in gauss, is used erroneously, although commonly, as the field strength; we will adopt this usage here. Note that quantitatively A is dependent on the g value. Thus, direct comparison of splitting constants without regard to their associated g values is improper.

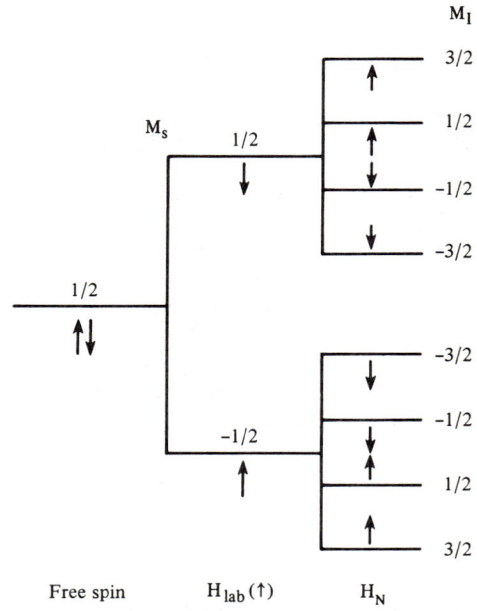

Fig. 7. Stick (splitting) diagram for an $S = \frac{1}{2}$, $I = \frac{3}{2}$ electron–nuclear system indicating orientations of the external and internal magnetic fields (moments) and the relative energies of the M_s, M_I states. There will be $2(1)(\frac{3}{2}) + 1$, or 4 transitions.

Splitting patterns can be fairly easily predicted and interpreted. If there are n nuclei with spin quantum number I, each exhibiting the same hyperfine coupling A, then each M_s orientation will be split into $2nI + 1$ states. If the nuclei are not coupled identically, then the number of lines is $n(2I + 1)$. This expression holds also for situations in which multinuclear (I not equivalent for all coupled nuclei) hyper- and superhyperfine splittings are involved. Isotropic spectra require either a specific symmetry of the electron-spin system as noted here, or a frequency of rotational reorientation rate large compared with the frequeny of the dipolar (anisotropic) term in the spin Hamiltonian. However, for spin systems associated with biological macromolecules, this tumbling rate is often not attainable. Thus, experimental determination of the separate contributions of dipolar and contact (A_0) interactions to A is not readily achievable.

2. Superhyperfine Splitting

The weaker coupling of other nuclei to the unpaired electron(s) is manifested by smaller splittings of the resonance lines associated with the principal M_I states (Figs. 6 and 7). This is illustrated in Fig. 8. The origin and character-

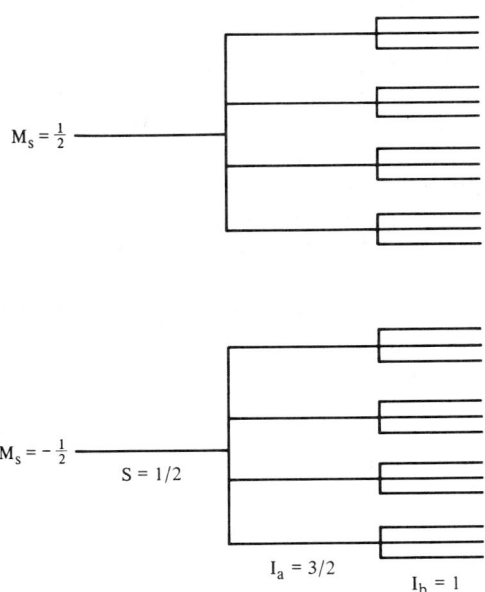

Fig. 8. Splitting (stick) diagram for an $S = \frac{1}{2}$, $I_a = \frac{3}{2}$, $I_b = 1$ system in which $A_a > A_b$. There will be $[2(1)(\frac{3}{2}) + 1] \cdot [2(1)(1) + 1]$, or 12 transitions.

istics of this coupling are identical to that previously described except that in Eq. (7) either A_0 is smaller, r is larger, or **r** is smaller (or a combination of these). These additional M_I states perturb the principal ones, as is shown in the figure. Again, there are $2I + 1$ additional states, thus an increase of $(2I + 1)$-fold in the number of resonance transitions (lines). Furthermore, since any given A value represents a unique contribution of dipolar and Fermi-contact interactions, comparisons of adjusted A values (in MHz or mK) may still be inappropriate. Calculation of A_0 from isotropic spectra addresses this difficulty.

III. Establishing and Detecting Electron Spin Resonance

A. Field Versus Frequency Sweep and the Choice of Resonance Energy

The absorption of energy by the electron spin occurs when $h\nu = g\beta H_{\text{eff}}$. Although this resonance condition could in theory be obtained by varying either ν or H_0, only fixed-frequency, variable-field ESR spectrometers are practicable. This is so since, at the fields readily available, the values of ν are in the range of 1–35 GHz, i.e., in the microwave region. The source of this radiant energy is a klystron tube which is monochromatic. Thus it differs from an rf coil, for example, which is tunable over a wide frequency range, or an ultraviolet or visible lamp, which is polychromatic. Thus, ESR spectrometers are restricted to a fixed-frequency design. Klystron tubes with outputs of ~9.5, ~24, and ~35 GHz have been used; these frequencies are referred to as X, K, and Q band. Low-frequency klystrons (1–4 GHz) have also been employed (S band). These tubes are tunable over a range of ~0.1% of their rated output (low-frequency generators are somewhat more flexible); this tuning allows for some variation in the design of the sample chamber (the cavity) and sample holder, and of the sample volume, geometry, and composition.

The choice of energy ($h\nu$) should be dictated by the experimental system and objective. The factors which choice of frequency affects are sensitivity, sample size, resolution, and saturation.

The sensitivity of the spectrometer does, in part, depend on the magnetic field used. Clearly, the absorption of energy by the sample depends on the number of electrons in the ground, $m_s = -\frac{1}{2}$ spin state (N^-). However, there is a concurrent stimulation of spins dropping *into* the ground state from the excited state (N^+). Thus, the net change n is given by

$$n = N^- - N^+ \tag{9}$$

The difference n is determined by a Boltzmann distribution, $N^+/N^- = e^{-\Delta E/kT}$; with $\Delta E = g\beta H_0$, this becomes $\exp(-g\beta H_0/kT)$. Thus, as H_0 becomes larger so does n, and for a given number of unpaired spins, the greater the absorption of energy.

An estimate of the value of n in a given field can be obtained using Eq. (10):

$$n = Ng\beta H_0/2kT, \tag{10}$$

where N is the concentration of spins (total) in the sample and k the Boltzmann constant, 1.38×10^{-16} erg/K. Thus, at a field of 3 kG (an X-band spectrometer operating with $\nu \simeq 9$ GHz, as discussed previously)

$$\exp(-g\beta H_0/kT) = N^+/N = 0.999$$

for $T = 298$ K and $g = 2$. That is, the difference between the ground and excited states (n) is approximately two parts in one thousand. At 12 kG (a Q-band instrument operating with $\nu \simeq 35$ GHz), Eq. (10) predicts that the spectrometer should be four times more sensitive than an X-band one. In practice, this increase in sensitivity is often not quantitatively realized because of differences in sample size, resolution, and relaxation.

Equation (10) also predicts that the sensitivity is inversely proportional to the temperature. This means that absorption observed at the temperature of liquid nitrogen (77 K) should be, and generally is, four times as intense as the same resonance observed at room temperature. A further increase in sensitivity is possible by approaching the temperature of liquid helium (4.2 K), although other factors, particularly *saturation,* then become important, as will be discussed in the following section.

Resolution is dependent on the resonance energy, since separation of *field-dependent* ESR transitions increases as H_0 ($h\nu$) increases. A field-dependent transition is one which arises from the interaction of the electron with a local field generated by the laboratory magnet. This excludes nuclear fields which exist independent of the applied field. Thus, by comparing spectra at two different values of $h\nu$, overlapping g values, which are field-dependent can be distinguished from overlapping resonances caused by hyperfine interactions, which are not. In effect, at higher frequencies and fields, the g-value spectrum is spread out. However, the g-value resolution at 35 GHz is often better than needed. At 12 kG ($g = 2$) and an inherent linewidth of 1 G, g-value differences of $\sim 0.02\%$ should be detectable. Since the linewidths of biological samples are commonly *greater* than 5–10 G, the resolution obtainable using a Q-band instrument is usually more than sufficient.

The sample size is dictated by the geometry of the resonance cavities of spectrometers operating at different frequencies. Cavity dimensions are inversely proportional to ν, thus an X-band cavity can accommodate a wide variety of sample holders (100–500 μl), whereas the cavity of a Q-band instrument is generally restricted to a relatively small capillary tube (10–50 µl). On the other hand, this limitation may be moot if the amount of sample is small anyway, or in single-crystal studies wherein the sample size is already fixed. Indeed, Q-band studies are often preferred when sample availability is limited. Cavities resonant at low frequencies (S band) tend to be large and inefficient. A refinement of an idea first proposed in 1940 has overcome this deficiency. The "loop-gap" resonator provides ~70 times the incident rf energy at the sample, effectively increasing the signal-to-noise ratio (Froncisz and Hyde, 1982). The other advantage of this design is its size. It accommodates only point samples (small volumes), thus, like the Q-band cavity, complements studies on sample-limited materials.

B. Modulation and the Second-Derivative Mode

As discussed earlier, absorption of microwave energy hv by the sample occurs when H_0, the magnetic field experienced by the electron, reaches a value which satisfies the relationship $hv = g\beta H_0$. Since the energy interval available from the incident electromagnetic radiation is fixed at $hv (=\Delta E)$, a spectrum is generated by steadily changing the differences between ground and excited spin states of the eletcron-spin systems by changing H_0, making the differences equal to hv for each system in turn. In going from lower to higher values of H_0, one brings systems into resonance which have steadily decreasing g values, i.e., as H_0 increases, that value of g necessary to satisfy $hv = g\beta H_0$ decreases.

In principle, the absorption of the microwave energy could be monitored as in any other absorption spectrophotometer, by sampling directly the difference in power between the incident and transmitted or reflected radiation. However, as outlined here, the net change in the populations of the ground and excited states during resonance is very small owing to the very small energy difference between these two states. Consequently, the net absorption is small and typically less than the random noise of an electronic circuit.

To allow for the necessary signal amplification, the dc absorption is converted to an ac one by modulating the external magnetic field H_0 with a secondary field H_m at a fixed frequency. The detector is driven at the same fixed frequency and is, therefore, sensitive only to (selects) input which varies with the frequency of modulation; essentially, the dc noise is ignored. How this modulation works is outlined briefly as follows.

As H_0 is swept through the g spectrum, the small modulating field H_m oscillates at a fixed frequency (typically 100 kHz) with an experimentally chosen amplitude. Thus, at a given value of H_0, the absorption of hv will vary sinusoidally with a frequency of 100 kHz and an amplitude determined by the absorption difference between the limits of the modulation field, $H_0 \pm H_m$. The 100-kHz detector circuit will respond only to this difference, not to the absorption itself. Amplification of this difference allows for a better signal-to-noise ratio, since only the 100-kHz noise in the circuit will be amplified; the low-frequency noise characteristic of the dc detector is lost.

The result of this modulation is that the "absorption" is displayed in the first-derivative mode. That is, the absorption difference between $H_0 \pm H_m$ represents the instantaneous slope of the absorption curve at H_0. As H_0 is swept through the absorption envelope, the change in slope per increment in H_0 is measured, and thus a first derivative is generated. This is illustrated in Fig. 9, which focuses on the three unique regions in the absorption and corresponding first-derivative curves. These are the two inflection points in the absorption curve, each of which is associated with a maximum (or minimum) in the derivative spectrum (the peak and trough), and the absorption maximum which corresponds to zero slope or the "crossover" point on the derivative curve. This latter point defines the position of the transition in the magnetic field.

The important variable in this method of signal detection is the magnitude of H_m. Based on this discussion and on Fig. 9, one can recognize that the larger H_m, the larger the potential difference between the absorption measured at the associated limits of the modulation field. Consequently, a large

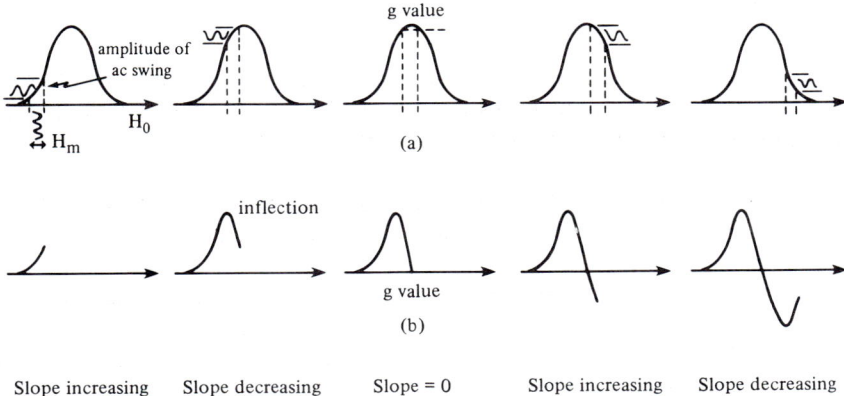

Fig. 9. (a) Frequency modulation of the ESR absorption envelop: H_m field is oscillating at a fixed frequency (100 kHz) causing 100-kHz oscillation of resonant absorption from the microwave field ($h\nu$); ac detector picks up 100-kHz oscillation, discards low-frequency dc noise. Relationships between absorption and first-derivative (harmonic) curves (b) are indicated.

ac swing would result and would yield more signal. However, one should also be aware of the limits to this conclusion, namely, at values of H_m comparable to the inherent absorption bandwidth (or greater), real differences might actually be obscured and the amplitude of the ac deflection reduced. Such overmodulation reduces the amplitude of the detected signal and, perhaps more important, distorts the first-derivative envelope, causing a loss of definition of the true slope of the absorption curve. Overmodulation causes a loss of spectral resolution and a decrease in the excursion of the first-derivative curve.

At very low modulation amplitudes (≤ 0.1 of the absorption linewidth), the derivative amplitude is directly proportional to the modulation. A maximum amplitude (peak height or peak to trough, Fig. 9) is obtained when H_m is approximately equal to the linewidth, although the *area* of the first-derivative envelope is proportional to the modulation amplitude even if the sample is overmodulated. Overmodulation is often used to improve the signal-to-noise ratio (because the higher-frequency noise is smoothed out) when making concentration measurements based on integrated area rather than on peak height, for example, or when attempting spectral simulations of specific hyperfine couplings. The areas obtained from overmodulated absorption envelopes contain appreciable contributions from unresolved components to lower and higher fields, however, so care must be taken to identify and eliminate such contributions.

C. Temperature, Power, and Saturation

The intensity (amplitude) of the first-derivative excursion of a resonance envelope can be increased in a number of ways by appropriate choice of instrument and instrument settings. Among these approaches are those that increase the difference in the population of ground and excited spin states, n, and thus the net flux of electron spins from the former to the latter state. H_0 and T are two operating parameters which change n, as indicated by Eq. (10). However, the intensity of the signal, I, is not only dependent on the value of n but also on the energy density or power of the incident microwave radiation. As the power increases, the transition probability increases, i.e., the signal becomes more intense. However, such simple behavior is not observed experimentally. While the signal intensity does in general respond as some function of the microwave power, at a certain level the signal actually starts to decrease, to broaden, and to eventually disappear altogether. This saturation broadening or power saturation is an important property of paramagnetic centers, since different types of systems, e.g., organic free radicals, metal ions, etc., saturate at different and characteristic levels of microwave power. This power required to saturate is a function of both the temperature and the magnetic field.

The saturation phenomenon is caused by the equalization of the ground and excited-state spin populations, since as $n \to 0$, I also $\to 0$. Thermal processes operate to maintain the equilibrium value n determined by the field strength, the g value, and the temperature. These processes "relax" the system, return it to equilibrium. The time constant for this thermal equilibration is the spin–lattice relaxation time T_1. This process allows for the dissipation of the energy absorbed by the electron spins as heat transferred to the medium surrounding these spins, reestablishing the normal Boltzmann distribution. This is the energy associated with the magnetization of the electron spins along the direction of the external magnetic field, that is, those spins aligned with and against this field. Thus, T_1 is also called the longitudinal relaxation time. Furthermore, T_1 is related to the inherent linewidth of a resonance line, ΔH, by

$$T_1^{-1} = 2g\Delta H/\hbar. \qquad (11)$$

This is also called lifetime broadening and applies to all members of a given spin system, i.e., it is homogeneous.

For the electron-spin system to relax efficiently (for T_1 to be short), the system must be coupled to the lattice; that is, there must be available in the lattice thermal (vibrational) energy levels exhibiting energy differences (quantized as phonons) similar to $g\beta H_0$. Since the thermal energy levels are temperature dependent, the marked temperature dependence noted for power saturation is not surprising. Thus, whereas very low temperatures increase n, they also tend to lengthen T_1. Under such conditions, $n \to 0$ even at very low microwave power, and consequently the anticipated increase in sensitivity owing to decreasing temperature [Eq. (10)] cannot be realized completely.

Although the spin–lattice relaxation process is what dissipates the energy absorbed by the electron spin in going from the ground to the excited state, and thus is the mechanism which reestablishes the Boltzmann distribution and determines the saturation behavior of the spin system, another relaxation mechanism is also of importance. This is the spin–spin relaxation process. The time constant for this process, T_2, is also called the transverse relaxation time, since T_2^{-1} is the rate at which the net magnetization of the x–y component of the spin system at resonance changes. As this component breaks up so that the individual spin vectors, while still in an "excited" T_1 state, exhibit a variety of x–y orientations, the laboratory field H_0 will be modified sightly by these randomly oriented magnetic dipoles. Each electron will be experiencing a slightly different resultant field and will thus be in resonance at a slightly different value of H_0, i.e., the resonance line will broaden. As these local fields become more random, the line broadens still further. This is a type of *secular* broadening and is related to the observed linewidth by

$$\frac{1}{T_2'} = \frac{1}{2T_1} + \frac{1}{T_2} \quad \text{and} \quad \frac{1}{T_2'} = \frac{g\beta \Delta H}{\hbar}. \tag{12}$$

Clearly, either T_1 or T_2 could be the determinant of the observed linewidth. In most systems of biological interest, however, T_2 makes the major contribution, since for these cases, $T_2 < T_1$. Important exceptions to this generality are found among transition metal ions in which very strong spin–orbital coupling provides an efficient spin–lattice relaxation mechanism. Such centers, particularly those containing Fe^{3+} (high and low spin) and Fe^{2+} (high spin), exhibit extreme lifetime broadening and are detectable by ESR only at temperatures of ~4 K.

To test whether saturation can occur at any temperature of interest, one plots the intensity (amplitude) of an ESR signal versus $p^{1/2}$. A straight line should be produced, since signal height is proportional to the square root of the microwave power. The deviation which usually occurs at higher microwave powers results from saturation. The use of lower temperatures to increase resolution (and intensity) of an ESR signal can also induce saturation if the same power level is maintained throughout a temperature study. Such saturation effects can be resolved by plotting the *area* of the resonance of interest against $1/T$, since the power absorbed by a sample is proportional to the difference in population in the two energy levels between which the transition takes place. This difference is governed by the Boltzmann equation, $\exp(-h\nu/kT)$. Since the area under an absorption curve is proportional to total spins (and thus to power absorbed), a graph of area versus $1/T$ should result in a straight line. Again, deviation from this behavior is indicative of saturation. For aqueous samples, power-saturation studies must be carried out with care, since some paramagnetic centers do not readily saturate at room temperature. It is possible to "cook" the sample while attempting to power saturate.

The decrease in linewidth with decreasing temperature will normally reach a limit below which a temperature decrease does not increase resolution. As noted, this temperature is the one at which the spin–spin relaxation has taken over from the spin–lattice relaxation as the dominant relaxation process. A plot of linewidth versus temperature can thus identify the "appropriate" temperature for spectral measurement.

IV. Applications of Direct (Continuous-Wave) ESR

A. Detection and Identification of Endogenous Paramagnetic Metal Centers and Flavin-Centered Free Radicals

Certainly, the most widely exploited area of biology with respect to the use of ESR has been that of bioinorganic chemistry, e.g., the study of metalloproteins. Table I (Fee, 1978) summarizes the characteristics of the main groups of transition-metal centers more commonly found in biomolecules; the flavin semiquinone is also noted. The identification of the metal ion responsible for a particular ESR transition(s) must be based first on correlation between the intensity of the ESR signal and the weight percent of a specific metal as it copurifies with the protein. That is, there is much that is too similar about the ESR spectral characteristics of, for example, molybdenum and iron–sulfur centers, to distinguish between them solely on the basis of simple cw ESR experiments. On the other hand, Cu and Mn commonly exhibit well-resolved HFS owing to the electron-spin–nuclear-spin interaction. This results in a four-line splitting for Cu ($I = \frac{3}{2}$) and six lines for Mn ($I = \frac{5}{2}$), although not all g values exhibit resolvable hyperfine couplings. For example, $A_{\parallel}^{Cu}$ is ca. 100–200 G, whereas $A_{\perp}^{Cu}$ is commonly ~ 0. ($A_{\parallel}^{Cu}$ refers to the nuclear hyperfine splitting in $g_{\parallel}$; $A_{\perp}^{Cu}$, the splitting in $g_{\perp}$.)

There are two relatively unambiguous ways to experimentally indicate if not determine the source of an observed ESR signal. The isotopic composition of the suspected metal can be changed by isolation of the protein from a biological source "fed" the chosen isotope. If the resulting ESR spectrum reflects the change in the nuclear spin I, then it may be inferred that that metal is the paramagnetic center. The classic example of this type of experiment is the preparation of ^{95}Mo-containing xanthine oxidase, described later. Similarly, ^{61}Ni ($I = \frac{3}{2}$) hydrogenase from *desulfovibrio gigas* has recently been prepared (Moura *et al.*, 1982). The natural abundance enzyme (^{58}Ni and ^{60}Ni, $I = 0$) exhibits an ESR spectrum characterized by $g = 2.02$, 2.23, and 2.31. None of these transitions exhibit any hyperfine structure. Hydrogenase enriched 60% with biosynthetically incorporated ^{61}N exhibits the same pattern of values, but both the $g = 2.02$ and 2.23 signals exhibit hyperfine splitting (the four-line pattern, $A \simeq 80$ G, expected for an $I = \frac{3}{2}$ is clearly resolved in $g = 2.02$), whereas the $g = 2.31$ resonance is broadened. This same type of experiment can be performed *in vitro* by resolving the metal(s) and reconstituting with the desired isotope. Such reconstitution is not possible with all proteins, however.

Another approach is to attempt to obtain the electron–nuclear double-resonance (ENDOR) spectrum. As will be described in Section V.B,

Table I Characteristics of Biological Metal and Flavin Prosthetic Groups

Prosthetic groups (electronic configuration)	S	Example(s)	g values[a] (range)
Mn(III) (3d^4)	2	Oxidized bacterial superoxide dismutase	ESR not observed
Mn(II) (3d^5)	5/2	Reduced bacterial superoxide dismutase concanavalin A	2–6
Fe(II) (3d^6)	Low spin, 0 High spin, 2	Deoxyhemoglobin	ESR not observed
Fe(III) (3d^5)	Low spin, 1/2 High spin, 5/2	Heme proteins Heme proteins Nonheme iron	0.5–3 2–6 0.5–10
Binuclear Iron			
Fe(III) · Fe(III) (3d^5,3d^5)	0	(Fe/S)$_2$ oxidized ferredoxins Purple acid phosphatase	No ESR No ESR
Fe(III) · Fe(II) (3d^5,3d^6)	1/2	(Fe/S)$_2$ reduced ferredoxins Purple acid phosphatase (reduced)	1.6–2.2
Co(III) (3d^6)	0	Cobalamin (B$_{12}$)	No ESR
Co(II) (3d^7)	Low spin, 1/2 High spin, 3/2	Cobalamin (B$_{12}$r) Cobalt-substituted proteins	2.0–2.3 1.8–6
Ni(III) (3d^7)	Low spin, 1/2	Carbon monoxide dehydrogenase Desulfovibrio gigas hydrogenase	2.0–2.3
Cu(II) (3d^9)	1/2	Copper proteins	2.0–2.3
Cu(I) and Zn(II) (3d^{10})	0	Copper and zinc proteins	No ESR
Binuclear Cu(II) · Cu(II)		Multinuclear copper proteins	No ESR
Mo(VI) (4d^0)	0	Xanthine oxidase (resting)	No ESR
(V) (4d^1)	1/2	Xanthine oxidase	1.95–2.03
(IV) (4d^2)	1	Xanthine oxidase	ESR not observed
Flavin semiquinone	1/2	Flavoproteins, free flavins	2.0
V (IV) as VO^{2+} (d^1)	1/2	Not physiological; useful as spin probe	1.94–1.98

[a] Ranges include g-value anisotropy. Adapted from Fee (1978).

ENDOR yields the Larmor precession frequencies of nuclei interacting with the electron(s). These frequencies for the metals listed in Table I are easily distinguishable (compare with Table XIII).

Another characteristic of certain metal ions and binuclear spin-paired clusters is that in some oxidation states $S = 0$, 1, or 2. That a diamagnetic spin-zero system does not exhibit an ESR spectrum is expected, but the ESR "silence" of spin-even systems should be noted. Spin-even systems are characterized by rather large zero-field splittings, so that the transitions allowed between the spin doublets are not readily accessible at the microwave and magnetic fields commonly employed. Furthermore, transitions within doublets are forbidden and not observed. Note, too, that both the coupling of some spin-paired systems and the saturation and relaxation of others are highly temperature dependent. Thus, the lack of an ESR spectrum under a given set of conditions is a negative experiment only and cannot be taken as evidence that inherent paramagnetism is lacking in the sample. On the other hand, the temperature dependence of the ESR spectra of certain types of metal centers has been fairly well characterized. Consequently, this type of behavior can be used to advantage in attempts at identification.

1. Coordination: Geometry and Ligation

There are four structural characteristics of an endogenous metal-ion center in a biological macromolecule that can be deduced theoretically from appropriate ESR experiments. These are spin state, oxidation state, coordination (ligand field) geometry, and nature of ligands. Clearly, these are completely interdependent characteristics. Thus, in general, no single ESR experiment can unambiguously answer a question about any one of these features.

The one aspect which can be inferred from the cw ESR spectrum is the overall geometry of the ligand field, i.e., axial or rhombic. That is, by assignment of the g tensors one can show the metal center to exhibit axial symmetry ($g_x = g_y \neq g_z$) or a rhombic distortion ($g_x \neq g_y \neq g_z$). This characterization, by itself, is not particularly informative because it does not define the nature of the distortion nor the extent to which the axial symmetry has been reduced from an octahedral field. Indeed, most axial spectra are associated with pseudosquare planar systems, which have weaker or nonexistent ligation in the fifth and sixth coordination positions. Furthermore, distortions from axial symmetry can be simple or complex. One example is a simple perturbation along one g tensor (coordinate axis), such as an odd component in the field (as in an N_3S chelate, for example). This type of change would not be a geometry change per se. On the other hand, a saddle distortion toward a tetrahedral field would constitute a major change in geometry, although it would not necessarily be reflected in $g_x \neq g_y$. Thus, distinctions between actual ligand orientations based on comparisons of the principal g values are not easy to make and require assignment of g values to specific orientations within the ligand field. This sort of assignment can be achieved using single-crystal ESR and, in some cases, ENDOR and the spin-echo technique on polycrystalline samples (Sections IV.D and V.B,C).

However, for a particular metal in a given oxidation and spin state, relationships between the principal spin-Hamiltonian parameters, A and g, and geometry and ligand type can be formulated. This is so because both A and g are related to the symmetry and energy levels of the electronic structure of the metal complex. For example, as noted, g deviates from the free-spin value because of coupling between S, L, and H_0. Thus, the spin-orbit coupling constant for a given ion in a particular ligand field has a direct bearing on both the anisotropy and the value of the g tensor(s). For Cu(II) in a tetragonal (square planar) field, we have

$$g_\parallel = g(1 - 4\alpha^2\lambda/\Delta_1) \qquad (13a)$$

and

$$g_\perp(g_{xx} = g_{yy}) = g(1 - \alpha^2\lambda/\Delta_2), \qquad (13b)$$

where $g = 2.0023$, α^2 is the coefficient of the d orbital contribution to the upaired electron (usually $d_{x^2} - d_{y^2}$); λ the spin-orbit coupling for the free ion (-830 cm^{-1} or kayser, K); and Δ_1 and Δ_2 the energy differences between the orbitals $d_{x^2-y^2}$ and d_{xy}, and $d_{x^2-y^2}$ and d_{xz}, d_{yz}, respectively.

An analogous pair of equations can be derived for the hyperfine constant:

$$A_\parallel = P[-k(4\alpha^2/7) + (g_\parallel - 2.0023) + 3(g_\perp - 2.0023)/7] \qquad (14)$$

and

$$A_\perp = P[-k(2\alpha^2/7) + \tfrac{11}{14}(g_\perp - 2.0023)], \qquad (15)$$

where the constants P and k are taken as 35 and 0.35 mK, respectively.

There are two ways of using this relationship experimentally. One, which will not be elaborated on here, is to attempt the solution of the actual electronic structure of a metal ion using measurements of g and A values to obtain values for Δ_1 and Δ_2 (in the example given). The other is simply to appreciate that both A and g reflect the sum of the properties of a metal center and attempt to delineate empirically how systematic changes in any of these factors affect either or both of these ESR parameters. Examples of this approach follow.

a. Cu(II) Proteins. (Peisach and Blumberg, 1974). Copper proteins have been divided into two groups based primarily on their visible absorbance near 600 nm. Those that have ϵ values in excess of $10^3 \, M^{-1}\,\text{cm}^{-1}$ are known as the "blue" copper proteins, whereas those that have lower extinctions are termed "nonblue." These groups are also characterized by having distinctly different $|A_{\parallel}^{Cu}|$ values, as is indicated in Fig. 10. Those with $|A_{\parallel}^{Cu}|$ values less than 10 mK (< 100 G) are denoted type 1 Cu(II); those with $|A_{\parallel}^{Cu}|$ more like that of typical Cu(II) complexes (15–20 mK) are designated type 2. Although the origin of these differences is now thought to be associated with the combination of effects of covalency (cysteine mercaptide ligation in type 1 sites) and geometry (these sites are tetrahedral), the effects on the A and g values of a systematic alteration in the ligand composition is still of

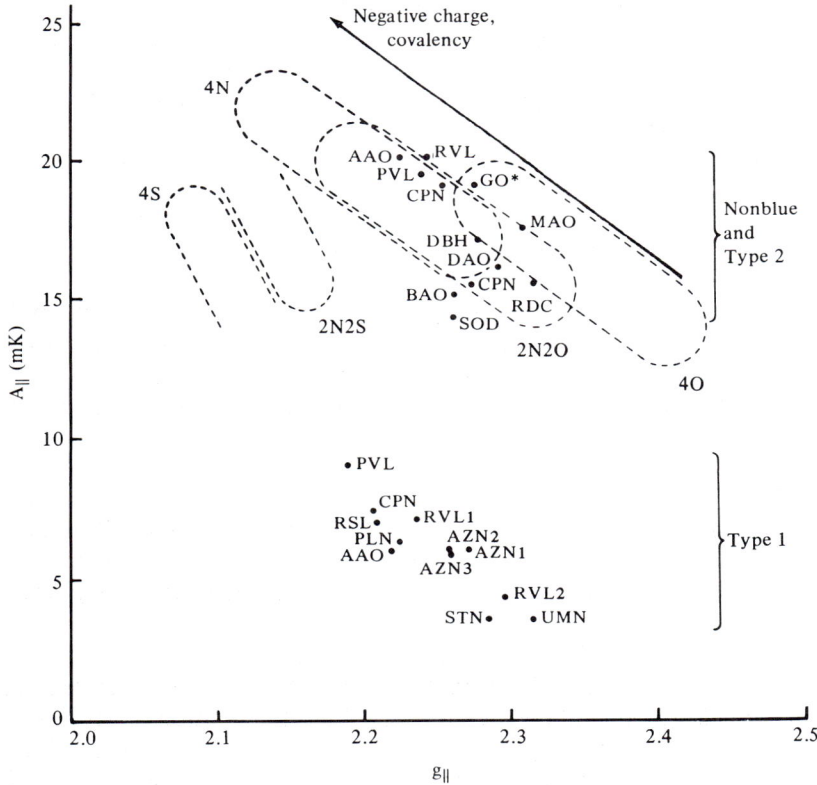

Fig. 10. Relationships between $A_{\parallel}$ and $g_{\parallel}$ for copper proteins (●) and simple Cu(II) complexes. Areas outlined are for the Cu(II) chelates with chemical ligation as shown. Note the consistent effects of increasing (relative) negative charge and covalency. Changes in $A_{\parallel}$ and $g_{\parallel}$ on ligand displacement occur along this diagonal. Galactose oxidase is indicated by (∗). (From Peisach and Blumberg, 1974.)

interest. Peisach and Blumberg documented these parameters for a series of (square planar) Cu(II) complexes with different combinations of oxygen, nitrogen, and sulfur ligation, e.g., O_4, NO_3, N_2O_2, N_3O, N_4, and N_2S_2. They then compared these types of sites to the type 1 and type 2 sites described here. Their analysis is summarized in Fig. 10.

The patterns observed confirmed that type 1 Cu(II) differed from type 2 Cu(II) in some manner independent of ligand type. As noted, this distinctive feature of the type 1 sites is their tetrahedral geometry. In addition, the analysis yielded useful empirical notions of how changes in charge and covalency, as well as ligand type, would be manifest in A and g values. These effects can be rationalized as the electron spending more time in ligand-like orbitals as the covalency (and negative charge) increases. Such an analysis also allows the prediction of the kind of effects exogenous ligand binding would have on an ESR spectrum with regard to changes in the metal-centered spin-Hamiltonian parameters. That is, addition of a ligand such as F^- to Cu(II) complexes could result in the displacement of an endogenous ligand which could be either more or less electron rich. If the ligand displaced, such as H_2O, is a weaker Lewis base (less negative), then F^- binding should decrease $g_\parallel$ and increase $A_\parallel^{Cu}$. If, on the other hand, F^- displaces a more electron-rich ligand, such as HO^- or a nitrogenous ligand, then $g_\parallel$ would increase and $A_\parallel^{Cu}$ decrease. Such behavior has been experimentally verified although, as with all relationships such as this one, other factors (e.g., changes in geometry) complicate the interpretation of the experimental results. Further aspects of ligand exchange will be addressed in the following.

b. Fe(III) Porphyrins. In addition to those for Cu(II), explicit relationships exist between g (and A) and λ for other transition metals. The manifestation of these relationships in the ESR spectra of Fe(III) porphyrins has been particularly informative with respect to the geometry and the spin state of these species.

The Hamiltonian of an electron-spin state can include a contribution from zero-field splitting of the nominally (in the absence of an external magnetic field) degenerate energy levels. This contribution arises directly from coupling (λ) or mixing of the orbital ground state with an excited state or states. For high-spin Fe(III), $S = \frac{5}{2}$, the electronic ground state is $^6A_1$. In a tetragonal field (i.e., an axially distorted octahedron as in an iron porphyrin), mixing occurs between this and the excited state, $^4A_2$, causing the splitting of the three electron-spin states, $M_s = \pm\frac{1}{2}, \pm\frac{3}{2}, \pm\frac{5}{2}$. Not surprisingly, the axial splitting parameter D is proportional to λ and inversely proportional to ΔE, the energy separation of the ground and excited states being mixed. The electron-spin Hamiltonian for this situation will include a field-independent term, $D[S_z^2 - \frac{1}{3}S(S+1)]$ [compare with Eq. (17)], that is, in the absence of H_0, $E_s = Dm_s^2$ or $\frac{1}{4}D$, $\frac{9}{4}D$, and $\frac{25}{4}D$ for $S = \frac{5}{2}$. Thus, the $m_s = \pm \frac{3}{2}$ and $\pm \frac{5}{2}$ spin states are $2D$ and $6D$ above the gound doublet, the $\pm \frac{1}{2}$ state, respectively (Fig. 11a).

When spin resonance is induced with H_0 aligned along the z axis (perpendicular to the porphyrin ring), it is this ground doublet which is excited; with $m_s = \pm \frac{1}{2}$, the transition energy is simply $2\beta H_0$, i.e., g_z $(g_\parallel) = 2$. The transition energies for the other two doublets are $6\beta H_0$ and $10\beta H_0$, respectively [Eq. (16) and Fig. 11], as follows:

$$E_s = 2\beta H_\parallel m_s + Dm_s^2. \tag{16}$$

On the other hand, when H_0 is in the porphyrin ring along $g_x = g_y$, only the $m_s = \pm \frac{1}{2}$ spin state is split, by $6\beta H_0$; that is, $g_\perp = 6$. (In general, $g_{max} = 2S + 1$ for systems exhibiting zero-field splitting.) This splitting pattern obtains when and if there is a strong axial crystal field, when $D \gg g\beta H$. Under these conditions, the z axis of quantization is independent of the orientation of H_0; in effect, there is an inherently stronger coupling between $m_s = \pm \frac{1}{2}$ (which has a large projection in the porphyrin plane) and H_0 when H_0 is in this plane. This stronger coupling is formally dealt with by assigning a large value of g to $g_\perp$. In summary, a high-spin tetragonal Fe(III) complex should have $g_\parallel \simeq 2$ and $g_\perp \simeq 6$ (Fig. 11a, b).

The introduction of rhombic distortion causes $g_x \neq g_y$ and necessitates the inclusion of an additional zero-field splitting parameter E, as $E(S_x^2 - S_y^2)$:

$$H_s = 2\beta \mathbf{H} \cdot \mathbf{S} + D_z[S_z^2 - \frac{1}{3}S(S+1)] + E(S_x^2 - S_y^2) \tag{17}$$

This factor causes the splitting of $g_\perp$ into $g_x = 6 + 24(E/D)$ and $g_y = 6 - 24E/D$ (to a first approximation), with E/D on the order of 10^{-3} (Fig. 11a, b). This distortion could be caused by a greater overlap of the metal d_{yz} orbital with an axial imidazole $p - \pi$ system in comparison with the d_{xz} one, for example. Asymmetric substitution in the pophyrin ring could also contribute to such rhombic distortion. That is, the splitting of the $g = 6$ resonance in high-spin Fe(III) porphyrins can be an exceptionally informative, albeit subtle, structural indicator.

Low-spin Fe(III)(S = $\frac{1}{2}$) naturally does not exhibit zero-field splitting [as Feher (1970) put it, "there is nothing to split off"], and thus lacks the $g = 6$ feature. On the other hand, nearly all biological, low-spin Fe(III) porphyrins exhibit a distinct rhombic distortion with $g_z \simeq 2$, $g_y \simeq 2.1-2.3$, and $g_x \simeq 2.4-2.8$ as typical g-value ranges (Fig. 11b). Since low-spin ($S = 0$) and high-spin ($S = 2$) Fe(II) are ESR silent, ESR spectra of iron porphyrins, in particular, can fairly reliably distinguish between both the spin and the oxidation state of biological iron. Also, since spin state is typically a temperature-dependent property, ESR can often be used to characterize the

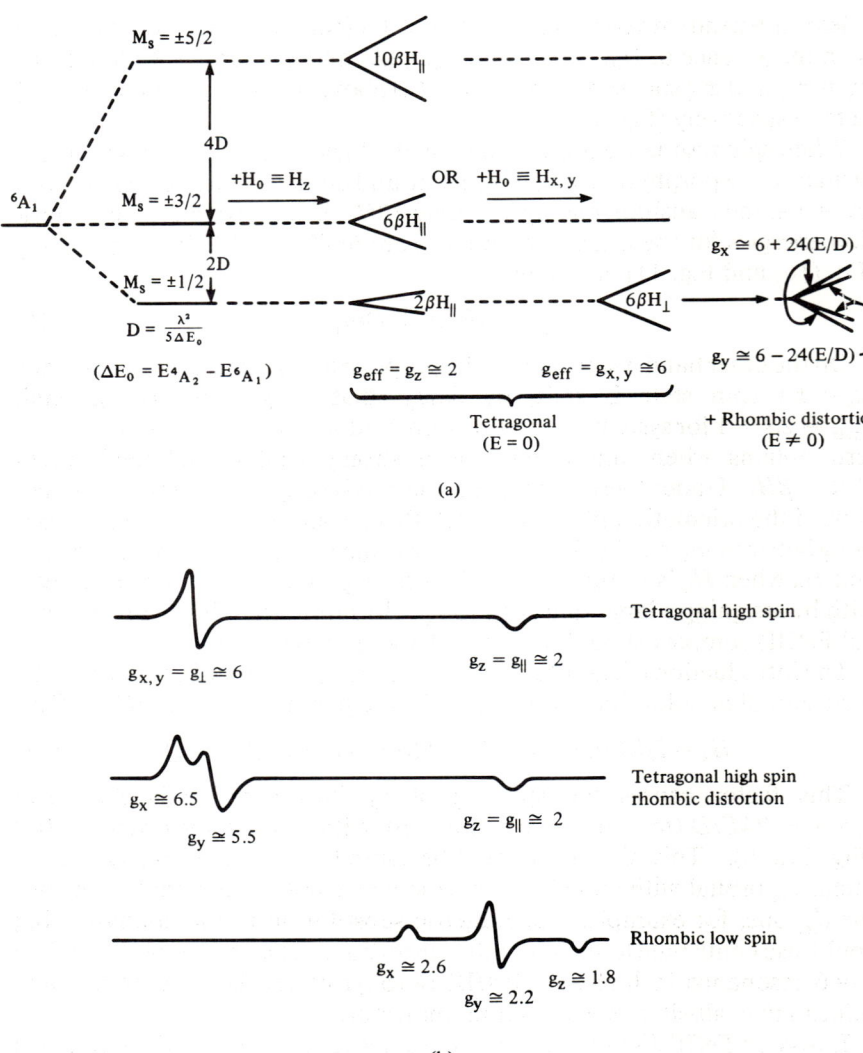

Fig. 11. ESR transitions in high- and low-spin Fe(III) porphyrins. (a) The effect of both the tetragonal (axial) D and rhombic E zero-field splitting of the ground $S = \frac{5}{2}$, $^6A_1$ state in a high-spin Fe(III) porphyrin is diagrammed, indicating the g_{eff} values for the ground-spin doublets associated with H_z and $H_{x,y}$. (b) Idealized first-derivative ESR spectra of both high- and low-spin Fe(III) porphyrins. The g values shown are typical for each spin system.

equilibrium constant for the high-spin–low-spin interconversion. A caution should be noted, however, about misinterpretation of g values. For example, with g_{max} given by $2S + 1$, for $S = \frac{3}{2}, g \simeq 4$. A completely rhombic $S = \frac{3}{2}$ system is isotropic with $g \simeq 4.3$. These two values could be confused quite easily. As emphasized in the introduction to this section, one must be aware of these types of spectrally similar paramagnetic centers.

c. Ligand SHFS. The assignment of protein ligation in a metalloprotein is based on a variety of information. As previously outlined, some idea of the ligand *set* can be gleaned from comparisons with model systems. Optical spectra, including absorbance and (M)CD, resonance Raman and Mössbauer spectroscopy, extended x-ray absorption fine structure (EXAFS), and, ultimately, x-ray crystallography all can provide significant insight into the nature of the ligand composition, perhaps even better than ESR can. ESR can provide valuable information, however, when SHF couplings to ligand nuclei can be resolved. The most direct way to do so is to detect SHFS in a cw spectrum and assign the splitting(s) to a particular ligand type (and, possibly, number) based on the number of resolved spectral features.

In general, this approach can detect only coordination by nitrogenous ligands, since among common biological ligand types N, O, and S, only N (in natural abundance) has a nuclear spin. Furthermore, ^{14}N coupling is often not resolvable in cw ESR spectra even if, in fact, it is present. As will be discussed in Section V, these unresolved SHFS can be detected by other ESR techniques. Nonetheless, cw ESR has been useful in characterizing the ligand types in certain proteins, particularly those containing Cu(II). An example of how both Cu(II) spin-Hamiltonian parameters and ligand SHFS are used to characterize the metal coordination in a copper protein is given here.

The ESR spectrum of galactose oxidase, a mononuclear Cu(II) enzyme, is axial, i.e., $g_x = g_y \neq g_z$ (Fig. 12) (Bereman and Kosman, 1977; Ettinger and Kosman, 1982). This is best seen in the 35 GHz spectrum, which spreads out the g field while (in general) losing HFS (H_0 dominates the local field at the nucleus). The Q-band spectrum is clearly axial. The values for $g_\parallel$ and $A_\parallel$ (Table II) are those of a type 2 Cu(II), although the characterization of galactose oxidase as a nonblue protein is somewhat misleading, since $E_{630 nm} \simeq 10^3 M^{-1} cm^{-1}$ (Ettinger and Kosman, 1982). These g and A values indicate that the Cu(II) is in an N_2O_2 chelate, which can have a net charge of zero (compare with Fig. 11). ESR data described next suggest that this may be correct.

The X-band spectrum exhibits clearly resolved SHFS in both $g_\parallel$ and $g_\perp$. Structure is apparent also in the Kivelson or "overshoot" line at high field. This feature represents a set of orientations between $H_0 \perp g_\parallel$ ($\parallel g_\perp$) and equilibrium constant for the high-spin – low-spin interconversion. A caution should be noted, however, about misinterpretation of g values. For example, with g_{max} given by $2S + 1$, for $S = \frac{3}{2}, g \simeq 4$. A completely rhombic $S = \frac{5}{2}$ system is isotropic with $g \simeq 4.3$. These two values could be confused quite easily. As emphasized in the introduction to this section, one must be aware of these types of spectrally similar paramagnetic centers.

c. Ligand SHFS. The assignment of protein ligation in a metalloprotein is based on a variety of information. As previously outlined, some idea of the ligand *set* can be gleaned from comparisons with model systems. Optical spectra, including absorbance and (M)CD, resonance Raman and Mössbauer spectroscopy, extended x-ray absorption fine structure (EXAFS), and, ultimately, x-ray crystallography all can provide significant insight into the nature of the ligand composition, perhaps even better than ESR can. ESR can provide valuable information, however, when SHF couplings to ligand nuclei can be resolved. The most direct way to do so is to detect SHFS in a cw spectrum and assign the splitting(s) to a particular ligand type (and, possibly, number) based on the number of resolved spectral features.

In general, this approach can detect only coordination by nitrogenous ligands, since among common biological ligand types N, O, and S, only N (in natural abundance) has a nuclear spin. Furthermore, ^{14}N coupling is often not resolvable in cw ESR spectra even if, in fact, it is present. As will be discussed in Section V, these unresolved SHFS can be detected by other ESR techniques. Nonetheless, cw ESR has been useful in characterizing the ligand types in certain proteins, particularly those containing Cu(II). An example of how both Cu(II) spin-Hamiltonian parameters and ligand SHFS are used to characterize the metal coordination in a copper protein is given here.

The ESR spectrum of galactose oxidase, a mononuclear Cu(II) enzyme, is axial, i.e., $g_x = g_y \neq g_z$ (Fig. 12) (Bereman and Kosman, 1977; Ettinger and Kosman, 1982). This is best seen in the 35 GHz spectrum, which spreads out the g field while (in general) losing HFS (H_0 dominates the local field at the nucleus). The Q-band spectrum is clearly axial. The values for $g_\parallel$ and $A_\parallel$ (Table II) are those of a type 2 Cu(II), although the characterization of galactose oxidase as a nonblue protein is somewhat misleading, since $E_{630 nm} \simeq 10^3 M^{-1} cm^{-1}$ (Ettinger and Kosman, 1982). These g and A values indicate that the Cu(II) is in an N_2O_2 chelate, which can have a net charge of zero (compare with Fig. 11). ESR data described next suggest that this may be correct.

The X-band spectrum exhibits clearly resolved SHFS in both $g_\parallel$ and $g_\perp$. Structure is apparent also in the Kivelson or "overshoot" line at high field. This feature represents a set of orientations between $H_0 \perp g_\parallel$ ($\parallel g_\perp$) and

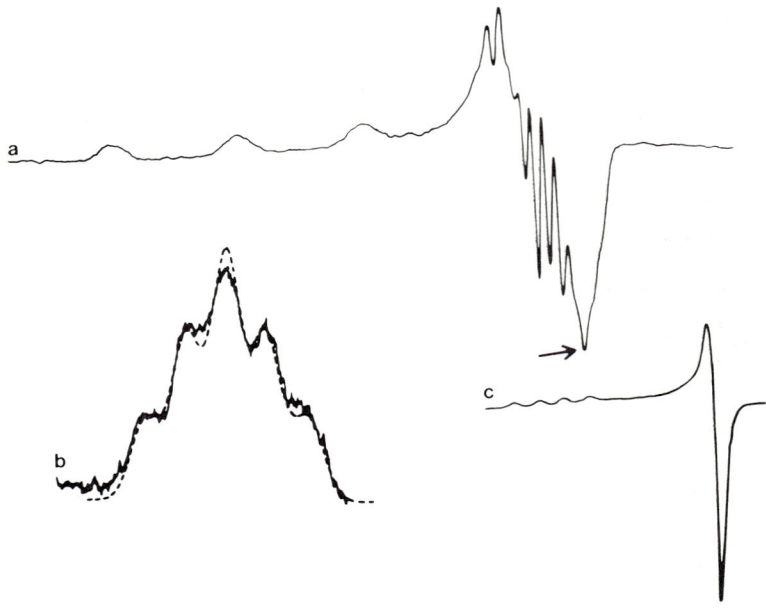

Fig. 12. ESR spectra of Cu(II) enzyme, galactose oxidase at 77 K. X- and Q-band spectra are shown in (a) and (c), respectively. The arrow indicates the position of the overshoot line. The $M = +\frac{3}{2}$ low-field line in the X-band spectrum at higher resolution is given in (b) to illustrate ^{14}N shfs.

$H_0 \parallel g_\parallel$ ($\perp g_\perp$) and appears when the values of $A_\parallel^{Cu}$ and $A_\perp^{Cu}$ are significantly different. As noted above, this is true for most Cu(II) systems in which $A_\perp^{Cu}$ is small, often ~0. Thus, as H_0 rotates out of $g_\perp$ (toward lower field), the contribution to A_{av} from $A_\parallel$ increases rapidly, thus actually spreading out the effective field at the electron (Fig. 13). Higher resolution of the $M_I = +\frac{3}{2}$ transition in $g_\parallel$ indicates that this contains five SHFS lines. This feature can be simulated using an $A_\parallel^N$ value of 14.5 for two equivalent nitrogen ligands (Table II) (Bereman and Kosman, 1977; Marwedel et al., 1981). Although such nitrogen SHFS has not been as well resolved by cw ESR in other native Cu(II) proteins, a number of artificial Cu(II) proteins exhibit similar couplings (compare with Section IV.C.1).

One particular endogenous ligand other than ^{14}N that can be detected by cw ESR is $H_2^{17}O$ ($I = \frac{5}{2}$). If a water molecule or molecules exchange at an inner coordination site of a metal ion with a residence time longer than the reciprocal of the frequency of observation, then significant hyperfine coupling may be apparent. Such coupling is seen as a broadening of a particular resonance line rather than as resolved SHFS. The effects of $H_2^{17}O$ or oxygens derived therefrom have been observed in ESR spectra for systems containing Cu(II), Mo(V), and Mn(II). Not only have these effects been useful in the elucidation of the coordination chemistry of these centers but they have also been related to the catalytic mechanisms of these centers, as will be discussed.

2. Ligand Exchange

As noted, the effects on metal g and A values by exogenous ligand coordination can be related to differences between the added and displaced endogenous ligand. Furthermore, the change in the metal chelate may improve the resolvability of endogenous ligand SHFS, either by increasing its magnitude or eliminating contributions to the linewidth (sample heterogeneity, lifetime and scalar effects, additional unresolved SHFS). Thus, ligand exchange not only can inform about the coordination chemistry of a metal center but may help in elucidating the nature of the endogenous ligands.

Table II Spin-Hamiltonian Parameters for Cu(II) Galactose Oxidase and Complexes

Cu spin-Hamiltonian parameters		Endogenous ligand nitrogen SHFS		Exogenous ligand SHFS	
Native enzyme[a]				In H_2 ^{17}O[d]	
$g_\parallel = 2.277$	$A_\parallel^{Cu} = 175.0$[c]	$A_\parallel^N = 14.5$			
$g_\perp = 2.055$		$A_\perp^N = 15.1$	$2^{14}N$		
		$A_O^N = 14.8$		$A_\parallel^O = 9$ } $1^{17}O$	
Imidazole complex[a]					
$g_\parallel = 2.254$	$A_\parallel^{Cu} = 167.5$	$A_\parallel^N = 12.1$		(includes exogenous	
$g_\perp = 2.041$		$A_\perp^N = 15.7$	$3^{14}N$	imidazole)	
		$A_O^N = 13.4$			
F^- complex[a]					
$g_\parallel = 2.305$	$A_\parallel^{Cu} = 159.7$	$A_\parallel^N = 11.2$		$A_\parallel^F = 41.0$	
$g_\perp = 2.050$		$A_\perp^N = 14.3$	$2^{14}N$	$A_\perp^F = 175.4$ } $1^{19}F$	
		$A_O^N = 13.3$		$A_O^F = 128.1$	
CN^- complex[b]				$^{13}CN^-$	
$g_\parallel = 2.233$	$A_\parallel^{Cu} = 159$	$A_\parallel^N = hr$		$A_\parallel^C = 96$	
$g_\perp = 2.050$		$A_\perp^N = 22.5$	$2^{14}N$	$A_\perp^C = (45)$[e] } $1^{13}C$	
		$A_O^N = 13.7$		$A_O^C = 49$	

[a] Bereman and Kosman (1977).
[b] Marwedel et al. (1981).
[c] Values of A in gauss; $A_\perp^{Cu} \simeq 0$ in all complexes.
[d] Melnyk and Ettinger (1984).
[e] Estimated value.

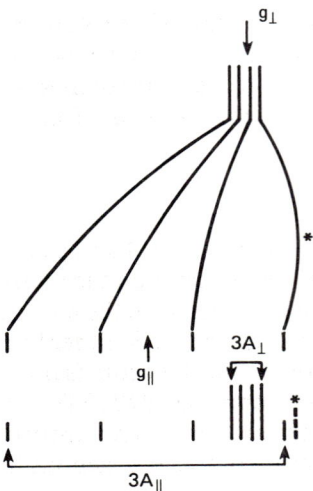

Fig. 13. A stick diagram for a tetragonal Cu(II) ESR spectrum showing the rotation of H_0 from $g_\perp$ into $g_\parallel$ (———). Note that the transition $A_\perp \rightarrow A_\parallel$ may change at a rate such that a false turnaround point (∗) occurs outside the actual limits of the spectrum. This point will have features intermediate between the $M_I = -\frac{3}{2}$ (high-field) transitions in $g_\perp$ and $g_\parallel$. It occurs when $g_\parallel > g_\perp$ and $A_\parallel \gg A_\perp$.

Anion binding to a number of Cu(II) proteins enhances the resolution of ^{14}N SHFS significantly. CN^- is particularly effective, in that unresolved SHFS from endogenous nitrogen ligands in the native protein is easily seen in the ESR spectra of the CN^- complexes. For example, although native bovine superoxide dismutase exhibits no ^{14}N SHFS, seven lines are seen in the low-field $M_I = +\frac{3}{2}$ transition upon CN^- coordination ($A^N \simeq 15$ G) (Haffner and Coleman, 1973c). This seven-line SHFS could indicate the presence of three nitrogen ligands to the Cu(II) in native SOD. However, a combination of ENDOR and electron spin-echo spectroscopy, and single-crystal cw ESR confirmed what the crystal structure revealed, namely, that the Cu(II) atoms in the protein are bonded to four histidine imidazoles. As detailed in Sections IV.D and V.C, CN^- bonding effectively displaces one of the coordinating imidazoles by realignment of the ligand field. This electronic effect was indicated by the conversion of the ESR spectrum from one exhibiting a rhombic distortion to an axial spectrum upon CN^- binding.

The effects of coordination of ligands such as F^- and imidazole to galactose oxidase are also striking (Fig. 14, Table II) (Bereman and Kosman, 1977). In both cases, well-resolved ^{14}N SHF structure is apparent in any or all of the transitions in $g_\parallel$, $g_\perp$, and the overshoot line at high field. That a seven-line spectrum in $g_\perp$ and the overshoot region is resolved in the presence of imidazole suggests that the endogenous N ligands are also imidazoles. This has been confirmed. The $^{14}F^-$ splitting in the fluoride complex is exceptionally well resolved and exhibits an extremely strong dipolar term in $g_\perp$. In none of these cases was ligand binding associated with a rhombic distortion as confirmed by spectra at 35 GHz.

H_2O coordinates to this exogenous ligand site as well with an $A_\parallel^{17O} \sim 9$ G (Melnyk and Ettinger, 1984). This water is displaced by CN^-. That CN^- binds to this site is indicated by the $A^{13}C$ value for $^{13}CN^-$ (~ 96 G) and the displacement by CN^- of the $^{19}F^-$ detectable by ESR, as in Fig. 14 (Bereman and Kosman, 1977; Marwedel et al., 1981). Inasmuch as a tyrosine phenol (phenoxide) may be the fourth ligand to the Cu(II) in galactose oxidase (Ettinger and Kosman, 1982), the expectation that this metal is in an N_2O_2 chelate is realized. Whether the water is coordinated as H_2O or HO^- is not known.

The coordination of ^{17}O derived from either $^{17}O_2$ or $H_2^{17}O_2$ has also been detected. For example, the multinuclear Cu(II) enzyme laccase catalyzes the four-electron reduction of O_2 to H_2O. The enzyme contains one type 2 Cu(II) site having an ESR spectrum which is resolved from that for the enzymes' type 1 Cu(II) at low field. The turnover by the enzyme of $^{17}O_2$ generates an intermediate with an ESR spectrum characterized by a broadening of the type $2M_I = +\frac{3}{2}$ low-field line (Brandén and Deinum 1977). This broadening is consistent with the coordination of a water molecule, $H_2^{17}O$, with $A \simeq 11$ G. This evidence suggests that an O_2-derived water molecule is coordinated at this site during turnover, although the binuclear Cu(II) pair in the enzyme is thought to be the site at which electrons are transferred to O_2.

Compound I, formed when peroxidases and catalases are treated with H_2O_2, has also been investigated with the use of ^{17}O (Roberts et al., 1981). Horseradish peroxidase compound I (HRPI) was generated using $H_2^{17}O_2$ (prepared by sodium reduction of $^{17}O_2$). Although the coupling of this nucleus to the unpaired spin density is unresolved in the large linewidths of the cw ESR spectrum, the nuclear Larmor precession frequency of the ^{17}O could be associated with the splitting of the lines in an ENDOR spectrum. As described in Section V.B, ENDOR is useful for the resolution of inhomogeneously broadened lines as occur when SHFS contribute to such linewidths. In HRPI—^{17}O, an ^{17}O ENDOR line was clearly evident, and from its position in the radiofrequency field the ^{17}O hyperfine splitting tensors could be calculated. These had axial symmetry ($A_x^{OT} \simeq A_y^{OT} \simeq 35$ MHz) and were supportive of an oxyferryl, $Fe^{IV} = O$, species characterizing the structure of compound I.

3. Redox Titrations and Kinetics

As the data in Table I indicate, the oxidation states of a given metal differ significantly with respect to their spin quantum number and state. Thus, changes in an ESR spectrum containing resonances from a particular metal (or other redox-active prosthetic group) can be expected to change as either reduction or oxidation of the system occurs. The ESR spectrum is relatively clean in that it contains relatively few features, most which are actually quite well resolved (compare with absorbance spectra). It is therefore an invaluable measure of the redox state of species in the sample which exhibit a detectable electron-spin transition in one or more oxidation states.

Redox titrations are performed by adding either oxidant or reductant in fractional equivalents to systems which have been prereduced or oxidized (Orme-Johnson and Beinert, 1969). In general, the systems are kept anaerobic; if O_2 is used as the oxidant, liquid volumes of O_2-saturated solutions are added as the oxidizing species. The ESR cells used for such titrations have at least four features: a stopcock and joint for vacuum connection or gas flush, a septum for injection of oxidant or reductant, a mixing bulb, and a ground-glass joint for coupling to a quartz ESR tube. Degassing can be achieved by application of freeze–thaw cycles, using vacuum, or by flushing via the septum, using a large-gauge needle. An additional component of the system can be a cell suitable for absorbance spectroscopy; thus, the ESR and absorbance spectra can be obtained on the same sample. An adaptation of this design is to carry out the titration in a closed vessel from which samples can be removed through a rubber septum with a gas-tight syringe and transferred to an argon or nitrogen-flushed ESR tube. The tube is then sealed under the inert gas. In addition, equipping the cell with calomel reference and indicating electrodes permits the measurement of the potential of the solution. Platinum gauze is commonly used for the latter electrode type. Lastly, the reduction can be effected by coulometry rather than electrochemically, by including a working electrode with current supplied by a potentiostat. In this system, however, protein concentrations used are normally below the

---

Fig. 14. ESR spectra of galactose oxidase ligand complexes. Top spectrum (a) is of imidazole complex at X band. In the center spectrum (b), the F^- complex is shown at X band (A) and Q band (B) to show that splitting in $g_\perp$ is owing to SHFS, not a rhombic distortion. At the bottom (c) is the X-band spectrum of the F^- complex at higher resolution with the analysis of the SHFS owing to ^{14}N and ^{19}F shown; $\perp$ refers to $g_\perp$, O refers to overshoot line. (From Bereman and Kosman, 1977. © 1977 American Chemical Society.)

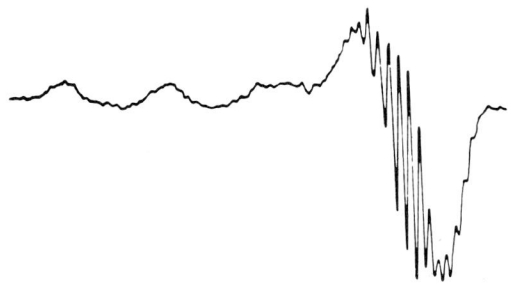

2582 2688 2744 2900 3006 3112 3218 3324
Magnetic field (G)
(a)

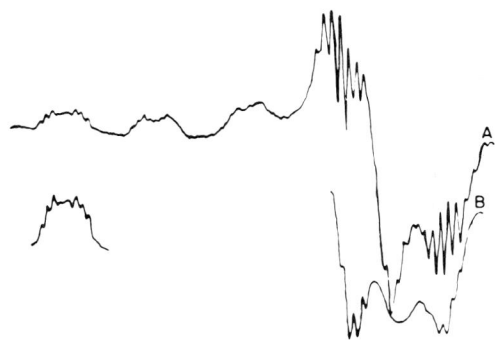

2476 2582 2688 2794 2900 3006 3112 3218 3324
Magnetic field (G)
(b)

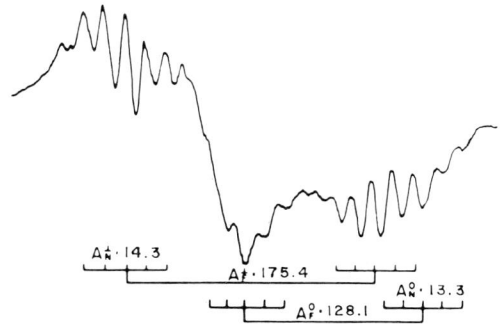

$A_N^+ \cdot 14.3$ $A_F^+ \cdot 175.4$ $A_N^0 \cdot 13.3$
$A_F^0 \cdot 128.1$

2986 3028 3070 3118 3160 3202 3244 3286 3328
Magnetic field (G)
(c)

limits of ESR detectability. Although thin-cell coulometric methods have been developed for combined absorbance–electrochemical measurement, at this point, coulometric–potentiostat methods have not been directly combined with ESR.

The kinetics of such redox reactions are often faster than can be measured without rapid mixing and observation. There are two basic ways by which the kinetic behavior of short-lived species can be assessed. These are continuous flow and stopped flow; both have been applied with some success in the area of biological ESR. Each has its own advantages and limitations.

a. Continuous-Flow Method (Borg, 1964; Piette, 1964). The advantage of the continuous-flow technique is also its disadvantage; it uses liquid samples at any temperature but often in prohibitively large quantities. In this method, the components of the system of interest are stored in separate reservoirs and are subsequently mixed in a continuous stream at a precise distance x from the center of the resonance cavity. By varying x or the flow rate, one can vary the time between mixing and observation. Thus, one can generate a profile of the concentration of the paramagnetic transient(s) as a function of time after mixing. A feature of this technique is that, since the concentration of species is held constant by the flow, one can record a complete spectrum of the radical, not simply measure the intensity of absorption at a single field setting. The flow can be driven by syringe or maintained by a pump; but in either case the volume of the system is typically quite large (1–2 ml aside from the reservoirs themselves). More important, the flow rate necessary (~ 10 ml/sec) to trap very short-lived species ($t_{1/2} <$ 100 msec) results in the consumption of large quantities of material (> 100 ml/flow). Add to this the fact that to generate the temporal profile of the radical(s) concentration requires several independent flows, and the disadvantage of this method is made clear. Continuous flow is not now the method of choice.

b. Stopped-Flow Methods (Bray and Petersson, 1961; Palmer and Beinert, 1964; Bray, 1961, 1964; Ballou, 1971; Ballou and Palmer, 1974; Klimes et al., 1980). Obviously, the stopped flow is a modification of the continuous flow methodology; one simply stops the flow and, at a given distance x from the mixing chamber, watches the corresponding radical (its concentration) decay. Although one could change both x and the flow rate to vary the time t (relative to mixing) at which the flow is stopped, it is usually just as easy mechanically, and clearly more economical of time and material, to make t as short as possible so that in one "flow" the whole temporal profile of the radical(s) can be established. This is the essence of stopped flow. Advances in mixing-chamber design have brought this mixing down to ~ 5 msec, a "dead time" only slightly longer than that encountered in other stopped-flow systems.

The method of choice for redox kinetics is that of rapid quenching by freezing the sample at precisely determined times after mixing. The spectrum of the frozen sample is subsequently recorded. This method combines the mixing concepts of the typical stopped-flow instrument — rapid flow and efficient mixing, e.g., short dead times — with the low temperatures often needed to detect the functional paramagnetic transition metals encountered in a variety of proteins. This technique was first developed and used successfully by Bray in the early 1960s (Bray, 1961, 1964; Bray and Petersson, 1961), and was further refined by Beinert (Palmer and Beinert, 1964) and subsequently by Ballou and Palmer (1974; Ballou, 1971). The elements of the systems used are essentially the same as in both continuous and stopped-flow methods, i.e., sample reservoirs driven by piston or pump, and a mixing chamber at a given distance from an injection port or nozzle. The reaction is "stopped" by spraying the reaction mixture into a cryogenic liquid, which is commonly isopentane precooled to ~ 130 K. The frozen pellets are packed into a quartz tube, and the spectrum of the sample can be recorded. This technique thus retains the primary advantage of continuous flow, namely, the ability to record the entire spectrum of the sample at any given time t after mixing. In addition, since the sample is temporally static, its behavior in the magnetic field can be more completely characterized. For example, the saturation properties of the sample can be determined over a range of temperatures down to that of liquid He. ENDOR spectra can be obtained and, theoretically, even electron-spin-echo spectra could be generated from the freeze-quenched material.

As in continuous flow, the time t is varied by either changing the distance between mixing chamber and nozzle or by changing the flow rate. In practice, since changes in the flow rate alter the mixing characteristics of the system and the size of the frozen pellets obtained, it is best to vary t by using reaction tubes of varying length. The shortest time accessible is ~ 5 msec, although the mixing itself is accomplished in ~ 2 msec; added to this is the actual quench time of the cryogenic bath, which is also ~ 5 msec. Much of the excellent kinetic data on xanthine oxidase transients discussed in the following were obtained using the rapid-freeze technique. Ballou detailed the design, construction, and use of the system now most commonly employed.

c. Xanthine Oxidase (Bray, 1975). The studies on the redox behavior of xanthine oxidase exemplify the applicability of ESR to this aspect of biochemistry. This enzyme, commonly isolated from cows' milk, catalyzes the oxidation of xanthine to uric acid; the reaction is, in fact, the hydroxylation of the C-8 position of the purine ring. This protein contains a variety of redox-active, and thus potentially paramagnetic, centers of different chemical types. A feature it shares in common with a few other enzymes (sulfite oxidase, aldehyde oxidase, nitrate and sulfate reductase) is the combination of nonheme iron, flavin, and molybdenum cofactors which together are responsible for the enzymatic activity. The structure of xanthine oxidase appears best described as a dimer of identical subunits, each of which contains 1 Mo, 1 FAD, and 2 Fe_2S_2 clusters. A pterin cofactor is also present. These two subunits appear to function independently of each other. Thus, each active site is capable of accepting or donating a total of six electrons during catalysis.

The Mo centers in xanthine oxidase are in the VI oxidation state in the completely oxidized enzyme; the enzyme is diamagnetic and exhibits no ESR spectrum in this form, as expected for the $d°$ Mo(VI) ion (compare with Table I). The Fe_2S_2 clusters are typical in that they appear to contain antiferromagnetically coupled, high-spin Fe^{3+} ($S = \frac{5}{2}$), and thus are also diamagnetic, as is the flavin, as FAD. The completely reduced enzyme exhibits the rhombic ESR spectrum characteristic of a reduced Fe_2S_2 cluster, with $g_{xx} = 1.899$, $g_{yy} = 1.935$, and $g_{zz} = 2.002$. Although observable in a broadened form at liquid-nitrogen temperatures, the spectrum is sharpened considerably at 4 K. In fact, below ~25 K a second species is observed in the ESR spectrum of the reduced enzyme, also attributable to a rhombic Fe_2S_2 cluster with $g_{xx} = 1.91$, $g_{yy} = 2.007$, and $g_{zz} = 2.12$. This is a unique Fe_2S_2 system, since only one of the principle g values is below the free-spin value (2.0023) (Lowe *et al.*, 1972).

These two clusters, termed Fe/S I and Fe/S II, respectively, are each one-electron redox centers. This can be determined by anaerobic titration of the oxidized enzyme with dithionite. Since one of the principal g values for each cluster is found at a unique magnetic field (at a given frequency), that is, $g_y = 1.95$ for Fe/S I and $g_z = 2.12$ for Fe/S II, the reduction of each cluster can be followed directly (Fig. 15a). Data can thus be obtained which show, qualitatively, that Fe/S II_{ox} has a more positive reduction potential than Fe/S I_{ox} (Olson *et al.*, 1974a).

This difference in e^- affinity of these two Fe/S centers is indicated kinetically, as well (Olson *et al.*, 1974b). Rapid-feeze experiments established the initial rate of reoxidation by O_2 of the substrate-reduced enzyme (Fig. 15b). In the initial ($t = 0$) spectrum, the ratio of the $g = 1.95$ and 2.12 signal-peak heights (e.g., Fe/S I/Fe/S II) is ~3.5. At 210 msec after mixing with O_2, this ratio is ~1.3, that is, the Fe/S I center is oxidized faster than is the Fe/S II one in these experiments. Chemically, this indicates that the reactivity (kinetic behavior) of these centers does reflect their thermodynamic properties ($E°'$), certainly a reasonable result but by no means one to take for granted.

That the Mo does not contribute to the ESR spectrum of the fully reduced enzyme shows that it is probably in the IV oxidation state. It is a spin-even

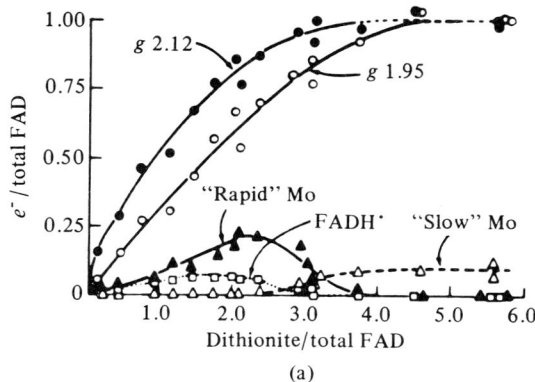

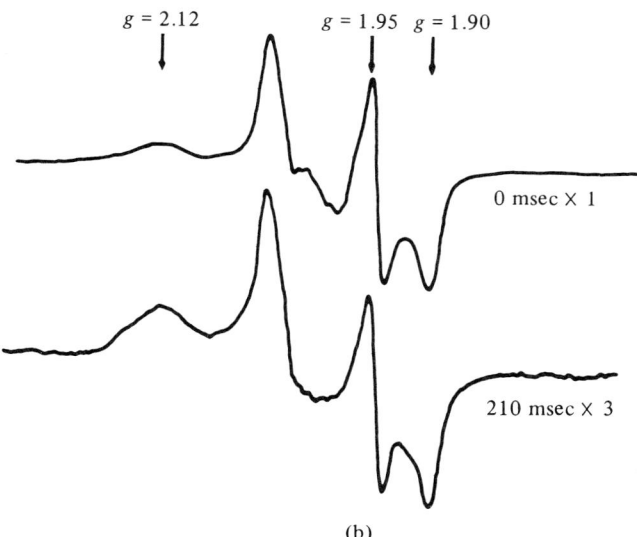

Fig. 15. Reductive ESR titration of xanthine oxidase and reoxidation of xanthine-reduced enzyme by O_2. (a) Integrated intensities of the paramagnetic reduction intermediates are plotted versus e^- equivalents from dithionite relative to total FAD: ●, $g = 2.12$; ○, $g = 1.95$; ▲, rapid Mo; △, slow Mo; □, FADH·. (b) Enzyme prereduced with xanthine is reacted with O_2 at pH 8.5 and 25°C and freeze quenched at time shown; the $t = 0$ spectrum is a sample in absence of O_2; the $g = 1.95$ signal is from Fe/S I, the $g = 2.12$ one from Fe/S II. (From Olson et al., 1974a, b.)

system ($S = 1$) and would not exhibit a detectable resonance. The two-electron reduction of the Mo (VI → IV) is expected to go through the paramagnetic ($S = \frac{1}{2}$) Mo(V) state. Indeed, reductive titration by dithionite does generate a signal attributable to this oxidation state (Fig. 15a). It never accounts for more than 25% of the total Mo and thus must have a more positive potential than Mo (VI); both species show less electron affinity than the two Fe/S centers. The FADH· intermediate is also evident in ESR spectra of partially reduced enzyme ($g = 2.0035$), although in relatively small amounts. In the reductive titration, the semiquinone accounts for, at most, ~5% of the total flavin. This shows clearly that this half-reduced cofactor has the most positive reduction potential of the six redox states (centers) in the active site. The relative electron-affinity constants for these centers are given in Table III (Olson, 1974a).

The absolute values of these potentials and their pH dependence have been reevaluated using a potentiometric titration protocol, as described earlier, in which both pH and electropotential were monitored (Barber and Siegel, 1982; Spence et al., 1982). Potentials were established first by addition of dithionite to oxidized enzyme with a series of overlapping potentials generated by addition of ferricyanide to the reduced enzyme. Mediators were used to insure electrode equilibration. Using this system, it is possible to fix the system at a given equilibrium potential and characterize the spectral properties of individual ESR-detectable components, including their saturation behavior. Thus, Barber et al. (1982) were able to evaluate the magnetic interactions between the centers in xanthine oxidase based on their relaxation behavior at various oxidation–reduction steps. As discussed more fully in Section IV.C, the magnetic interaction between two paramagnetic centers can include both exchange and dipolar terms. The dipolar term is distance and angle dependent, thus resolution of its contribution can be used to evaluate the distance between and orientation of two interacting paramagnetic centers. Following an excellent discussion of their analysis of the saturation data, Barber et al. (1982) estimate that the distances between the redox center in the xanthine oxidase subunit are as given here:

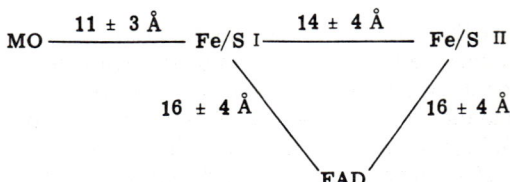

Of the four redox centers which exhibit ESR spectra, the most informative mechanistically is the Mo(V). In addition to the redox behavior outlined here, the ESR spectra also help to characterize the enzyme–substrate inter-

Table III Relative Redox Potentials for the Prosthetic Groups in Xanthine Oxidase[a]

	Relative redox potential (mV)	
Species	Alone	Xanthine bound
Fe/S II_{ox}/Fe/S II_{red}	0	0
Fe/S I_{ox}/Fe/S I_{red}	−24	−24
FAD/FADH·	−60	−60
FADH·/FADH$_2$	60	60
Mo(VI)/Mo(V)	−60	120
Mo(V)/Mo(IV)	−31	57

[a] From Olson et al. (1974a).

action. First, the Mo(V) signal is not represented by a single spectrum; during turnover, there are two distinct Mo(V) species formed in sequence, termed the very rapid and the rapid signals. Early literature refers to these two as the γ,δ and α,β signals, respectively (Bray, 1975). These two resonances differ in three important ways: rate of formation and decay, principal spin-Hamiltonian parameters, and the presence of superhyperfine splitting. That the two species are temporally related is indicated by their formation half-times, 5 and 15 msec, and decay half-times, 40 and 500 msec with maximal signal intensity at 15 and 65 msec, respectively.

The enzyme as normally isolated contains a natural abundance mixture of $I = 0$ Mo isotopes (−94, −96, and −98) and $I = \frac{5}{2}$ isotopes (−95 and −97). The former constitute 75% of the total, and thus the Mo(V) ESR spectra are essentially devoid of metal hyperfine splittings, although the contribution from the latter isotopes can be detected at high signal levels.

The analysis of the ESR spectra of the Mo(V) center, and the characterization of this site, was in large measure carried out by Bray and Meriweather (1966), who compared the natural abundance Mo enzyme with an enzyme prepared from a cow which had received an injection of ^{95}Mo (as sodium molybdate). For reference, 184 mg of ^{95}Mo were used for a 546-kg cow. They included a tracer of ^{99}Mo, and by determining the specific activity of the xanthine oxidase could quantitate the isotopic composition of the enzyme independent of the ESR measurements. This stands as one of the most elegant experiments in isotopic substitution yet reported. Their analysis yielded the g and A values given in Table IV. Interesting is the similarity of these values to those for complexes between Mo(V) and various thiols

Table IV Spin-Hamiltonian Parameters for the Mo(V) States in Xanthine Oxidase[a]

Very rapid signal	$A^{95}Mo^b$	$A^{17}O^c$	$A^{13}C^d$
$g_x = 1.949$	$A_x = 37$	13.6	2.9
$g_y = 1.955$	$A_y = 24$	14.0	3.7
$g_z = 2.025$	$A_z = 41$	13.4	3.1
Rapid signal		$A'H^e$	
$g_\perp = 1.971$	$A_\perp = 28$	16	
$g_\parallel = 1.990$	$A_\parallel = 67$	13	

[a] Values of A in gauss.
[b] From Pariyoduth et al. (1976).
[c] From reaction in $H_2^{17}O$; exchanges with $H_2^{16}O$ during turnover. (From Gutteridge and Bray, 1980.)
[d] From reaction with 8-[^{13}C]-xanthine. (From Tanner et al., 1970.)
[e] Proton from C_8—H of xanthine; exchanges with H_2O. (From Gutteridge et al., 1978.)

(Pariyaduth et al., 1976). EXAFS studies have shown that in the oxidized enzyme the Mo atom contains one terminal sulfur (Mo=S) as well as several Mo—SR ligands (Cramer et al., 1981).

Significantly, the rapid or α, β—, Mo(V) signal is different in detail from the very rapid one. As noted in Table IV, this Mo(V) exhibits an axial spectrum ($g_x = g_y$ within limits of resolution at X-band frequencies and corresponding magnetic fields). In addition, a doublet splitting ($I = \frac{1}{2}$) is observed.

The origin of the doublet splitting of $g_\perp$ and $g_\parallel$ could be investigated by determining the frequency dependence of the magnitude of this splitting. However, SHFS is generally not seen in 35-GHz spectra, for example, thus this approach often yields ambiguous results. Isotopic substitution is a more direct approach. In fact, generation of the very rapid and the rapid species in D_2O results in the initial appearance of the typical rapid signal, which then loses the SHF structure contributed by the $I = \frac{1}{2}$ nucleus. This confirms that this nucleus is ^1H, and it is in exchange with solvent within the lifetime of the rapid species. By determining the extent of exchange at various pH values, the pK_a of the conjugate acid involved was found to be ~8. Even more precise information is obtained when 8-deuteroxanthine is used as substrate; the $I = \frac{1}{2}$ splitting in the rapid signal is lost. Thus, the ^1H nucleus responsible is from the C-8 position of xanthine (the reaction center) (Gutteridge et al., 1978).

2. ELECTRON SPIN RESONANCE

Where this proton is in the rapid species is a matter of some debate. Two possibilities exist. First, the ^{1}H nucleus, as a hydride, could be coordinated directly to the Mo(V). However, A_H in $g_\parallel$ and $g_\perp$ are nearly the same; that is, the SHFS is nearly isotropic. This would be unusual if the ^{1}H coordination, itself, were highly anisotropic. The second possibility is that the nucleus as a proton is bound to one of the ligands to the Mo(V) and in this way interacts with the unpaired spin density through a dipolar interaction or, via spin delocalization, into the ligand itself.

In fact, use of isotopic substitution in conjunction with ESR has largely resolved this question. Initial studies involved the appearance of ^{17}O SHFS from $H_2^{17}O$ into the very rapid Mo(V) ESR spectrum (Gutteridge and Bray, 1980, Fig. 16) and its exchange with $H_2^{16}O$ (Bray and Gutteridge, 1982). This rapid-freeze ESR experiment involved the use of three syringes and two mixers and is a good example of this protocol (Gutteridge and Bray, 1980). Another aspect relating to the use of ^{17}O in such experiments is also dealt within this report in some detail. That is that $(H_2)^{17}O$ is generally available enriched to only 50–60%, thus spectra must be corrected for the fact that two species (at least) are present, $I = 0$ and $I = \frac{5}{2}$. Nonetheless, the spectra clearly indicate the additional spectral features to be expected from ^{17}O SHFS. The simulations in Fig. 16 used the spin-Hamiltonian parameters in Table IV. Also listed is the result of a complementary experiment in which the substrate xanthine is labeled at C_8 with ^{13}C. A weak SHFS is also contributed by this nucleus. The presence of the sulfur ligand is indicated by EXAFS, as noted (Cramer et al., 1981). In addition, reactivation of a cyanide-treated protein preparation with [^{33}S]-sulphide yields an active enzyme enriched in this isotope ($I = \frac{1}{2}$). The very rapid Mo(V) ESR spectrum exhibited by this labeled enzyme showed the SHFS expected from this nucleus (Fig. 17). Simulation of the observed spectrum indicated that the $A^{33}S$ tensors were highly anisotropic and were not colinear with the Mo(V) g tensors (Fig. 17). In contrast, the rapid spectrum for the [^{33}S]-protein exhibited a much smaller coupling to the sulfur, $A \simeq 3.6$ G, which was isotropic.

A model for the very rapid and the rapid Mo(V) complexes is shown here (Bray and Gutteridge, 1982). The structure of the very rapid is based on the anisotropic coupling of the ^{33}S and isotropic coupling of ^{17}O and ^{13}C (compare with Table IV). This intermediate is thought to result from the nucleophilic attack of an Mo–O$^-$ ligand at the C_8 of xanthine.

Very rapid   Rapid

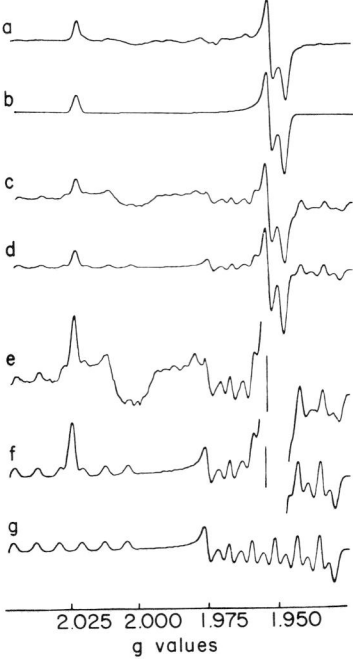

Fig. 16. ^{16}O and ^{17}O ESR spectra of the very rapid species in the xanthine oxidase reaction: (a) very rapid signal in $H_2{}^{16}O$; (b) simulation; (c) in $H_2{}^{17}O$, 50.4% enrichment; (d) simulation; (e) spectrum c at higher gain; (f) simulation; (g) simulation assuming 100% enrichment of $H_2{}^{17}O$. Simulations are based on values in Table IV. From Gutteridge and Bray, 1980.)

The structure of the rapid species is based on the exchange of the ^{17}O out of the system, the reduced and isotropic coupling of ^{33}S, and the appearance of an isotropically coupled ^1H. The imaginative use of isotopic substitution is illustrated here, and it must be stressed that the interpretation of the ESR data would not have been possible without the spectral simulations. At the same time, to do these experiments at all required the ability to prepare by redox titration or rapid-quench kinetics the ESR-detectable species, in particular oxidation states at a specific point along the reaction coordinate. Without these protocols, none of the experiments described or alluded to could have been performed.

d. ESR and Metal Metabolism. In early years of ESR development, interest in signals from *in vivo* systems was high. Although this use of ESR has waned, with the increased viability of cells in primary culture, the time seems ripe to think of significant *in vivo* ESR experiments using cellular material. As will be noted, even an endogenous protein radical such as is found in the Fe-dependent ribonucleotide reductase can be observed within *E. coli* and a mouse cell line, 3T6. It is in the study of metal metabolism that ESR may be most useful, since essential metals such as Fe and Cu undergo

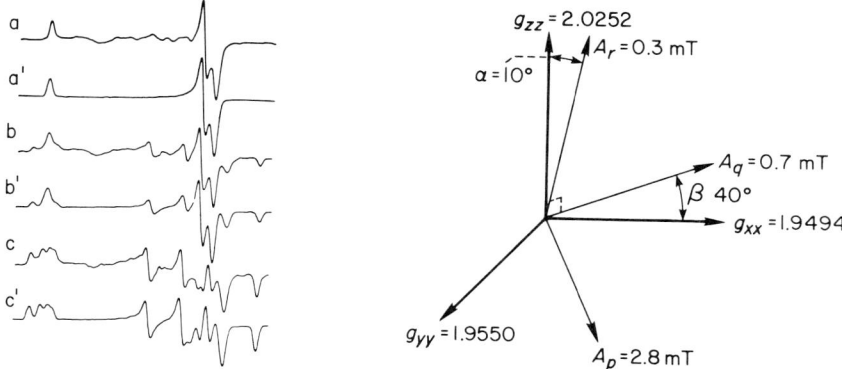

Fig. 17. Very rapid ESR signal from [^{33}S]xanthine oxidase. In the left panel are shown the experimental and simulated spectra: (a) normal enzyme and (a') simulation; (b) [^{33}S]enzyme; (c) difference spectrum of (a) and (b), corresponds to 100% enrichment; (c') simulation; (b') simulation based on sum of fractional contributions from (a') and (c'). In the right panel are shown the g and A^{33}S tensors used for these simulations with 1 mT = 10 G. (From Malthouse et al., 1981.)

redox cycling as part of their uptake, intercellular and transepithelial transport, and intracellular utilization. Since Fe(II)/Fe(III) and Cu(I)/Cu(II) differ so markedly in their ESR characteristics (compare with Table I), changes in the oxidation state of these metals within a physiologically active cell system should be readily detected by ESR.

One simple example of this concept is provided by the study of Fe uptake by microorganisms such as *Fusarium* and *Neurospora* (Ecker et al., 1982). Such eukaryotes as well as certain bacteria secrete Fe(III) chelating agents—siderophores or deferrichromes—which scavenge the metal from the environment. These organisms apparently have specific membrane receptors which recognize the metal chelate. Once the metal chelate is bound the metal can enter the cell with or without the siderophore. If the chelator does enter the cell, it is thought of as a true iono(ferro)phore. ESR is useful to compare the rate of Fe disappearance from the medium per se with the rate of its actual release within the cell. These rates can be distinguished by ESR directly, since on entry of the Fe(III) into the cytosol of the cell, it is reduced to Fe(II). If, however, it enters the cytosol bound to the siderophore, then reduction will occur only on dissociation of the metal from the chelating agent. Thus, ^{59}Fe disappearance from the medium may be more rapid than Fe(III) disappearance from the cell culture.

In fact, by monitoring the $g = 4.3$ signal characteristic of ferrichrome and ferrichrome A, Ecker *et al.* were able to show that *U. sphaerogena* uses ferrichrome as a true ionophore; that is, the removal of ^{59}Fe-ferrichrome from the medium was faster than the reduction of the metal in culture, as determined by ESR. In contrast, ^{59}Fe-ferrichrome A uptake and Fe(III) reduction were synchronous, showing that for this ferrichrome reduction was a prerequisite for iron uptake (Ecker *et al.*, 1982).

B. Other Endogenous Paramagnetic Centers

1. The Iron-Dependent Ribonucleotide Reductase

Unpaired electrons may be localized on organic cofactors other than the isoalloxazine ring. An apparently novel radical species is one associated with the ribonucleotide reductases from *E. coli* and mouse fibroblast 3T6 and *L*-cells (Gräslund *et al.*, 1982). This radical is generally observed as a doublet species. Each of these enzymes has two subunits; one contains an antiferromagnetically coupled high-spin Fe(III) pair, the B2 subunit of the *E. coli* reductase, and the M2 subunit of the mammalian enzyme (Atkin *et al.*, 1973). A description of how the properties of the Fe(III) and radical centers were elucidated will provide a good illustration of how ESR can be useful.

The *X*-band ESR spectra of *E. coli* and mouse 3T6 cell ribonucleotide reductases are shown in Fig. 18. The saturation behavior at 77 K is given in Fig. 19 and shows that, at this and higher temperatures, saturation is apparent only above ~2 mW, although the mouse enzyme exhibits anomalous saturation behavior in this region. Below 77 K, the signal is more readily saturated. Although line broadening occurs as the temperature is increased, a spectrum is readily obtainable at 25°C in liquid solution; the doublet splitting persists under these conditions. With the enzyme from *E. coli*, isotopic replacement with ^{57}Fe or ^{56}Fe, or D_2O from H_2O (for sample preparation only) has no effect on the spectrum. In the *Q*-band spectrum ($\nu = 35$ GHz) the major doublet persists. The data all indicate that the resonance originates from an unpaired electron in an organic molecule which is coupled to a nuclear spin, $I = \frac{1}{2}$ (Ehrenberg and Reichard, 1972).

The radical is not centered on the iron; Mössbauer spectroscopy and magnetic susceptibility measurements show that the two iron atoms are high-spin Fe(III) centers which are antiferromagnetically coupled and thus diamagnetic (below 195 K) (Atkin *et al.*, 1973). The existence of the radical depends on the presence of the metal, however; also, a known free-radical scavenger such as hydroxylamine, which does not remove the metal, eliminates the electronic properties associated with the radical, including the ESR spectrum (Ehrenberg and Reichard, 1972).

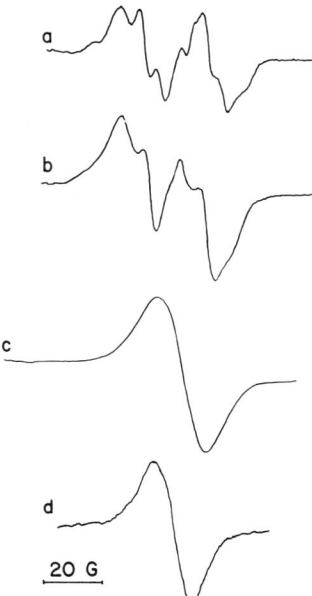

Fig. 18. ESR spectra of ribonucleotide reductase. Enzyme was isolated from mouse 3T6 cells (a and c) and *E. coli* (b and d). Cells in spectra c and d were grown in the presence of [β,β-^{2}H$_2$]tyrosine. X-band spectra ($g = 2.0047$) were recorded at 77 K except for b, which was taken at 32 K. (From Sjöberg et al., 1977; Grässlund et al., 1982.)

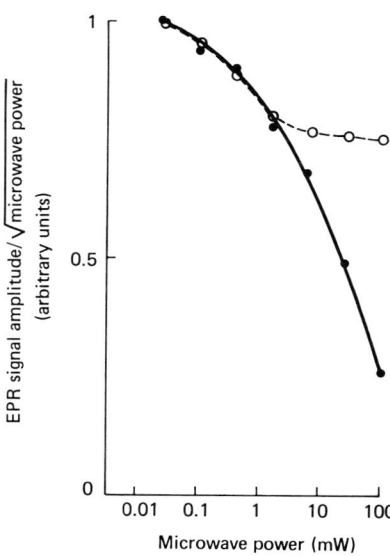

Fig. 19. ESR saturation behavior of *E. coli* (●) and mouse 3T6 cell (○) ribonucleotide reductase at 77 K. (From Grässlund et al., 1982.)

To establish the identity of the radical, two elegant experiments were performed. A strain of *E. coli* which produces 5–10% of its soluble protein content as reductase (the "overproducing" reductase in Fig. 18b) was grown in D_2O and the ESR spectrum of the *cells* recorded; the comparison of the cells grown in H_2O with those grown in D_2O is shown in Fig. 20 (Sjöberg *et al.*, 1977). Significantly, the doublet collapses to a single line, as would be expected if

$$\text{>C−H were replaced by >C−D.}$$

Note that because of the differences in g_N the hyperfine coupling of D should be $\sim \frac{1}{6}$ that of H, a coupling which would not be resolved.

In a second experiment, *E. coli* were again grown in D_2O, but with the addition of specific nondeuterated amino acids; the "D_2O" spectrum persisted for all amino acids added with the exception of tyrosine. When tyrosine was added, the normal doublet spectrum was observed (Sjöberg *et al.*, 1977).

In a complementary experiment, growth of both *E. coli* and mouse 3T6 cells in H_2O in the presence of specifically deuterated tyrosine localized the radical species still further (Gräslund *et al.*, 1982). The collapse of the doublet is caused by incorporation into the reductase of β, β'-dideuteriotyrosine (Fig. 18c, d). Similar experiments were carried out using 3,5-[$^2H_2$]-tyrosine. Although the doublet spectrum persists, the fine structure apparent in the ^1H spectra is lost. This indicates that substantial spin density is located in the aromatic ring, particularly at C_3 and C_5.

The analysis of the spectra, particularly the β—CH coupling, has recently been carried out (Sahlin *et al.*, 1982). This coupling depends on two factors: the through-bond spin polarization of electrons in the C_1—C_β and C_β—H bonds, and the through-space spin–spin interaction. The former is usually neglected for β protons (although there are cases for which this is a poor assumption). The hyperfine coupling is then given by the through-space interaction $B = \rho B'' \cos^2 \theta$, in which B'' is normally set equal to 50 G. The angle θ is the dihedral angle between the C_β—H bond and the p_z orbital containing the unpaired electron-spin density ρ, as shown here.

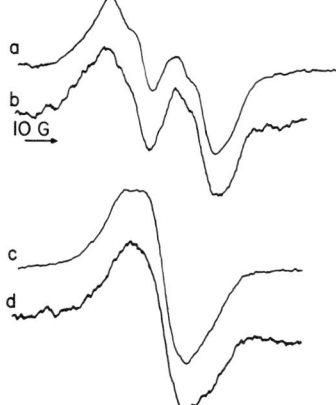

Fig. 20. ESR spectra of *E. coli* (a and c) and partially purified B_2 protein (b and d). Cells were grown in H_2O (a) and D_2O (c) and the protein isolated from the two cell cultures, respectively. X-band spectra were recorded at 95–97 K. From Sjöberg et al., 1977.

The hyperfine coupling to C_3 and C_5 ^1H can be related to the spin density at those carbons, as well, since $A_H = \rho Q_H$, where $Q \approx 25$ G. (This represents a through-bond spin-polarization mechanism, e.g., the type of coupling neglected in the foregoing.)

Using these relationships, the set of parameters shown in Table V were determined (Sahlin et al., 1982). The *E. coli* data is less precise, since the B_2 coupling is not resolved ($B_2 \lesssim 6$ G), providing only an upper limit on θ_2 (between 90 and 120°). However, the two radicals are remarkably similar in structure, differing only in a 10° twist-angle difference around the C_1-C_β bond.

The different saturation behavior of M2 above 2 mW could well be related to this conformational difference. An interpretation of the saturation data

Table V Hyperfine Coupling Parameters in Fe-Dependent Ribonucleotide Reductase[a]

Parameter	E. coli	Mouse 3T6
B_1	~18.8 G	18.2 G
θ_1	~0°	10°
B_2	~5 G	7.5 G
θ_2	~120°	130°
ρ_{C_1}	0.37	0.37
$A_{H_{3,5}}$	~6 G	~6 G
$\rho_{C_{3,5}}$	0.25	0.25

[a] From Sahlin et al. (1982).

for the 3T6 enzyme is that there is a small population of conformations similar to that in the *E. coli* protein which saturate similarly. The data show, however, that the major fraction of the free-radical sites remain unsaturated at 100 mW at 77 K. The relaxation process(es) necessary to achieve this result can be associated with the Fe(III) pair. Indeed, the coupling between these centers is typically temperature dependent. Significantly, even the M2 protein (from 3T6 enzyme) can be saturated at 20 K, as would be consistent with its relaxation mechanism dominated by an interaction with the iron centers. This enzyme continues to be a fascinating study in the properties of a distinct type of paramagnetic center.

2. Substrate Free-Radical Intermediates

Free-radical intermediates occur in a number of biological redox reactions, particularly those involving O_2 or H_2O_2 and heme proteins. The role of such reactions involving peroxidatic-type mechanisms in the metabolism of carcinogens (aromatic polyamines), prevention of polyspermy (Boldt *et al.*, 1981; Josephy *et al.*, 1982), and prostaglandin biosynthesis (Kalyanaraman *et al.*, 1982) has been the subject of considerable research and deserves comment. ESR has been useful here in both the detection and assignment of these radicals.

One recent report illustrates this research area well. This work involved the characterization of a free radical formed during the O_2-dependent autoinactivation of prostaglandin synthase, a putative hemoprotein (Kalyanaraman *et al.*, 1982). Some of the characteristics of this radical species, detected in freeze-quenched samples of O_2-saturated ram seminal-vesicle microsomes in the presence of arachidonic acid, are given in Table VI. The g value is significantly different from a variety of oxygen-centered radicals, such as $HO^{\cdot}$ and $HO_2^{\cdot}$ ($g = 2.025$ and 2.016, respectively). On the other hand, the g values and saturation behavior are similar to those for the radicals derived from H_2O_2-treated methemoglobin and myoglobin. In any event, the saturation observed at $-196°C$ does indicate that the radical is organic centered, since most paramagnetic transition metals saturate readily only at much lower temperatures. Although neither the methemoglobin nor myglobin radicals have been characterized structurally, it is likely the spin is in a π system, perhaps in the porphyrin ring. Another intriguing possibility is that the species seen in the prostaglandin synthetase is a tyrosine free radical. Peroxidatic reactions are known to cause the formation of o,o-dityrosine via a tyrosine free-radical intermediate (Gross and Sizer, 1959). Furthermore, the g value reported here is remarkably similar to that of the well-studied tyrosine free radical in the iron-dependent ribonucleotide reductases ($g = 2.0047$) which have been reviewed.

Table VI ESR Parameters of Putative Heme Protein Organic Radicals

Radical	g (K)	$P_{max}{}^a$ (mW)	Linewidth,^{a} peak to trough (G)
Ram seminal vesicle	2.0038 (296) 2.0054 (106)	~250	25 25
Methemoglobin	2.0033 (296) 2.0048 (106)	100	18
Metmyoglobin	2.0036 (90)		10–16

^{a} At 77 K. From Kalyanaraman et al. (1982).

The mechanism of action of certain enzymes in which the B_{12} coenzyme serves as cofactor has been elucidated by the use of ESR. In particular, ESR experiments demonstrated the presence of organic free radicals during turnover generated via homolytic bond fission of a carbon—cobalt or a C—H bond. The nature of these radicals was determined by a combination of g value, linewidth, and saturation analysis, and by assignment of hyperfine couplings. Isotopic substitution has also proven useful. Kinetic studies indicate the competence of these radicals as intermediates. Thus, these systems provide a variety of examples of how ESR can be used in the study of biologically interesting systems.

ESR resonances have been observed in (frozen) solutions of dioldehydrase (Finley et al., 1973; Valinsky et al., 1974a,b), glyceroldehydrase (Cockle et al., 1972), ethanolamine ammonia lyase (Babior et al., 1972, 1974), and the B_{12}-dependent ribonucleotide reductase (Hamilton and Blakley, 1969; Hamilton et al., 1972; Orme-Johnson et al., 1974). The intensity of these resonances depends on the concentration of available substrate(s). These spectra are not attributable to the coenzyme itself, since B_{12} or 5'-deoxyadenosylcobalamine (DBCC) is not paramagnetic [perhaps low-spin Co(III)]. However, what is known as Co(II)-cobalamine, or $B_{12(r)}$, which lacks the carbon-cobalt bond of the native coenzyme, does exhibit a characteristic axial ESR spectrum ($g_x = g_y = 2.27$, $g_z = 2.022$). A typical spectrum is shown in the low-field portion of Fig. 21. Minimally, therefore, the reactions catalyzed by these enzymes involve rupture of the bond between the 5'-carbon of the deoxyadenosyl ligand and the cobalt; that a Co(II) species is formed rather than Co(I) requires that homolytic bond fission

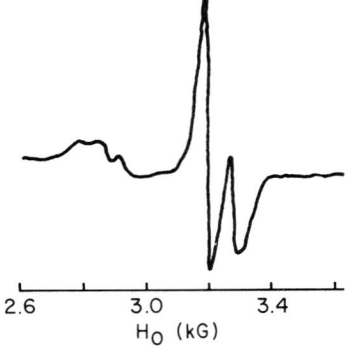

Fig. 21. ESR spectra of ethanolamine ammonia lyase in the presence of L-2-aminopropanol. The ESR spectra were taken at 98 K, at 9.162 GHz and 1.0 mW. (From Babior et al., 1974.)

occurs, and consequently a second unpaired electron must result. In fact, such a radical species has been detected, as illustrated in Fig. 21.

The properties of this spectral "doublet," which is observed in these four enzymes, are summarized in Table VII (Schepler et al., 1975). Although an assignment of this radical species has been elusive, the strategies employed in an attempt to do so are instructive.

That the doublet is attributable to a single electron center is established readily by comparing the separation of the two resonances at different microwave frequencies. There is no difference in this parameter at 9 and at 35 GHz; this frequency-independent nature shows that neither g-value anisotropy nor two (unrelated) g values are responsible for the spectra observed.

Table VII Properties of Substrate Spectral Doublet in B-12 Dependent Enzymes[a]

Parameter	Ribonucleotide reductase	Glycerol dehydrase	Ethanolamine ammonia lyase	Diol dehydrase
Integrated intensity [(% of Co(II)]	8.5%	50%	60%	0.76
g_0	1.965	1.944	1.99	1.96
Linewidth (G)	37	39	41	43.6
g_1	2.032	2.035	2.04	2.05
Linewidth (G)	30	36	54.5	38.6
g_1/g_0 line height ratio	2:1	4.7:1	$\approx$2.2:1	4.7:1
Microwave power	2.7 mW	Not given	1 mW	1 mW
Temperature dependence of signal	12 K (splitting decrease)	Not studied	8–243 K (no effect)	100–214 K (no effect)
Optical evidence for B_{12_r}		+	+	+
Kinetic competence	+	+	+	+
Spin density on substrate	−(?)	Not given	+	+

[a] From Schepler et al. (1975).

The linewidths themselves are suggestive of the unpaired electron being contained within an organic molecule rather than centered on a metal ion. The spin relaxation differences which are responsible for these linewidth effects can be gauged by saturation experiments. A typical result (for ribonucleotide reductase) is that at 12 K, the power at half-saturation, $P_{1/2}$, for the two resonance lines is 23 mW ($g = 2.036$) and 2.5 mW ($g = 1.965$) (Sando et al., 1975). The slopes of the lines in the saturation region are similar, indicating that the spin packets which make up the resonance envelopes have similar relaxation rates. The $P_{1/2}$ values for the doublet are minimally one tenth of those needed to cause appreciable saturation of the Co(II) signal and are similar to those for the organic (tyrosine) radical in the Fe-dependent ribonucleotide reductases. Thus, the saturation study further distinguishes between the two types of resonances, but also indicates that the doublet spectrum does not arise from a metal-centered electron spin.

The doublet nature of the spectrum is suggestive of a hyperfine interaction with a single nuclear spin, $I = \frac{1}{2}$, thus selective deuterium substitution is a reasonable way to elucidate the electron-spin distribution. As noted in Table VII, the results have been inconsistent in that there is no single pattern of isotopic effect. This perhaps is not surprising, for although all four enzymes use B_{12} as cofactor, the reactions catalyzed are distinctly different, particularly the reductase one. As is obvious from the table, the assignment of the radical in this latter enzyme has proven difficult. This system will be discussed in more detail. The data from the other three enzymes indicate that the electron spin contributing to the doublet spectrum is at least in part localized in the substrate, but little ESR evidence exists that places spin on the deoxyadenosyl moiety. Thus, the stable intermediate formed following cleavage of the cobalt-deoxyadenosyl-5′-carbon bond results from transfer of a hydrogen atom from substrate to cofactor. This process can be pictured (Sando et al., 1975):

$$R-CH_3 + \overset{CH_2-dA}{\underset{Co}{|}} \rightleftharpoons \underset{A}{\overset{R-CH_3 \quad {}^*CH_2-dA}{Co(II)}} \rightleftharpoons \underset{B}{\overset{R-CH_2^* \quad H-CH_2dA}{Co(II)}}$$

Presumably, B is the predominant species, based on the results summarized in Table VII.

As noted, ribonucleotide reductase is distinct from the other three B_{12} enzymes which exhibit ESR spectra in the presence of substrates. Aside from and related to experimental differences is the fact that this enzyme's reaction consumes reducing equivalents in the form of the dithiol ↔ disulfide redox reaction. The biological reductant is thioredoxin, but various 1,3- and 1,4-dithiols can substitute (at higher concentrations); prereduced enzyme can also support turnover, suggesting that a dithiodisulfide redox pair exists in the enzyme.

As shown in Table VII, isotopic substitution has not unambiguously determined the source of the doublet spectrum exhibited by this enzyme system. However, fast-freeze kinetic experiments as well as analog studies suggest that in this enzymic process, at any rate, the doublet species is not catalytically relevant (Orme-Johnson et al., 1974).

An interpretation of the doublet spectrum which may also be relevant to these freeze-quench experiments is that it is caused by an exchange coupling of Co(II) and organic-centered unpaired electrons (Schepler et al., 1975). This can be compared to an "AB" system of coupled spins commonly encountered in NMR, two coupled nuclei which have similar chemical shifts (g values). This analysis can explain a number of puzzling and seemingly unrelated spectral phenomena, i.e., the $\sim 2:1$ intensity ratio of the resonance lines of the doublet, the doublet splitting, and intensity dependence on temperature. Briefly, the typical doublet spectrum including the $g \simeq 2.3$ region may not represent simply B_{12r} and the "organic radical" but, in fact, a system in which the two electrons experience an electrostatic-exchange coupling J which is equivalent to the energy separation of the radical doublet. By taking into account the anisotropic nature of the cobalt g and hyperfine values, the doublet spectra of the four B_{12} enzymes discussed here can be simulated quite successfully (Schepler et al., 1975).

C. Spin Labels

1. Metals as Spin Labels: VO^{2+}, Cu(II), Co(II), and Mn(II)

Relatively few proteins are inherently paramagnetic or functionally generate such species. However, many proteins either contain a diamagnetic metal ion or specifically bind a variety of transition-metal ions. Thus, the substitution of a suitable paramagnetic ion can introduce an ESR-detectable probe into the apoprotein. Such probes can reflect aspects of the coordination geometry and ligand type by the spin Hamiltonian parameters or superhyperfine couplings they exhibit. In addition, these parameters may be responsive to functional perturbations of the protein, e.g., ligand or substrate binding, cooperative effects, conformation, etc. That is, they can provide just the type of information as was described in the preceding sections (Sections IV.A,B).

The VO^{2+}, Cu(II), Co(II), and Mn(II) ions have been the most useful as ESR probes, although Mn(II) has been used principally in NMR relaxation experiments in which the metal ion enhances the relaxation of protein or substrate nuclei in a time- and distance-dependent fashion. This is discussed in Chapter 1 and will not be considered here. Co(II) is an exceptionally useful probe in that it exhibits a rich envelope of electronic transitions in the visible and near infrared. In some respects, it is the optical properties of this metal ion which have been exploited to good advantage. Nonetheless, as an ESR spin label, Co(II) has proven useful, also, as will be noted here and in Sections IV.D and V.B.

Many proteins which normally contain Zn(II) as a prosthetic group will also exhibit comparable affinity for Cu(II) [or Co(II)] (Ochiai, 1977). Thus, Cu(II)-serum albumin (Peters and Blumenstock, 1967), carbonic anhydrase (Taylor and Coleman, 1973), and carboxypeptidase (Brill and Kirkpatrick, 1967), among others, were prepared. The latter two enzymes, with Cu(II), exhibited no activity. Nonetheless, the paramagnetic properties of these Cu(II) proteins were useful in assessing aspects of metal coordination and ligand binding. For example, two equivalent nitrogen ligands to the metal were inferred from the ESR spectrum of Cu(II) carboxypeptidase A and carbonic anhydrase, as well as three nitrogen ligands in Cu(II) alkaline phosphatase (Taylor and Coleman, 1972).

Exogenous anion binding to these Cu(II) centers can also be followed by ESR. These pseudo-Cu(II) proteins were used as models for the type 2 Cu(II) found in true copper proteins (compare with Fig. 10) and for imidazole coordination as investigated by the electron-spin-echo modulation technique (Section V.C). That many of the enzymes involved in polynucleotide formation and proteolysis, e.g., DNA and RNA polymerases and restriction endonucleases, appear to be Zn(II)-containing proteins suggests that further use of Cu(II) as a spin probe may be warranted. However, Co(II) may be a more appropriate choice; not only are its electronic properties well characterized and sensitive to perturbation, but, notably, it appears to be a better substitute physiologically. Co(II)-substituted enzymes are generally active, thus this substitution is not lethal to a living organism (Ochiai, 1977). As a result, *E. coli* Co(II) DNA polymerase can be prepared *in vivo* by growing this prokaryote in the absence of zinc with cobalt supplementation (Speckhard *et al.*, 1977). This exemplifies a logical use of cell culture in ESR, namely, to prepare specifically labeled material. More use of this approach should be made.

One interesting aspect of the use of Co(II) as a spin probe is that as a d^7 system it can exist in either high-spin ($S = \frac{3}{2}$) or low-spin ($S = \frac{1}{2}$) states. The high-spin state exhibits extreme lifetime broadening, thus its ESR spectrum is readily seen only at liquid-helium temperatures; g values extend from 2 to 6. The low-spin ($S = \frac{1}{2}$) state exhibits a typical axial ESR spectra with $g_\parallel = 2.00$ and $g_\perp = 2.26$. That the latter is undoubtedly 6–8 coordinate (either octahedral or a tetragonal distortion of that field) whereas the former is likely to be tetrahedral provides the basis for interpretation of effects on the ESR spectrum caused by ligand binding to the Co(II) center. For example, Co(II)-carbonic anhydrase appears to be an $S = \frac{3}{2}$ system for the reasons noted here (Haffner and Coleman, 1973a). On the binding of one equivalent of CN^-, the Co(II) remains high spin, presumably tetrahedral; CN^- apparently first displaces a water molecule. The second equivalent of CN^-, however, causes the Co(II) to go low spin and at least six coordinate. Since the Co(II) ESR spectra do not exhibit the well-resolved ^{14}N SHFS often seen in Cu(II) spectra, the endogenous coordination of the Co(II)—biscyanide complex is not known. In the Cu(II)—carbonic anhydrase CN^- complex only two ^{14}N are coordinated (Haffner and Coleman, 1973b). Possibly in both the Co(II) and Cu(II) proteins, CN^- either displaces one of the three endogenous imidazoles directly or causes a rearrangement of the ligand field. This would result in one of these protein groups becoming an axial (weakly coupled) ligand. This type of effect occurs upon CN^- binding to the Cu/Zn superoxide dismutase, as discussed later.

Cu(II) as a probe of the Fe(III)-binding proteins, transferrin (Zweier and Aisen, 1977) and conalbumin (Zweier, 1980), has also been quite informative. Not only is the nature of the endogenous ligands of interest, but the equivalence of these proteins' two metal sites and their requisite binding of counter anions, CO_3^{2-} or oxalate, was investigated using Cu(II) as a spin label. Both Cu(II) proteins exhibit axial ESR spectra ($g_\parallel \simeq 2.3$, $g_\perp \simeq 2.06$) with SHFS ($A_\parallel^N \simeq 9$ G) associated with the coupling of one ^{14}N ligand. Spin-echo experiments show this nitrogen to be from a histidine imidazole (discussed later).

Both $g_\parallel$ and $A_\parallel$ are sensitive to ligand binding; $g_\perp$, typically, is less so. In general, $g_\parallel$ increases whereas $A_\parallel$ decreases (from ~ 150 to 140 G) upon CO_3^{2-} bonding; oxalate has a somewhat larger effect. These effects are consistent with an increase in negative charge in the ligand field (compare with Fig. 10), suggesting that anion binding displaces a neutral ligand, presumably H_2O. Two tyrosine phenols complete the equatorial ligation at the Cu(II) site. In both proteins, inequivalence of the two sites is clearly evident in the ESR spectra. In Cu(II) transferrin, this inequivalence can be detected in hybrid complexes which contain one each of carbonate and oxalate. The ^{14}N SHF triplet from the carbonate site is considerably narrower than that observed for the biscarbonate species, thus indicating heterogeneity at these two sites. These sites in conalbumin are somewhat more dissimilar in that they can be assigned distinct g and A values.

Vanadyl (VO^{2+}) also has been valuable as a probe of the metal-binding sites of proteins involved in iron metabolism, perhaps even more so (Campbell and Chasteen, 1977; Chasteen et al., 1977; Chasteen, 1981; Chasteen and Theil, 1982). As a metal spin probe it has several experimental advantages over Cu(II), for example: in large measure, these advantages are owing to the fact that VO^{2+} has an orbitally nondegenerate ground state. With no electronic excited states nearby in energy, the VO^{2+} ion and its complexes exhibit well-resolved ESR spectra both in liquid solutions at room temperature and in frozen glasses. Furthermore, the HFS is anisotropic and thus is sensitive to orientation, conformation, and rotational relaxation. The VO^{2+} ion also has a well-characterized coordination chemistry, whereas VO^{2+} is itself fairly stable. Most important, VO^{2+} does bind specifically to certain sites in a variety of biological systems.

In general, VO^{2+} complexes in frozen glasses exhibit axial symmetry, particularly at X band, with $g_\| \simeq 1.94-1.98$ and $g_\perp \simeq 1.98$ (Chasteen, 1981). As the ligation changes from $O_4 \rightarrow N_2O_2 \rightarrow N_4 \rightarrow S_2O_2$, $g_\|$ increases while $A_\|$ decreases (from ~ 200 to 140 G). A corresponding change is seen in g_0 (from ~ 1.97 to ~ 1.98) and A_0 (~ 105 to ~ 90 G). Note that these effects are similar to those observed for Cu(II) complexes (Fig. 10). The complexes of VO^{2+} are generally square pyramidal or bipyramidal with the V=O bond along the g_z ($g_\|$) axis. The nuclear hyperfine interaction ($I = \frac{7}{2}$) is largest in this orientation. Isotropic spectra exhibit the eight lines expected of this nuclear HFS (Fig. 22a). The nuclear-spin states are identified assuming that $A < 0$. The spacings between the lines increase as H_0 increases owing to the second-order effects caused by the large value of $|A|$ (compare with Appendix

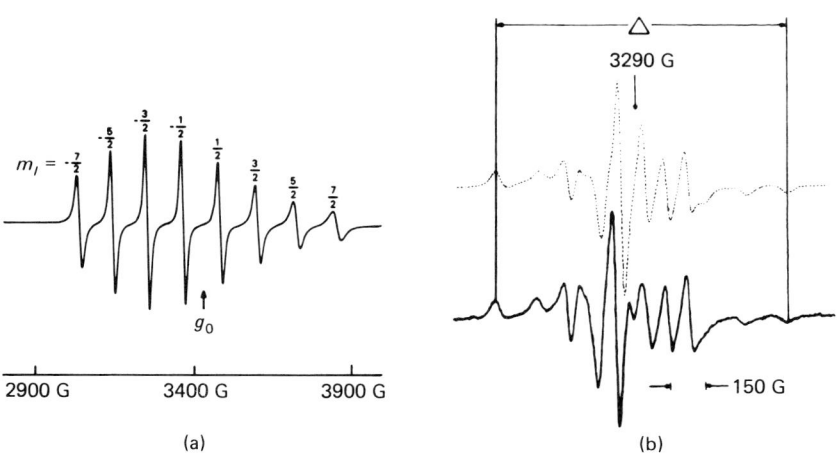

Fig. 22. Isotropic (a) and slow-motion (b) VO^{2+} X-band ESR spectra: (a) is the spectrum of VO^{2+} $(H_2O)_6$ at 298 K; (b) VO^{2+} (acetonylacetone)$_2$ in supercooled toluene (174 K); the dotted line is a simulated spectrum using $\tau_c = 4.2 \times 10^{-9}$ sec/rad. (From Chasteen, 1981.)

C). The increasing linewidth noted is the result of incomplete averaging, since the transitions at higher field are the most anisotropic.

The anisotropy of the g and A values becomes apparent in the slow-motion regime as both parallel and perpendicular features are in evidence (Fig. 22b). As noted for the nitroxide spin label in Section IV.C.2, information about the rotational correlation time τ_c can be obtained from the maximum splitting in the spectrum, in this case Δ. For VO^{2+} in the region $8 \times 10^{-10} \lesssim \tau_c \lesssim 7 \times 10^{-8}$ sec, we have

$$\tau_c = a(1 - S)^b, \tag{18}$$

where $S = \Delta/7A_\parallel$; Δ is the observed spread of the outer lines, $7A_\parallel$ the rigid-limit separation for the VO^{2+} complex, and $a \simeq 1.1 \times 10^{-11}$ sec/rad, $b \simeq -1.9$.

The rigid-limit (powder) spectrum is generally that exhibited by VO^{2+} proteins (Fig. 23). This is so because, in general, the value of τ_c for the VO^{2+} ion is the same as for the protein. The Stokes–Einstein relationship [compare with Eq. (23)] shows that for most proteins, $\tau_c \geq 10^{-8}$ sec/rad, well within the slow-motion limit. Although this limits the amount of information obtained about motion per se (in the sense of an order parameter, as discussed later), the high resolution of VO^{2+} spectra in both aqueous and frozen samples remains valuable. For example, the heterogeneity of the

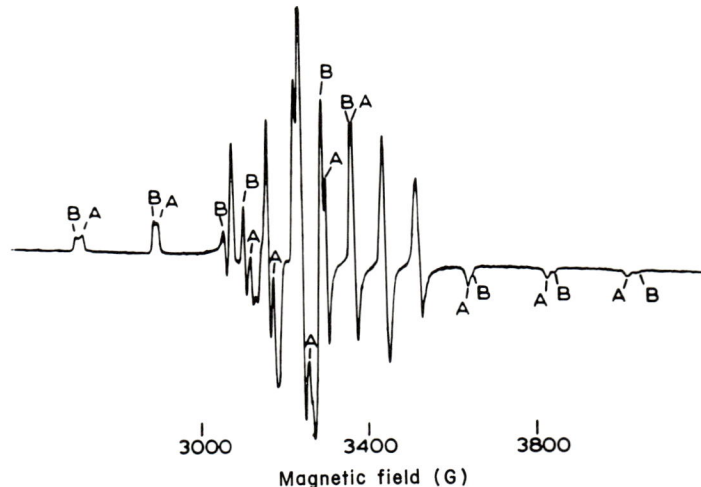

Fig. 23. X-band ESR spectrum of bis VO^{2+} human transferrin at 77 K. Two sets of resonance lines are indicated corresponding to the two metal-binding sites, A and B. (From Campbell and Chasteen, 1977.)

iron-binding sites in human transferrin to which VO^{2+} binds is readily apparent in the ESR spectrum (Fig. 23). The origin of the differences could be owing to a number of factors, but by assessing the effects of changing the counter anion required for promotion of VO^{2+} binding [as is required for Fe(III) binding, also], Campbell and Chasteen (1977) were able to elucidate these differences. Since the effects on $A_\parallel$ (transferrin) caused by ligand substitution (thioglycolate → lactate → malonate, for example) were linearly correlated to the change in $A_\parallel$ for the corresponding ligand (nonprotein) complexes, it appeared that the ligands' effects on the protein-bound VO^{2+} were direct, i.e., the ligands were within the innersphere coordination shell, bridging the ion to the protein. Although this was clear for site B (Fig. 23), it was less definitive for site A. These results appear complementary to those outlined here for the Cu(II)-transferrin system. New work on the ENDOR (Mulks *et al.*, 1982) and spin-echo modulation properties (compare with Section V.B,C) of VO^{2+} model and protein complexes appears warranted to fully exploit the utility of this ion as a biophysical spin probe.

A unique aspect of vanadyl as a probe is its interaction with membrane components, particularly (Na,K)-ATPase. Actually, the inhibiting ion is not VO^{2+} (IV), but V (V) as VO_4^{3-}, (ortho)vanadate, at physiological pH. This ion is a PO_4^{3-} analog, explaining its activity (Chasteen, 1981). Vanadate incubated with cells (erythrocytes, adipocytes, yeast, etc.) appears within the cytosol as VO^{2+}, normally in complex with thiol-containing species, e.g., glutathione, or macromolecular ones. The differences in these two types of complexation are owing to the τ_c values for the associated complexes (Fig. 24) (Cantley and Aisen, 1979; Degani *et al.*, 1981). The VO^{2+} bound to protein (presumed to be hemoglobin, Fig. 24a) is in the slow-motion limit, whereas the glutathione complexes are nearly isotropic (Fig. 24b). The biochemical response to VO^{2+} by mammalian cells is also of considerable interest, since vanadyl (or vanadate) has insulinlike effects, i.e., stimulation of glycolysis and energy metabolism and inhibition of lipolysis (Degani *et al.*, 1981). The biochemistry of vanadium has been little characterized and represents a fertile area for new research.

Mn(II) as a spin probe presents certain aspects not shared by the other metals discussed. First, as an $S = \frac{5}{2}$, $I = \frac{5}{2}$ system with $A^{Mn} \simeq 90$ G, the spectrum is spread out over the g field, particularly for systems that exhibit large deviations from cubic symmetry. The six-lined isotropic spectrum observed at X band results from the electron–nuclear hyperfine coupling with a near superposition of the six electron-spin-state transitions. However, the spin Hamiltonian for Mn(II) includes orientation-dependent zero-field splitting (ZFS), that is,

$$\mathcal{H} = g\beta H_0 \cdot S + D[S_{z^2} - \tfrac{1}{3}S(S+1)] + E[S_{x^2} - S_{y^2}], \tag{19}$$

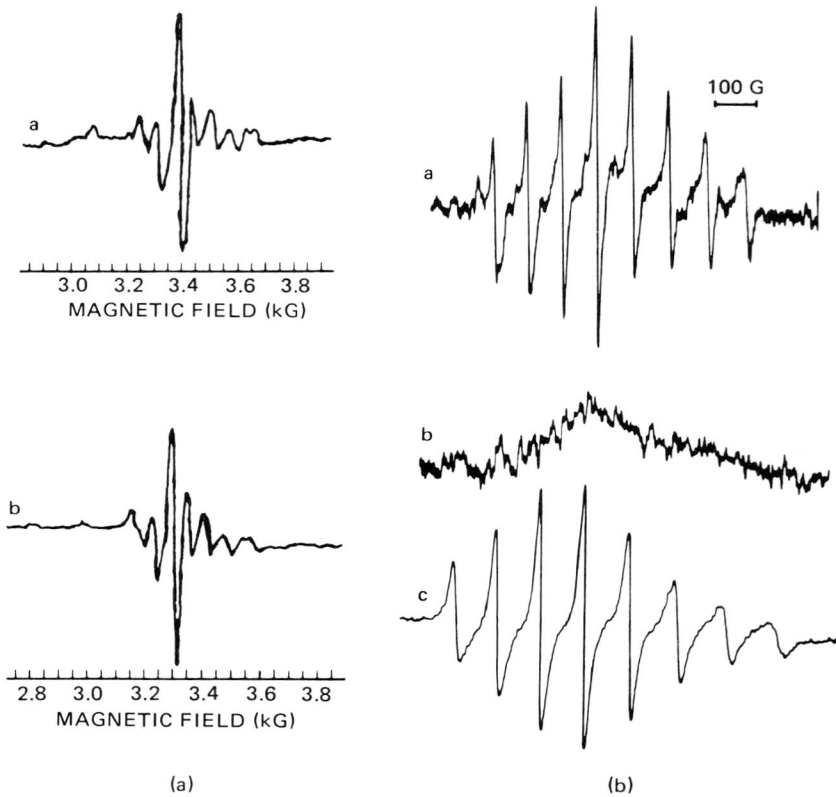

Fig. 24. X-band ESR spectra of VO_4^{3-}-treated cells: (a) human erythrocytes (50% v/v) incubated with 1.3 mM Na_3VO_4 for 1 h at 37°C; spectrum a was taken at 246 K, spectrum b at 77 K; (b) rat adipocytes (~6 × 10⁶) incubated with 1 mM $NaVO_3$ (meta vanadate) for 90 min at 37°C; spectrum taken at ambient temperature. (A) cells alone; (B) medium alone; (C) medium alone treated with 10 mM glutathione. [(a) from Cantley and Aisen, 1979; (b,c) from Degani et al., 1981. © 1981 American Chemical Society.]

where D and E are the axial and rhombic distortion parameters of the zero-field interactions. If the ZFS ($\Delta\omega$, in rad/sec) is smaller than $1/\tau_c$, the rotational relaxation rate, then the terms in D and E are effectively averaged to zero. If $\Delta\omega > 1/\tau_c$, as is often case for Mn(II) bound to proteins ($\tau_c \simeq 10^{-8}$ sec), then the anisotropy of the ZFS will become evident in the spectra, again, if the metal's ligand field is distorted ($D \neq 0$, $E \neq 0$). There is the additional feature that nominally forbidden $\Delta M_I = \pm 1$ transitions are often observed. Thus, the combination of multiple allowed and forbidden transitions, as well as possibly large and anisotropic ZFS offer both a challenge to the ESR spectroscopist and an opportunity to use this complexity to advantage.

A last, and sometimes limiting, feature of Mn(II) ESR spectra is the electron-spin relaxation time τ_s. This is generally very short, leading to appreciable lifetime broadening, when the ion is accessible to solvent, providing an efficient relaxation mechanism. It is this property which makes Mn(II) such a good enhancer of NMR relaxation rates. Since the separation of the spin states (their eigenvalues) is field dependent, the coupling of the lattice and nuclear spins to the m_s transitions can be damped, i.e., the value of τ_s lengthened. In fact, Q-band ESR spectra are usually sharper (less lifetime broadening) than those at X band. In addition, at the higher field, the six $m_s = \pm\frac{1}{2}$, m_I transitions at $g \simeq 2$ are better resolved since this electron Zeeman splitting will be dominant (at moderate values of the ZFS parameters, that is, if the system is not significantly distorted). The ΔM_I (forbidden) transitions will also be reduced, since their probability is inversely proportional to the magnetic field. Consequently, protein-bound Mn(II) ESR spectra taken at 35 GHz are even more useful (relative to X band) than for most other spin systems discussed here.

The importance of Mn(II) as a probe is because of its ability to substitute for Mg(II) in dNTP and NTP-utilizing systems, particularly those dependent on ATP. The metal can itself be an obligatory cofactor or prosthetic group, as in glutamine synthetase and pyruvate carboxylase. The former enzyme's interaction with various substrates and substrate analogs has been particularly well-characterized using both NMR and ESR approaches (Villafranca *et al.,* 1976). The X-band spectrum of the enzyme-Mn(II) species (Fig. 25a) is essentially isotropic, as can be seen by the lack of apparent transitions at lower and higher fields from the $g = 2$ sextet of lines. This indicates that D and E are relatively small. The broadness of the lines is made evident by comparison with spectrum b, which shows the effect of the binding of a substrate (transition state) analog, methionine sulfoximine. The considerable sharpening of the lines is owing to a longer τ_s caused by a shielding of the Mn(II) from solvent protons. With the resolution of the forbidden transitions (Fig. 25d), the value for the axial splitting parameter D can be calculated from the intensity ratio (IR) of the forbidden to the allowed lines:

$$\text{IR} = \frac{8}{15}\left(\frac{3D}{4g\beta H_0}\right)^2 \frac{1 + S(S+1)}{3M(M-1)}\left[I(I+1) - m^2 + m\right], \qquad (20)$$

where $M = m = \frac{1}{2}$. The value of D can be compared with that for model structures, indicating the nature of the electronic structure of the Mn(II) center under study.

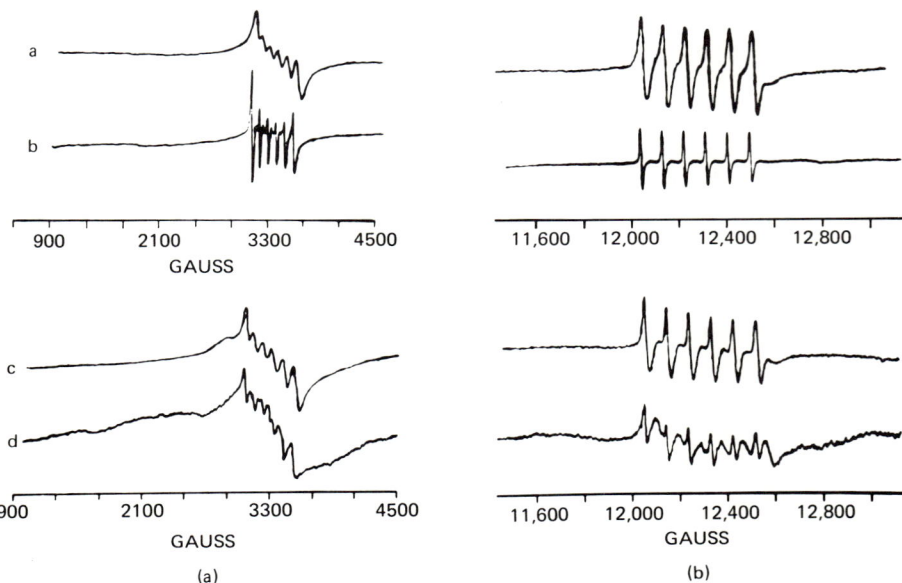

Fig. 25. ESR spectra of glutamine synthetase Mn(II) complexes at 1 °C with concentrations of enzyme subunits and Mn(II) ~0.5–0.7 mM at (a) 9 and (b) 35 GHz. The spectra are curve a, enzyme · Mn(II); curve b, enzyme · Mn(II) · methionine sulfoximine; curve c, enzyme · Mn · glutamate; curve d, enzyme · Mn · ATP. (From Villafranca et al., 1976. © 1976 American Chemical Society.)

The addition of glutamate and ATP cause distinctly different changes in the X-band ESR spectra. Glutamate has little effect on the linewidth while inducing a small distortion from the near-cubic symmetry noted. This is seen in the broad features just outside of the $m_s \pm \frac{1}{2}$ envelope (Fig. 25c, d). ATP causes an even greater distortion; the additional unresolved electron-spin transitions are readily apparent. Here, too, there is little change in linewidth. Neither glutamate nor ATP have an appreciable effect on the electron-spin relaxation rates or the processes which determine them.

The Q-band spectra reveal the same differences between the effects of methionine sulfoximine, glutamate, and ATP (Fig. 25) while illustrating the advantages of the higher field discussed here. The linewidths are reduced by half, and the $m_s \pm \frac{1}{2}$ transitions dominate the spectra; only small-intensity, resolved, forbidden transitions are observed. The anisotropy of the glutamate and, in particular, the ATP complexes is very evident, and the increase in τ_s in the analog complex is striking. In fact, the higher overall resolution at Q band shows a decrease in the linewidths in the ATP complex, as well, indicating that ATP binding causes both a shielding of the Mn(II) from H_2O and the rhombic distortion noted.

A second type of spin-probe experiment using Mn(II) is the double label one (Section IV.C.3). As will be discussed, the interaction between two paramagnetic centers is characterized by an electron-exchange (overlap) term, J, and a dipolar coupling, D. If there is significant exchange, a number of additional lines may be observed. For high-spin Mn(II), if $J \gg A$, eleven lines will be present [as if there were $2n(S + 1)$ species] separated by $\frac{1}{2}A$, or ~45 G. With $J \simeq A$, the spectrum becomes even more complex (compare with Fig. 40). Of course, if $A \gg J$, there is essentially no electron coupling, and the spectrum observed contains the normal six lines. In the case of S-adenosylmethionine synthetase, the two metal sites [which accept either Mn(II) or Mg(II)] are close together (Markham, 1981). That is, the ESR spectra of the Mn_2-enzyme in the presence of imidotriphosphate (PPNP) and S-adenosyl-L-methionine (AdoMet) exhibit a complex of poorly resolved lines around $G = 2$, which is not observed when less than a stoichiometric amount of metal is bound (Fig. 26). The spacing of these lines is ~45 G or roughly what would be expected for a spin exchange comparable

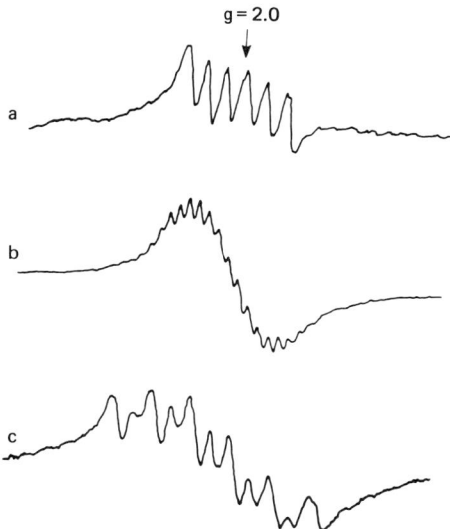

Fig. 26. ESR spectra of Mn(II) complexes of S-adenosylmethionine synthetase at 4°C with 1–3 mM enzyme subunits, 0.4–1.0 mM Mn(II). Spectra (a) and (b) are at 9 GHz: spectrum a, enzyme · Mn complex; spectrum b, enzyme · Mn · imidotriphosphate · S-adenosylmethionine · K^+ complex; spectrum c is of the latter complex at 35 GHz. (From Markham, 1981.)

with or larger than the hyperfine coupling. That there are clearly more than eleven lines indicates that, in fact, J may be similar to A. The 35-GHz spectrum lacks some of these lines; it has the eleven lines expected for $J > A$, with a spacing of 45 G. Although a distance calculation cannot be done based on the exchange integral (unlike the dipolar term, as will be discussed later), the results indicate that the two Mn(II) are within the same molecular orbital and probably share a common, bridging ligand. Although the temperature dependence of the intensity could help estimate the magnitude of the coupling, the spectrum was too poorly resolved to make precise intensity measurements on spectra obtained from 300 to 10 K. However, the behavior observed (increasing intensity) ruled out antiferromagnetic coupling of the two ions (Markham, 1981).

The coordination chemistry of Mn(II) bound to a protein or biological solute can also be assessed. As with other paramagnetic ions, ligand SHFS, resolved in either cw or pulsed ESR, or ENDOR, is the most definitive measure of coordination type and number. The use of ^{17}O in the interaction of ADP with Mn(II), and the MnADP complex with a fragment from myosin, subfragment 1, which possesses a metal—nucleotide binding site is an example of this approach (Webb *et al.,* 1982). Although the ^{17}O SHFS is not resolved, the line broadening observed can be used to evaluate the metal coordination. Using 35 GHz for the reasons detailed, Webb *et al.* demonstrated ^{17}O-ADP ligation in this way (Fig. 27a). As seen, ^{17}O on the β-phosphate (the terminal phosphate) causes a broadening of the Mn(II) subfragment 1 spectrum. The binding of $H_2\,^{17}O$ can also be detected. Generation of a difference spectrum which takes into account the number of $^{17}O - H_2O$ atoms (molecules) indicates that there are *minimally* two water molecules bound as well (Fig. 27b). It is possible, therefore, to suggest that of the six ligand sites in the Mn(II) coordination complex, two are occupied by H_2O, one by a β-phosphate oxygen, and three by, presumably, protein ligands. Again, spectral studies on these systems using ENDOR or spin-echo spectroscopy would be valuable.

2. Nitroxides

Relatively few biological macromolecules are or contain paramagnetic centers. However, the development of the concept of the spectral "reporter group" early on encompassed ESR as a potential spectroscopic probe of biological structure. The extensive and diverse use of the technique of spin labeling has grown out of these early efforts.

The nitroxide is by far the most exploited of those reagents classed as spin labels. A generalized structure of the paramagnetic center of this label type is shown in Fig. 28. The anisotropy of the g and A tensors and the spectra

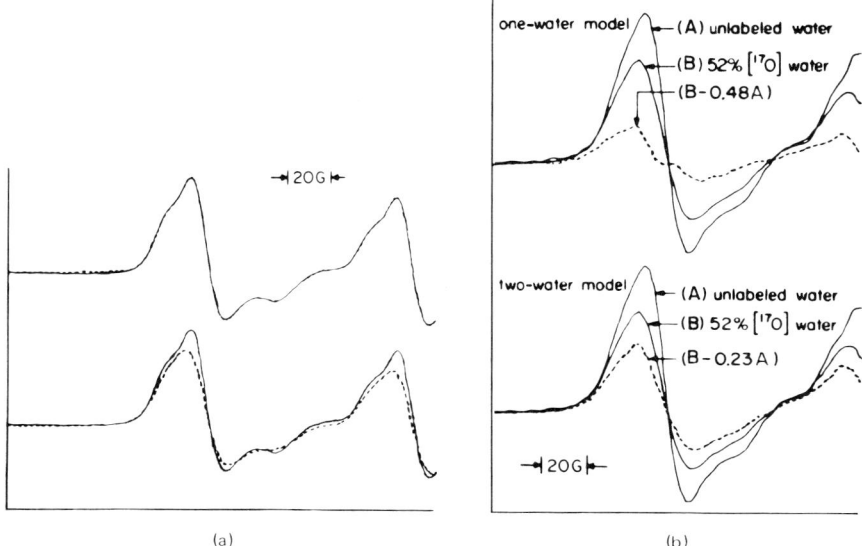

Fig. 27. ESR spectra of myosin subfragment 1 complexes with Mn(II)-ADP labeled with ^{17}O, at 35 GHz: (a) the α- and β-labeled ADP derivatives (- - -), upper and lower curves, respectively, are compared to unlabeled ADP (———). The two low-field M_I transitions ($+\frac{5}{2}, +\frac{3}{2}$) are shown. (b) The lowest field line is shown for the unlabeled Mn-ADP complex in $H_2^{16}O$ and $H_2^{17}O$ (50% enrichment). The difference (- - -) is calculated assuming one $H_2^{17}O$ or two $H_2^{17}O$ bound. A smooth spectrum is obtained with $2H_2^{17}O$ indicating this to be the minimum number coordinated. (From Webb *et al.*, 1982.)

associated with the three principal orientations are indicated there as well. As outlined in Sections II and IV, except in single crystals, these principal tensors will be manifest as an average or a sum of all orientations, depending on the rotational correlation time of the nitrogen center; i.e., the spectrum can vary from truly isotropic to the powder spectrum of a frozen glass (Fig. 28). Thus, the degree of spectral averaging (from completely anisotropic to isotropic) is determined by the freedom and anisotropy of motion of the nitroxide label. If the motion is isotropic and fast relative to the frequency equivalent of the anisotropy in the spin-Hamiltonian, i.e., $\Delta A \simeq 0.7 \times 10^8$ sec^{-1} and $\Delta g \beta H_0 \simeq 0.3 \times 10^8$ sec^{-1}, then an isotropic spectrum results. This motional regime occurs when $\tau_c \simeq 10^{-11}$ sec. At values larger than this, 10^{-8} sec $> \tau_c > 10^{-11}$ sec, the observed spectral anisotropy can be interpreted in terms of the motional freedom or immobilization of the spin label.

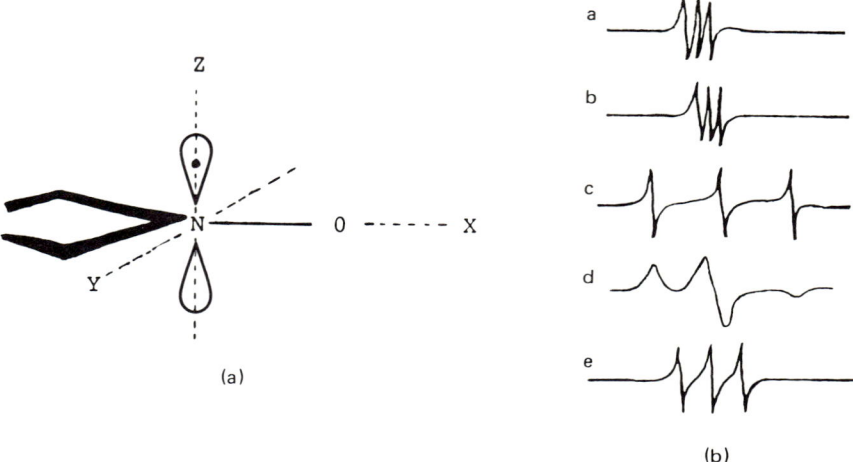

Fig. 28. Tensorial axes in (a) common nitroxide and (b) associated spectral anisotropy at 9.5 GHz. In a–c, orientation of H_o is indicated. The motional limits of the nitroxide are represented in d and e. Values given for g and A are typical; note that both sets vary (particularly a) among nitroxides: for curve a, $H_o \| x$, $g_{xx} \simeq 2.009$, $A_{xx} \simeq 6$ G; for curve b, $H_o \| y$ $g_{yy} \simeq 2.006$, $A_{yy} \simeq 6$ G; for curve c, $H_o \| z$, $g_{zz} \simeq 2.002$, $A_{zz} \simeq 32$ G; curve d, powder (rigid glass); curve e, isotropic solution: $g_o \simeq 2.006$, $A_o \simeq 14.7$ G.

Thus, the nitroxide spin label as a probe relies primarily though not exclusively on the anisotropy of the g and A tensors, i.e., the differences in these in magnitude, and the label's rotational correlation time (frequency) in comparison with these differences. Spin labels are primarily probes of motion; in the vocabulary of the biological chemist, motion implies conformation (changes), order, and fluidity. Within the constraints posed by this fundamental sensitivity to motion, the spin label can also be used to assess dimension (distance). On the other hand, as a true structural probe, the spin label works best as a perturbant of the relaxation rates of neighboring nuclei. The use of spin labels as origins of nuclear relaxation-rate enhancement is discussed in Chapter 1.

The utility of the spin label as a probe results from two factors. First, introduced into a biological system it generally exhibits a unique ESR spectrum, i.e., there is no background. Second, although the nitroxide itself is essentially invariant (e.g., Fig. 29), the organic framework of which the paramagnetic center is a part can be tailored to conform to nearly any structural design. As such, labels have been synthesized as enzyme substrates, inhibitors, and affinity labels; probes of conformational changes in

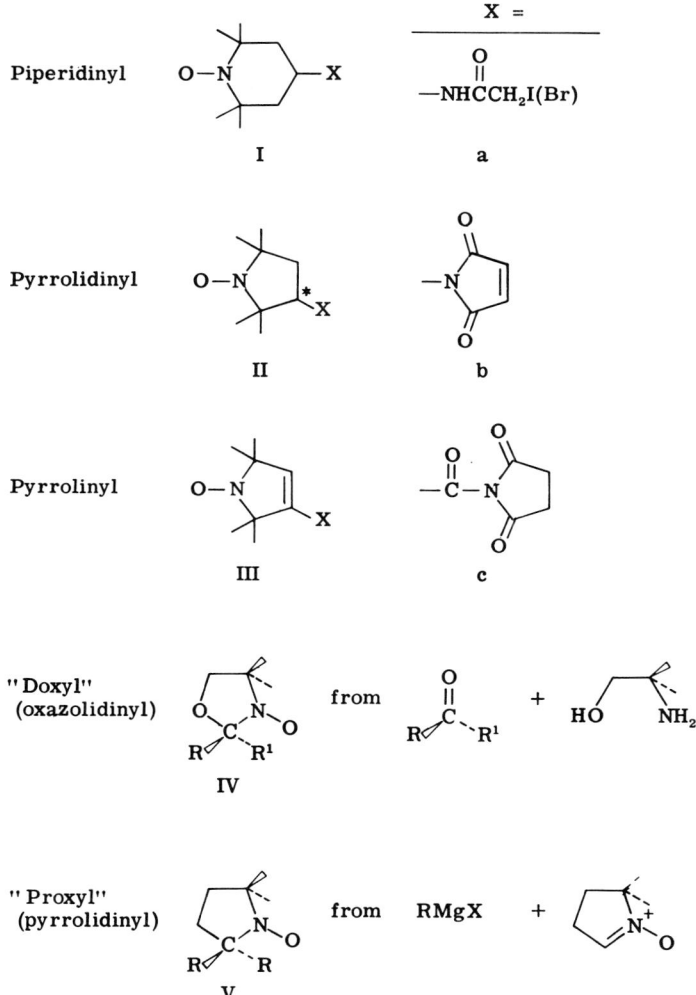

Fig. 29. Structures of common nitroxide (nitrogen heterocycle *N*-oxyl) spin labels and reactive side chains for covalent modification. Note that II has an asymmetric center (∗); in rigid environments, it will assume two orientations. Compounds IIa and IIb are often designated ISL and MSL, respectively (see text). In all cases, ⤙ represents gem-dimethyl substitution.

proteins, nucleic acids, and lipids; probes of protein–protein, protein–nucleic acid, protein–lipid, and lipid–lipid interactions; and probes of the structure and fluidity of biological and model membranes. This panoply of labels clearly cannot be reviewed here. The remainder of this section will present selected examples of spin labels which illustrate both the applicability and the use of this type of probe with emphasis on those examples which illustrate the fundamental aspects of spin-label behavior.

The simplest use of the spin label is as an indicator of macromolecular conformation change. Such experiments are carried out without regard to quantitation of the change in the structural constraints placed on the nitroxide, reflected in the changes in the ESR spectra. For example, the deoxy-oxy transition of hemoglobin can be followed by the spectral change exhibited by the maleimide or iodoacetamide labels MSL and ISL, used as covalent modifying reagents of the β-93 cysteine-SH (McConnell and McFarland, 1970). Activation of chymotrypsinogen can be followed by the change in spectral behavior of Ia when it is covalently linked to the zymogen methionine 192 (Fig. 30) (Kosman, 1972). The activation of pepsinogen labeled via reaction with IIIc at 2-3 lysine ϵ-NH_2 groups near the N terminus has been followed similarly (Fig. 31) (Twining et al., 1981). When covalently linked to chymotrypsinogen, the motion of the label becomes more constrained by the conformation change(s) associated with activation; thus, the peak height (line height) decreases. In the second example, the pepsinogen sequence to which the labels are covalently linked is released during activation proteolysis, thus the motion of the label becomes less restricted. The increase in the peak height reflects this behavior.

These changes in peak height can be related to rotational correlation times under specific conditions. These conditions are that the motion is isotropic and is in the fast-tumbling regime, 10^{-11} sec $< \tau < 3 \times 10^{-9}$ sec (for a nitroxide at resonance in an $\sim$3-kG effective magnetic field). With these constraints the line broadening (and peak height) can be readily related to the dominant relaxation process involving the rate and amplitude of motion; this is the transverse relaxation time T_2. The resonance lines associated with the three S, I transitions ($M_I = +1, 0, -1$, from low to high field, respectively) are differentially affected by motional broadening. Thus, to calculate τ_c values from line shape, the line shapes (heights) of all three must be taken into account:

$$\tau_c = \Delta H(0) \left[15.236 \left(\frac{B^2}{4} \right)^{-1} \right] \left[\sqrt{\frac{h(0)}{h(-1)}} + \sqrt{\frac{h(0)}{h(+1)}} - 2 \right], \quad (21)$$

$$\tau_c = \Delta H(0)[28.567(B\Delta\gamma H_0)^{-1}] \left[\sqrt{\frac{h(0)}{h(-1)}} - \sqrt{\frac{h(0)}{h(+1)}} \right], \quad (22)$$

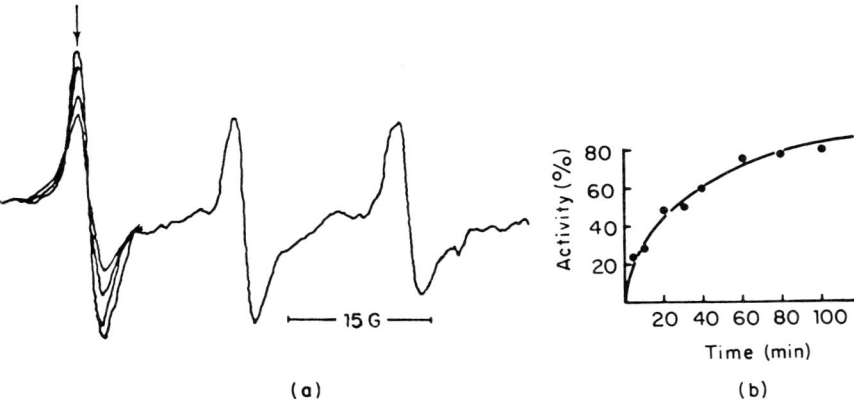

Fig. 30. Activation of spin-labeled chymotrypsinogen. (a) The 9.1-GHz ESR spectrum of chymotrypsinogen covalently labeled by spin label Ia, and the alteration of this spectrum in the presence of 1% w/w trypsin. The arrow indicates the decrease in peak height associated with the activation process. The other resonance lines change to a somewhat lesser degree. Conditions: [Ctgn-Ia] = 1 × 10^{-5} M, μ = 0.1 M (CaCl$_2$), pH 7.8, 0.01 M PO$_4^{2-}$ at 30°C. (b) The correspondence of the decrease in the indicated resonance peak height and the appearance of chymotryptic activity. The former is represented by the trace (———), which is a reversed experimental curve. Portions were removed from this sample and assayed for protease activity. The values are represented by the closed circles (●). (From Kosman, 1972. © 1972 Academic Press Inc. (London) Ltd.)

where $\Delta H(0)$ is the linewidth of the $m_I = 0$ transition at half height; $h(0, \pm 1)$ the line heights (peak to trough) of appropriate transitions; $B = \frac{4}{3}\pi[A_{zz} - A_{xx(yy)}]$ with A in megahertz (axial symmetry is assumed); and $\Delta\gamma = (\beta/\hbar)[g_{zz} - \frac{1}{2}(g_{xx} + g_{yy})]$. Two independent calculations of τ_c (in seconds) can be obtained from Eqs. (21) and (22). In theory, they should yield comparable values. That they may not is related to the fact that they respond differentially to changes in the hyperfine (the B term) and g tensors (the $\Delta\gamma$ term). Lack of correlation between these two calculations is most often caused by a breakthrough of the constraints, i.e., the system is undergoing anisotropic motion or slow motion or both. Another factor in the use of these equations to calculate τ_c values is the choice of spin-Hamiltonian parameters. Values for g and A do vary from nitroxide to nitroxide, and both exhibit some dependence on solvent polarity, H-bonding, etc. It is for these reasons that spin labels are so often used only as reporter groups rather than true probes of structure and molecular order.

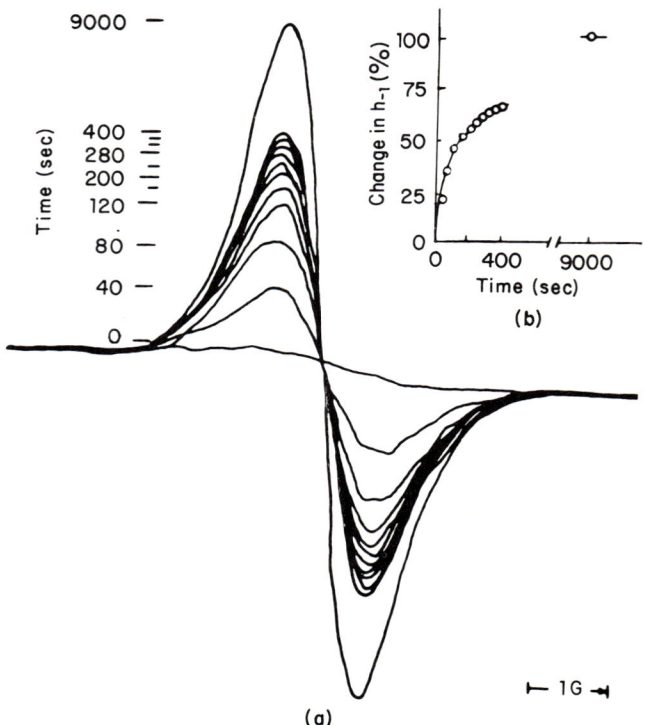

Fig. 31. Activation of spin-labeled pepsinogen. (a) The 9.1-GHz ESR spectrum of pepsinogen labeled with IIIc was followed at the low-field nitrogen hyperfine line ($M_I = +1$) at pH 2.38. Data in (b) represent the change in peak height versus time. This rate is *slower* than activation (see text). (From Twining *et al.*, 1981. © 1981 American Chemical Society.)

A related use of the spin label without regard to quantitation of the motional behavior is as a kinetic probe. By following the time dependence of a specific spectral change under defined conditions of concentration, rate constants for the related biological process can be determined. For example, in the zymogen-activation examples noted, the rate of change of peak height(s) yields the rate of structural change associated with activation. This may or may not equal the rate of activation (expressed as enzyme activity), depending on the temporal and structural relationships between the change in the motion of the label and the event which produces an active enzyme. In the chymotrypsinogen example, the ESR and activation events are controlled by the same rate-limiting steps, since the changes are experimentally synchronous (Fig. 30) (Kosman, 1972). In pepsinogen activation, on the other hand, activation precedes the spectral change, since the latter is apparently caused by the release of the activation peptide (to which the labels are linked) rather than the conformational change actually associated with activation per se (Twining et al., 1981). The parallel physiologically relevant experiments must be done along with the ESR ones to uncover these types of relationships.

Spin labels are, however, well suited to kinetic experiments. Being stable, they are unlike many biologically interesting paramagnetic species, which are relatively short-lived and have long since decayed within the time required to manually fill a quartz tube or flat cell, position it in the resonance cavity, tune the instrument, and record the spectrum. In fact, this is exactly how the zymogen-activation experiments were carried out. Even faster reactions can be monitored using a stopped-flow method even simpler than the freeze-quench technique outlined in Section IV.A.3. This involves the use of an aqueous (flat) quartz cell with built-in mixing jets. The dead volume of such a cell is $\sim 500\ \mu l$; that is the volume needed per flow to renew the solution in the cell. However, because of the nature of the cell, very fast flow rates are not possible; the dead time is limited to at best ~ 100 msec. The spectrometer is set at a field position chosen to be a maximum of signal intensity (for sensitivity) or extent of signal-amplitude change. This setting is determined by first preparing initial and final-state samples and comparing their respective ESR spectra. The reactant solutions can be driven by hand, piston, or pump, and the spectrometer can be triggered manually or electronically. Data can be displayed on the recorder (for $t_{1/2} \geq 0.5$ min limited by pen time constant, sample concentration, and noise) or collected by computer. An example of data acquired from direct-recorder tracings is in Fig. 32, which shows the pH dependence of the pseudofirst-order rate of labeling human hemoglobin A with ISL (Fig. 29). The indicated pK_a is probably that of the $\beta 93$ cysteine-SH groups, which are alkylated by the reagent.

As the motion of the nitroxide is restricted further, the label enters the intermediate and slow-motion regime, 3×10^{-9} sec $< \tau_c < 2 \times 10^{-7}$ sec. In the slow-motion limit, the nitroxide's spectrum is that of a rigid powder (Fig. 28). No direct calculation of τ_c from spectral parameters is possible. To extract motional parameters from such spectra simulation is necessary, since the positions and line shapes, in addition to the line heights and widths, depend on τ_c. Such simulations have provided empirical calibrations of spectra which provide an estimate of τ_c assuming isotropic motion. Such spectra can also be reproduced by slowing down the isotropic motion of a

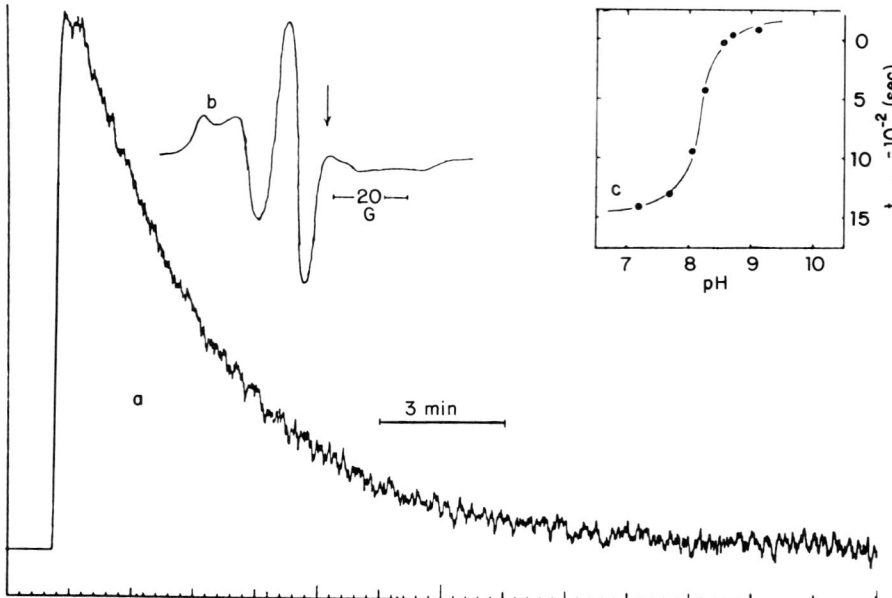

Fig. 32. The pH dependence of alkylation of β-chains of human hemoglobin A by spin-label IIa. (a) Time course of alkylation following decrease in signal amplitude of $M_I = -1$ transition of free label (3403 G at 9.5 GHz, compare with b). Conditions: [β chains] = 1.75 mM, [IIa] = 22 μM, 30 mM borate, pH = 9.30 at 24°C. (b) ESR spectrum of labeled HbA; arrow indicates position of H_o employed in A. (c) The pH dependence of $t_{1/2}$ values derived from pseudofirst-order tracings, as in a. Note that dependence exhibits a fair degree of cooperativity ($n \simeq 1.7$). (From unpublished data, D. J. Kosman).

label in solutions of increasing viscosity or decreasing temperature or both (Fig. 33) (Hsia and Piette, 1969), as indicated by the Stokes–Einstein relationship:

$$\tau_c = (4a^3/3kT)\eta. \qquad (23)$$

The former approach has yielded the following empirical equations, assuming an inherent linewidth of 3 G and a simple random-motional model:

$$\tau_c (5.4 \times 10^{-10})[1 - (A'_{zz}/A^R_{zz})]^{-1.36} \qquad (24)$$

and

$$\tau_c = a'_m [(\Delta H_m/\Delta H^R_m) - 1]^{b'm}, \qquad (25)$$

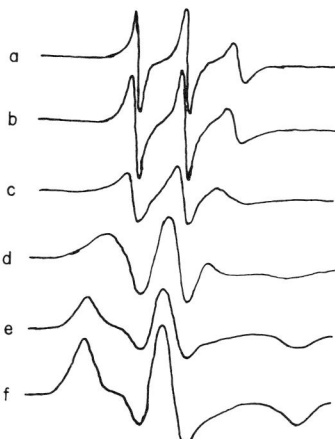

Fig. 33. τ_c calculated from the Stokes–Einstein equation, $r = 5$ Å: 1.5, 4.0, 6.9, 20.0, 76.6, and 115.0 nsec for a–f, respectively. Values for η (P) and T (K) are: curve a, 0.116 and 283, curve b, 0.367 and 293, curve c, 0.523 and 293, curve d, 1.58 and 283, curve e, 5.77 and 273, and curve f, 9.42 and 273. The nitroxide was 1-oxyl-2,2,6,6-tetramethyl-4-piperidinol at 9.5 GHz. (From Hsia and Piette, 1969.)

where $m = h$ (high-field) or l (low-field) line (compare with Fig. 34); R refers to the rigid limit (powder spectrum); A'_{zz} is the measured separation of the low- and high-field lines in the sample; and ΔH_m the measured linewidths of the respective spectral features:

$$a'_h = 2.12 \times 10^{-8}, \quad b'_h = -0.778;$$
$$a'_l = 1.15 \times 10^{-8}, \quad b'_l = -0.943.$$

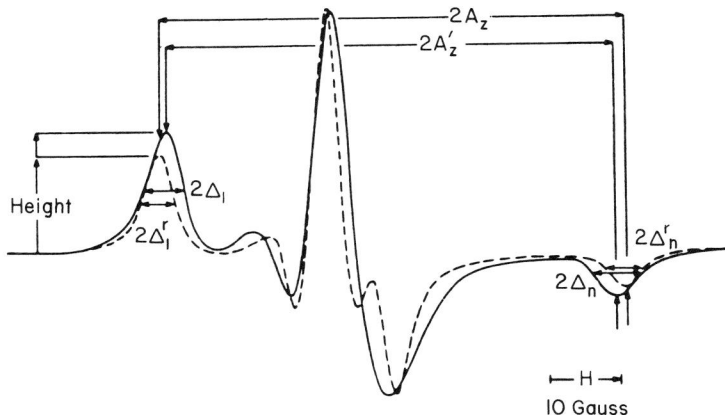

Fig. 34. Powder (rigid limit) spectrum (---) and slow-motional spectrum (——) indicating parameters used in Eqs. (24) and (25): $A'_{zz} = A'_z$; $A^R_{zz} = A_z$; $\Delta H m = \Delta_m$; $\Delta H^R_m = \Delta^r_m$, where $m = h$ or l. (From Marsh, 1981.)

Again, note the constraints placed on the use of these equations by the underlying assumptions. Furthermore, the choice of A_{zz}^R is critical. As mentioned here, A values are sensitive to environment and can vary independently of τ_c.

Despite these constraints, either or both A'_{zz} or ΔH_m can and have been used at least to monitor structural change if not to calculate (estimate) τ_c values. That is, the spin label in slow to intermediate motion can indicate conformation (change), but the experimental parameter used is separation or linewidth of the spectral low- and high-field lines rather than line heights.

One of the first attempts to use a spin label in the slow-motion regime as a probe of structure involved a series of spin-labeled 2,4-dinitrophenyl haptens of the generalized structure (Hsia and Piette, 1969), as shown here:

These haptens were used to probe the combining site of a rabbit anti-DNP IgG by comparing spectra of a given hapten–antibody complex to empirical spectra, as in Fig. 35. Figure 36 gives a sense of how the spectra respond to changes in τ_c (Likhtenshtein, 1976). The parameter ΔH in this plot is given by

$$\Delta H = (H_i^\tau - H_i^I)/(H_i^R - H_i^I), \qquad (26)$$

where the notation is as given for Eqs. (24) and (25), τ refers to a given sample, I to the isotropic spectrum, and H is the position in the magnetic-field gradient of the respective transitions. Based on such empirical comparisons, the τ_c value for each hapten was estimated and plotted versus the value of d (Fig. 35). As the data indicate, the label "emerges" from the

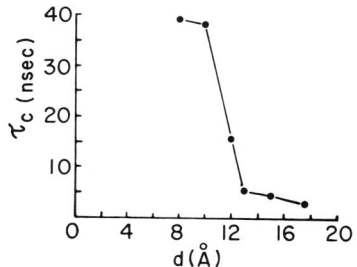

Fig. 35. Plot of τ_c versus d for DNP spin-labeled haptens bound to rabbit anti-DNP antibody; τ_c calculated from A_{zz} values by comparison with empirical data, as in Fig. 33. (From Hsia and Piette, 1969.)

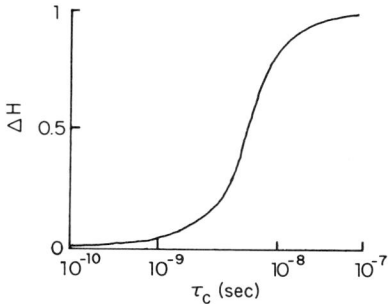

Fig. 36. Relationship between τ_c and the normalized separation of the low-field ($M_I = +1$) transition in rotational state, τ, and the fast-rotation limit. See Eq. (26). (From Likhtenshtein, 1976.)

intermediate-motion region ($\tau_c \simeq 4 \times 10^{-8}$ sec) when $d \simeq 10\text{--}12$ Å. At a separation of the combining hapten (the DNP group) and the bond linking the nitroxide to the spacer arm greater than 13 Å, the nitroxide ring exhibits fast rotation ($\tau_c \simeq 5 \times 10^{-9}$ sec). These data were interpreted to suggest that the depth of the antibody combining site was ~10 Å. More recent x-ray crystallographic results show the hapten combining site in a mouse myeloma antibody to be a cavity about 12 Å deep (Padlau et al., 1976).

A rather novel use of the nitroxide as an ESR reporter group is as an absolute pH indicator (Keana et al., 1982). The basis for this use is the sensitivity of A_N toward the microenvironment; factors which diminish the contribution of the resonance form $\overset{+}{N}-\bar{O}$ to the electronic structure of the nitroxide, will decrease A_N. This would occur with electron withdrawal or, similarly, by protonation at a site adjacent to the nitroxide nitrogen, as in reagents A and B, shown here:

Reagent A   Reagent B

These labels suffer from the rather limited pH range over which they respond. Also, as intravesicular pH probes they work best at the pH values at which they are charged. At higher pH values (as they ionize to neutral, conjugate-base species), the membrane becomes permeable to them. Further development incorporating permanent charges and other (higher) pK_a values should alleviate these limitations. This approach to measurement of intravesicular (and, perhaps intracellular) pH could rival the use of pH-dependent fluorescent and NMR probes.

The discussion and examples so far have focused on or assumed isotropic motion of the nitroxide. However, this motion need not be isotropic. Measurements of the angular amplitude (which specified the direction) and the rate of anisotropic motion have proven extremely valuable in the study of membrane structure and the interaction between membrane components.

Motion may be fast although anisotropic. The anisotropy of the motion of spin labels in oriented bilayers or membrane dispersions can be used to calculate an order parameter S for the label environment. This can be done most readily for fast motion. In brief, S is a measure of the angular amplitude of rotation of the nitroxide, i.e., the extent to which the principal tensorial axes overlap in the rotational-state population. When $S = 1$, the motion is purely anisotropic; there is no "wobble" in the molecular rotation. When $S = 0$, rotation is isotropic, i.e., the principal axes are reoriented several times in one "rotation," and the spectrum is isotropic as a result.

In a random dispersion of membrane bilayers (as vesicles or fragments) nitroxide motion depends on the order parameter, i.e., all label reorientation takes place within the bilayer. The spectra contain contributions from both $A_\parallel$ (A_{zz}) and $A_\perp$ ($A_{xx} \simeq A_{yy}$) as outer ($A'_\parallel$) and inner ($A'_\perp$) hyperfine splittings (Fig. 37). Although $A'_\parallel \simeq A_\parallel$ for $1 \geq S \geq 0.1$, $A'_\perp$ deviates significantly from $A_\perp$. These lines are sensitive to differential line broadening caused by rotational relaxation effects. By using $A'_\parallel \simeq A_\parallel$, we get

$$S = 3(A_\parallel - A_0)/[2(A_{zz} - A_{xx})], \qquad (27)$$

where A_0 is the isotropic splitting in the bilayer system. In addition, this value of S can be normalized by a polarity factor, as discussed here. Note that if motion is slow, $A'_\parallel$ will tend to be large anyway (see previous discussion) and thus lead to an overestimate of S. This is an important consideration in the evaluation of these types of spin-label data and explains why S is less easily determined for cases in which slow motion obtains.

In oriented thin films or bilayers, two orientations of the sample are possible with respect to the magnetic field H_0: H_0 normal ($\perp$) to the plane of the bilayers, and H_0 in this plane. $A_\parallel$ and $A_\perp$ refer to the measured hyperfine splittings in these two orientations, respectively. If the Z-axis tensor of the nitroxide is perpendicular to the lipid bilayer (parallel to the long molecular axis of the lipid chains, then

$$S_{33} \text{ (or } S_{zz}) = \frac{(A_\parallel - A_\perp)}{(A_{zz} - A_{xx})} \frac{A_0 \text{ (crystal)}}{A_0 \text{ (bilayer)}}. \qquad (28)$$

If this axis is not coincident with the long axis of the chains (not normal to the bilayer), then

$$S_{\text{mol}} = 2S_{33}/(3 \cos^2 \alpha - 1), \qquad (29)$$

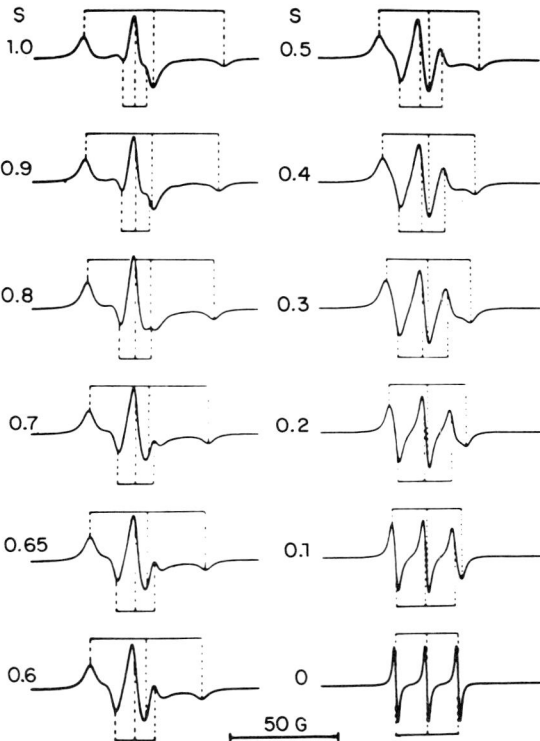

Fig. 37. Simulated ESR spectra of randomly oriented lipid spin labels with different values of S_{mol}. Dotted lines indicate positions of absorption features $A_{\parallel}$ (equivalent to A'_z in Fig. 34) and $A_{\perp}$ used in the simulations. (From Griffith and Yost, 1976.)

where α is the angle between these tensorial and molecular axes. The term A_0 (crystal)/A_0 (bilayer), which includes the isotropic hyperfine splittings of the label in the two environments indicated, corrects for polarity differences between membrane systems. S_{33} and S_{mol} are used to calculate order parameters for fatty-acid and phospholipid spin labels VI and VII.

$$H_3C-(CH_2)_m \overset{\overset{\displaystyle O\diagdown_C\diagup N\diagdown O}{\big|}}{} (CH_2)_n-CO_2H$$

VI (m, n)

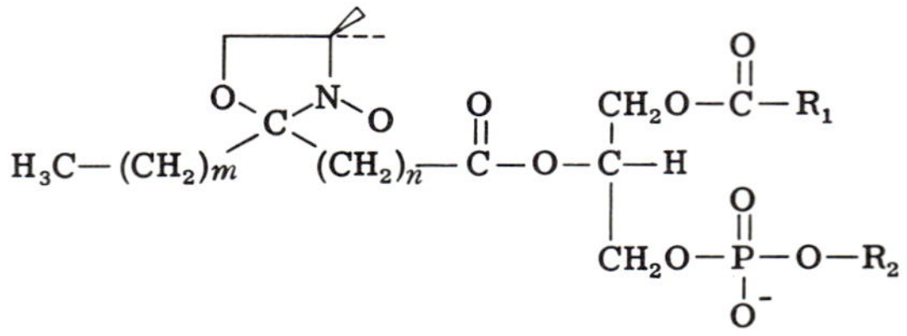

166   DANIEL J. KOSMAN

VII (m, n)

If another tensorial axis coincides with the long molecular axis, e.g., the nitroxide y axis, then motion averages A_{zz} and A_{xx} (rather than A_{xx} and A_{yy}) in the plane perpendicular to the long axis (parallel to the bilayer). This is characteristic of the steroid spin labels, as shown here:

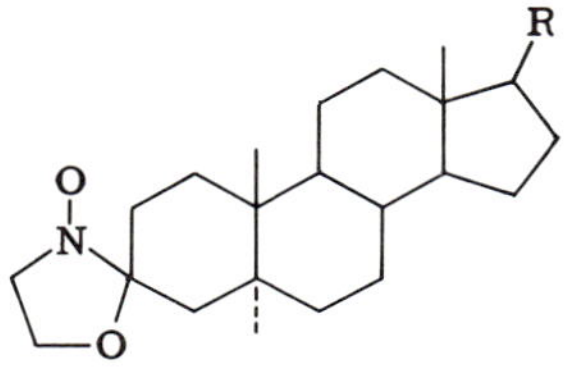

R = (a)  Cholestane SL
(b) —OH Androsterol SL
(c) —H Androstane SL

The order parameter for such species is

$$S_{mol} = \frac{A_\| - A_\perp}{A_{yy} - \frac{1}{2}(A_{zz} + A_{xx})} \frac{A_0 \text{ (crystal)}}{A_0 \text{ (bilayer)}}. \tag{30}$$

Numerically, these two order parameters are

$$S_{mol} = \frac{A_\| - A_\perp}{A_0 \text{ (bilayer)}} \times 0.5407 \tag{31}$$

and

$$S_{mol} = \frac{A_\| - A_\perp}{A_0 \text{ (bilayer)}} \times (-1.131). \tag{32}$$

Note that for steroid labels [Eq. (30)], $(A_\| - A_\perp) < 0$ since the measured hyperfine coupling will be larger when H_0 is parallel to the bilayer (perpendicular to the long molecular lipid axis). In this orientation, H_0 is coincident with the nitroxide Z axis. Order parameters have been used to monitor the effects of temperature (membrane fluidity and phase), lipid and steroid composition, and protein content. For example, in bilayers of egg phosphatidylcholine, S_{mol} varies from 0.55 to 0.80 (using cholestane spin label VIIIa) as the cholesterol content is increased from 0 to 50 mol % (Marsh, 1981).

There is some controversy about the reliability of order parameters based on spin labels and their use to describe membrane structure (Taylor and Smith, 1981). Some of the problems have been mentioned, namely, the overestimate of S caused by slow motion and the importance of the correct choice of A_0 (bilayer). Both of these problems are most common to the fatty-acid and phospholipid spin labels VI and VII, particularly when S_{mol} values for labels where $m > n$ are compared with labels where $m < n$. As the label moves down the chain deeper into the lipid bilayer, the polarity decreases as does A_0 (Table VIII) (Fretten et al., 1980). Structurally, there are an increasing number of rotational states of the chain at methylene carbons distant from the fixed-head group; thus, the label experiences less motional constraint.

An additional factor is the perturbation of the bilayer by the nitroxide (doxyl ring) itself. A static tilt in the nitroxide orientation is induced by the close parking and rigidity of the bilayer near the polar-head groups. This is alleviated further into the bilayer. Thus, there is a steady decrease in the ESR-derived order parameter in the region of the top of the chain which is not deduced from ^{2}H-NMR measurements, for example. This tilt can be estimated and corrected for (Gaffney and McConnell, 1974). Nonetheless, the most reliable membrane spin-labeling experiments are those employing the steroid labels or those fatty-acid and phospholipid derivatives for which

Table VIII Isotropic Hyperfine Constants (A_0) for Stearic Acid Spin Labels in Chromaffin Granule Membranes: Effect of Decreasing Polarity[a]

Spin label VI (m,n)	$A_0 (G) = \frac{1}{3}(A_\parallel + 2A_\perp)$	
	Membrane	Aqueous
(13,2)	15.2	15.6
(12,3)	15.1	15.7
(9,6)	15.0	15.7
(7,8)	14.4	—
(5,10)	14.0	—
(1,14)	14.0	15.7

[a] From Fretten et al. (1980).

$n > m$, i.e., those that have the doxyl group in the lower portion of the chain(s).

This perturbation of a membrane or bilayer by the nitroxide is indicated by the effects on the temperature dependence of the phase transition T_m. For example, T_m values for spin-labeled analogs of 1-palmitoyl-2-stearoyl-phosphatidyl choline (PSPC), VII (12,3) and (1,14), where R_1 = palmitoyl and R_2 = choline, have been compared by differential scanning calorimetry with the values for normal phospholipids (Chen et al., 1982). The two spin-labeled derivatives exhibited T_m values in bilayers 18 and 16°C, respectively, which were below that of PSPC but essentially the same values as a branched-chain derivative, 16-methyl-PSPC. As indicated, the effect of the doxyl ring, which is likened to the introduction of branching or a double bond, is larger the closer the nitroxide is to the phospholipid head group.

On the other hand, the disadvantage in having the reporter group deep within the membrane is that it may not provide as relevant structural information. This can be inferred from Fig. 38, which shows the ESR spectra of doxylstearic acids (VI, $n = 5$, 12, and 16, respectively) incorporated into rabbit small-intestinal brush-border membrane vesicles (Fig. 38a) and a dispersion of lipids extracted from this membrane (Fig. 38b) (Hauser et al., 1982). Only the 5- and 12-doxylstearic acids exhibit anisotropic motion, which can be analyzed by Eq. (29) to yield an order parameter S_{mol}. The 16-doxyl derivative exhibits rapid ($\tau_c \simeq 10^{-9}$ sec), essentially isotropic motion and thus may not be as sensitive to physiologically relevent membrane perturbations. Clearly, in the design of any experiment employing a spin-

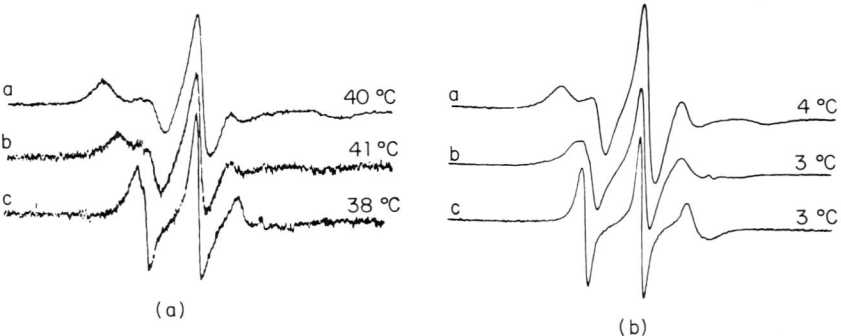

Fig. 38. X-band ESR spectra of doxylstearic acids (VI, $n = 5$, 12, and 12: spectra a, b, and c, respectively) in rabbit small-intestinal brush-border membrane vesicles (set a) and a dispersion of lipids extracted from this membrane (set b). (From Hauser et al., 1982. © 1982 American Chemical Society.)

labeled analog, the choice of analog must be dictated by a specific objective, and the analysis of the results must take into account the effect the label chosen may have on the data obtained in relation to that objective.

3. ESR Double Labeling

A complement to the other types of ESR experiments discussed in this chapter and to the paramagnetically enhanced nuclear relaxation described in Chapter 1, is the general technique of ESR double labeling. That is, two paramagnetic centers are brought together within a biomolecule. The resulting electron–electron interaction is assessed by ESR and used to evaluate the parameters r, $\mathbf{r}$, and ν_c, which are the distance and the radius vector relating the two spins and the frequency of collision at that distance. Obviously, this technique can provide structural information directly; since it can also reflect translational motion (as implied by the collision parameter ν_c), movement between specific sites can also be assessed. (If the spin centers are fixed with respect to translation, ν_c is irrelevant; ν_c^{-1} can be taken as the lifetime of the biradical complex at the exchange distance r.) These effects, of course, are in addition to those related to the anisotropy, the amplitude, and the rate of overall spin-center motion.

There are three general types of double-labeling schemes: biradicals, in which the labels are part of the same reagent molecule; double labeling per se, in which two independent paramagnetic centers are constrained to a specific region of a macromolecular system; and the electron–electron interaction between monoradicals of a specific type brought together by translation in a fluid medium on the time scale of the ESR measurement.

In parallel with the nature of S, I interactions [Eq. (7)], electron–electron coupling can involve both dipolar and "contact" components. The former is related to the zero-field splitting of a triplet state ($M_s = 1$, as it is for a diradical), i.e.,

$$D = 3g^2\beta^2/2r^3. \tag{33}$$

Consequently, if the spectrum can be resolved into (and assigned to) the principal components of this tensor, or if the value E is taken to be zero, D can be estimated from the extreme features of the spectrum (Fig. 39) (Luckhurst, 1976). This assumption is implicit in Eq. (33), and borne out by experiment (Table IX). A more quantitative approach is to simulate the spectrum using appropriate choices for the hyperfine, g, and D tensors, as well as residual (inherent) linewidths and τ_c. This has recently been described for the diradical, tetraphenyl(bis)verdazyl in oriented bilayers of dimyristoyl phosphatidylcholine using a Fortran simulation, TRISUPRO (Meier *et al.*, 1982).

The "contact" interaction is better described as electron–electron exchange, that is, it is the energy of exchange, $JS_1 \cdot S_2$, where J determines the singlet–triplet separation of the biradical. Briefly, when $J \gg A_0$, the spins are completely correlated, and only triplet ($\Delta M_I = \pm 1$) transitions are allowed. A dinitroxide, for example, would exhibit a five-line spectrum (Fig. 40). When $J \ll A_0$, there is no spin exchange, and the diradical exhibits a normal three-line spectrum (Fig. 40) associated with singlet transitions. Note that dipolar effects are not considered here. The spectra are considerably more complex when $J \simeq A$, since both singlet and triplet transitions are possible. A further complication is the linewidth effects associated with slow and intermediate motion, which will tend to obscure the additional

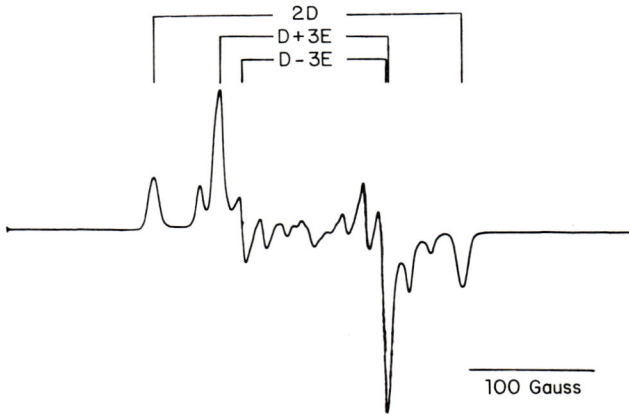

Fig. 39. X-band ESR spectrum of the biradical XI (Table IX) in a toluene glass at 77 K with the principal components of the zero-field-splitting tensor, D and E, indicated. ^{14}N hyperfine is observed on only the $D + 3E$ pair; A_n in the other two directions is small; E is also very small (compare with Table IX). (From Luckhurst, 1976.)

2. ELECTRON SPIN RESONANCE

Table IX Zero-Field Splitting Parameters for Some Nitroxide Biradicals

Radical	D (MHz)	E (MHz)	Reference
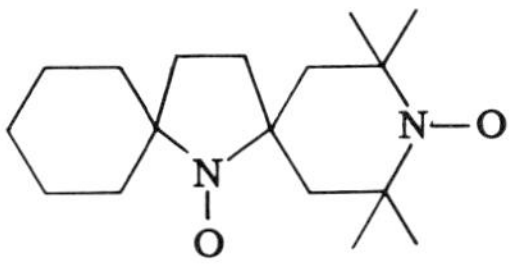 IX	644	50	Rassat (1972, private communication) Luckhurst (1976)
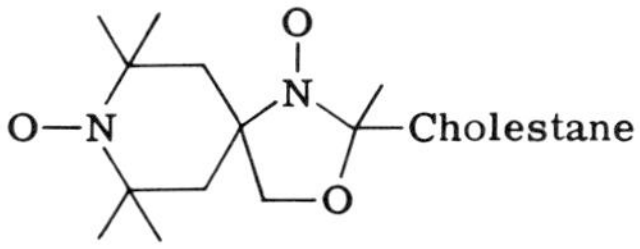 X	636	—	Keana and Dinerstein (1971)
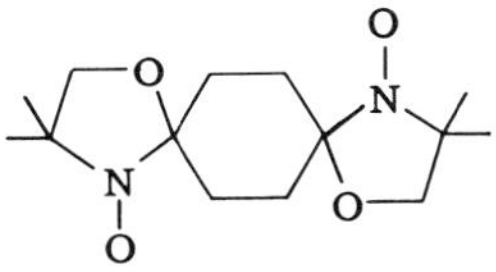 XI	336	9	Luckhurst and Poupko (1974)

splittings. In any event, exchange coupling has not been extensively used as a probe of biomolecular structure. In most cases reported, distance estimates have been obtained and spectra simulated assuming J to be small.

The addition of a nitroxide spin label to a protein containing a paramagnetic transition metal has been used as a probe of exogenous ligand coordination. Ferric, Fe(II) (high spin), Co(II) and Co(III), and Mn(II) have served as endogenous paramagnetic centers. A variety of covalent and noncovalent nitroxide probes have been employed, and, in general, the nitroxide line heights are measured to estimate the Me → nitroxide distance, based on a strictly dipolar interaction. More recently, in a study of the binding of a metyrapone analog, RNO (as shown here),

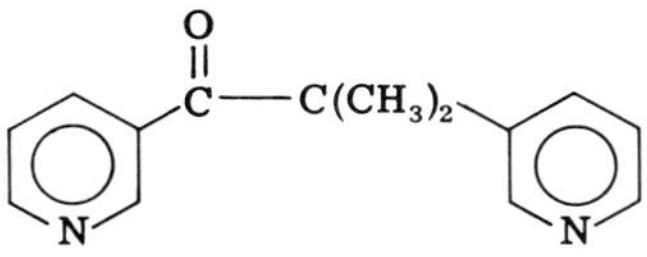

Metyrapone 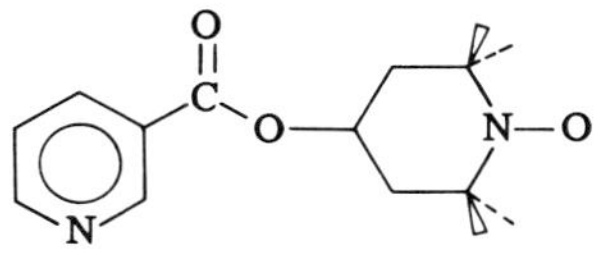   RNO

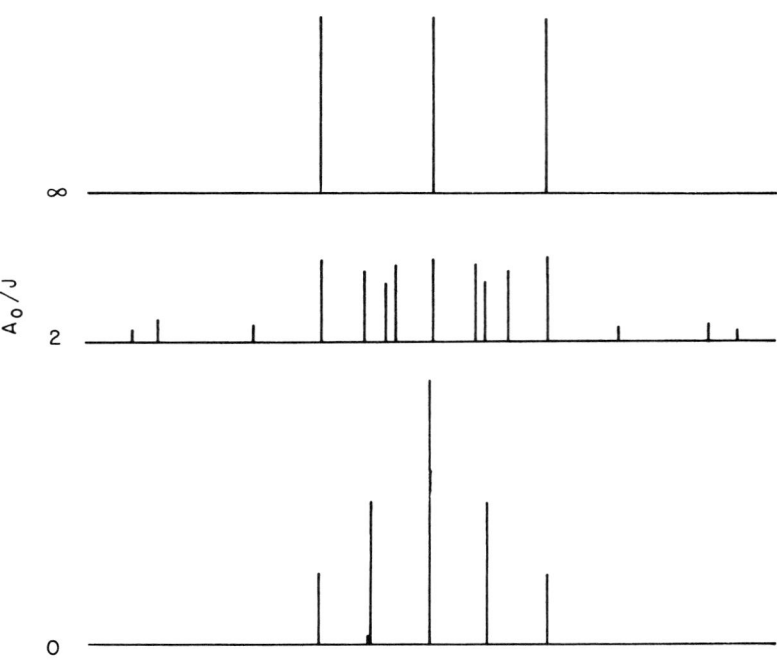

Fig. 40. Theoretical stick diagrams for a nitroxide biradical at various values of A/J. (From Luckhurst, 1976.)

to cytochrome P-450, a Fortran IV simulation program was used to derive "best fit" values for the distance between the Fe(III) and the nitroxide and the location of the nitroxide (the electron as a point dipole) in the g tensor coordinate frame of the Fe(III) (Mock *et al.*, 1982). The result is shown here:

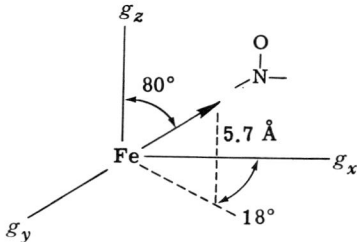

Again, the exchange integral was set equal to zero. This assumption was supported by the similarities in the linewidths of the Fe(III) transitions with either metyrapone or the spin label bound, i.e., there was no apparent exchange broadening (Fig. 41). If $r \simeq 5.7$ Å as indicated, the p_z orbital of the

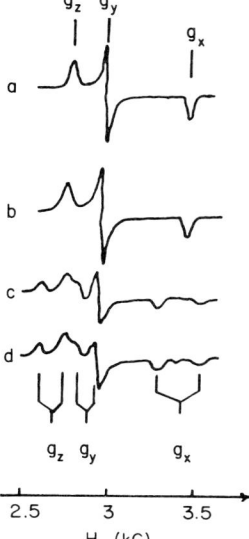

Fig. 41. ESR spectrum of *P. putida* cytochrome *P*-450 cam at 103 K and 9.16 GHz: Curve a, simulated spectrum of camphor-free cytochrome; curve b, metyrapone-bound cytochrome; curve c, simulated spectrum of RNO-bound cytochrome; curve d, simulated spectrum of 9:1 mixture of b and c, corresponds to experimental spectrum of RNO-bound cytochrome. (From Mock *et al.*, 1982.)

nitroxide must be orthogonal to the unpaired electron-spin density centered in the Fe(III).

The third type of double-labeling experiment is conceptually somewhat different from the first two in that it specifically addresses the question of the frequency of electron spin–spin encounter, assuming such encounters to be at a van der Waals radius. The electron–electron exchange is a measure of the collision frequency of two labels and thus the rate of lateral diffusion of the labels in a mobile phase, as in a lipid bilayer. Spin labeling has been used to assess the exchange and segregation of membrane components in this way.

The exchange frequency is given by

$$W_{ex} = J/h, \tag{34}$$

which can be analyzed further in a two-site model for the two extreme cases, $W_{ex} \ll A_n (J \ll A_n)$ and $W_{ex} \gg A_n (J \gg A_n)$, for which the linewidths are given, respectively, as $\Delta v = 2W_{ex}$ and $\Delta v = A_n^2/W_{ex}$.

Thus, in the absence of contributions from *dipolar* broadening in these limits, W_{ex} can be measured from the linewidths. This is usually possible if the label is in the fast-motion region, since dipolar contributions (which are anisotropic) are averaged out (to zero). In the region of intermediate exchange ($W_{ex} \simeq A_n$), dipolar broadening and exchange have opposite effects on the spectrum, the former moving the sidebands outwards (compare with

Fig. 39), whereas exchange moves them inwards (Fig. 40). Thus if dipolar effects are dominant, increasing the concentration of the label should push the outer spectral features apart, for example.

The frequency of collision is given by $v_{coll} = 3W_{ex}$. Two models have been developed to relate the label-concentration dependence of the spectral effects (i.e., linewidth analysis) to the diffusion constant of the label in the membrane. One is based on homogeneous two-dimensional diffusion [Eq. (35)] (Träuble and Sackmann, 1972), the other on a nearest-neighbor exchange in a hexagonal lattice [Eq. (36)] (Devaux et al., 1973):

$$v_{coll} = [8dc/(F \cdot \lambda)] \times D_{diff} \times c \tag{35}$$

and

$$v_{coll} = (12/\lambda^2) \times D_{diff} \times c, \tag{36}$$

where F is the area per lipid molecule; λ in Eq. (35) is the length of one "jump," whereas in Eq. (36) it is the lattice constant a. Equations (35) and (36) are related, since $F = \sqrt{\frac{3}{2}} \cdot a^2$; λ^2 in Eq. (36) has been replaced by $12\sqrt{\frac{3}{2}}F$, which is numerically similar to $F \cdot \lambda$ [Eq. (35)]. Using these equations, diffusion coefficients for several steroid and phospholipid spin labels in a variety of membrane systems were calculated. Two examples are given here:

$D_{diff} \simeq 10^{-8}$ cm²/sec in dipalmitoyl-phosphatidylcholine bilayers
(Träuble and Sackmann, 1972)

$D_{diff} \simeq 6 \times 10^{-8}$ cm²/sec in sarcoplasmic reticulum membrane vesicles
(Scandella et al., 1972)

4. Spin Trapping

As discussed earlier (Section IV.A.3), paramagnetic transients are often too short-lived to detect using conventional mixing techniques. Being short-lived, they are rarely present at appreciable concentrations. Although recent reaction techniques such as freeze quenching have been useful in a variety of situations, even these fail if the concentrations of intermediates are below the detectable limits of ESR spectrometers.

When faced with this type of problem, the kineticist commonly attempts to trap such intermediates as are indicated by the overall reaction. This approach has been adopted for the ESR experiment, as well, albeit with relatively limited success. This section will discuss briefly the technique of spin trapping.

Spin-trapping agents are diamagnetic reagents which react with a free radical to generate a relatively stable nitroxide. These agents are essentially all nitroso or nitrone compounds of the general structures $R'N=O$ and $R'R''=\overset{+}{N}-O^-$, respectively. Their reactions with a free radical are

$$R^{\cdot} + R'N=O \longrightarrow R'-\underset{\underset{R}{|}}{N}-\overset{\cdot}{O}$$

and

$$R^{\cdot} + R'CH=\underset{\underset{R''}{|}}{\overset{+}{N}}-O^- \longrightarrow R'-\underset{\underset{R''}{|}}{CH}-\overset{\overset{R}{|}}{N}-\overset{\cdot}{O}$$

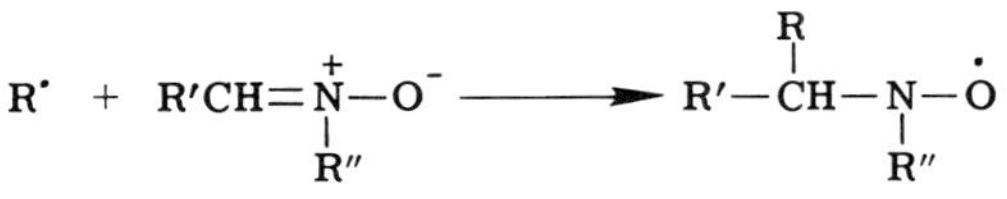

The most commonly used spin traps (they are commercially available, see Appendix A) are

$(CH_3)_3C-N=O$

2-Methyl-2-nitrosopropane
(nitro-t-butane, NtB)

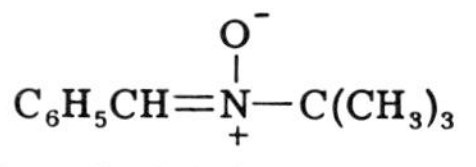

$C_6H_5CH=\overset{\overset{O^-}{|}}{\underset{+}{N}}-C(CH_3)_3$

Phenyl N-t-butyl nitrone
(PBN)

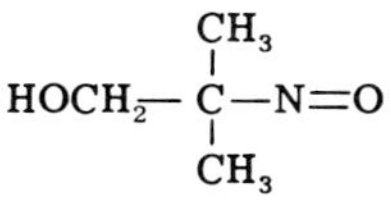

$HOCH_2-\underset{\underset{CH_3}{|}}{\overset{\overset{CH_3}{|}}{C}}-N=O$

2-Methyl-2-nitrosopropanol
(hydroxynitroso-t-butane
HONtB)

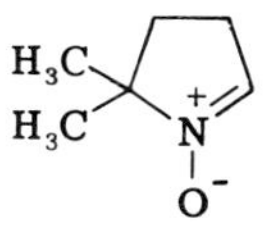

5,5-Dimethyl-1-1-pyrrolidine-
N-oxide (DMPO)

The use of the spin trap in the study of biological systems has largely focused on the detection of the oxygen-centered radicals generated by the odd-electron intermediates of O_2 reduction, the superoxide and hydroxyl radicals, O_2^- (HOO·) and HO·. As with the use of spin traps generally, assignment of a paramagnetic signal to a particular adduct depends on the interpretation of the ESR spectrum of the adduct. The most readily distinguishable characteristic of the adduct (nitroxide) spectra is the presence of SHFS owing to the ^{1}H in the radical portion of the adduct. (In principle, ^{13}C or ^{17}O could be employed as well to help assignment of carbon- and oxygen-centered radicals, respectively.) These couplings are generally small, since the ^{1}H atoms in R^1 are at some distance from the unpaired electron.

Values for the spin-Hamiltonian parameters for O_2^- (HOO·) and HO· adducts with PBN and DMPO are given in Table X (Harbour et al., 1974). The adducts are not stable. They disproportionate readily and are susceptible to redox quenching. These reactions are generally pH dependent. For example, the $t_{1/2}$ for the O_2^-—DMPO adduct is 80 sec at pH 6 and ~ 35 sec at pH 8; yes, that's right, seconds (Buettner and Oberley, 1978). Thus, although ESR-detectable adducts have been demonstrated during light-induced chloroplast reduction of O_2 (Harbour and Bolton, 1975), microsomal turnover of nitrophenyl compounds (Sealy et al., 1978), O_2^- generation by the xanthine–xanthine oxidase reaction (Buetner and Oberley, 1978), and by the Fenton chemistry postulated to generate HO· when Fe(II) bleomycin is oxidized by O_2 (Oberley and Buettner, 1979), in all of these cases the adducts may not be substrate dependent. Irradiation of DMPO generates an oxazirane (which can be formed from reaction with H_2O_2 also) (Bonnet et al., 1959); this can undergo ring opening and oxidation, producing a putative hydroxyl spin adduct (Finkelstein et al., 1980; Tero-Kubata et al., 1982), as shown here:

$$\underset{O^-}{\underset{|}{N^+}}\diagup\!\!\!\diagdown \xrightarrow[\text{or } h\nu]{H_2O_2} \underset{O}{\underset{|}{N}}\diagup\!\!\!\diagdown \xrightarrow{H_2O} \underset{OH}{\underset{|}{N}}\diagup\!\!\!\diagdown OH^- \xrightarrow[-H^+]{-e^-} \underset{O^-}{\underset{|}{N^+}}\diagup\!\!\!\diagdown OH$$

Additionally, in phosphate buffer Fe^{2+} catalyzes the apparent decomposition of PBN with the subsequent generation of a benzoyl spin adduct (Tero-kubata et al., 1982):

$$\begin{array}{c} \text{H} \quad \text{O} \\ | \quad\; | \\ C_6H_5-C-N-C(CH_3)_3 \\ | \\ C_6H_5-C=O \end{array}$$

The g and $A_{N,H}$ values of this radical (2.0055, 16.0, and 4.35) are close enough to that of the HO· spin adduct (Table X) to suggest that spin-trapping experiments involving Fe^{2+}-dependent redox chemistry may have been misinterpretated (Tero-Kubata et al., 1982).

On the other hand, the sulfur trioxide radical anion SO_3^- has been cleanly

Table X A and g Values for Spin-Trap Adducts[a]

	Trapping agent						
	PBN			DMPO			
Radical	g	A_N[b]	A_H	g	A_N	A_H	A_H^{γ}[c]
O_2^- ($HO_2\cdot$)	2.0057	14.8	2.75	2.0061	14.3	11.7	1.25
$HO\cdot$	2.0057	15.3	2.75	2.0060	15.3	15.3	0.61, 0.25

[a] Adapted from Harbour et al. (1974).
[b] Values of A in gauss.
[c] γ protons in pyrrolidine nitroxide; ^1H on $-O_2\cdot$ and $-O\cdot$ in fast exchange with solvent contribute to linewidths only.

trapped by DMPO (Mottley et al., 1982). This radical apparently is formed in the prostaglandin synthase-dependent (bi)sulfate oxidation using aerobic arachidonic acid or 15-hydroperoxyarachidonic acid, or H_2O_2 directly as oxidant. The horseradish peroxidase–H_2O_2-dependent reaction was used as a model reaction. By correlating O_2 uptake with the presence or absence of DMPO, inhibition of the $SO_3^{\div}$ production with indomethacin and other controls, the unique ESR spectrum observed was clearly not a "false positive." This spectrum was identical to that of the SO_3^- · adduct formed photochemically, was well resolved, and is distinct from those associated with the side reactions noted ($A_H = 16.06$, $A_N = 14.7$ G, $g = 2.0056$; compare with the values in Table X). This work is a good model for anyone contemplating using the spin-trapping technique.

In summary, unless used with care, the spin-trapping technique is susceptible to false positives. Furthermore, the failure to detect a spin adduct is ambiguous; as with any negative result, no definitive conclusions can be reached about the possible intermediacy of a free-radical species. Thus, although it is a technique that may be useful in specific, well-defined situations, spin trapping has not contributed significantly to the elucidation of biological structure or function.

D. Single-Crystal ESR (Chien and Dickinson, 1981)

This section is not concerned with diamagnetic host crystals doped with a paramagnetic impurity. What has proven informative to the biochemist is the analysis of ESR spectra of single crystals of biological macromolecules (primarily proteins) which contain paramagnetic centers. (Depending on one's perspective, this might be considered a paramagnetically doped crystal!) Although other ESR techniques have been employed in the study of such crystals (see Section V.B), only X-band continuous-wave spectra will be discussed here.

The advantage of obtaining ESR spectra from single crystals is that a much more complete set of structural data can be obtained. Certainly the most productive use of single-crystal ESR has been in the study of metalloproteins which possess paramagnetic metal centers. In such systems, the relationships between the crystallographic axes and g and A tensors can be determined, thus providing real insight into the metal-ligand (coordination) field within the protein matrix.

Two studies illustrate the utility of single-crystal studies. Fee and his coworkers have determined the values for the principal magnetic tensors in bovine superoxide dismutase (SOD), a Cu_2Zn_2 protein in which Cu—Zn pairs are located in identical subunits, $M_r \simeq 16$ K (Lieberman et al., 1982). The Cu—Zn binuclear pairs are bridged by a protein imidazole. Whereas the Cu(II) sites in the native protein exhibit a rather rhombic g tensor, the CN^- complex is axial with a significant alteration in the A tensors as well (Table XI). Although the crystal structure of native enzyme has been determined to 2 Å, that of the CN complex has not, and thus the alterations in the ligand field indicated by the changes in the g and A tensors have been of some interest with respect to the coordination chemistry of the catalytic Cu(II).

The rhombic distortion of the Cu—Imidazole$_4$ chelate at the Cu(II) sites in SOD involves the movement of one ligand N above its original position in the square-planar conformation. This motion tips all three g tensors so that $g_{x(or\,y)}$ will point between the square-planar and rhombic positions of that N while the other tensor $g_{y(or\,x)}$ will tip toward the N trans to the distortion; g_z will remain roughly perpendicular to the original plane. The single-crystal ESR analysis suggests that the correct assignment in native SOD is

i.e., g_y is "in between" the new and old positions of the $N_{His\,46}$ whose rhombic distortion has moved it out of the plane perpendicular to this page. $N_{His\,61}$ is the bridging ligand.

In the CN complex, the g_z tensor is rotated either 30 or 60° from its position in the native protein. This uncertainty arises from the lack of crystallographic data on the complex and the fact that SOD crystallizes in a

Table XI Principal Magnetic Tensor Values of ^{63}Cu—Zn Superoxide Dismutase[a]

Tensor	Native	CN$^-$ complex
g_x	2.03 ± 0.01	2.05 ± 0.01
g_y	2.09	2.05
g_z	2.26	2.205
A_x (MHz)	156.9 ± 10	70.1 ± 10
A_y	103.7 ± 10	70.1 ± 10
A_z	426 ± 3	6.9 ± 3

[a] From Lieberman et al. (1982).

unit cell containing four Cu atoms (two SOD dimers) having four asymmetric (crystallographically inequivalent) units. Nonetheless, the data show that the axial displacement involves the realignment of this tensor. The new orientation results in the g_z tensor directed either toward N$_{His\,118}$ or between N$_{His\,44}$ and N$_{His\,118}$. In the former conformation, N$_{His\,118}$ would be considered an axial (apical) ligand; in the latter, either N$_{His\,44}$ or N$_{His\,118}$ would not be a ligand at all. That is, through either a realignment of the principal axes or ligand displacement, one of the imidazoles is removed from equatorial coordination to the Cu(II), as shown here:

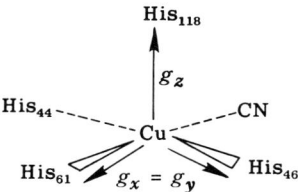

Since the bridging imidazole (His 61) remains intact, it cannot be the displaced ligand. The loss of one coordinating N as indicated by these data is confirmed by other continuous-wave ESR, spin-echo, and ENDOR experiments (discussed later). Recent refinement of the x-ray structural data indicate a 37° rotation of g_z upon CN$^-$ coordination, resulting in the effective replacement of His 44 (not His 118) as an equatorial ligand (Tainer et al., 1983).

The care with which this study was carried out deserves comment. The crystal was placed in a capillary which was mounted on a goniometer. Fifteen-degree precession photographs were taken to determine the orientation of the crystal relative to the goniometer, and these data were fed into a computer which directed the magnet rotation setting. For the ESR experiments, the crystal was transferred to the ESR goniometer system, which was mated to the waveguide terminating in a "home-built" quartz, gold-lined, rectangular cavity. Spectral simulations were performed with similar care employing second-order corrections on the A tensors and line-broadening effects of the nuclear Zeeman fields. Care was also taken to insure the correct assignment of crystallographic, noncrystallographic ("local"), and principal tensorial axes.

The other recent report concerned a study of single crystals of Co(II)-substituted myoglobins (Mb) (Hori *et al.*, 1982). The Fe(III)—Mb crystals grown by Kendrew provided the first crystals on which biological single-crystal ESR spectra were obtained and analyzed. As noted in section IV.C.1, Co(II) has been a useful spin probe, replacing diamagnetic metals in metalloproteins, e.g., Co(II) replaces high-spin Fe(II) (spin even, $S = 2$) in the porphyrin ring of Mb. Both the Co(II) and Fe(III) protein crystallizes in the $P2_1$ space group as monoclinic crystals.

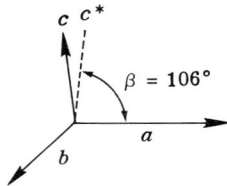

There are two Mb molecules in the asymmetric unit related by a twofold symmetry axis. The symmetry of the crystallographic axes has been determined, as has the local symmetry which relates the protein molecules to each other in the asymmetric unit. The question the ESR spectroscopist poses is: What relationships do the principal magnetic tensors of the paramagnetic centers have with the crystallographic and local symmetry axes. In answering this question, the spectroscopist can orient these tensors relative to the metal center itself.

The Mb crystal has well-defined faces; in particular, the $a-b$ face can be readily identified. Mounting the crystal in a cylindrical cavity with this face horizontal and rotating the cavity in the magnetic field (or the field around the cavity) sweeps the magnetic-field tensor in the $a-b$ plane. If the spectrum is recorded at incremental values of H_0 with respect to the a axis, *both* orientations of the protein, and thus the heme (Fe or Co), in the symmetric unit will be observed. A plot of either g^2 or A^2 derived from these spectra versus this angle yields two curves (Fig. 42).

There are two important elements in such curves. First, they can indicate whether one (or more) of the principal tensors is or is not in the crystallographic plane being sampled, in this case, $a-b$. For example, g_{xx} for oxyCoMb

Table XII g and A^{Co} Tensors and Their Directions in oxyCoMb[a]

g^b	Angle (°) to			A^{Co} (G)	Angle (°) to		
	a	b	c^a		a	b	c^a
$g_x = 2.056$	56	48	60	$A_x = 18.1$	85	50	40
$g_y = 2.011$	66	−42	58	$A_y = 11.1$	68	46	−52
$g_z = 2.003$	43	−85	−47	$A_z = 7.3$	23	−70	79

[a] Mb reconstituted with Co protoporphyrin. Adapted from Hori et al. (1982).
[b] These g tensors are not centered on the Co since ~90% of the spin density is on the dioxygen. The x-y-z notation is used here arbitrarily.

is found to be 2.056 (Table XII). The value of g_x^2, 4.227, is not attained when H_0 is in a-b, nor in b-c^* or a-c^* either (Fig. 42). Thus, g_x does not lie in any of the crystallographic planes, i.e., it cannot be coincident with any of the crystallographic axes. Second, there are apparent in Fig. 42 two angles at which the spectra became coincident, that is, angles of H_0 relative to a (a-b) or b (b-c^*) when the interaction between H_0 and the two heme groups in the asymmetric unit is symmetrical. This occurs when H_0 is either perpendicular or parallel to the twofold axis which relates these two centers. Thus the orientation of this axis with respect to the crystallographic axes can be determined by ESR. The g^2 variation in the a-c^* plane is unique here because there is only one spectrally distinct metal. This indicates that the a-c^* plane is a plane of symmetry relating the two hemes, that is, they project on this plane equally.

The curves in Fig. 42 are associated with

$$g_{obs}^2 = g_{zz}^2 \cos^2\theta + g_{xx}^2 \sin^2\theta \cos^2\phi + g_{yy}^2 \sin^2\theta \sin^2\phi, \quad (37)$$

where θ and ϕ are assigned as

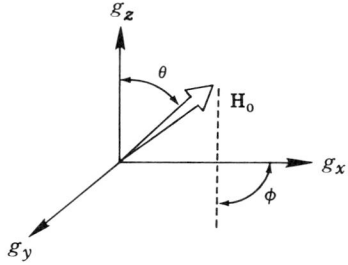

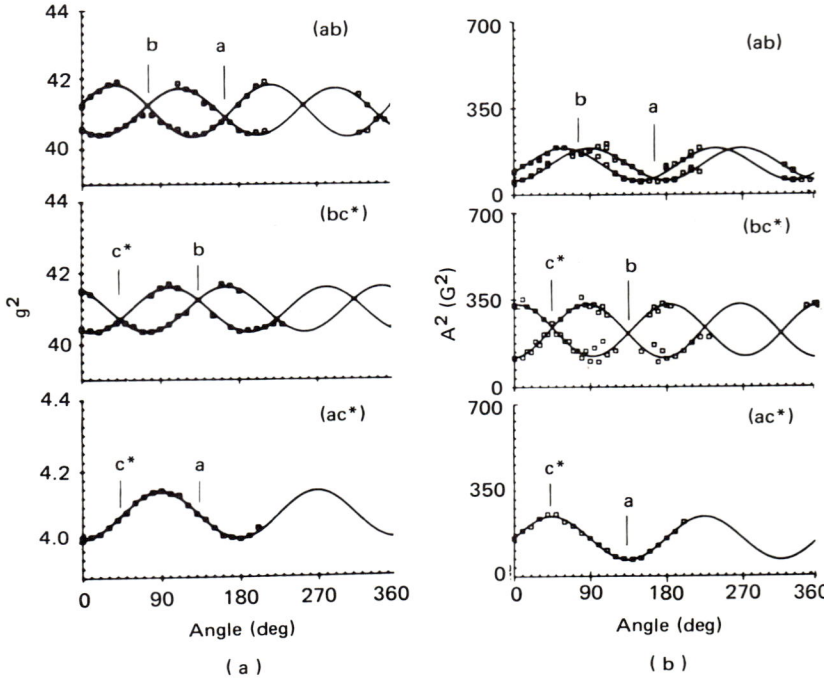

Fig. 42. Angular variations of g^2 (a) and A^2 (b) in oxyprotoporphyrin CoMb single crystals at 293 K. Data were collected as H_0 we swept through the $(a-b)$, $(b-c^*)$, and $(a-c^*)$ planes, respectively. (From Hori et al., 1982.)

The data are analyzed as follows: g values were collected at specific angles of H_0 relative to the crystallographic axes a, b, and c^* (Fig. 42). The observed values are related to the g tensors by Eq. (37). The direction cosines which relate the g (and A) tensors to the crystallographic axes can thus be calculated (Table XII). Knowledge (from the x-ray structure) of the orientation of the heme relative to these axes permits the assignment of these g (and A) tensors to specific orientations relative to the heme.

Data as in Table XII are often presented in a stereographic diagram (Fig. 43). This diagram is generated by projection of a point denoting the intersection of a radius line (axis or magnetic tensor) drawn out from the center of a sphere with the sphere's surface (Fig. 43a). This projection on the diametral plane of the sphere is generated by a line drawn from the point on the surface to the bottom (or top) pole of the sphere. The two values (angles) θ and ϕ uniquely define the orientation of the vector. A stereographic dia-

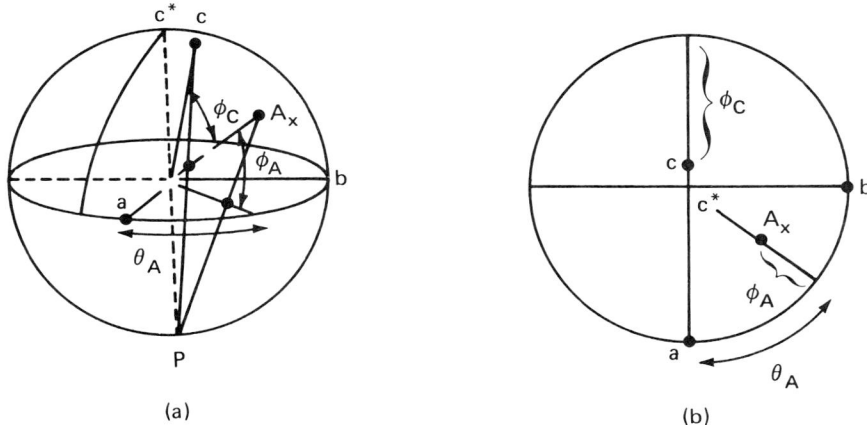

Fig. 43. Generation of a stereographic representation of crystallographic and spin-Hamiltonian tensor elements from single-crystal studies. In (a), the crystallographic axes a, b, and c are placed within a spherical coordinate frame. The a–b plane is a projection surface and is illustrated in (b). The coordinates θ, ϕ (in radians), for a given element are given by the radial line passing through the projection point [c, A_x in(b)] in the a–b (diametral) plane generated by drawing a line from the point of intersection of the element with the spherical surface [c, A_x in (a)] to the bottom (or top) pole P.

gram can be generated using some of the data in Table XII, as shown in Fig. 43b.

V. Resonance as a Perturbation

A. Resolution of Inhomogeneously Broadened Lines

Resonance lines which arise from identical paramagnetic centers whose magnetic environment fluctuates with time are homogeneously broadened. Saturation of such centers causes line broadening and a decrease in amplitude. Inhomogeneously broadened lines, however, arise from a set of spin packets, each of which is experiencing a different, but constant, local magnetic field. Consequently, these spin packets saturate independently of one another, and the resonance line is not broadened by excess microwave power. The effects and uses of power saturation were discussed in greater detail in Section III.C.

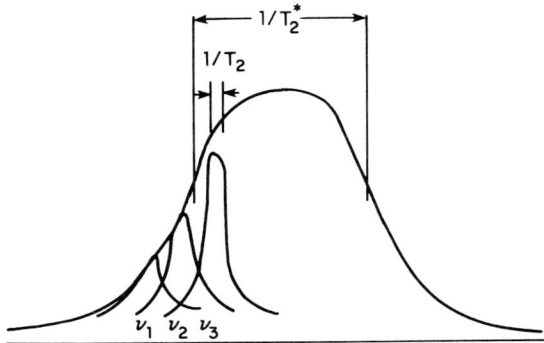

Fig. 44. An inhomogeneously broadened resonance line represented by the precessional frequencies v_i of the individual spin packets which contribute to the resonance. The linewidths of the whole line $(1/T_2^*)$ and spin packets $(1/T_2)$ are indicated.

Inhomogeneously broadened lines contain significant information in the form of the energies (frequencies) of the local fields modulating the behavior of each spin packet. For a given electron-spin transition, this information corresponds to the nuclear spins coupling to the electron-spin states; if these hyperfine interactions are similar in magnitude to the inherent linewidth (caused by secular or lifetime homogeneous line-broadening mechanisms), then they will go unresolved. This is illustrated in Fig. 44. Such inhomogeneously broadened lines generally have a Gaussian shape, described by the function $\exp[-0.69(T_2^*/T_2)^2]$, where T_2^{-1*} and T_2^{-1} are the linewidths at half height for the whole (inhomogeneously broadened) line and the individual spin packets, respectively (Fig. 44).

Two resonance techniques exist for the resolution of these small nuclear hyperfine interactions: electron–nuclear double-resonance (ENDOR) and the electron-spin echo. They yield the same information, namely, the nuclear frequencies of nuclei coupled to the electron spin. Since the magnetogyric ratios are known for all stable nuclei, the determination of the nuclear frequencies at a particular magnetic field serves to identify the nuclei types which are coupled to the unpaired electron(s). Both techniques are based on the same phenomenon, namely, the mixing of nuclear-spin and electron-spin transitions in an electron-spin system at resonance.

B. Electron–Nuclear Double Resonance

The electron–nuclear double-resonance (ENDOR) technique uses partial power saturation and an applied radio-frequency (rf) field to stimulate electron-spin transitions. The experiment can best be described by reference to a simple example (Fig. 45). In this system a nuclear spin, $I = \frac{1}{2}$, is coupled to an electron-spin doublet ($S = \frac{1}{2}$). The resultant energy levels are indicated.

Note the effects on the level energies caused by the interaction between the externally applied field and the *nuclear* moment [the *nuclear* Zeeman term in Eq. (7)]. This term results in $E_{a \to b} \neq E_{c \to d}$. This difference can be calculated and is $2g_{N_i}\beta_N H_0$. Being able to measure this energy difference

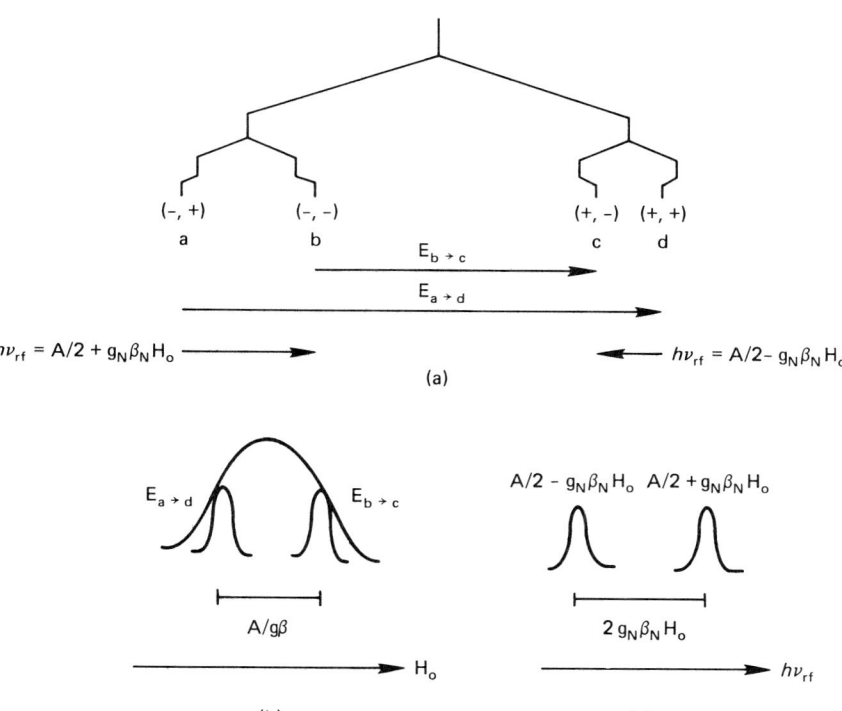

Fig. 45. (a) Energy diagram for an $S = \pm\frac{1}{2}$, $I = \pm\frac{1}{2}$ system, including the nuclear Zeeman splitting ($-g_N\beta_N I \cdot H_0$). The electron-spin transitions $E_{a \to d}$ and $E_{b \to c}$ are indicated. (b) The inhomogeneously broadened resonance line of which these two transitions are a part is shown as generated by the typical variable-field experiment. (c) The ENDOR spectrum is shown, as generated by monitoring the partially saturated $a \to d$ transition while sweeping through an rf field around $h\nu_{rf} = \frac{1}{2}A$. The "+" and "−" ENDOR frequencies are characteristic of individual nuclei; g_N can be readily calculated from the value of the ENDOR splitting. In this example $\frac{1}{2}|A| > g_N\beta_N H_0$ and $A > 0$.

means one can determine the value of g_{N_i} and thus the nuclear species N_i. The ENDOR experiment (as does the electron-spin-echo one) yields the values for these differences.

Consider the $E_{a \to d}$ arrow in Fig. 45. This represents the excitation of one of the spin packets in an inhomogeneously broadened line, chosen by fixing the field and microwave frequency of the experiment. The pair of transitions associated with this $I = \frac{1}{2}$ system is indicated as well in the variable-field experiment typical of the electron-resonance measurement; characteristically, the hyperfine splitting is $A/g\beta$ (Fig. 45).

The ENDOR experiment proceeds as follows. The transition $a \to d$ is at resonance and is partially saturated by the appropriate choice of microwave power based on measurements as outlined in Section III.C. This involves, of course, a decrease in the number of spins in state a relative to d. An rf field at right angles to H_0 is now swept through a frequency range centered around $A/2h$. When $hv_{rf} = A - g_N\beta_N H_0$ (the energy difference between states d and c) a *nuclear*-spin transition will be stimulated which will reduce the population of state d ($I = +\frac{1}{2} \to I = -\frac{1}{2}$). This has the effect of reducing the saturation of the electron-spin transition $a \to d$, causing a stimulation of absorption of microwave energy by the electron-spin packet. It is this ESR absorption which is recorded in the ENDOR experiment and is pictured in Fig. 45c.

When $hv_{rf} = A/2 + g_N\beta_N H_0$ (the energy difference between states a and b), another absorption of microwave energy will be observed. This is the result again of a "relief" in the saturation of the electron-spin system, this time owing to the nuclear-spin transition $a \to b$. This relief is caused by the connecting link between states c and d ($S = +\frac{1}{2}$) that is provided by the $a \to b$ nuclear excitation and relaxation processes. Thus, the ENDOR experiment involves detection of electron-spin transitions, as in a simple ESR measurement. However, the transitions are generated by varying an rf field, and that is how the ENDOR spectrum is presented as in Fig. 45c.

For the example given, the two absorption lines are separated in units of hv by $2g_N\beta_N H_0$; from this one can calculate the nuclear moment of the modulating nuclei, and thus identify it. Conversely, at a given experimental field, the frequencies of likely nuclei can be calculated before the measurement is made, and the ENDOR spectrum obtained can subsequently be interpreted by comparison.

ENDOR can simplify complex ESR spectra. This can again be best understood by an example. Consider a system which contains two nuclei, $I = \frac{1}{2}$, which are identical in that they are characterized by the same hyperfine coupling value. This is illustrated in Fig. 46. As pictured, in the ESR spectrum itself these two nuclei give rise to three lines. However, in the ENDOR spectrum only two transitions are observed (stimulated) associated

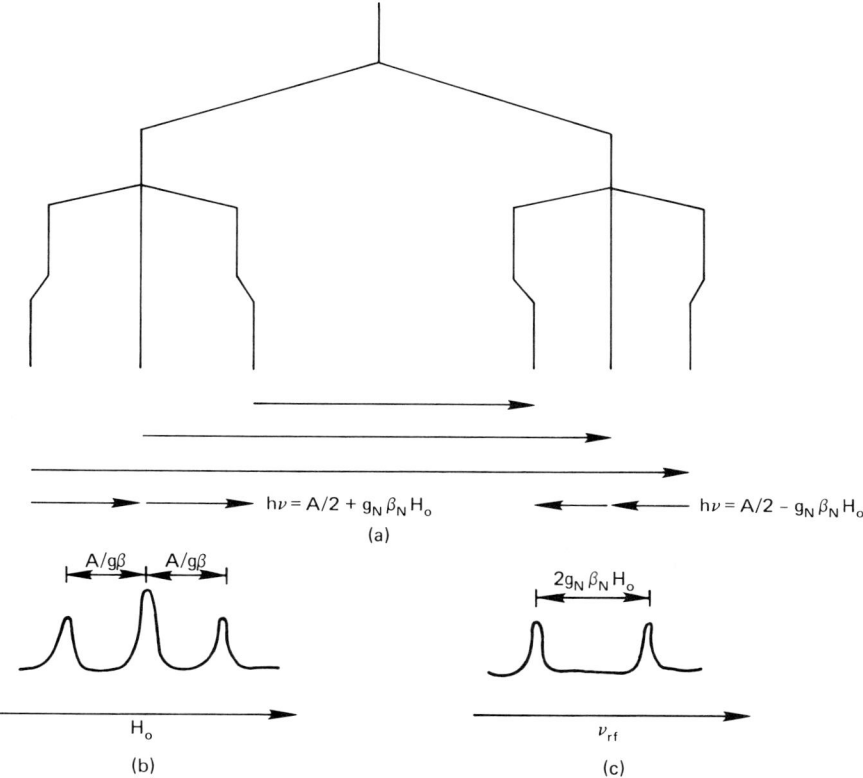

Fig. 46. (a) Energy diagram for $S = \pm\frac{1}{2}$ coupled to two identical $I = \pm\frac{1}{2}$ nuclei. The three different electron-spin transitions are noted, as are the two nuclear-spin transitions. These two spin-transition types are illustrated as: (b) the direct ESR experiment (variable H_0, constant microwave frequency); (c) the ENDOR experiment (variable ν_{rf}, constant H_0 and microwave frequency). As in Fig. 45, $\frac{1}{2}|A| > g_N \beta_N H_0$ and $A > 0$.

with the modulation of the electron's local field, caused by the two orientations of the nuclear moments precessing around the nuclear Zeeman field. In general, in the absence of nuclear quadrupole splitting there will be $2N$ ENDOR lines, where N is the number of different coupled nuclei types or different hyperfine couplings. For example, the ESR spectrum of a system containing one ^{14}N ($I = 1$), one ^{19}F ($I = \frac{1}{2}$), and three inequivalent ^{1}H ($I = \frac{1}{2}$) would be expected to have 48 lines ($3 \times 2 \times 2 \times 2 \times 2$). The ENDOR spectrum, however, could have as few as ten lines, two pairs separated (in units of $\beta_N H_0$) by g_{14_N} and g_{19_F}, respectively, and three pairs (centered at ν_{1H}) separated by different values of $|A|$ in megahertz.

These latter three pairs of ENDOR lines for ^{1}H are centered around the free spin for this nucleus because for ^{1}H, $\nu_N > \frac{1}{2}|A|$, in most cases. Some values of ν_N are listed in Table XIII. ^{1}H differs from most other biologically relevant nuclei in this fashion. ^{19}F, although characterized by having a large ν_N, also can exhibit large hyperfine couplings because of the spin polarization of s electrons by unpaired electron density in fluorine p orbitals. Thus, $|A_{19_F}|$ values as large as 500 MHz have been measured (Bereman and Kosman, 1977), although this is by no means common.

The field dependence of ν_N indicated by the values in Table XIII shows that the separation of the ENDOR lines (if $\frac{1}{2}|A| > \nu_N$) or their position (if $\nu_N > \frac{1}{2}|A|$) depends on the g_{eff} of the ESR transition being observed, i.e., the magnetic field H_0. Clearly, this behavior must be considered in both the interpretation of ENDOR spectra and in the choice of resonance field. Increasing the field could improve the resolution by separating resonance lines or could help in assignment by inverting the relative magnitudes of ν_N and $\frac{1}{2}|A|$. Note that $|A|$ values to a first approximation are independent of H_0. It is worth mentioning that since ENDOR lines are inherently narrower than the inhomogeneously broadened ESR transitions (since they are only homogeneously broadened), much more precise A values are obtainable.

Since ENDOR spectra are generated by frequency sweep (as opposed to magnetic-field sweep, as for conventional ESR), it is convenient to express

Table XIII Nuclear Frequencies ν_N at 3390 G (9.5 GHz, X Band) and 12,408 G (35 GHz, Q Band)

Nucleus	ν_N (MHz)	
	at 3390 G	at 12,488 G
^{1}H	14.4	53.2
^{2}H	2.22	8.16
^{13}C	3.63	13.4
^{14}N	1.04	3.83
^{15}N	1.46	5.39
^{19}F	13.6	50.0
^{51}V	3.97	14.0
^{55}Mn	3.58	13.2
^{57}Fe	0.47	1.72
^{59}Co	3.42	12.6
^{63}Cu	3.83	14.1
^{65}Cu	4.10	15.1

the energy levels as in Fig. 47 in terms of the frequencies of the transition energies. For the $S = \frac{1}{2}, I = \frac{1}{2}$ system described here, the spin-Hamiltonian is

$$\mathcal{H} = g\beta \mathbf{H} \cdot \mathbf{S} - g_N \beta_N \mathbf{H} \cdot \mathbf{I} + hA\mathbf{S} \cdot \mathbf{I}, \tag{38}$$

where β_N is the nuclear magneton (5.05×10^{-24} erg/G) and A the isotropic coupling (A_0) in hertz. The term $-g_N \beta_N \mathbf{H} \cdot \mathbf{I}$ is the nuclear Zeeman interaction omitted in Eq. (7). The energy levels of the spin states are given (in hertz) by

$$\frac{E}{h}(m_S, m_I) = \frac{g\beta H m_S}{h} - \frac{g_N \beta_N H m_I}{h} + A m_S m_I. \tag{39}$$

This is commonly expressed in terms of the frequencies of the electron and nuclear transitions, i.e.,

$$E/h = \nu_e m_S - \nu_N m_I + A m_S m_I, \tag{40}$$

where $\nu_e = g\beta H/h$ and $\nu_N = g_N \beta_N H/h$.

The splitting diagram in Fig. 45 can be redrawn in these terms, as shown in

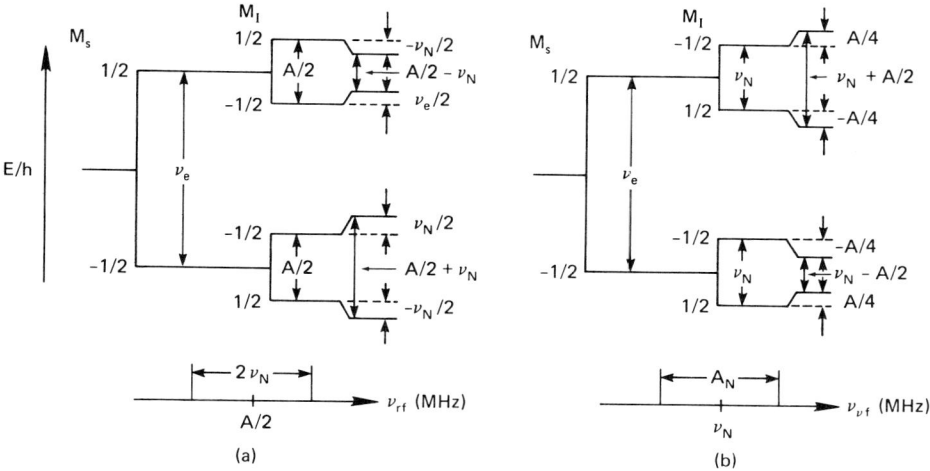

Fig. 47. Splitting diagrams for an $S = \frac{1}{2}, I = \frac{1}{2}$ electron–nuclear-spin system including the nuclear Zeeman energy contribution, ν_N. How the relative signs and magnitudes of the hyperfine and nuclear precessional frequencies affect the ordering of the energy levels are indicated. (a) The splitting of the $M_2 = \pm \frac{1}{2}$ states is dominated by A_N, where $A_N > 0, \frac{1}{2}A_N > \nu_N$. The ENDOR splitting is thus $2\nu_N$. (b) The splitting of the $M_s = \pm \frac{1}{2}$ states is dominated by ν_N, where $A_N < 0, \frac{1}{2}A_N < \nu_N$. The ENDOR splitting is thus A_N.

Fig. 47a. ESR transitions will be stimulated when $v_{rf} = \frac{1}{2}|A| - v_N$ and $v_{rf} = \frac{1}{2}|A| + v_N$; thus, the ENDOR spectra will consist of two lines separated by $2v_N$ centered around $\frac{1}{2}|A|$, as shown here. Note that the order of the m_S, m_I states depends on the sign of A. In addition, the relationship between the resonance lines in the ENDOR spectrum to v_N and A depends on the relative magnitudes of v_N and $\frac{1}{2}|A|$. Thus, for example, if $A < 0$ and $v_N > \frac{1}{2}|A|$, the energy levels are ordered as in Fig. 47b. In this situation, ESR transitions are stimulated when $v_{rf} = v_N - \frac{1}{2}A$ and $v_N + \frac{1}{2}A$. Consequently, the ENDOR spectrum will consist of two lines separated by $|A|$ centered around the "free-spin" nuclear frequency v_N.

A complication in the ENDOR spectrum arises for nuclei $I \geq 1$ because of the nuclear-quadrupole coupling tensor **Q** which is caused by the nonspherical charge distribution in such nuclei (**P** is also used to denote this tensor, reserving **Q** as the symbol for the quadrupole moment). First-order ESR spectra do not reflect this term in the spin-Hamiltonian [Eq. (7)]

$$\mathcal{H} = g\beta \mathbf{H} \cdot \mathbf{S} - g_N \beta_N \mathbf{H} \cdot \mathbf{I} + h \cdot \mathbf{S} \cdot \mathbf{A} \cdot \mathbf{I} + h\mathbf{I} \cdot \mathbf{Q} \cdot \mathbf{I}, \quad (41)$$

since it is usually much smaller than the others. ENDOR, however, can resolve this contribution. For example, if A_{zz} and Q_{zz} are defined along a principal axis and $|A| > |Q|$, a first-order axial-spin-Hamiltonian can be written:

$$\mathcal{H} = g\beta H_{zz} S_{zz} - g_N \beta_N H_{zz} I_{zz} + A_{zz} S_{zz} I_{zz} + Q[I_{zz}^2 - \tfrac{1}{3} I(I+1)], \quad (42)$$

with eigenvalues

$$E/h = v_e m_S - v_N m_I + A m_S m_I + Q_{zz}[m_I^2 - \tfrac{1}{3} I(I+1)]. \quad (43)$$

The ENDOR frequencies for some $S = \frac{1}{2}$ systems are:

for $I = \frac{1}{2}$ (two ENDOR lines), $\quad v = |\tfrac{1}{2} A_{zz} \pm v_N|$;

for $I = 1$ (four ENDOR lines), $\quad v = |\tfrac{1}{2} A_{zz} \pm v_N \pm Q_{zz}|$;

for $I = \frac{3}{2}$ (six ENDOR lines), $\quad v = |\tfrac{1}{2} A_{zz} \pm v_N|, v = |\tfrac{1}{2} A_{zz} \pm v_N \pm 2Q_{zz}|$.

In general, the observed ENDOR frequencies are given by

$$v = |\tfrac{1}{2} A_{zz} \pm v_N \pm Q_{zz}(2m_I - 1)|, \quad (44)$$

where $-I + 1 \leq m_I < I$.

In this discussion, A and Q were assumed positive. In practice, the *relative* signs of A and Q can be determined by determining which nuclear transitions when pumped stimulate a particular ESR transition. This is illustrated in Fig. 48, which shows the energy levels for an $S = \frac{1}{2}$, $I = 1$ system. If A, $Q > 0$, stimulation of the low-field $m_I = +1$ ESR transition occurs at the highest ($v_1 = |\tfrac{1}{2} A_{zz} + v_N + Q_{zz}|$) and lowest ($v_4 = |\tfrac{1}{2} A_{zz} - v_N - Q_{zz}|$) ENDOR frequencies. If $A > 0 > Q$, however, saturation of this line gives the inter-

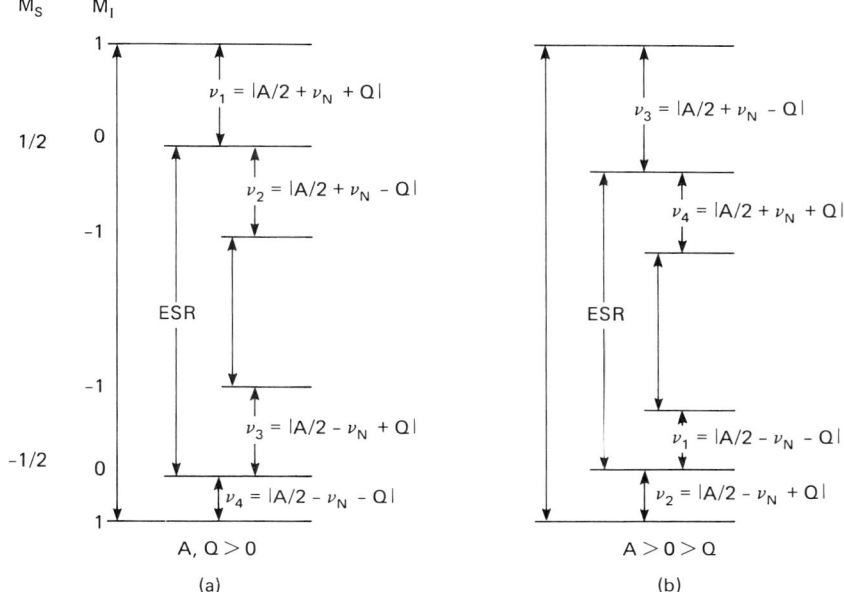

Fig. 48. Effect of nuclear quadrupole coupling on splitting diagram of an $S = \frac{1}{2}$, $I = 1$ electron-spin–nuclear-spin system and the relative signs of A_N and Q_N. (a) The nuclear hyperfine and quadrupole couplings are both positive. Stimulation of the $M_I = +1$ ESR transition yields the extreme ENDOR frequencies v_1 and v_4. (b) Q is negative, thus stimulation of the $M_I = +1$ ESR transition yields the intermediate frequencies v_2 and v_3.

mediate frequencies, $v = \frac{1}{2}A_{zz} \pm |v_N - Q_{zz}|$. If the sign of either A or Q is known, this analysis can help determine the sign of the other.

In summary, ENDOR is a highly valuable resonance technique. Although the instrumentation is somewhat more sophisticated than that employed in direct ESR experiments, it is available commercially. The coupling of the rf and microwave radiation to the sample necessitates a special cavity design, although even the conventional rectangular cavity can be readily modified for ENDOR by the insertion of an rf coil. However, there is also the experimental difficulty of being able to selectively saturate one part of an inhomogeneously broadened line. "Spin diffusion," which in effect causes saturation of other spin packets, broadens the ENDOR line. This can usually be overcome by correct selection of temperature, since the diffusion process is temperature independent. T_1 processes, which are temperature dependent, can bring neighboring spin packets into thermal equilibrium at rates fast enough to nullify the unwanted contributions of spin diffusion. Of course, as T_1 becomes very short, the microwave power required to effect saturation can become unobtainable (or undesirable). This obviously represents an experimental difficulty which for some systems is not surmountable.

Another aspect of ENDOR applied to biological samples is the one associated with the experimental conditions often required to detect the inherent paramagnetism of such samples. Because of the relaxation properties of, in particular, biologically relevant transition-metal ions, most experiments are conducted on frozen glasses in which the paramagnetic centers generally exhibit a (large) degree of anisotropy. Note, however, that neither the broad lines nor the anisotropy characteristic of such systems prohibits a successful ENDOR experiment. At a given microwave frequency and Zeeman field, only one spin packet is stimulated by the incident rf field. Clearly, this anisotropy can actually be of some use in that the nuclei and their associated hyperfine couplings, which modulate each principal g tensor, can be assigned and thus, theoretically, a quite detailed picture of the metal complex may be obtainable.

Biological ENDOR has enjoyed a well-deserved renaissance since 1975. The technique has been valuable in the characterization of the Mo—Fe centers in nitrogenase (Hoffman *et al.*, 1982a,b); ferric heme in myoglobin (Scholes *et al.*, 1982); cobalt hemoglobin (Höhn and Hüttermann, 1982); Cu(II) centers in superoxide dismutase (Van Camp *et al.*, 1982), and cytochrome *c* oxidase (Stevens *et al.*, 1982); vanadyl complexes (Mulks *et al.*, 1982); and photosynthetic reaction centers in bacteria (Lendzian *et al.*, 1981). A few examples illustrating the use of ENDOR follow.

Both ^{57}Fe and ^{95}Mo ENDOR (Hoffman *et al.*, 1982a,b) of *Azotobacter vinelandii* nitrogenase MoFe protein have been analyzed. The ESR spectrum is rhombic: $g_3 = 4.32$, $g_2 = 3.65$, and $g_1 = 2.0$. The versatility of ENDOR is made apparent by comparison of the ENDOR spectra in Fig. 49, taken at the extreme g values, g_3 and g_1. As noted here, the ENDOR experiment essentially "sees" a frozen glass as a single crystal in that the field can be set along a given g tensor if the tensor is at the edge of the resonance envelope, as g_3 and g_1 are. At g_3, six well-defined ^{57}Fe peaks are observed. With $\nu_{Fe} = 0.220$ MHz at 1600 G (at g_3), the splitting between the pairs of ENDOR peaks for each of the six Fe atoms (assuming they are inequivalent) should be 0.440 MHz, and the pairs should be centered around A_{g_3}. This rather small splitting can go unresolved and, commonly, ENDOR transitions do not have equal intensity. Consequently, the twelve lines expected from six inequivalent Fe atoms are not resolved. Nonetheless, the A_{g_3} values can be determined for these six atoms (Table XIV).

The A_{g_1} values are markedly different, as is indicated by the spectra; these are also listed in the table. The higher field used (3350 G) resolves the

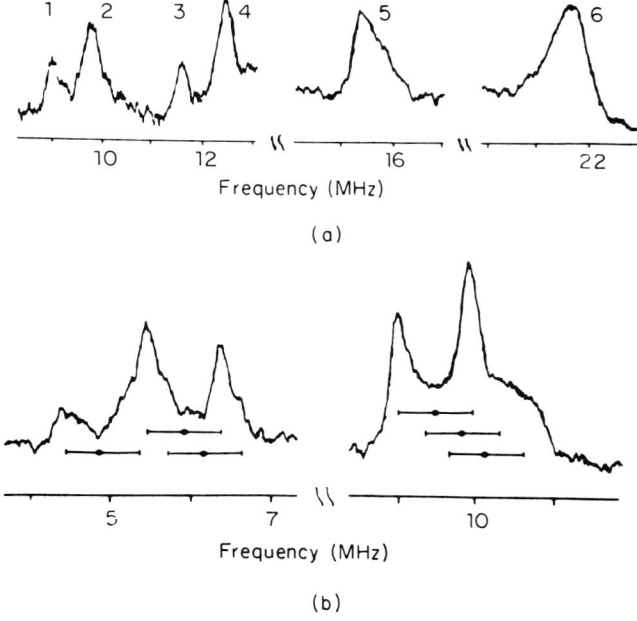

Fig. 49. ^{57}Fe ENDOR spectra of MoFe protein from *Azotobacter vinelandii* at 2 K. Spectra were taken in segments as indicated and represent the accumulation of 5000 scans. (a) $H_0 = 1600$ G at $g_3 = 4.32$; (b) $H_0 = 3350$ G at $g_1 = 2.0$. The six contributing ^{57}Fe centers are numbered in (a). In (b), the six pairs of lines separated by $2v_{^{57}Fe}$ (0.94 MHz) are indicated. (From Hoffman *et al.*, 1982b. © 1982 American Chemical Society.)

Table XIV ^{57}Fe Hyperfine Tensors (MHz) in Azotobactor Vinelandii MoFe Protein Fe Centers[a]

Fe center	$A (g_3 = 4.32)$	$A (g_2 = 3.65)$	$A (g_1 = 2.0)$
1	8.3	11.5	9.9
2	9.0	12.6	9.9
3	10.7	11.5	11.8
4	11.5	15.3	18.9
5	14.3	18.2	19.9
6	20.5	17.0	20.6

[a] From Hoffman *et al.* (1982b).

nuclear Larmor frequency ($v_{Fe} = 0.47$ MHz). The assignment of A value sets to Fe sites is unsupported by experiment. However, the data clearly show the complexity of the Fe sites in the MoFe protein, since they are all characterized by markedly different metal hyperfine interaction tensors.

^{14}N and ^{15}N ENDOR spectra of single crystals of aquomethyoglobin illustrate the assignment of nitrogen frequencies and the value of isotopic substitution (Scholes et al., 1982). In these experiments, [^{15}N] protoporphyrin IX was prepared in bacterial cultures and reconstituted with apomyoglobin. Thus, ENDOR spectra were markedly better resolved in that the heme nitrogens (^{15}N, $I = \frac{1}{2}$) contributed pairs of ENDOR lines only, rather than the four lines associated with ^{14}N, $I = 1$. Furthermore, ^{15}N exhibits no quadrupolar coupling. The surprising result was that in the ^{15}N heme spectrum there were eight lines, not the four that would have been expected for two pairs of equivalent heme (pyridine) nitrogens (Fig. 50). The paired values for the ENDOR frequencies for the heme ^{15}N and proximal histidine (^{14}N$_\epsilon$) are given in Table XV for two orientations of H_0 in the heme plane, and with H_0 along the heme normal.

Given the completeness of the data set obtained including the comparisons of ^{14}N and ^{15}N as noted (as well as the differences in signs for the hyperfine and g tensors), an excellent fit to a standard spin Hamiltonian was possible. An example of how a consistent analysis is developed is related to the discussion about the relative signs of A and Q. If $A_{14_{N\epsilon}} > 0$ in the heme normal, as would be expected if unpaired electron density is in an N2s orbital (as this work established), and $Q < 0$, then the lower-energy ENDOR pair v^- can be predicted to have a ca 0.07 MHz larger splitting than the v^+, or higher energy, pair. As the data in Table XV show, this is borne out experimentally. As a result of being able to assign magnitudes and signs to specific A tensors, and because these tensors could be correlated with the crystallographic axes and the heme planes, Scholes et al. (1982) can describe the electronic asymmetry of the Fe(III) porphyrin explicitly. This contrasts with the interpretation of the ENDOR pairs for the nitrogenase MoFe protein, whose assignment to specific hyperfine tensors and Fe centers was somewhat speculative (Hoffman et al., 1982b). The difference, in part, results from the greater order of the true crystal compared with the "crystallike" frozen glass, which makes the angular relationships of specific transitions to structural axes much better resolved. Indeed, the ENDOR experiments of Scholes et al. indicate that the Fe(III) in aquometmyoglobin is ~ 0.02 Å closer to one heme nitrogen than its diagonally opposite partner, which results in a larger Fermi contact term. This fine study exemplifies the utility of ESR in determination of biological structure generally, and the value of single-crystal, isotopic substitution, and ENDOR techniques in particular.

Another example illustrates the use of ENDOR to describe the spatial

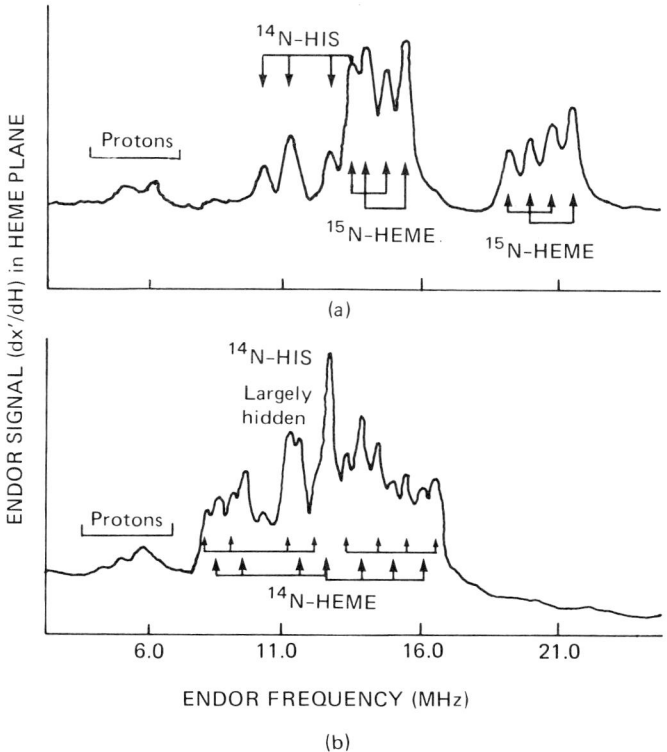

Fig. 50. Nitrogen ENDOR spectra for (a) [^{15}N]-heme at 9.09 GHz and (b) [^{14}N]-heme at 8.94 GHz, protoporphyrin aquometmyoglobin single crystals at 2.1 K. Both spectra were recorded near $g = 5.9$ with the magnetic field in the heme plane along the $N_1 - N_3$ pyridine nitrogen diagonal. (From Scholes *et al.*, 1982. © 1982 American Chemical Society.)

arrangements of protons in the vicinity of a paramagnetic center, in this case deoxy and oxy Co(II) hemoglobin (CoHb) (Höhn and Hüttermann, 1982). The deoxy derivative exhibits a highly axial ESR spectrum, as indicated by the data in Table XVI. The ENDOR assignments were generated by analysis of spectra, as in Fig. 51, which gives the ENDOR spectrum of deoxy CoHb in H$_2$O with H_0 along g_1. As is customary for ^{1}H ENDOR spectra, the lines are referenced to v_N, the "free-spin" value for the nuclear Larmor precession at the field of the experiment. This is because, for weakly coupled protons, $v_N > |\frac{1}{2}A|$ with $A < 0$, as discussed here. Thus, the pairs of lines in Fig. 51 are separated by $|A|$, in megahertz.

Since the crystal structures of the heme proteins myoglobin and hemoglo-

Table XV Measured ENDOR Frequencies (MHz) for [^{15}N]-Protoporphyrin-[^{14}N]-Aquometmyoglobin ([^{14}N$_\epsilon$]-His)

Sample	$H_0 \parallel$ heme normal[a] ($H_0 = 3.24$ kG, $g = 2.00$)	$H_0 \parallel N_1-N_3$[a] ($H_0 = 1.10$ kG, $g = 5.91$)	$H_0 \parallel N_2-N_4$[a] ($H_0 = 1.09$ kG, $g = 5.97$)
[^{15}N]-Heme	N$_1$; N$_3$ $\begin{array}{l}3.57 \leftarrow\\ \rightarrow 3.87\\ 6.29 \leftarrow\\ \rightarrow 6.55\end{array}$ N$_2$; N$_4$ (2.79)[b]	N$_2$ $\begin{array}{l}\rightarrow 13.42\\ 13.98 \leftarrow\\ \rightarrow 14.71\\ 15.36 \leftarrow\end{array}$ N$_4$ (1.36)	N$_3$ $\begin{array}{l}\rightarrow 13.38\\ 14.02 \leftarrow\\ \rightarrow 14.72\\ 15.36 \leftarrow\end{array}$ N$_1$ (1.35)
[^{14}N$_\epsilon$]-His	$\begin{array}{l}3.10 \leftarrow\\ 5.16 \leftarrow\\ 6.43 \leftarrow\\ 8.42 \leftarrow\end{array}$ (1.99)	N$_3$ $\begin{array}{l}\rightarrow 19.23\\ 19.98 \leftarrow\\ \rightarrow 20.70\\ 21.48 \leftarrow\end{array}$ N$_1$ (1.56) $\begin{array}{l}10.22 \leftarrow\\ 11.18 \leftarrow\\ 12.59 \leftarrow\\ 13.42 \leftarrow\end{array}$ (1.05)	N$_2$ $\begin{array}{l}\rightarrow 19.61\\ 20.58 \leftarrow\\ \rightarrow 21.15\\ 22.02 \leftarrow\end{array}$ N$_4$ (1.55) $\begin{array}{l}11.50 \leftarrow\\ 13.38 \leftarrow\end{array}$ (1.03)

[a] H_0 along N_1-N_3 heme diagonal; similarly for $H \parallel N_2-N_4$. Field position and g values given in parentheses. Adapted from Scholes et al. (1982).
[b] Values (MHz) in parentheses calculated for $2g_N\beta_N H_0$ at resonance field employed.

bin are similar in the vicinity of the porphyrin rings, the geometry of the former was used to calculate an expected hyperfine coupling for protons closest to the Co(II). This estimate was based on the relationship

$$A_N = (g\beta g_N \beta_N/r^3)\rho(3\cos^3\phi - 1), \tag{45}$$

which is the dipolar term of the hyperfine interaction in Eq. (7); ϕ is the angle between the field and the radius line between the electron and nuclear dipoles; ρ was taken as 1.0 for deoxy CoHb, since the unpaired electron density resides fully on the metal in this state. The protons on the distal side (on the side of the O_2 binding) were too far away to couple significantly to the electron (maximum $A_H \leq 0.5$ MHz). On the other hand, the F-8 histidine protons at N_ϵ and C_ϵ

Table XVI ¹H ENDOR Couplings (MHz) for Cobalthemoglobin (CoHb) and Oxycobalthemoglobin (oxyCoHb) in H_2O and D_2O[a]

g value (resonance field, kG)	Sample	
	CoHb (line pairs as in Fig. 51)	
$g_1 = 2.32$ (2.44)	0.40 (AA′)	
	1.35 (BB′)	
$g_2 = 2.29$ (2.98)	0.45 (AA′)	
	1.35 (BB′)	
$g_3 = 2.03$ (3.36)	1.10 (BB′)	
	oxyCoHb	
	H_2O	D_2O
$g_1 = 2.078$ (3.28)	2.45	2.50
	5.90	—
$g_2 = 2.005$ (3.40)	0.40	0.40
	1.30	1.25
	3.10	3.20
	5.60	—
$g_3 = 1.982$ (3.44)	0.45	0.45
	1.25	1.25
	2.90	2.90
	5.70	—

[a] Höhn and Hüttermann, 1982.

were calculated to have A_H values in g_1/g_2 and g_3 of 0.4/1.4 and 1.0 MHz, respectively, in good agreement with experimental data (Table XVI). Since the ENDOR experiment cannot, in general, yield information about the *number* of identical nuclei, one cannot tell whether both of these protons are, in fact, coupled and contributing to the spectrum observed. To the extent that the direct ESR spectrum does indicate the number of nuclei of a particular type, it has an advantage over ENDOR.

The ESR spectrum of oxyCoHb is not axial, thus the frozen glass can be treated much like a crystal in the ENDOR experiment (Table XVI). The analysis by comparison of calculated with experimental A_H values as given here is complicated by the fact that in oxyCoHb (as in oxyCoMb, see Section IV.D), more than 90% of the spin density ρ is on the dioxygen, and two orientations of the O_2 are possible. However, A_H values for both orientations I and II were calculated assuming that the proton(s) on the distal His (E7) were coupled to the spin density on the dioxygen. In fact, both I and II

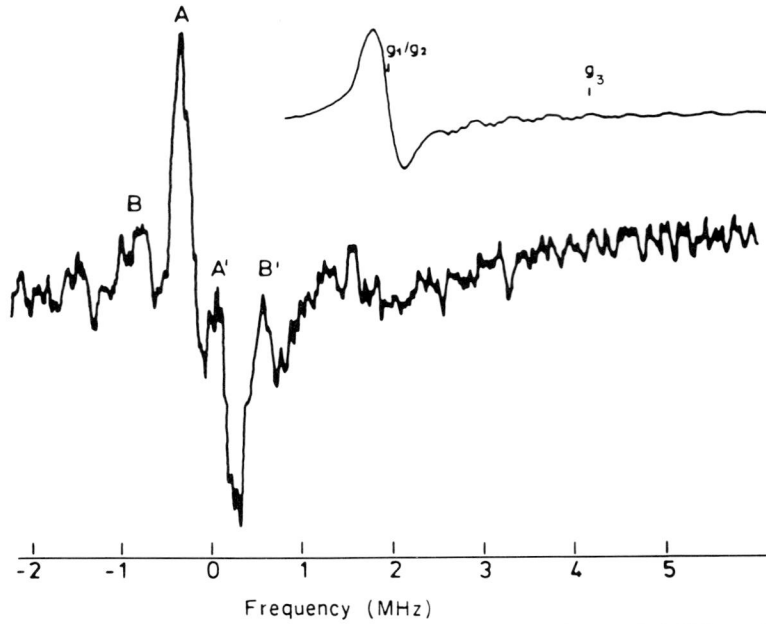

Fig. 51. ¹H ENDOR spectrum of deoxycobalt hemoglobin at ~9.4 GHz and ~10 K. The field was at $g_1 = 2.32$ (2.44 kG). Spectral lines are referenced to v_N for free-proton frequency in the Zeeman field. Pairs of lines, AA′ and BB′, are mirrored around this frequency, separated by the value A, with the positions of each of a pair of lines $v_\pm$ given by $|v_N \pm \tfrac{1}{2}A|$. The insert shows a sample ESR spectrum. (From Höhn and Hütterman, 1982.)

orientations gave the same strong 5.9 and 5.6-MHz couplings along g_1 and g_2 for the His (E7) N_{ϵ_2} proton, and this coupling disappeared in D_2O as expected for this H-bonded exchangeable proton. The other hyperfine couplings can be assigned to the methyl-group protons of Val (E11) which point toward the O_2. Since the bulk of the spin density is on the ligand in the dioxygen complex, protons on the proximal side do not contribute a hyperfine interaction.

The single-crystal ESR experiments on CoMb outlined in Section IV.D and these on CoHb are complementary and illustrate the unique insights provided by each type of technique. Single-crystal ESR can provide unambiguous assignments of g and dominant A tensors with respect to some defined structural axes of the protein environment. However, the direct ESR spectrum does not resolve the superhyperfine couplings which are lost in the large linewidths. Thus, direct ESR cannot deduce aspects about the nuclear environment outside of the "nearest-neighbor" sphere of the spin density (typically, the inner coordination sphere of a paramagnetic metal ion). While ENDOR on frozen glasses cannot make tensorial assignments, it *can* detect and, in favorable cases, assign small hyperfine couplings to specific nuclei, generally those at a distance greater than a bond length or a van der Waals radius. Another technique which can detect such hyperfine couplings is the spin-echo envelope-modulation approach, which will be presented next.

C. Electron-Spin Echo (Mims and Peisach, 1978, 1981)

The electron-spin echo also begins by the excitation of a single spin packet within an inhomogeneously broadened ESR resonance line. The electron-spin echo differs, however, in that it is, in effect, a transverse T_2 relaxation measurement; that is, it represents the modulation of the decay of the x–y component of the electron-spin magnetization following the spin packet's absorption of microwave energy. The information derived from the experiment is contained within the pattern(s) of modulation of this spin–spin relaxation process. As will be outlined here, this pattern of modulation is attributable to weakly coupled nuclei ($I \neq 0$) whose hyperfine couplings are too small to be resolved by direct (continuous-wave) ESR. As in the ENDOR experiment, it is the *frequency* of the modulation which serves to identify these weakly coupled nuclei.

Consider the behavior of a spin packet, precessing at a frequency v_0 in a magnetic field H_0. All other spin packets, in this and other resonance lines, will be precessing with frequencies greater or less than v_0; thus a pulse of microwave energy hv_0 will be absorbed by this spin packet only. If such a pulse is of sufficient power and duration (typically 200 W and 20 nsec), the spins in the packet will be tipped into the x–y plane; that is, the spin magnetization will nutate around the microwave field H_1 by 90°.

The experiment now becomes one of monitoring what happens to this collection of coherent spins as a function of time following this 90° pulse. Since the energy bandwidth of the exciting radiation is not infinitely narrow, the packet will consist initially of spins experiencing a variety of local magnetic fields. This will be apparent after the pulse is terminated, because spins at higher field will be precessing faster, at slightly higher frequencies than those at slightly lower field. Thus, the spin packet will "fan out" in the x–y plane; the magnetic-moment vector of the spin assembly, **M**, will break up. The time constant for this is the transverse relaxation time, and in the spin-echo experiment, the length of time for which **M** has a nonzero value (until the spin moments are completely randomized) is termed the phase memory. The loss of precessional coherence within the spin packet is also associated with time-dependent processes, i.e., those dynamic factors which contribute to homogeneous line broadening (such as fluctuations in the local nuclear fields). Put another way, the "faster" spins are not always faster.

Consider for the moment the idealized situation in which the precessional frequency of one half of the spins is $2v_0$ and that of the other half is $\frac{1}{2}v_0$. Thus, the magnetization vector would fluctuate with a frequency v, that is, attain its initial amplitude every $1/v$ sec. As long as the resonance (precessional) frequency is not altered by a fluctuation in local magnetic fields, this peak amplitude magnetization will persist as an "echo" of the initial state. This echo is experimentally measured by nutating the precessing moments around H_1 again, employing a second microwave pulse of sufficient power and duration to effect a 180° "flip." In effect, this "slows" down the more rapidly precessing spins and "speeds" up the slow ones. This shortens the time required to generate the echo, and the echo produced will be rotated by 180°. At the moment when the "slow" spins catch up to the "fast" ones (in our simple example, when they both have frequency v_0), current will be generated in the microwave detector circuit (tuned to v_0) by the dynamo principle; in essence, the microwave bridge is in "resonance" with the sample. The timing of this echo response is pictured in Fig. 52.

As noted, the precessional frequencies of the individual moments are not constant. Random fluctuations in the local fields experienced by each spin cause the spin packet to lose the coherence it possessed at the time of the initial pulse. Thus, as τ increases, the echo signal generated decreases. In

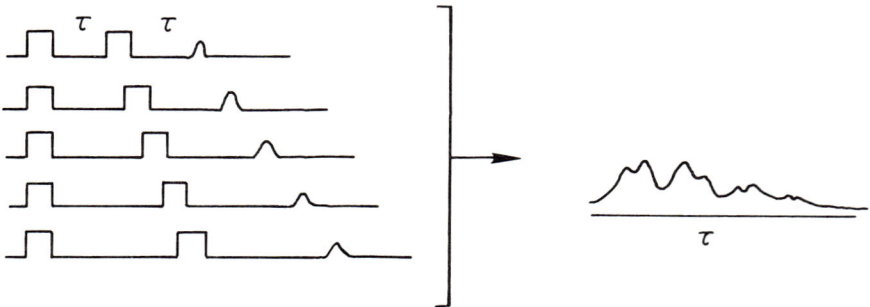

Fig. 52. Diagrammatic representation of generation, collection, and display of electron-spin echoes. The time τ between the first and second (90° and 180°) microwave pulses is made progressively longer (from ~50 nsec to ~3 μsec) by ~20-nsec increments. The "echo" generated at each time is detected by the crystal detector, and the detector output is stored. At the end of this series of microwave pulses, the stored detector output is fed into a recorder (or oscilloscope) and displayed as a function of the time τ.

biological samples containing paramagnetic metal ion centers, the phase memory is typically 3 μsec measured from the first pulse. Although of practical importance (short-phase memories limit the amount of usable data derivable from the experiment), this "unmodulated" decay is devoid of information. Rather, it is the pattern of echo amplitudes as a function of τ which, when observed, characterizes the electronic environment of the spin packet. Specifically, the oscillations in the echo amplitude, caused by *regular* fluctuations in the local fields, will exhibit a periodicity related to the precessional frequencies of weakly coupled nuclei (Fig. 52). Thus, what is termed the envelope modulation function of the spin-echo decay is generated by combinations of nuclear frequencies. As in the ENDOR experiment, the identification of the nuclei types contributing to this modulation process simply involves the matching of the frequencies observed in the modulation envelope to the frequencies expected of various nuclei in the magnetic field of the experiment.

The periods of nuclear precession obtaining at the magnetic fields of the experiment (~ 3 kG) are of magnitude 50–750 nsec. The shortest time of τ usually employed is 20 nsec, and the value is increased incrementally in a series of pulses until the echo signal is no longer observable. Thus, although the electron spins are going through hundreds of precessional cycles between each pulse, the nuclei with even the shortest periods are sampled more than once *within one* of their precessional cycles. This sampling of the echo during one nuclear precessional period will reflect the difference between the nuclear moment adding to and subtracting from the Zeeman field. This fluctuation is what causes the precessing electron to speed up and slow down, respectively, which changes the degree of spin coherence at any time τ, as indicated in the modulation of the spin-echo decay envelope. The envelope-modulation function is, therefore, a curve representing the fluctuating local magnetic fields experienced by an electron spin. In essence, the spin-echo technique samples a number of S,I states much as is done in the ENDOR experiment. In fact, the frequencies resolvable from modulation functions are just those determined in ENDOR experiments. A quantum-mechanical analysis of the mixing of the electron and nuclear states makes this apparent (Mims and Peisach, 1978). What such an analysis also indicates is that for the modulation to be observed, the electron–nuclear hyperfine interaction ($S \cdot A \cdot I$) must be similar in magnitude to the nuclear Zeeman and quadrupole terms. These latter values for ^{14}N, for example, are ~ 1 MHz and ~ 2 MHz (at 3 kG), respectively, corresponding to a hyperfine coupling of ~ 0.5 G.

Figure 53 presents the envelope modulation patterns obtained for two Cu(II) complexes (Mondovi *et al.*, 1977). In the first, Cu(II)-diethylenetriamine, only ^1H modulation is observed, with a period of ~ 70 nsec. The

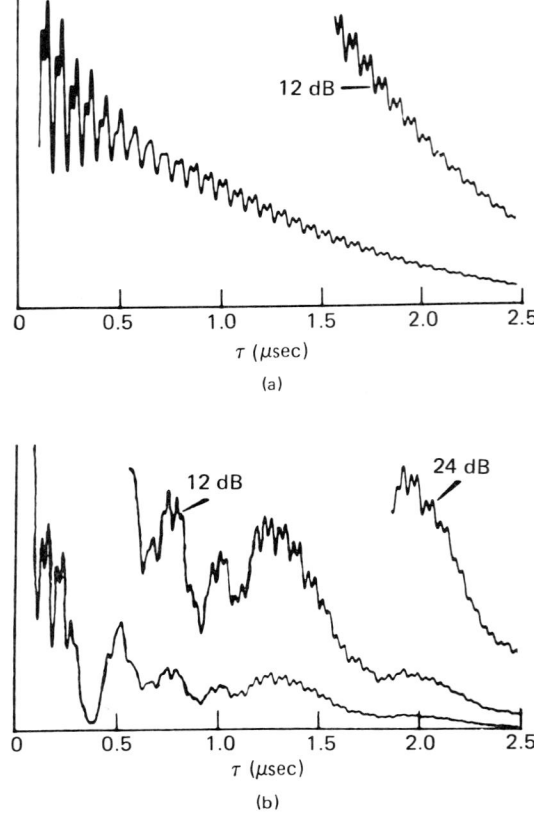

Fig. 53. Envelope modulation spectra for Cu(II)-diethylene triamine (a) and Cu(II)-diethylene triamine imidazole (b). The spectra were recorded with $H_0 \simeq 3.2$ kG at ~ 9.3 MHz and 4.2 K. Complexes were 10 mM in Cu(II) in 50% (v/v) glycerol–water, pH 8.3. (From Mondovi *et al.*, 1977. © 1977 American Chemical Society.)

coordinating ^{14}N do not contribute to the envelope since the $\mathbf{S} \cdot \mathbf{A} \cdot \mathbf{I}$ term (~ 40 MHz) far exceeds the nuclear terms in the spin Hamiltonian. The other pattern is of the same complex in which an imidazole has replaced H$_2$O as an equatorial ligand. The high-frequency ^{1}H precession is still evident as is a lower-frequency modulation attributed to ^{14}N. Again, the coordinating pyridine nitrogen in the imidazole ring is too strongly coupled to modulate the spin-echo decay. However, the distal, noncoordinating, pyrrole nitrogen apparently does contribute to the envelope-modulation function. The period at ~ 3 kG is ~ 0.7 μsec. Thus, among amino-acid ligands of biological significance, the imidazole ring of the histidine side chain is unique in its pattern of modulation (Mims and Peisach, 1978).

The low-frequency modulation of heavier nuclei such as ^{14}N can be emphasized in a three-pulse, or stimulated-emission, sequence. This involves a second 180° pulse at a time T following the first 180° pulse (Fig. 54). In essence, the period between τ and T allows for the dephasing of the ^{1}H fields, leaving only the lower-frequency nuclear moments to contribute to the fluctuations in the H_N fields experienced by the electron spin. While the three-pulse envelope for Cu-diethylenetriamine is smooth (the unmodulated phase memory), that for Cu-diethylenetriamine imidazole contains the 0.7-μsec period attributable to ^{14}N. This envelope can be transformed into a frequency spectrum and simulated with appropriate choice of hyperfine coupling and nuclear quadrupole moment. This is illustrated in Fig. 55 (Mims and Peisach, 1979a).

The details of the three-pulse technique and the transform deserve further comment. The nuclear precession modulating the electron spin is caused by a magnetic field, $H_{\text{eff}} = H_{\text{lab}} \pm \frac{1}{2} Ah/g_N\beta_N$. For the distal nitrogen in a coordinating histidine imidazole, the pyrrole nitrogen, the two components of H_{eff} at the nucleus are of the same absolute magnitude. Thus, when $H_{\text{eff}} = H_0 - \frac{1}{2}Ah/g_N\beta_N$ (when $m_s A < 0$) the field experienced by the nucleus is essentially zero, and the modulation of the spin echo is caused by zero-field quadrupolar coupling associated with the values ν_+, ν_-, and ν_0. Thus, for a given m_s state (depending on the sign of A), a set of modulation frequencies will be detected corresponding to these quadrupolar ones. For this nitrogen in the imidazole ring, these are given in Table XVII.

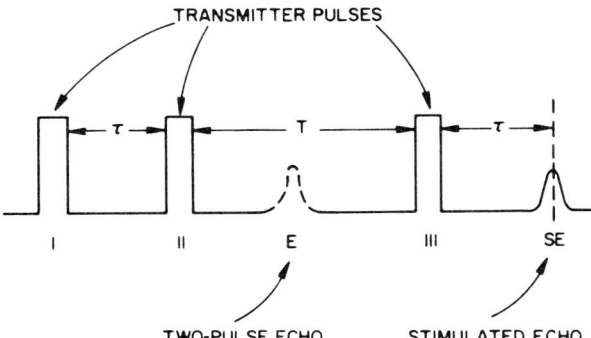

Fig. 54. Three-pulse or "stimulated" spin-echo sequence. The interval τ is held constant and the interval T gradually lengthened, as shown. The sample hold circuit is gated in synchronism with the stimulated echo. (From Mims and Peisach, 1978.)

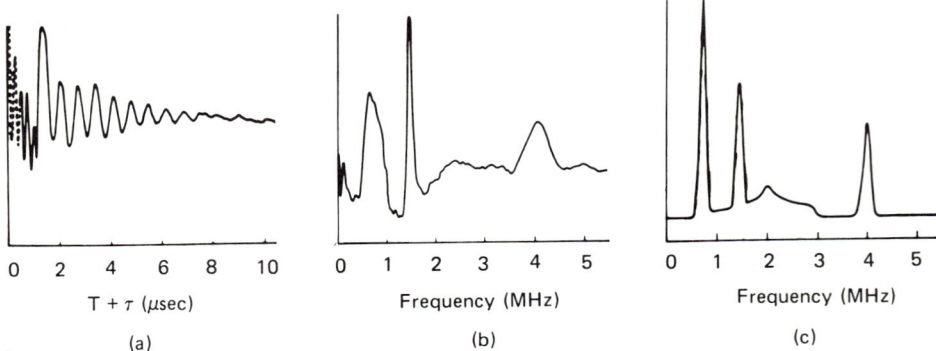

Fig. 55. Three-pulse modulation envelope (a), Fourier transform (b), and simulation (c) for Cu(II)-diethylenetriamine-imidazole. Modulation envelope obtained at 3.17 kG and 9.192 MHz with $\tau = 370$ nsec. The simulation used $A = 1.80$ MHz and the nuclear quadrupole frequencies v_+, v_-, and v_0 for the pyrrole (protonated) nitrogen of imidazole (Table XVII). (From Mims and Peisach, 1979a.)

Of some interest, the data show that these frequencies are quite sensitive to the state of protonation of the pyrrole nitrogen as well as to substitution in the ring. This characteristic makes possible evaluation of subtle bonding patterns in such a ligand. The other m_s state (associated with $H_{eff} = H_{lab} + \frac{1}{2}Ah/g_N\beta_N$) will generate a different nuclear precession dominated by the hyperfine coupling A. Thus, this frequency will yield the value of this interaction.

Returning now to the three-pulse spectrum and its Fourier transform shown in Fig. 55, the frequencies observed at 0.6–0.8 and 1.5 MHz can be assigned to v_-/v_0 and v_+, respectively. The breadth of the line at ~0.7 MHz

Table XVII Imidazole Quadrupole Frequencies[a]

Frequency	:N⌐N-H	:N⌐N⁻
v_+	1.417	2.579
v_-	0.719	2.344
v_0	0.698	0.235

[a] From Hunt et al. (1978).

indicates that either or both v_- and v_0 may be shifted from their value(s) in free imidazole by ligation to the metal. The sensitivity of these frequencies to structural perturbation will be illustrated by the following example. The frequency at 4 MHz can be assigned to the energy state in which H_{lab} and $\frac{1}{2}|A|(h/g_N\beta_N)$, are additive; a value for $|A|$ can be derived from this peak. The determination of all these frequencies, however, relies on a successful simulation, since for none of the experimental frequencies do v_+, v_-, v_0, or A represent independent variables. The simulation of the transform shown in Fig. 55 thus constitutes a fit to the experimental data. The quadrupole frequencies used were those for free imidazole with $|A| = 1.8$ MHz. Values for a characteristic and thus identifying nuclear frequency and A can be extracted only indirectly by simulation of three-pulse spectra; this constitutes a difference between the spin echo and ENDOR as a probe of small, hyperfine frequencies directly.

The spin-echo technique has been most useful in establishing imidazole coordination of transition metals in metalloproteins. As noted, the noncoordinating pyrrole nitrogen contributes a characteristic quadrupolar modulation. Thus, the two- and three-pulse modulation envelopes and Fourier transform of the latter have been obtained for a number of proteins and peptides (Burger et al., 1981; Freedman et al., 1982) including the antineoplastic drug bleomycin (Freedman et al., 1982), containing the metals Co(II), Cu(II), or Fe(II). In comparisons of blue and nonblue Cu(II), spin echo has shown that both types of sites contain imidazole coordination. Thus Cu(II) sites in laccase, ceruloplasmin, stellacyanin, superoxide dismutase, and galactose oxidase all exhibit the modulation pattern characteristic of imidazole coordination (Mims and Peisach, 1976a, 1979a; Mondoví et al., 1977; Kosman et al., 1980; Fee et al., 1981).

The data for galactose oxidase, a nonblue Cu(II) protein, is presented in Fig. 56 (Kosman et al., 1980). For comparison, the same data for the blue Cu(II) protein, stellacyanin, is provided as well (Mims and Peisach, 1976a, 1979a). The salient features of these spectra are as follows:

(1) The greater "depth" of the modulation for galactose oxidase indicates that there are probably two coordinating imidazoles compared with the one in stellacyanin, as established by x-ray crystallography. The depth is related to the number of modulating nuclei and their distance(s) from the paramagnetic center.

(2) The hyperfine coupling used to fit the data was 1.8 MHz for both proteins.

(3) The quadrupolar frequencies used, however, were different. Those for stellacyanin were (for v_+, v_-, v_0) 1.42, 0.72 and 0.70 MHz [identical to free imidazole and Cu(II) DET Im (Table XVII and Fig. 55)], whereas for

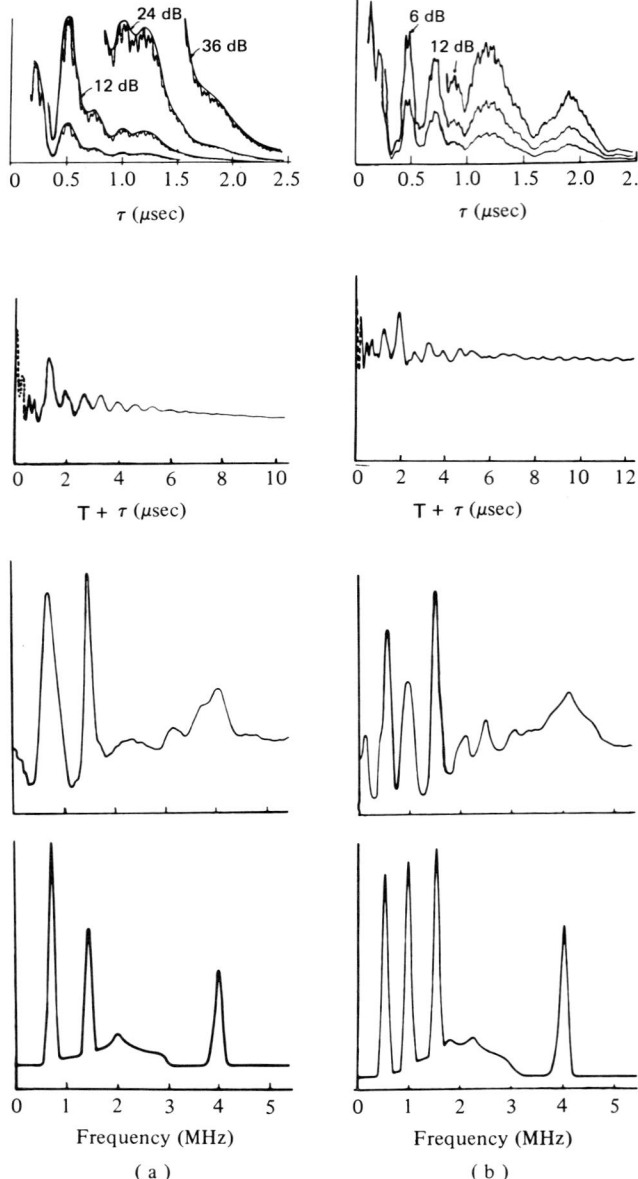

Fig. 56. Spin-echo spectra for stellacyanin (a) and galactose oxidase (b). From top to bottom: two-pulse modulation envelope; three-pulse stimulated-echo modulation; Fourier transforms of three-pulse data; simulation of frequency spectrum. For both simulations, $A = 1.8$ MHz. The values for v_+, v_-, v_0 were: for stellacyanin, 1.42, 0.72, and 0.70 MHz; for galactose

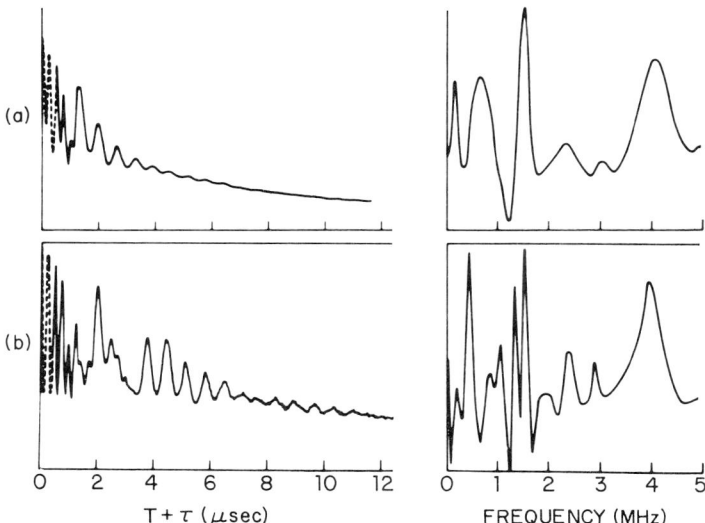

Fig. 57. Three-pulse modulation envelopes for zinc-free bovine superoxide dismutase ($Cu_2/\square$) (a) and the holoprotein (Cu_2/Zn_2) (b), respectively. The left panels are the envelopes in the time domain ($\tau = 372$ nsec); the right panels are the Fourier-transformed spectra in the frequency domain. Note the relatively normal frequency spectrum in (a) with lines at 0.7, 1.5, and 4.0 MHz, whereas that for (b) exhibits a number of additional lines. (From Fee *et al.*, 1981.)

galactose oxidase these frequencies were 1.54, 1.0, and 0.54 MHz, respectively. The increase in v_+, and splitting of v_-/v_0 in galactose oxidase is in the direction of the unprotonated, imidazolate form (Table XVII). This suggests that the pyrrole proton(s) are H-bonded to (a) protein group(s).

This effect, although more marked, is seen in the Fourier transforms of the three-pulse spectra of bovine superoxide dismutase (SOD, Fig. 57) (Fee *et al.*, 1981). The zinc-free form (2 Cu) exhibits a normal frequency spectrum. The alteration in the spectrum caused by reconstitution with Zn presumably results from the perturbation of the quadrupolar frequencies of the pyrrole nitrogen of His_{61}, the bridging ligand. This ligand is in the imidazolate form

---

oxidase, 1.54, 1.0, and 0.54 MHz, respectively. (Stellacyanin data from Mims and Peisach, 1979a; Mims and Peisach, 1976a; galactose oxidase data from Kosman *et al.*, 1980. © 1976, 1980 American Chemical Society.)

in the holo protein. Also, with regard to the discussion in Section IV.D concerning the single-crystal studies of the SOD—CN complex, CN binding to the Cu_2 and holo protein had effects in the three-pulse modulation envelope indicating that this bridging ligand remained intact and was not displaced.

This is an appropriate point to complete the model for CN^- annation to bovine SOD. In another excellent study, Van Camp et al. analyzed the ^{14}N and ^{15}N ENDOR spectra from $C^{14}N^-$ and $C^{15}N^-$—SOD as well as native enzyme (Van Camp et al., 1982). ENDOR lines separated by the Larmor precession frequency for ^{14}N were observed around 20 MHz as well as ~4 MHz. These lines were characteristically better resolved in the CN^- complexes. The differences between $C^{14}N^-$ and $C^{15}N^-$ were manifest in the low-frequency region, that is, they were attributed to the SHFS of the CN^- nitrogen. The ^{14}N ENDOR lines that were independent of this isotopic

Table XVIII ENDOR Frequencies and Hyperfine and Quadrupole Couplings (MHz) for Cu/Zn Superoxide Dismutase and CN^- Complex[a]

Frequency (MHz)	Coordinating ^{14}N		Distal ^{14}N[b]	
	A^N_{zz}	P^N_{zz}	$A_\parallel$	$A_\perp$
SOD				
→ 17.7	38.7	0.86		
→ 19.3				
→ 21.1				
SOD-$C^{14}N$				
→ 17.7	37.0	≪ 1.6	1.6	1.6 (Im)
→ 19.3			5.7	3.9 (CN)
→ 21.51	47.8	1.62		
→ 23.09				
→ 24.55				
→ 26.53				

[a] Adapted from Van Camp et al. (1982).
[b] Values used in simulation of low-frequency spectral region with couplings for protein imidazole (Im) and CN^- nitrogens in the orientations as noted.

substitution could therefore be assigned to the distal, noncoordinating nitrogens of the protein imidazoles, those nuclei which are responsible for the echo-envelope modulation. The ENDOR analysis yielded the frequencies and hyperfine and quadrupolar couplings for these species (Table XVIII).

Three features of these data are of particular note. First is the well-resolved four-line spectrum centered around 23.5 MHz characteristic of ^{14}N spectra given by the sums and differences of the quadrupole and nuclear Zeeman fields. Second, the values of A_{zz}^N (15.5 and 12 G, respectively) can be used to simulate the low-field $M_I = +\frac{3}{2}$ line in the CN$^-$ ESR spectrum, which exhibits the seven-line SHFS (Fig. 58.) An excellent fit is obtained assuming 2N ($A_\parallel$ = 47.77 MHz) and 1N ($A_\parallel$ = 37.1 MHz) (Fig. 58). Third, the low-frequency parameters can be used to recreate the spin-echo frequency spectrum which exhibited lines at 0.55, 0.8, 1.35, and 3.5 MHz for the SOD—CN$^-$ complex. The simulation yielded values of 0.5, 0.9, 1.4, and 3.74 MHz, a very satisfactory fit considering the resolution and simplifying assumptions made. As is indicated by the data which are clearly consistent with the cw and spin-echo ESR results touched on in preceding sections, the interaction of the imidazoles in the native enzyme with unpaired spin is complex, e.g., these four nitrogens are electronically nonequivalent. In the CN$^-$ complex, however, not only is the ESR spectrum axial but the nitrogen coordination is more regular, with two equivalent ^{14}N and one with a 25% weaker coupling. The CN$^-$ is in the plane and presumably trans to this latter, more weakly coupled imidazole.

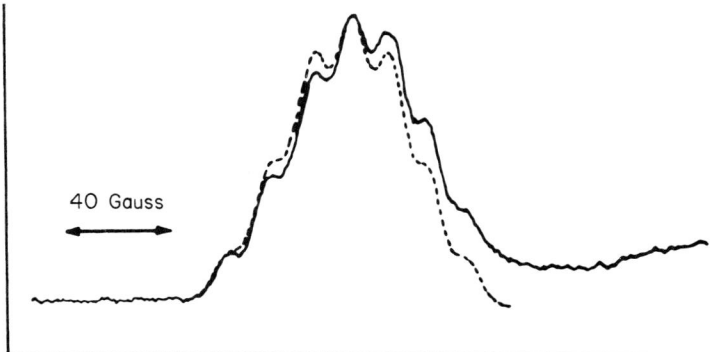

Fig. 58. Comparison of the experimental ESR spectrum for CN—SOD complex (———) of the low-field ^{65}Cu hyperfine peak with a simulated ESR spectrum (– – –) using two nitrogens with hyperfine coupling of A = 47.77 MHz (15.48 G for g = 2.208) and one nitrogen with coupling A = 37.1 MHz (12.02 G). The simulation used a Gaussian line shape with intrinsic linewidth of 12 G at half-height. (From Van Camp et al., 1982.)

The coordination chemistry of a metal site can be characterized by the spin-echo method as well. As in the CN^--binding experiment noted for SOD, exogenous ligand coordination to the Cu(II) in galactose oxidase has also been informative (Kosman et al., 1980). For example, binding of imidazole is not only detected by direct ESR (see Section II.A) but also is seen in the modulation envelope. As expected, the modulation "deepens" upon coordination of a third imidazole. Comparison of two-pulse envelopes of one, two, and three imidazole complexes, as in Fig. 59, indicates the qualitative differences due to variation in the number of modulating nuclei. The use of Cu(II) as a spin-label probe of the metal-binding sites [to which Zn(II) is normally coordinated] in serum albumin (Mims and Peisach, 1976a) and carbonic anhydrase (Kosman et al., unpublished) is illustrated here (Section IV.D).

The nature of anion binding by metals bound to both transferrin (Zweier et al., 1979) and conalbumin (Zweier et al., 1982) has been studied effectively by the spin-echo method. Both proteins bind Cu(II) as well as Fe(III); Cu(II) has proven to be a useful probe of these sites (Section IV.D). Direct ESR has shown that the metal is coordinated to at least one nitrogenous ligand (Zweier and Aisen, 1977; Zweier, 1980). In fact, the modulation envelope exhibits the shallow patterns with peaks at 0.2, 0.5, 0.7, 1.0, and 1.2 μsec that are characteristic of the coupling of a single ^{14}N nucleus. In binding metal ions, both proteins also bind the anions carbonate and oxalate, although the two sites exhibit some specificity. The binding of oxalate can be assessed directly by the spin-echo method because of the weak hyperfine coupling which this technique can detect when [^{13}C]—oxalate is used as ligand (90% isotopic purity). The two-pulse spectra for the [^{12}C]— and [^{13}C]—oxalate—transferrin complexes are given in Fig. 60 along with the envelope for the Cu(II) [^{13}C]—oxalate complex.

The visual comparison of the two transferrin-derived envelopes do not reveal the effect of the isotopic substitution. By taking the ratio of the modulation amplitudes for the [^{13}C] relative to the [^{12}C]—oxalate spectra, however, the ratio pattern obtained is clearly related to the spectrum for Cu[^{13}C]—oxalate. In effect, by taking the ratio of the two envelopes, the contribution of the ^{14}N is removed from the periodicity observed. This is a reasonable manipulation, since a modulation pattern is given by the product of the modulation patterns of the contributing nuclei considered separately. The period for [^{13}C]—oxalate coupled to the Cu(II) is about 150 nsec, which corresponds to a modulation frequency of 6.5 MHz. This is larger than the 3.0-MHz Larmor precision frequency expected for ^{13}C at the field employed (2830 G); thus, there must be a Fermi contact interaction experienced by the nucleus equivalent to the Zeeman field. The actual magnitude of this coupling has not been determined, although NMR and

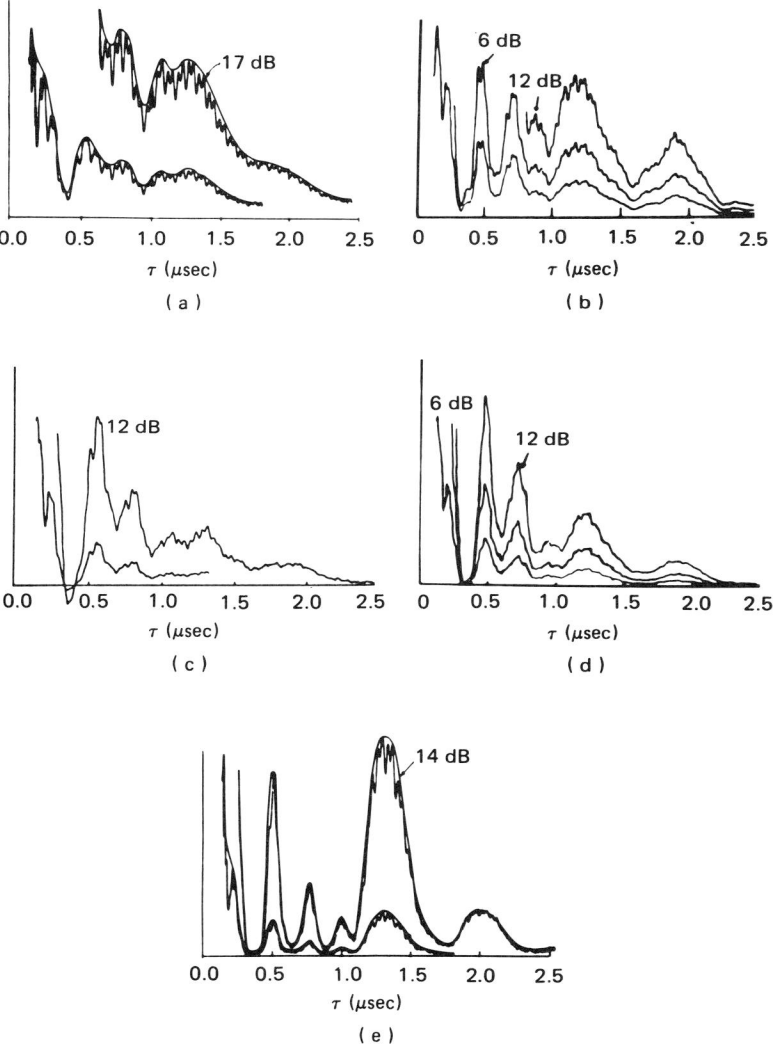

Fig. 59. Two-pulse modulation envelopes for one-, two-, three-, and four-imidazole Cu(II) complexes. These are (a) Cu(II) bovine serum albumin; (b) galactose oxidase; (c) Cu(II) human carbonic anhydrase B; (d) galactose oxidase-imidazole complex; and (e) Cu(II)—tetraimidazole. (From Mims and Peisach, 1976a; Kosman *et al.,* 1980; Kosman *et al.,* unpublished).

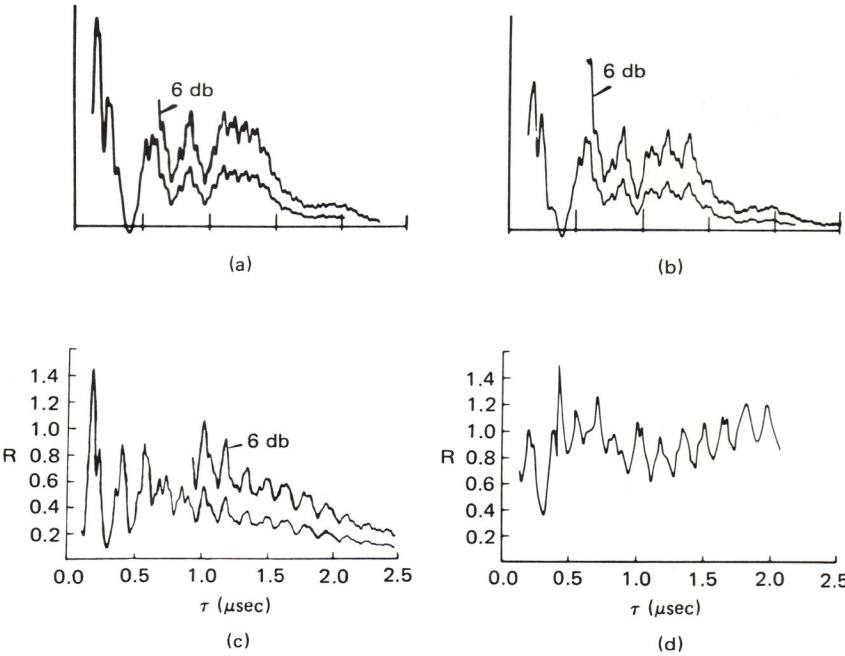

Fig. 60. Two-pulse modulation envelopes for (a) Cu(II), human transferrin [^{12}C]-oxalate; (b) Cu(II), transferrin [^{13}C]-oxalate complex; (c) Cu(II)-[^{13}C]-oxalate complex; and (d) The ratio of modulation envelopes (b)/(a). (From Zweier *et al.*, 1979.)

direct ESR experiments are consistent with an upper limit to this contact hyperfine coupling of 3 MHz or 1 G.

Curiously, $^{13}CO_3^{2-}$ does not introduce an additional element into the modulation envelopes for Cu(II) transferrin or conalbumin (Zweier *et al.*, 1979, 1982). This would result if the ligand were not directly coordinated to the metal (i.e., outer sphere or in a weak, apical coordination site). However, there is considerable evidence that CO_3^{2-} is directly coordinated, although it is significant that a reasonable upper limit to A/h for $^{13}CO_3^{2-}$ in Fe^{3+}—transferrin is ~0.031 MHz (0.09 G), based on an estimate of the ^{13}C NMR linewidth broadening observed (Harris *et al.*, 1974). The spin–echo results indicate that CO_3^{2-} and oxalate coordinate to the metal centers in somewhat different ways. [^{13}C]—oxalate produces an enchanced effect in the modulation envelope, probably as a result of a larger contact interaction.

The spin-echo method is not confined to studies on Cu(II)-containing proteins. A series of measurements on a number of Fe(II) porphyrin complexes and heme proteins indicates some additional experimental aspects of this pulsed ESR technique (Peisach et al., 1979). Peisach et al. attempted to delineate the ^{14}N modulation frequencies observed in these Fe(II) complexes and to assign them to nitrogens in either the porphyrin or axially coordinating nitrogenous ligands. Of particular interest were these latter interactions, since the identification of the axial ligands in cytochrome P-450, for example, is important to the understanding of this enzyme's structure-to-function relationships. To distinguish between modulation caused by the coupling of porphyrin nitrogens and that caused by the axial ligand(s), ^{14}N— and ^{15}N—imidazole were both used, and these frequencies were compared with ligand-free systems. This is reminiscent of the effective use of ^{15}N substitution in the elucidation of the ENDOR spectra for aquometmyoglobin, discussed in Section IV.B.

The effect on the modulation function of this substitution is to eliminate the zero-field quadrupole couplings seen when $H_{\text{eff}} \simeq 0$, i.e., when $H_0 \simeq |A|h/2g_N\beta_N$, and the contribution of the quadrupole coupling in general. This difference is illustrated in Fig. 61, which represents the energy-level

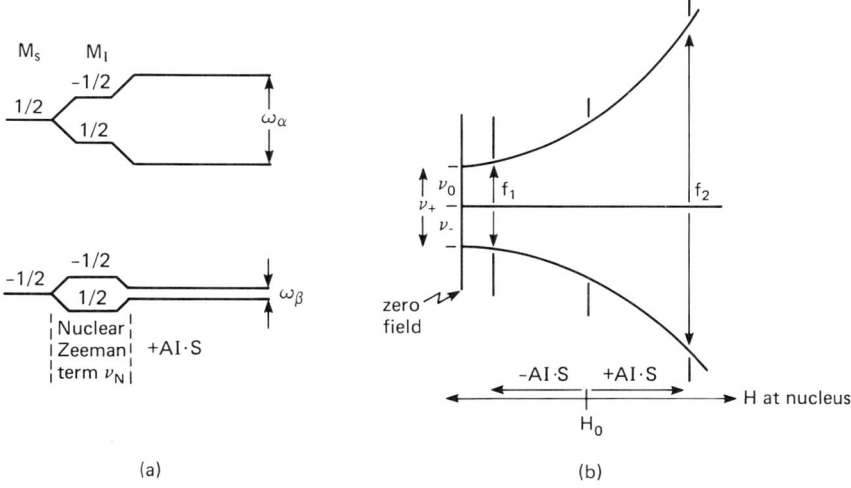

Fig. 61. Energy-level schemes for $S = \frac{1}{2}$, $I = \frac{1}{2}$ (a) and $S = \frac{1}{2}$, $I = 1$ (b) electron-spin–nuclear-spin systems. In (a), the frequencies ω_β and ω_α represent the low- and high-frequency modulation seen in ^{15}N spin-echo modulation envelopes (periods of ~4 and 0.35 nsec, respectively). In (b), if $H_0 \simeq \frac{1}{2}|A|H/g_N\beta_N$, then the frequency f_1 will be v_+ and f_2 a strong, higher-frequency line in the transform for an ^{14}N system. (From Mims and Peisach, 1978.)

schemes for $S = \frac{1}{2}, I = \frac{1}{2}$ (^{15}N) and $S = \frac{1}{2}, I = 1$ (^{14}N) systems. The former was discussed in Section V.B, ENDOR (compare with Fig. 48). The values ω_α, ω_β in Fig. 61 and the sums and differences of them may be seen as frequencies in a Fourier transform of the modulation envelope. The low-frequency modulation ω_β is associated with the spin state when the nuclear Zeeman term, $\omega_I = g_N \beta_N H_0/\hbar$, is nearly canceled by the contact term $AI \cdot S$. For the time evolution of the spin-echo amplitudes in the three-pulse regime, the modulation envelope is given by

$$E_{\text{mod}}(\tau, T) = 1 - \tfrac{1}{2} k \sin^2 \tfrac{1}{2}\omega_\alpha \tau [1 - \cos \omega_\beta (\tau + T)]$$
$$+ \sin^2 \tfrac{1}{2}\omega_\beta \tau [1 - \cos \omega_\alpha (\tau + T)], \tag{46}$$

where k is a term describing the depth of the modulation.

What is clear from Eq. (46) is that at certain times τ, $\omega_\alpha \tau = 2n\pi$ (i.e., if τ is a multiple of the period of the frequency $\omega_\alpha/2\pi$), the term in $\cos \omega_\beta (\tau + T)$ will be missing from the envelope. Similarly, if $\omega_\beta \tau = 2n\pi$, the term in $\cos \omega_\alpha (\tau + T)$ will be missing. Since the three-pulse envelope is generated by fixing τ and incrementally increasing T (compare with Fig. 54), the choice of τ can be used to reduce or enhance the contribution of a particular superhyperfine transition to the spectrum.

For example, consider Fig. 62 (Peisach et al., 1979). The three-pulse spectrum of ^{15}N–Im–heme–RSH with $\tau = 140$ nsec (a) exhibits envelope amplitudes missing when τ is increased to 300 nsec (b). The transform of the former yields a frequency spectrum with strong components at 3.5 and 6.8 MHz (Table XIX). The same pair of three-pulse protocols using ^{14}N–Im–heme–RSH resulted in a similar spectrum at $\tau = 140$ nsec (see Fig. 62) but the $\tau = 300$ nsec data exhibited a periodicity that transformed into modulation frequencies of 0.4, 1.7, and 2.1 MHz (Table XIX). Replacing imidazole by pyridine in the heme complex yielded the data given in Table XIX, namely, the 140 nsec spectrum contained (at least) the 6.7-MHz coupling whereas the 300 nsec one contained a set of three couplings, comparable to those seen with imidazole. The data obtained for cytochrome P-450 are also listed in the table.

These results can be interpreted consistently as follows. The lines seen at 140 nsec must be from the porphyrin, since they are not affected by the isotopic substitution. That two lines are observed can be ascribed to the large contact term, 4.2–4.5 MHz, which dominates the field seen by the nucleus ($H_{\text{nuc}} = \tfrac{1}{2}|A|h/g_N\beta_N \pm H_0$, where $\tfrac{1}{2}|A| \simeq 2.2$ MHz and the Zeeman field $\simeq 1$ MHz). Thus, the two lines correspond to the frequencies f_1 and f_2 seen in Fig. 61. At longer times, τ, these frequencies are suppressed, but lower-frequency modulation (associated with even longer periods) can be enhanced. In fact, although $\tau = 300$ nsec effectively suppresses the ^{14}N

Table XIX ^{14}N Modulation Frequencies for Fe(II)—Heme Complexes[a]

Complex[b]	τ (nsec)	Modulation frequencies (MHz)	A ($g = 2$) (MHz)	^{14}N assignment
[^{15}N]—Im·heme·RSH	140	3.5, 6.8	4.2	Porphyrin
[^{14}N]—Im·heme·RSH	140	3.5, 6.8 (5.15)[c]	4.3	Porphyrin
	300	0.4, 1.7, 2.1	2.0	Imidazole N$_1$ (pyridine)
Py · heme · RSH	140	6.7 (5.2)[c]	4.5	Porphyrin
	300	0.5, 2.1, 2.6	2.0	Pyridine
Cytochrome P-450	360	0.7, 1.5, 4.1	1.8	Imidazole N$_3$ (pyrrole)

[a] Adapted from Peisach et al. (1979).
[b] Im is imidazole; heme, from hemin chloride; RSH, β-mercaptoethanol; and Py, pyridine.
[c] Values in parentheses attributed to axial ligand.

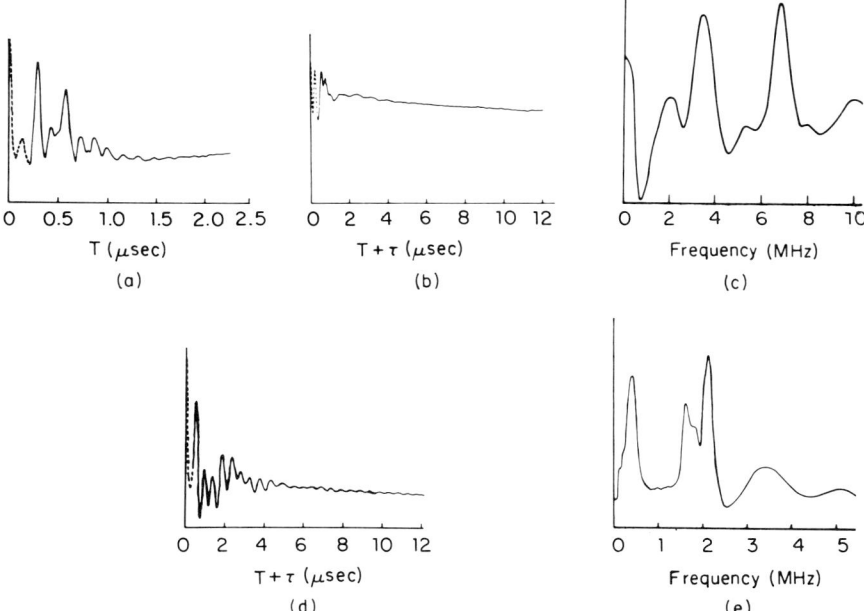

Fig. 62. Three-pulse modulation envelopes and Fourier transforms for [^{15}N]-imidazole heme mercaptide (a–c) and [^{14}N]-imidazole heme mercaptide (d–e) complexes. The value for τ was 140 nsec in (a) and (c), 300 nsec in (b), (d), and (e). $H_0 \simeq 3$ kG at ~ 9.4 MHz. (From Peisach et al., 1979.)

modulation due to the porphyrin, it enhances that due to the coordinating imidazole and pyridine ligands. Thus, the frequencies observed at this interval can be assigned to these ligand nitrogens.

Two experimental observations helped to distinguish between N_1 (pyridine) and N_3 (pyrrole) imidazole nitrogen coupling. First, note that pyridine itself gave a set of data comparable to that for [^{14}N]–imidazole. This by itself suggests that it is the pyridinelike nitrogen in imidazole that is coupled and is presumably the coordinating atom. Consistent with this are the frequencies themselves, which can be assigned to the nuclear quadrupole frequencies $v_+ = 2.1$, $v_- = 1.7$, and $v_0 = 0.4$ MHz for the [^{14}N]–Im complex and the corresponding set for the Py complex. These are more characteristic of the unprotonated nitrogen (the pyridine nitrogen) in imidazole (Table XVII). Thus, the axial nitrogenous ligand is rather weakly coupled to the spin on the iron; consequently, the distal pyrrole nitrogen in the imidazole does not contribute to the modulation field at all.

This differs from those cases discussed previously and for the cytochrome P-450 heme system. As the data in Table XIX show, the coupled ^{14}N generates frequencies that can be ascribed to the pyrrole nitrogen in a coordinating imidazole. Thus, in the enzyme, the axial histidine is much more tightly coupled than in those models which also contain a transverse mercaptide ligand. This presents a paradox because a variety of evidence indicates that the sixth ligand in the enzyme is a cystienyl sulfur, which would be expected to exert a trans effect on the coordination of the nitrogen ligand. Although this is observed in the models, this expectation is not realized in the enzyme. Perhaps bonding and steric constraints imposed by the protein account for this difference.

Most recently, deuterium (d) substitution in substrates for metalloenzymes has been used in conjunction with the spin-echo technique to characterize substrate-metal geometry in the enzyme active site. Groh et al. (1983) have determined the spatial relationship between the substrate, 22(R)-hydroxycholesterol-22-d, and cyt P450$_{scc}$, the enzyme responsible for the side-chain cleavage involved in the conversion of cholesterol to pregnenolone, the precursor for all steroid hormones. The modulation associated with the 22-d was clearly evident in the three-pulse echo decay of the enzyme in the presence of the labeled substrate. Fourier transformation yielded the frequency of this modulation (1.9 MHz) to be compared to the expected nuclear frequency for deuterium at the magnetic field of the resonance stimulated (1.93 MHz at 2950 G). Based on a choice of isotropic hyperfine coupling ($A_0 = 0.1$ MHz) and quadrupolar coupling ($Q = 0.05$ MHz), the distance between the deuterium and the Fe(III) in the porphyrin was calculated to be 4 Å. An indication of the significance of this result was the observation that 22(S)-hydroxycholesterol-22-d, the non-substrate enantiomer, did not give rise to this low-frequency deuterium modulation although it binds to the enzyme in a competitive fashion. The deuterium must be at a distance greater than 6 Å from the iron in this complex. In summary, specific isotopic substitution together with analysis of echo modulation patterns offers significant potential in the study of protein–ligand interactions.

The spin-echo technique has been exploited in another type of measurement, that of the electric-field effect on unpaired electron-spin density as characterized by g-value shifts, the electron Stark effect. This effect is associated normally with changes of optical spectra as a result of the interaction of the absorbing molecule with an electric field. The effect is ascribed to the displacement of the electron distribution in the molecule caused by the external field. Since it is the molecular electric field(s) experienced by an electron which gives the unpaired electron its characteristic g value(s), an external electric field would be expected to cause shifts in g values, as well. Note that for shifts in g to be observable, the paramagnetic center must be noncentrosymmetric. Thus, a first level of interpretation of any *observable* g shift relates to the symmetry of the paramagnetic center. The measurement and usefulness of the electron Stark effect will now be outlined briefly (Mims, 1974; Mims and Peisach, 1978).

Consider the sequence of events diagrammed in Fig. 63. The spin-echo

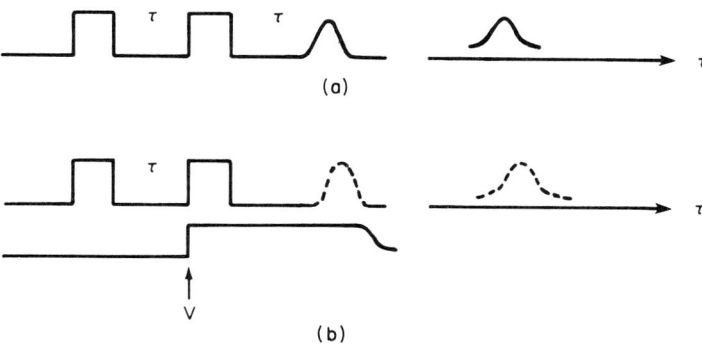

Fig. 63. (a) The two-pulse microwave sequence is shown and the echo signal generated at one time τ. (b) The same time τ, between the first and second pulses, is employed, but an electric field is turned on with the second pulse. It remains on until the echo signal is detected, but is turned off before the next pulse cycle. In (b), the electric field is shown as having "delayed" the echo, thus at time τ there is no echo signal. The g-value shift (and thus the change in electron precessional frequency) induced by the field has caused this alteration in the echo amplitudes at specific times τ_i.

cycle is initiated as has been described. Simultaneously with the second (180°) pulse, an electric field is switched on and remains on until after the appearance of the echo signal. This field is typically 100 kV/cm. The presence of the electric field causes a change in the local (orbital) field experienced by the precessing electron spin and thus the electron's precessional frequency. Consequently, the process which after a time τ brings the spin packet into convergence (the echo) is perturbed. For example, suppose of two spin packets, the faster were speeded up and the slower slowed still further. At certain times τ, a null point in the echo amplitude would be expected because the two spin packets would be out of phase by exactly π. The shift in frequency associated with these times, τ_n, is given by (Peisach and Mims, 1973; Mims and Peisach, 1976b) $\Delta\omega = (2n + 1)\pi/2\tau_n$.

The shift in frequency is related to the shift in g by the basic equation for the Zeeman of the electron. Rather than attempting to find these specific null times, one can simply define some finite value of the echo amplitude (e.g., $A_\tau^{(E)}/A_\tau = 0.5$, the "half-fall" value where $A_\tau^{(E)}$ is the amplitude in the presence of the electric field) and determine the "half-fall" time $\tau_{1/2}$ at which this value occurs. The frequency shift at $\tau_{1/2}$ is $\pi/6\tau_{1/2}$. $\Delta\omega$ has also been represented by f or Δf, and, as such, used to define a shift parameter σ (or S). This is defined as $\sigma = d/6f(\tau V)_{1/2}$, where d is the thickness of the sample and $(\tau V)_{1/2}$ the product of time and voltage which yields the "half-fall" of the echo amplitude. This indicates that varying the voltage at a fixed τ will cause a variable shift and thus a varying echo-amplitude attenuation. As indicated, however, the product $(\tau V)_{1/2}$ is determined at a fixed voltage.

The detection of a g-value shift, although indicative of noncentrosymmetry, is the minimum information derivable from an LEFE ("linear" electric field effect) experiment. In a metal (or other paramagnetic) center which exhibits g-value anisotropy, the potential for shifts will be determined by two factors: the orientation of the magnetic and electric fields relative to one another and to the principal g tensor(s); and the symmetry of each g tensor, that is, whether or not an axis contains an odd component (is not itself centrosymmetric). This latter feature is characteristic of tetrahedral geometry, for example, or D_{2h} or D_{4h} systems in which a significant ligand difference exists on any particular axis in the two spatial directions (±displacement). Thus, by physically changing the alignment of the electric and magnetic fields relative to one another (generally ∥ and ⊥) and, at a fixed microwave frequency, by appropriate choice of magnetic field (to select spin packets of specific g value) one can locate in a ligand field the major "odd component(s)," that ligand element or geometric distortion which in greatest measure contributes to the noncentrosymmetry of the paramagnetic center.

LEFE has been useful, certainly, in the context of the time of its application. For example, the continuing effort to determine what constituted the fundamental difference(s) between blue and nonblue (types 1 and 2) Cu(II) in various proteins was aided by LEFE experiments (Mondoví et al., 1977). Subsequently, actual structures became available which now obscure to some extent the contributions made by earlier spectroscopic techniques. Type 1 Cu(II) was thought to be characterized by either or both mercaptide ligation or a tetrahedral ligand field. The latter by itself is centrosymmetric, whereas, the former would contribute an odd component which would generate a g-value shift in certain orientations of the electric and magnetic fields. No such coordination elements have been ascribed to type 2 Cu(II) centers, which are thought to be comparable to simple N,O-containing Cu(II) complexes.

The LEFE clearly indicates these differences (Fig. 64). A nonblue Cu(II) center (galactose oxidase) exhibits a very small, nearly field- and orientation-independent g- shift (Kosman et al., 1980). This is similar to hexaquocopper (II). On the other hand, a type 1 blue Cu (II) center, as in stellacyanin, exhibits a much larger shift which is highly dependent on the relative orientation of the electric and magnetic fields and on the relative orientation of the magnetic field and g tensors (Mims and Peisach, 1979b). The larger shift can be ascribed to the polarizability of the mercaptide sulfur. That the maximum shift occurs in $g_\parallel$ when $E \perp H_0$ (and thus $E \perp g_\parallel$) suggests that the polarizable (odd component) is in $g_\perp$. This suggestion has proven consistent with the crystal structure of stellacyanin (Gray and Solomon, 1981). Furthermore, the type 1 Cu(II) in laccase and azurin also exhibit comparable shifts, whereas those for the Cu(II) in galactose oxidase (and hexaquo copper) are much smaller. The use of shift values to distinguish type 1 (blue) and type 2 (nonblue) Cu(II) is attractive, as indicated by the data in Table XX. There seems to be a good correlation between relative $g_\parallel/A_\parallel$ (compare with Section IV.A.1) and σ values for type 1 and type 2 proteins, that is, the same factor(s) which determines the unusual $A_\parallel$ values for the type 1 sites contributes to the large LEFE. However, as the data in the table show, the molar extinction of the ~600-nm band is a somewhat less distinguishing feature of these sites inasmuch as $\epsilon_{630\ nm}$ for galactose oxidase is much more similar to that for the blue copper proteins than to aqueous Cu(II) (Ettinger and Kosman, 1982). It is of some interest that the LEFE of the Cu_A in cytochrome c oxidase yielded an anomalous LEFE which could not be interpreted in the simple context developed here (Mims et al., 1980).

Another system which has been exploited by the LEFE technique has been the low-spin Fe(III) heme complex (Peisach and Mims, 1973; Mims and Peisach, 1976b). Briefly, maximum shifts occur when the electric field is aligned with or has a component in the direction of the coordination axis of the ligand field with the largest difference between the transverse ligands.

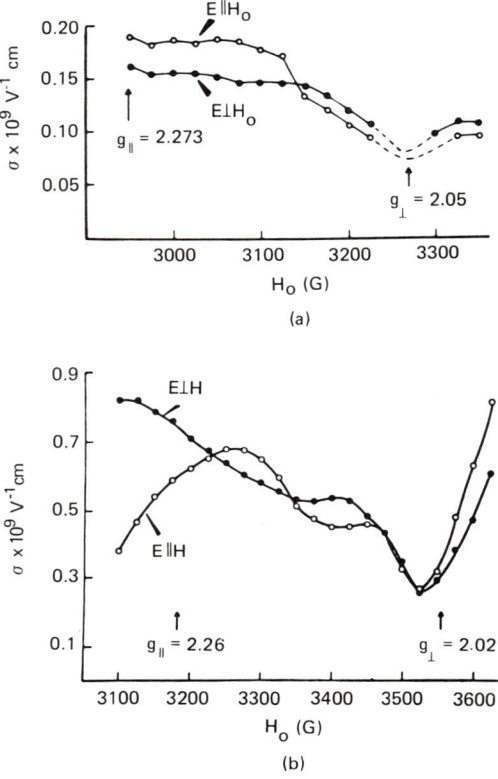

Fig. 64. LEFE curves at ~9.4 GHz for galactose oxidase (a) and stellacyanin (b), with the relationships between the electric field E and magnetic field H indicated. Shift parameter σ determined as described in the text. [(a) from Kosman *et al.*, 1980; (b) from Mims and Peisach, 1979b.]

Thus, in an Fe porphyrin, the maximum difference would be between the axial ligands, e.g., mercaptide versus imidazole. Indeed, as the data in Table XXI indicate, this expectation is realized. These limited data show some anomalies, however, particularly in the context of the modulation frequencies presented in Table XIX. Although pyridine and imidazole exhibit a similar contact interaction, imidazole clearly has a larger effect on the ligand field, i.e., it is more like RS^- than is pyridine. A possible explanation for this is that the coordinated imidazole has partially deprotonated, that is, it is an imidazolate anion. Again, the modulation frequencies do not indicate this, unless, of course, they are for the distal nitrogen (unprotonated) and not the coordinating one. This type of anomaly exists also for cytochrome P-450 in

Table XX LEFE Maximum Low-Field-Shift Parameters with $E \perp H_0$, $g_\parallel/A_\parallel$ Values, and Extinction Coefficients for Some Protein Cu(II) Sites

Protein	$\sigma \times 10^9$ (Vcm^{-1})	$g_\parallel/A_\parallel$ (mK^{-1})	λ_{max} (nm)	ϵ (M^{-1} cm^{-1})	Reference
Blue Cu(II)					
Azurin	0.70	0.38	625	3500	Mims and Peisach (1979b)
Stellacyanin	0.85	0.65	604	3820	Mims and Peisach (1979b)
Decuprolaccase	1.25	0.37	614	5700	Mondoví et al. (1977)
Nonblue					
Galactose oxidase	0.15	0.12	630	1015	Kosman et al. (1980)
[Cu(II)(H$_2$O)$_6$]	0.15	0.11[a]	780[a]	15[a]	Mims and Peisach (1979b)

[a] From G. DaCosta and D. J. Kosman, unpublished.

that while the modulation frequencies and contact term show the putative imidazole to be a "normal" ligand, this heme complex exhibits the smallest shift.

Perhaps the best indication that the spin-echo technique and the related LEFE one are useful is that the data they generate give rise to difficult yet fundamental questions. The spin-echo technique is complementary to ENDOR in that it can resolve hyperfine interactions that go unresolved even in the ENDOR line envelope. The extension of spin echo to single-crystal studies should prove useful in the assignment of specific nuclei in the protein matrix to observed superhyperfine couplings. There is every indication that

Table XXI Maximum Shift Parameters (σ) for Fe(III) Heme Complexes[a]

Complex	$\sigma \times 10^9$ V^{-1} cm	
	$H_0 \parallel E_0 \parallel g_3$[b]	$H_0 \perp E_0 (\parallel g_1/g_2)$[b]
Myoglobin · N$_3^{-c}$	2.3	2.35
Py · Heme · RSH[d]	2.1	0.55
Im · Heme · RSH[c]	1.4	0.55
Cytochrome P-450	1.0	0.50

[a] Adapted from Peisach and Mims (1973).

[b] g_3 is normal to porphyrin ring; g_1/g_2 are in porphyrin plane.

[c] Transverse ligand is imidazole.

[d] Py is pyridine; heme, hemin chloride; RSH, β-mercaptoethanol; and Im, imidazole.

ENDOR and pulsed ESR are the two "growth" areas in biological electron spin resonance.

D. Saturation Transfer

The absorption of energy by a spin packet from the microwave field H_1 causes a realignment of the spin moment in H_0. Relaxation is the time-dependent process of returning these dipoles to their ground state in which the Boltzmann distribution of dipole orientations is reestablished. This reorientation can be electronic, that is, associated with longitudinal (T_{1e}) and transverse (T_{2e}) relaxation processes, or motional, associated with the rotational correlation time τ_2. (The subscript 2 is used here rather than "c" since this is the notation customary in the saturation-transfer literature.) The combination of both types of relaxation processes has been described as the random walk of a drunken sailor (τ_2) coupled to the sailor's tendency to stumble and fall periodically (T_{1e}). The mean square angular displacement of a spin before it *electroncially* relaxes is $\theta^2(T_{1e})$. It is related to T_{1e} and τ_2 by

$$\theta^2(T_{1e}) = 2T_{1e}/3\tau_2, \quad (47)$$

$$\tau_2 = \tfrac{2}{3}T_{1e}/\theta^2(T_{1e}). \quad (48)$$

For nitroxides, $T_{1e} \simeq 10^{-5}$ sec, thus measurement of $\theta^2(T_{1e})$ yields a value for the rotational correlation time τ_2.

The significance of this relationship is that the use of the nitroxide as a probe of macromolecular motion can be extended below the rigid limit ($\tau_2 \sim 10^{-7}$ sec) available to line-shape analysis of the first-derivative ESR spectra. Conventional ESR is sensitive to nitroxide motion only within a period T_{2e} (related to the inhomogeneous linewidth) given by

$$\tau_2 \leq \tfrac{8}{3}T_{2e}^3(A_{zz} - A_{xx})^2. \quad (49)$$

Saturation-transfer (ST) ESR extends this sensitivity into the millisecond range:

$$\tau_2 \leq \tfrac{8}{3}T_{1e}T_{2e}^2(A_{zz} - A_{xx})^2, \quad (50)$$

because of the 300-fold difference between T_{1e} and T_{2e} ($\sim 10^{-8}$ sec). The indirect evaluation of $\theta^2(T_{1e})$ by saturation-transfer spectroscopy is the value of this technique.

Saturation transfer can be described as the apparent diffusion of spins out of a spin packet at resonance under saturating conditions. This "diffusion" is manifest as an absorption of energy detected *out of phase* with the field H_m which modulates the external field H_0. In fact, this out-of-phase absorption is owing entirely to the *failure* of the spin system to keep up with the modulation field. The intensity of this out-of-phase absorption is maximal when $\omega_m T_{1e} \simeq 1$, that is, when the frequency of the modulation field is approximately equal to the longitudinal electron-relaxation rate (frequency). It is the *shape* of this out-of-phase absorption which is related to τ_2; the sensitivity of spectral line shape to τ_2 is greatest when $\omega_m \tau_2 \simeq 1$. Since most experiments are performed with $\omega_m \simeq 50-100$ kHz, the technique is most sensitive to changes in rotational rates around 10^5 sec^{-1}.

The analysis of saturation-transfer spectra, yielding motional information, is almost entirely carried out by reference to a "model" system. The regions of a typical nitroxide spectrum which are most sensitive to changes in θ can be theoretically evaluated, however. The spectra in Fig. 65 illustrate this (Thomas *et al.*, 1976). V_0 is the simple absorption spectrum of an isotropic, rigid-limit nitroxide population; V_1 the common first-harmonic (derivative) spectrum. The lower trace is the derivative of the spin Hamiltonian used to generate the absorption envelope V_0 with respect to θ, the external magnetic-field orientation in the molecular coordinate system. As is clear from this simulation, the regions of the *absorption* envelope V_0 *between* turning points are most sensitive to rotational diffusion (changes in θ). Consequently, the common first-derivative absorption spectrum V_1 is not an appropriate display for ST–ESR. Rather, either the first harmonic of the dispersion signal U_1, or more typically, the second harmonic of the absorption V_2 are used. The ST–ESR spectra (out of phase, as denoted by the *prime*) in these two displays for a model system, hemoglobin labeled with maleimide nitroxide Ib (compare with Fig. 29), are presented in Fig. 66 (Thomas *et al.*, 1976). The values for τ_2 shown in the figure are calculated using the Stokes–Einstein relationship; the nitroxide is taken as rigid within the protein matrix.

(Dispersion spectra are not commonly used because few spectrometers or their operators are equipped to generate them. Dispersion is related to the slight *frequency* shift that occurs in the cavity when the microwave field passes through an absorbing sample, similar to the refractive index increment of a solution. Whereas absorption spectra record the effective change of the Q of the cavity—its transparency—dispersion spectra are generated by the frequency shift at resonance. These two responses are associated with the imaginary (Q) and the real (Δf) components of the rf magnetic susceptibility. Since most spectrometers are operated so that the detector *follows* the microwave frequency, frequency changes cannot be measured. However, an investigator planning substantial use of ST–ESR should consider acquiring a microwave bridge possessing a reference arm usable for both absorption and dispersion.)

The spectra in Fig. 66 demonstrate what was simulated in Fig. 65, namely,

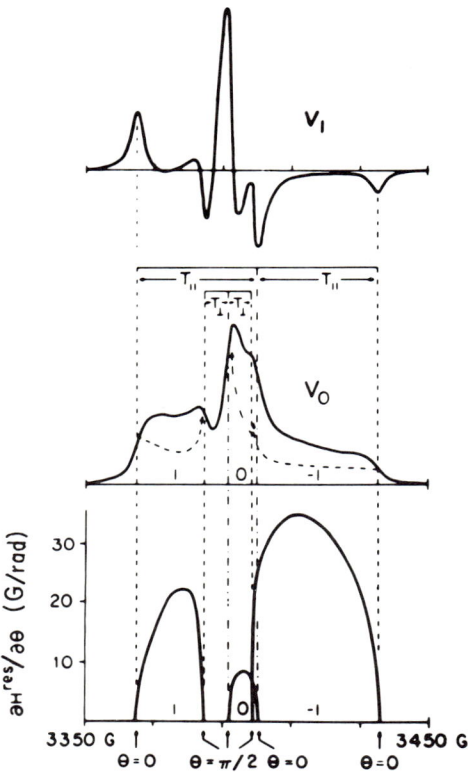

Fig. 65. Simulated ESR spectra and derivative of an isotropically oriented, motionless nitroxide population at 9 GHz. The top spectrum V_1 is the normal first derivative (harmonic) of the absorption spectrum shown below (V_0). The $T_\parallel$ and $T_\perp$ hyperfine tensors are indicated here, as well as the M_I transition regions (± 1, 0). The lower trace is the derivative of V_0 with respect to θ, the angle between the magnetic-field direction and the nitroxide magnetic tensor at excitation. The sensitivity of this derivative (its amplitude) is greatest between the turning points of the V_1 spectrum, rendering the latter relatively useless as a diagnostic of slow diffusion. (From Thomas *et al.*, 1976.)

that the regions between turning points—those regions between the center and the extreme wings—are most sensitive to changes in motion within this regime of τ_2. Because of this, the use of this type of reference system in the evaluation of ST–ESR data for other systems involves comparison of ratios of line heights. Specifically, the rations of extreme features at low (L'' to L') and high (H'' to H') field, respectively, give a good correlation to τ_2 in a system undergoing isotropic motion (Fig. 67).

The higher resolution afforded by Q-band (35 GHz) ESR of the spectral turning points associated with the principal values should enable similar evaluation of anisotropic motion (Hyde and Johnson, 1981; Johnson et al., 1982). At 9 GHz, only the low- and high-field turning points are resolved and amenable to analysis (e.g., as done empirically in Fig. 67). For example,

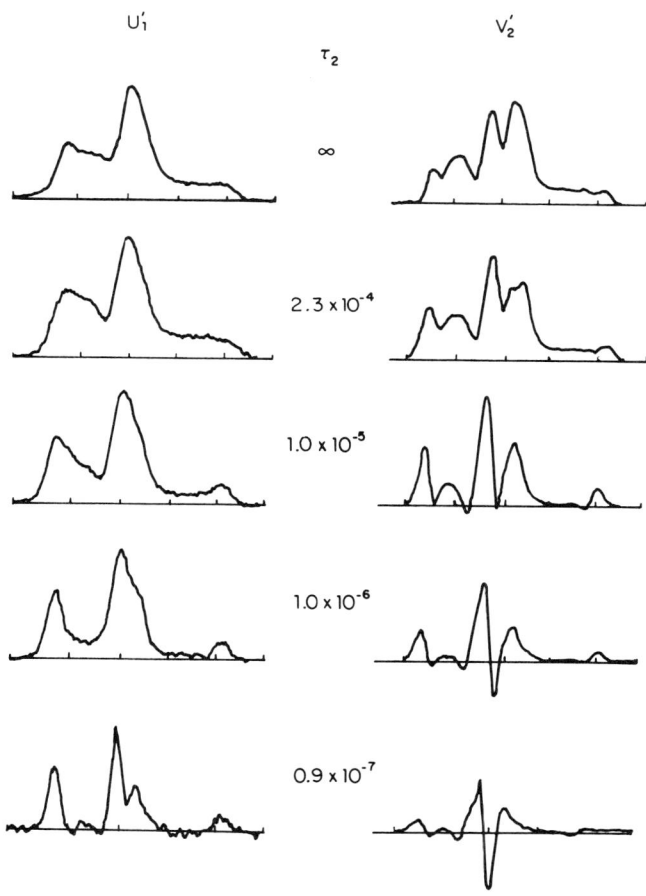

Fig. 66. Saturation transfer for maleimide (Ia)-hemoglobin in mixtures of glycerol-H_2O; τ_2 values were calculated using the Stokes–Einstein relationship, assuming a rigidly bound nitroxide. The U'_1 series includes the (90°) out-of-phase, first-harmonic, dispersion spectra; these are designated by the prime, the subscript, and U, respectively; V'_2 refers to the out-of-phase second-harmonic absorption (V) spectrum. The frequency was 9 GHz. (From Thomas et al., 1976.)

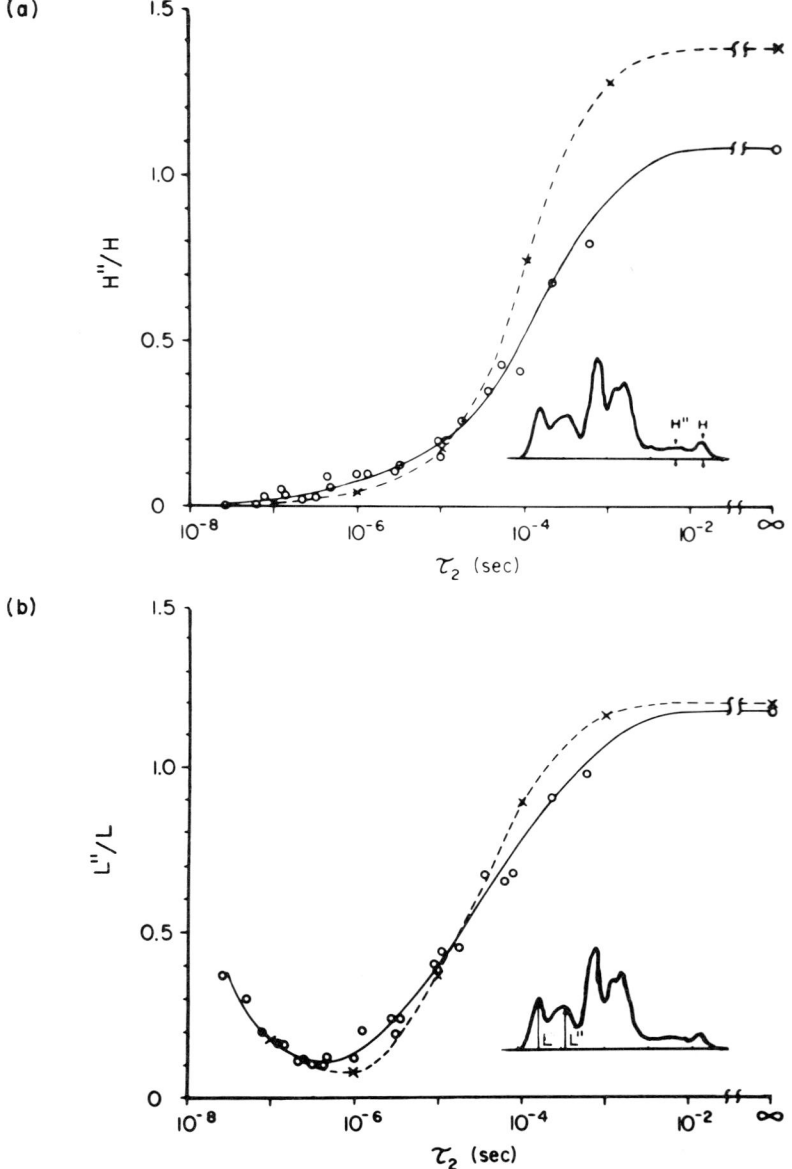

Fig. 67. Line height ratios of reference spectra used for τ_2 determination. The two common linewidth ratios derived from model spectra, as in Fig. 66, are displayed as a function of τ_2. (———, ○) is the experimental curve, (---, ×) the theoretical, assuming isotropic Brownian diffusion. These ratios afford the best correlation between experiment and theory. (From Thomas et al., 1976.)

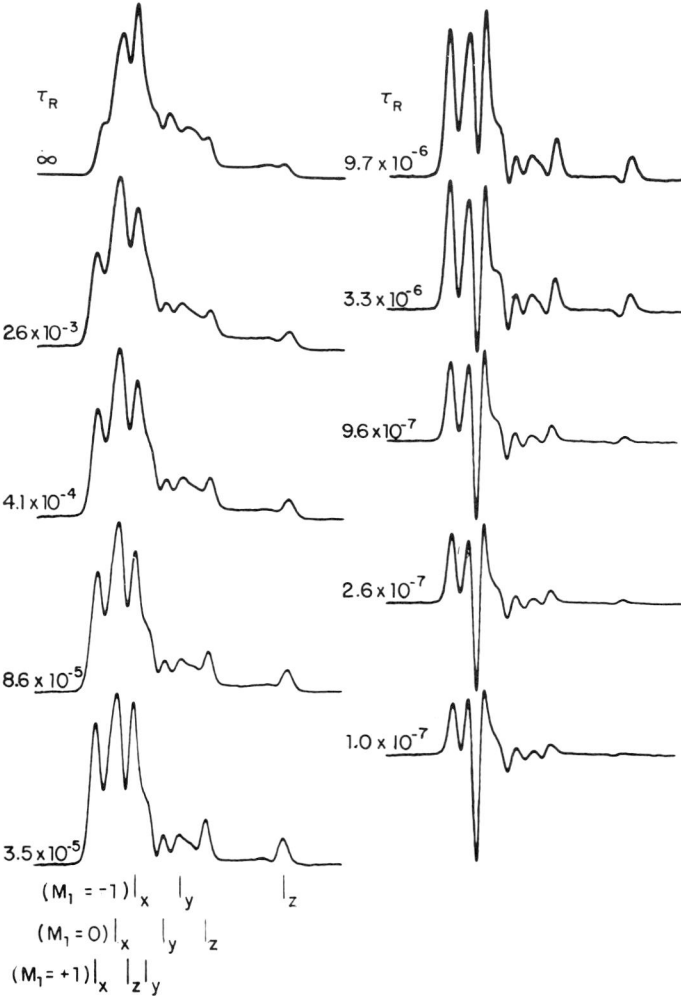

Fig. 68. Saturation-transfer (ST)–ESR spectra (V_2') of maleimide (Ia)-hemoglobin at 35 GHz. Solutions were prepared and τ_2 values calculated as in Fig. 66. The turning points in X, Y, and Z for each of the ^{14}N M_I spin states are shown; nearly all of them are clearly resolved at this field. (From Johnson and Hyde, 1981.)

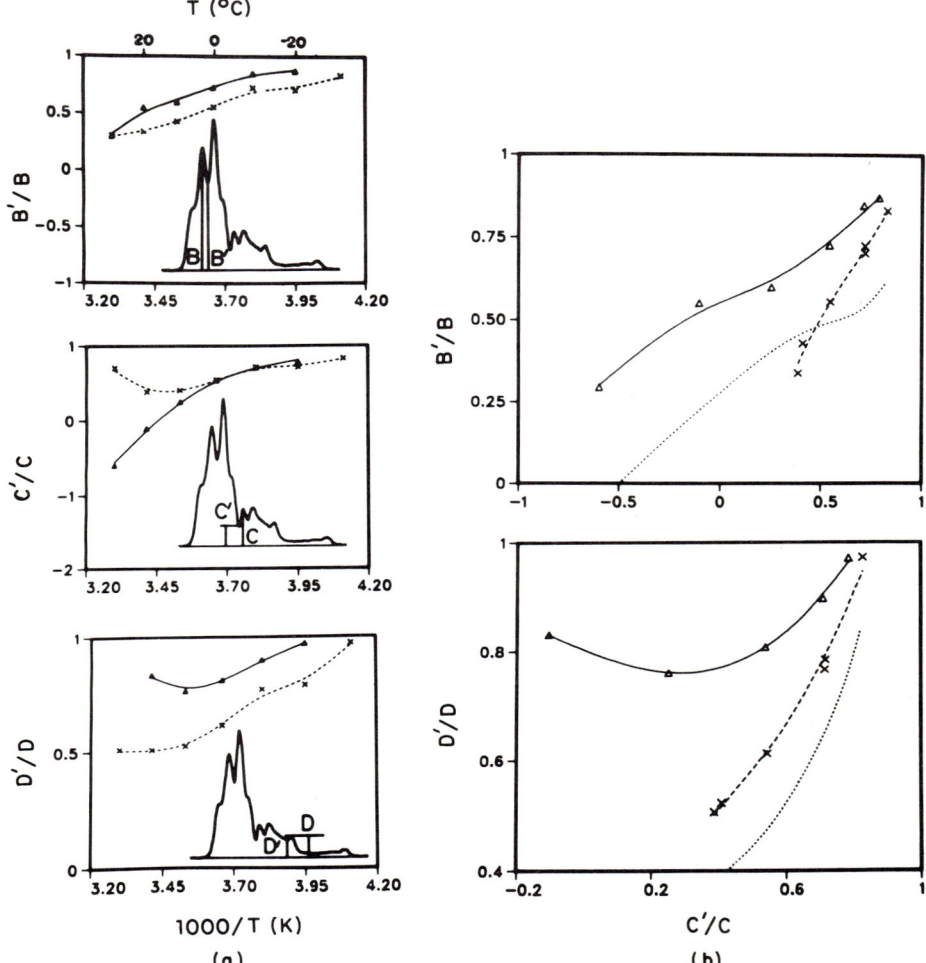

Fig. 69. Line height ratios of 35 GHz ST-ESR spectra (V_2') of bilayer dispersions of cholestane (Δ, ———) and fatty acid (x, - - -) spin labels in dipalmitoylphosphatidylcholine. (a) Selected ratios are plotted versus temperature; The respective spectral features are indicated in the insert. (b) Two of these ratios are plotted versus the third. Also included are data for the isotropic diffusion model (hemoglobin-Ia, · · ·, from Fig. 68). These plots show that the Y-axis motional anisotropy of the cholestane label is distinguishable from the Z-axis anisotropy of the fatty acid one. Both can be distinguished from an isotropic system. (From Johnson et al., 1982.)

the same reference model has been studied at the higher frequency (Fig. 68) (Hyde and Johnson, 1981). Since the nitroxide anisotropy inherent in g_{zz} and g_{xx}/g_{yy} is resolved at 35 GHz, anisotropic motion can be detected and characterized. As an example of this anisotropy, the axial motions of the fatty-acid [VI(12,3)] and steroid (VIIIa) spin labels in bilayer dispersions of dipalmitoylphosphatidylcholine and cholesterol have been studied (Johnson *et al.,* 1982). The former is a good model for z-axial diffusion, whereas the cholestane spin label is appropriate for the characterization of *y*-axial motion. The motion of these labels was restricted by temperature modulation of the bilayers; as at 9 GHz, the spectra were analyzed by line-height ratios (Fig. 69a). The ratios of similar spectral features can be determined from the spectra of the isotropic-motion model (Fig. 68). Comparisons of two sets of these ratios for these three systems are shown in Fig. 69b.

The choice of C′/C as a "reference" set is arbitrary. However, the choice of D′/D is related to its location near a *z*-axis turning point, and thus this ratio should be particularly sensitive to loss of correlation between the nitroxide *z* axis and the magnetic field. Indeed, the D′/D versus C′/C plot does show that the two types of axial motion are distinguishable from one another, and both are distinct from purely isotropic rotational diffusion. Inasmuch as saturation transfer has been viewed primarily as a probe of membrane structure, and nitroxides, in general, are intrinsically more valuable as reporter groups because of the anisotropy of their behavior, the ability of 35-GHz ESR to take advantage of both these characteristics clearly warrants its use in future spin-label studies. At the same time, the full potential of ST-ESR has not been realized; its usefulness encompasses far more than simple spin-label studies. The reader should be alert to the application of this technique in the study of a variety of biological systems.

VI. An ESR Perspective

There is no question that ESR has made a major contribution to the accumulation of data on and analysis of the structure–function relationships in biological systems. No area in biological chemistry has gone untouched by this technique, although it is clear that ESR has had a predominant impact in the area of biological oxidation–reduction—particularly metallo- and flavoproteins—and in the general application of the nitroxide spin-label technique. With the evolution of the spin-echo and saturation-transfer techniques, a renaissance of ENDOR, and the availability of single crystals and crystallographic data, as well as refinements in data acquisition and analysis, ESR has become an even more useful biophysical probe. In addition, advantage should be taken of our ability to use cells in defined media to synthesize isotopically pure reagents, [^{15}N]-heme, for example, or to incorporate a substitute metal, Co(II) for Zn(II). Such systems may also provide the quantities of material often needed for the ESR experiments proposed.

There are drawbacks to its indiscriminate use, of course, some of which have been mentioned in this chapter. Furthermore, ESR spectra are not sensitive to all perturbations nor as sensitive to some perturbations as are other spectral techniques. An example of this was described in Section IV. Generally, optical measurements (absorbance, (M)CD, and fluorescence) are more sensitive to overall symmetry–environmental alterations than is ESR. Further, there is always the problem of correlating changes in ESR spectra with real physiological–biostructural events. Thus, while ESR has established itself as a viable and, in some instances, unique biophysical technique, the true measure of its success and its correct application must be in the context of independent analyses of related, linked biochemical phenomena. The reader should view this chapter only in the context of this volume as a whole and not see ESR as a necessarily unique choice from a menu of independent experimental techniques.

Appendix A: ESR Hardware, Glassware, and Supplies

1. Spectrometers

There is only one company currently marketing ESR hardware in the United States, Bruker/IBM: Bruker, Manning Park, Billerica, MA 01810; or IBM Instruments, P. O. Box 322, Danbury, Conn 06810.

JOEL also carries a full line of ESR spectrometers but they no longer compete in the US market: JOEL Ltd., 1418 Nakagami Akishima, Tokyo, Japan 196. Although Varian spectrometers may be available, production of them was terminated in late 1983.

These manufacturers can supply the consoles, microwave bridges, magnet systems, and cavities for continuous-wave ESR and ENDOR at 9.5 and 35 GHz. Spectrometers operating at 3.8 (S band) and 24 (K band) GHz and pulsed spectrometers have been assembled from individual components and are not readily commerially available. For example, S-band measurements can be obtained using a Varian console and magnet along with a Micro New Instruments Co. 3.8 GHz microwave bridge. Bruker, however, appears eager to provide a much more extensive line of speciality ESR hardware.

2. Quartz Glassware

Although the manufacturers listed above carry certain types of quartzware (including dewars for low temperature work), the following are also excellent

sources of a wide variety of quartz glassware: James F. Scanlon Co., 3428 Baseline Avenue, Solvang, CA 93403; Wilmad Glass Co., U. S. Route 40 and Oak Road Buena, NJ 08310.
Wilmad is also a good source of pens and paper for ESR recorders.

3. Cryogenics

ESR manufacturers also have available low-temperature dewars, heat exchangers, and temperature controllers. Other sources of low-temperature support, particularly in the region below 25 K, are: Air Products & Chemicals, Inc., P. O. Box 538, Allentown, PA 18105; Andonian Cryogenics, Inc., 26 Farwell Street, Newtonville, MA 02160; Janis Research Co., Inc.; 22 Spencer Street, P. O. Box 487, Stoneham, MA 02180; Oxford Instruments, 3 New England Executive Park, Burlington, MA 01803.

4. Spin Labels

Many spin labels and spin-label precursors are commercially available from the following: Aldrich Chemical Co., 940 W. St. Park Avenue, Milwaukee, WI 53233; SYVA, 3221 Porter Drive, Palo Alto, CA 94304; Eastman Organiz Chemicals, 343 State Street, Rochester, NY 14650; Frinton Labs, P. O. Box 301, Grant Avenue, S. Vineland, N. J. 08360.

5. Spin Traps

Aldrich and Eastman both supply the common spin-trapping reagents.

Appendix B: ESR Software

The collection, analysis, and simulation of ESR spectra have increasingly added to the effectiveness of the ESR technique. These operations enhance sensitivity and improve effective resolution, have permitted the development of pulse sequences and the spin-echo technique, and have made possible more precise analysis and interpretation of a wide variety of ESR data. The ESR spectrometer is significantly upgraded as a research tool when interfaced to a dedicated minicomputer with background and mainframe programs available for multi parameter spectral simulation.

An early appraisal of the role of the dedicated minicomputer in ESR will be found in the following: Klopfenstein, C., Jost, P., and Griffith, O. H. (1972). The dedicated computer in electron spin resonance. *In* "Computers in Chemical and Biochemical Research," vol. I (C. Klopfenstein and C. Wilkins eds.), pp. 175–221. Academic Press, New York.

ESR manufacturers all now provide hardware–software packages for use specifically with the spectrometers they sell, although any micro or mini can easily be interfaced. The computer support provided by the ESR manufacturer has become a most competitive aspect of spectrometer merchandising. The acquisition of Bruker by IBM Instruments has had a significant impact; software is being consultant-generated for a number of the applications covered in this book. Complementary sources of software are: DEC Users Society, Maynard MA 01754; Nicolet Instrument Corp., Madison, WI 53711; and OLIS (On-Line Instrument Systems), Jefferson, GA. For software written primarily in Fortran, the Quantum Chemistry Program Exchange (Department of Chemistry, Rm. 204, Indiana University, Bloomington, IN, 47401) is replete with programs for both ESR and NMR. One can become a member for a small fee and receive the Quarterly Bulletin. Catalogs and guides and indexes for the catalogs are also available. Program documentation and listings are available at cost.

The ESR and some computer-hardware manufacturers can supply the software for interfacing the instrument with the computer for data accumulation, digitizing, and analysis. Simulations are less commonly carried out on a microcomputer, since it is likely to be limited by speed (particularly if the program is written in BASIC or a comparable language) and memory. This limitation is becoming less severe, however, as 16- and 32-bit microprocessors are adopted. Two common choices are the DEC/MINC 1123 and the IBM-PC. The Apple II+ suffers from its size and speed, although the IIe with 128K RAM represents a significant improvement. However, if the computer is to be used for simulation, one should consider carefully the advantages of the newer microprocessors.

A useful reference to interfacing micros is Rich, E. S., and Wampler, J. E., (1982). EPR spectrometer control and data acquisition using a general-purpose IEEE-488 bus interface and structured software. Amer. Lab. 14, 17–28. The software for this application was written in HP-85 BASIC, although the interface is designed to work with any IEEE-488 bus microcomputer. The program is structured like SPECOS, a general-purpose program designed by Professor Wampler and referenced therein.

A limited number of programs have occasionally appeared listed in the literature. For example:

(1) Freed, J. H. (1976). In "Spin Labeling: Theory and Applications" (C. J. Berliner, ed.). Academic Press, New York. pp. 121–129 has listed a program (FORTRAN IV) for the spectral simulation of slow tumbling (isotropic) of nitroxides.

(2) Lowe, D. J. (1978). Electron paramagnetic resonance in biochemistry. Biochem. J. **171**, 649–651, describes a program available in both BASIC and FORTRAN IV for spectral simulation of spin $=\frac{1}{2}$ systems in frozen aqueous samples. This program is available upon request.

(3) Twilfer, H., Gersonde, K., and Christahl, M. (1981). Resolution enhancement of EPR spectra using Fourier transform technique. *J. Mag. Reson.* **44,** 470–478, describes a program (FORTRAN IV) for resolution enhancement, integration, and differentiation by Fourier transform. This program is available on request.

(4) van Veen, G. (1978). Simulation and analysis of EPR spectra of paramagnetic ions in powders. *J. Mag. Reson.* **30,** 91–109, describes in some detail the calculation of resonance positions with respect to H_0; a program (FORTRAN IV) is available for simulation of axial spectra.

(5) Vistnes, A. I. (1978). A new method for simulation of EPR spectra with particular application to nitroxide spin labels. *J. Mag. Reson.* **29,** 495–507, outlines the flow diagram for the program and discusses its subroutines.

(6) Belford, R. L., and Duan, D. C. (1978). Determination of nuclear quadrupole coupling by simulation of EPR spectra of frozen solutions. *J. Mag. Reson.* **29,** 293–307, presents a simulation technique for resolving the quadrupole couplings usually detected only by "double-spin–flip" protocols, i.e., ENDOR and the electron-spin echo. Use is made of an unpublished program written by T. M. Northern (Ph. D. Thesis, University of Illinois, Champaign/Urbana, IL, 1976). White, L. K., Ph. D. Thesis, University of Illinois, Champaign/Urbana, IL, 1975, is also a useful reference. Professor Belford has contributed several EPR spectral simulation routines and is an excellent resource person.

A useful review of the theoretical bases for generating simulations complete with an extensive bibliography is Taylor, P. C., Baugher, J. F., and Kriz, H. M. (1975). Magnetic resonance spectra in polycrystalline solids. *Chem. Rev.* **75,** 203–240.

A rich source of software and computational methodology is generally the Journal of Magnetic Resonance. A monthly perusal of this publication is well worth the effort. Also, as has been noted the Quantum Chemistry Program Exchange has listed a number of simulation programs. For example, see Lozos, G. P., Hoffman, B. M., and Franz, C. G. (1974). SIM14/SIM14A: Simulation of powder EPR spectra. *QCPE* **11,** 265.

Appendix C: Calibration of ESR Spectra

The scan range, field, and frequency of an ESR spectrometer can be calibrated using the combination of an NMR gauss meter (which measures the frequency of ^{1}H resonance of water) and the frequency meter (which is an in-line resonant cavity tunable over the range of klystron output). Hewlett-Packard markets a widely used frequency meter; AEG (Oak Ridge, TN 37830) manufactures an appropriate gauss meter.

Neither piece of equipment is necessary for calibration, however. The spectrometer field can be calibrated using a variety of standards, e.g., diphenylpicrylhydrazyl (DPPH, $g = 2.0036$) or vanadylacetylacetone [VO(acac)$_2$ in benzene, $g = 1.986$, located between the fourth and fifth lines of the 8-lined spectrum, $I = \frac{7}{2}$] are often used, although the latter requires a correction factor owing to the slight field dependence of the hyperfine coupling (second-order correction). The scan range can be calibrated using VO(acac)$_2$, as well, since the separation of the $M_I = +\frac{1}{2}$ and $-\frac{1}{2}$ lines in this spectrum is 108.0 G. Both standards are available from either Aldrich or Eastman Chemical Co. Several other calibration standards have been suggested (Fee, 1978).

The determination of the magnitude of the A and g values for an ESR spectrum is normally straightforward, particularly if the hyperfine coupling(s) is large (> 25 G). If A is small, however, second-order corrections are necessary to derive accurate values. The equation employed to calculate the position of a line in an isotropic spectrum is

$$H_0 = H_m + AM_I + \frac{A^2}{2H_0}[I(I+1) - M_I^2] + \frac{A^3}{4H_0^2}[\cdots \qquad (51)$$

In this expression, H_0 represents the center of the particular set of lines of interest, and it is this H_0 value which will later be determined to calculate the correct g value; H_m is the field position of the line of interest; A the hyperfine coupling to be determined; and I the spin quantum number of the coupled nucleus. As an example of the use of this equation, consider the VO^{2+} ion, which is valuable as an ESR probe as outlined in Section IV.C. If the A value were approximately 100 G, then the last of the terms in Eq. (51) would be only 0.1 G at 9.1 GHz and could be neglected. We then can write a sample expression for the positions for the eight lines in the isotropic ESR spectrum for VO^{2+}, thus

$$H_{7/2} = H_0 - \tfrac{7}{2}A - \tfrac{7}{2}K, \qquad (52)$$

$$H_{5/2} = H_0 - \tfrac{5}{2}A - \tfrac{19}{2}K, \qquad (53)$$

$$H_{3/2} = H_0 - \tfrac{3}{2}A - \tfrac{27}{2}K, \qquad (54)$$

$$H_{1/2} = H_0 - \tfrac{1}{2}A - \tfrac{31}{2}K, \qquad (55)$$

$$H_{-1/2} = H_0 + \tfrac{1}{2}A - \tfrac{31}{2}K, \qquad (55a)$$

$$H_{-3/2} = H_0 + \tfrac{3}{2}A - \tfrac{27}{2}K, \qquad (54a)$$

$$H_{-5/2} = H_0 + \tfrac{5}{2}A - \tfrac{19}{2}K, \qquad (53a)$$

$$H_{-7/2} = H_0 + \tfrac{7}{2}A - \tfrac{7}{2}K, \qquad (52a)$$

where $K = A^2/2H_0$. Subtracting pairs of these equations yields four independent determinations of the A value:

$$H_{1/2} - H_{-1/2} = A,$$
$$H_{3/2} - H_{-3/2} = 3A,$$
$$H_{5/2} - H_{-5/2} = 5A,$$
$$H_{7/2} - H_{-7/2} = 7A.$$

Note that the only pair of lines which is separated by A is the inner pair; the position and separation of these two lines are often the most imprecise.

By adding pairs of these equations, (52) and (52a), (53) and (53a), and rearranging terms, equations can be obtained which are of the form

$$(55) + (55a) = H_{1/2} + H_{-1/2} = 2H_0 - (62/2)K.$$

Thus

$$(H_{1/2} + H_{-1/2})/2 = H_0 - (62/4)K$$

or

$$H_0 = (H_{1/2} + H_{-1/2})/2 + (31/4)(A^2/H_0).$$

Representing $(H_{1/2} + H_{-1/2})/2$ as the average of the two field positions in gauss, $H_{4,5}$, gives

$$H_0 = H_{3,6} + (27/4)(A^2/H_0),$$
$$H_0 = H_{2,7} + (19/4)(A^2/H_0),$$
$$H_0 = H_{1,8} + (7/4)(A^2/H_0).$$

The evaluation of H_0 is thus an iterative process. One first assumes that H_0 is $H_{a,b}$ and uses that H_0 to calculate a new H_0, and so forth. Normally, convergence will result after two or three cycles.

As noted, these second-order calculations for H_0 are not needed if A is large. For example, if $A = 50$ G the error in H_0 can be estimated. At a field of 3 kG the K term would be approximately 5 G, or about 0.2% of H_0.

Equations similar to all of those given here can be written for $A_\parallel$ and $g_\parallel$ ($K = A_\parallel^2/2H_0$) and for $g_\perp$ and $A_\perp$ [$K = (A_\parallel^2 + A_\perp^2)/4H_0$].

Bibliography

Electron Spin Resonance

1. Bersohn, M., and Baird, J. C. (1966). "An Introduction to Electron Paramagnetic Resonance." Benjamin, New York. Relatively descriptive, very readable.
2. Poole, C., Jr. (1967). "Electron Spin Resonance-A Comprehensive Treatise on Experimental Techniques." Wiley (Interscience), New York. Very useful for details about all aspects of instrumentation.

3. Carrington, A., and McLachlan, A. (1967). "Introduction to Magnetic Resonance." Harper and Row, New York. Treats both ESR and NMR.
4. Pake, G. (1967). "Paramagnetic Resonance." Harper and Row, New York. Treats the theory, quite mathematical.
5. Ingram. D. J. E. (1967). "Biological and Biochemical Applications of Electron Spin Resonance." Plenum, New York. Very good for the novice in magnetic resonance.
6. Alger, R. (1968). "Electron Paramagnetic Resonance Techniques and Applications." Wiley (Interscience), New York. A second choice after Poole.
7. Feher, G. (1970). "Electron Paramagnetic Resonance with Applications to Selected Problems in Biology." Gordon and Breach, New York. Has very useful sections on theory.
8. Wertz, J., and Bolton, J. (1972). "Electron Spin Resonance-Elementary Theory and Applications." McGraw-Hill, New York. Wertz and Bolton write very clearly; excellent mix of theory and practice.
9. Swartz, H. M., Bolton, J. R., and Borg, D. C. (Eds.) (1972). "Biological Applications of Electron Spin Resonance." Wiley-Interscience, New York. Remains an excellent reference book.
10. Atherton, N. M. (1973). "Electron Spin Resonance." Wiley, New York. An exceptionally well-written, complete book; good for both theory and practice, will not overwhelm the less physically oriented.

ESR of Transition Metals

1. Abragam, A., and Bleaney, B. (1970). "Electron Paramagnetic Resonance of Transistion Metal Ions." Clarendon Press, Oxford. Very complete, very useful references; accessible to both biologist (with reasonable effort) and chemist.
2. Brill, A. S. (1977). "Transition Metals in Biochemistry." Springer-Verlag, New York. This monograph is full of readily accessible information pertaining to biological metal coordination and spectral properties. Focus is primarily on Cu and Fe. Very useful reference.

Spin Labeling

1. Berliner, L. J. (1976/1979). "Spin labeling: Theory and Applications," Vol. I and II. Academic Press, New York. These volumes are both excellent.
2. Lichtenshtein, G. I. (1976). "Spin Labeling Methods in Molecular Biology." Wiley (Interscience), New York. Useful since draws examples from Russian literature; underscores fact that large amount of spin-labeling research has been done in Russia and has gone largely unreported in the Western scientific literature.
3. Jost, P. C., and Griffith, D. H. (1978). The Spin-Labeling Technique. *Meth. Enzymol.* **69,** 369–418. A concise yet very useful introduction, using membrane studies for examples, but providing much practical information.
4. Berlinger, L. J. (1978). Spin Labeling in Enzymology: Spin-Labeled Enzymes and Proteins. *Meth. Enzymol.* **69,** 481–480. A companion to the Jost, Griffith chapter, written in much the same informative manner. A "how-to" description of carrying out single-crystal studies on a spin-labeled protein is a highlight.
5. Marsh, D. (1981). Electron Spin Resonance: Spin Labels. *In* "Membrane Spectroscopy" (E. Grell, ed.). Springer-Verlag, New York. A superbly written chapter covering all aspects of the use of spin labels to probe membrane structure.
6. Schreier–Muccillo, S., Polnaszek, C.F., and Smith, I. C. P. (1978). *Biochim. Biophys. Acta.* **515,** 375–436. A thoughtful, honest appraisal of the spin-label technique applied to membranes.

Spin Trapping

1. Evans, C. A. (1979). *Aldrichimica Acta,* **12,** 23–29. Brief but useful introduction to the technique.
2. Perkins, M. J. (1980). Spin trapping. *In* "Advances in Physical Organic Chemistry" (V. Gold and D. Bethell, eds.). Academic Press, New York. Good review of the chemistry of traps, with limited information about biological applications per se.
3. Janzen, E. G. (1979). Spin Trapping. *In* "Free Radicals in Biology"(W. A. Pryor, ed.), Vol. IV. Wiley (Interscience), New York. A good review focusing on biological applications by a principal developer of the technique.

Double-Resonance Techniques (ENDOR)

ENDOR is discussed in Feher (1970), Ingraham (1967), Atherton (1973), and Abraham and Bleaney (1970) in this bibliography (General References). In addition, the following are excellent references:

1. Kevan, L., and Kispert, L. D. (1976). "Electron Spin Double Resonance Spectroscopy." Wiley (Interscience), New York. The place to start if you want to develop a real insight into these techniques, including ELDOR (electron–electron double resonance, in which a second microwave field is used).
2. Dorio, M. M., and Freed, J. H. (Eds.) (1979). "Multiple-Electron Resonance Spectroscopy." Plenum, New York. A complete overview of the theory and practice of multiple ESR, including ENDOR, ELDOR, and TRIPLE (electron–nuclear–nuclear resonance in which two rf fields are used to stimulate two different nuclear transitions simultaneously).

Relaxation Techniques

1. Kevan, L., and Schwartz, R. N. (1979). "Time Domain Electron Spin Resonance." Wiley (Interscience), New York. A book for the cognocenti, perhaps, but excellent chapters on saturation recovery and spin-echo techniques. The book is important because ESR methodology is going in this direction.
2. Mims. W. B. (1972). Electron spin echoes. *In* "Electron Paramagnetic Resonance" (S. Geschwind, ed.). Plenum, New York. An early but complete discussion of the spin–echo technique. Precedes much of the experimental application of the technique but still has useful theoretical and instrumental background.

Saturation Transfer

A general introduction to the concepts and application of this technique is afforded by the following:

1. Hyde, J. S. (1978). Saturation-Transfer Spectroscopy. *In Meth. Enzymol.* **69,** 480–511.
2. Dalton L. R., Robinson, B. H. Dalton, L. A., and Coffey, P (1976). Saturation Transfer Spectroscopy. *In Adv. Magn. Reson.* **8,** 149–259.

Saturation-transfer spectroscopy represents a rather subtle use of an ESR spectrometer (particulary of the microwave bridge). For those unfamiliar with it (as this author still is!) having some direct contact with the technique probably would be more useful than simply reading through these references. See also Kevan and Schwartz (1979) in the bibliography (Relaxation Techniques).

References

Atkin, C. L., Thelander, L., Reichard, P. and Lang, G. (1973). Iron and free radicals in ribonucleotide reductase. *J. Biol. Chem.* **248**, 2464–2469.

Babior, B. M., Moss, T. H., and Gould, D. C. (1972). The mechanism of action of ethanolamine ammonia lyase. *J. Biol. Chem.* **247**, 4389–4392.

Babior, B. M., Moss, T. H., Orme-Johnson, W. H., and Beinert, H. (1974). The mechanism of action of ethanolamine ammonia lyase. *J. Biol. Chem.* **249**, 4537–4544.

Ballou, D. P. (1971). Instrumentation for the study of rapid biological oxidation-reduction reactions by EPR and optical spectroscopy. Ph.D. thesis, University of Michigan, Ann Arbor, Michigan.

Ballou, D., and Palmer, G. A. (1974). Practical rapid quenching instrument for the study of reaction mechanisms by electron paramagnetic resonance spectroscopy. *Anal. Chem.* **46**, 1248–1253.

Barber, M. J., and Siegel, L. M. (1982). Oxidation-reduction potentials of molybdenum, flavin, and iron-sulfur centers in milk xanthine oxidase. *Biochemistry* **21**, 1638–1647.

Barber, M. J., Salerno, J. C., and Siegel, L. M. (1982). Magnetic interactions in milk xanthine oxidase. *Biochemistry* **21**, 1648–1656.

Bereman, R. D., and Kosman, D. J. (1977). Stereoelectronic properties of metalloenzymes. 5. Identification and assigment of ligand hyperfine splittings in the electron spin resonance spectrum of galactose oxidase. *J. Am. Chem. Soc.* **99**, 7322–7325.

Boldt, I., Schuel, H., Schuel, R., Dandekar, P. V., and Troll, W. (1981). Reaction of sperm with egg-derived hydrogen peroxide helps prevent polyspermy during fertilization in the sea urchin. *Gamete Res.* **4**, 365–377.

Bonnet, R., Clark, V. M., and Todd, A. (1959). Formation of bicyclic oxaziran from a Δ-pyrroline and from the corresponding nitrone. *J. Chem. Soc.,* 2102–2104.

Borg, D. C. (1964). Continuous flow methods adopted for EPR apparatus, in "Rapid Mixing and Sampling Techniques in Biochemistry," pp. 135–141. Academic Press, New York.

Brandén, R., and Deinum, J. (1977). Type 2 copper (II) as a component of the dioxygen reducing site in laccase: Evidence from EPR experiments with ^{17}O. *FEBS Lett.* **73**, 144–146.

Bray, R. C. (1961). Sudden freezing as a technique for the study of rapid reactions. *Biochem. J.* **81**, 189–193.

Bray, R. C. (1964). Quenching by squirting into cold immiscible liquids. *In* "Rapid Mixing and Sampling Techniques in Biochemistry," pp. 195–204. Academic Press, New York.

Bray, R. C. (1975). Molybdenum iron-sulfur flavin hydroxylases and related enzymes. *In* "The Enzymes"(P. D. Boyer, ed.), Vol. 12, pp. 299–357, 3rd ed., Academic Press, New York.

Bray, R. C., and Gutteridge, S. (1982). Numbers and exchangeability with water of oxygen-17 atoms coupled to molybdenum (V) in different reduced forms of xanthine oxidase. *Biochemistry* **21**, 5992–5999.

Bray, R. C., and Meriwether, L. S. (1966). Electron spin resonance of xanthine oxidase substituted with Mo-95. *Nature London* **212**, 467–469.

Bray, R. C., and Pettersson, R. (1961). ESR Measurements. *Biochem J.* **81**, 194–195.

Brill, A. S., and Kirkpatrick, P. R. (1967/1971). *In* "The Enzymes" (J. A. Hartsuch, and W. N. Lipscomb, eds.), Vol. 3., pp. 1–56. Academic Press, New York.

Buettner, G. R., and Oberley, L. W. (1978). Considerations in the spin trapping of superoxide and hydroxyl radical in aqueous systems using 5,5-dimethyl-l-pyrroline-l-oxide. *Biochem. Biophys. Res. Commun.* **83**, 69–74.

Burger, R. M., Adler, A. D., Horwitz, S. B., Mims, W. B., and Peisach, J. (1981). Demonstration of nitrogen coordination in metal-bleomycin complexes by electron spin echo envelope spectroscopy. *Biochemistry* **20**, 1701–1704.

Campbell, R. F., and Chasteen, N. D. (1977). An anion binding study of vanadyl (IV) human serotransferrin. *J. Biol. Chem.* **252,** 5996–6001.

Cantley, L. C., Jr. and Aisen, P. (1979). The fate of cytoplasmic vanadium. *J. Biol. Chem.* **254,** 1781–1784.

Chasteen, N. D. (1981). Vanadyl (IV) EPR spin probes: Inorganic and biochemical aspects. *Biol. Magn. Reson.* **3,** 53–119.

Chasteen, N. D., and Theil, E. C. (1982). Iron binding by horse spleen apoferritin. A vanadyl (IV) EPR spin probe study. *J. Biol. Chem.* **257,** 7672–7677.

Chasteen, N. D., White L. K., and Campbell, R. F. (1977). Metal site conformational states of vanadyl (IV) human serotransferrin complexes. *Biochemistry* **16,** 363–368.

Chen, S. -C., Sturtevant, J. M., Conklin, K., and Gaffney, B. J. (1982). Calorimetric evidence for phase transitions in spin-label lipid bilayers. *Biochemistry* **21,** 5096–5101.

Chien, J. C. W., and Dickinson, L. C. (1981). EPR crystallography of metalloproteins and spin-labeled enzymes. *Biol. Magn. Reson.* **3,** 155–211.

Cockle, S. A., Hill, H. A. O., Williams, R. J. P., Davies, S. P., and Foster, M. A. (1972). Detection of intermediates during the conversion of propane-1,2-diol to proprionaldehyde by glyceroldehydrase. *J. Am. Chem. Soc.* **94,** 275–277.

Cramer, S. P., Rajagopalan, K. V., and Wahl, R. (1981). Molybdenum sites of sulfite oxidase and xanthine dehydrogenase. A comparison by EXAFS. *J. Am. Chem. Soc.* **103,** 7721–7727.

Degani, H., Gochin, M. Karlish, S. J. D., Shechter, Y. (1981). Electron paramagnetic resonance studies and insulin-like effects of vanadium in rat adipocytes. *Biochemistry* **20,** 5795–5799.

Devaux, P. Scandella, C. J., and McConnell, H. M. (1973). Spin-spin interactions between spin-labelled phospholipids incorporated into membranes. *J. Magn. Reson.* **9,** 474–485.

Ecker, D. J., Lancaster, J. R., Jr., and Emery, T. (1982). Siderophere iron transport followed by electron paramagnetic resonance spectroscopy. *J. Biol. Chem.* **257,** 8623–8626.

Ehrenberg, A., and Reichard, P. (1972). ESR of the iron-containing protein B2 from ribonucleotide reductase. *J. Biol. Chem.* **247,** 5485–5489.

Ettinger, M. J., and Kosman, D. J. (1982). Chemical and catalytic properties of the mononuclear, type 2 Cu(II) enzyme, galactose oxidase. *In* "Bioinorganic Chemistry" (T. S. Spiro, ed.), Vol. 5. Wiley (Interscience), New York.

Fee, J. A. (1978). Transition metal electron paramagnetic resonance related to proteins. *Meth. Enzymol.* **69,** 512–528.

Fee, J. A., Peisach, J., and Mims, W. B. (1981). Superoxide dismutase: Examination of the metal binding sites by electron spin echo spectroscopy. *J. Biol. Chem.* **256,** 1910–1914.

Feher, G. "Electron Paramagnetic Resonance with Applications to Selected Problems in Biology." Gordon and Breach, New York.

Finkelstein. E., Rosen, G. M., and Rauckman, E. J. (1980). Spin trapping. Kinetics of the reaction of superoxide and hydroxyl radicals with nitrones. *J. Am. Chem. Soc.* **102,** 4994–4999.

Finley, T. H., Valinsky, J., Mildvan, A. S., and Abeles, R. H. (1973). ESR studies with dioldehydrase. *J. Biol. Chem.* **248,** 1285–1290.

Freedman, J. H., Pickart, L., Weinstein, B., Mims, W. B., and Peisach, J. (1982). Structure of the glycyl-L-histidyl-L-lysine-copper(II) complex in solution. *Biochemistry* **21,** 4540–4544.

Fretten, P., Morris, S. J., Watts, A., and Marsh, D. (1980). Lipid-lipid and lipid-protein interactions in chromaffin granule membranes. A spin label ESR study. *Biochim. Biophys. Acta* **598,** 247–259.

Froncisz, W., and Hyde, J. S. (1982). The loop-gap resonator: a new microwave lumped circuit ESR sample structure. *J. Magn. Reson.* **47,** 515–521.

Gaffney, B. J., and McConnell, H. M. (1974). The paramagnetic resonance spectra of spin labels in phospholipid membranes. *J. Magn. Reson.* **16**, 1–28.

Gräslund, A., Ehrenberg, A., and Thelander, L. (1982). Characterization of the free radical of mammalian ribonucleotide reductase. *J. Biol. Chem.* **257**, 5711–5715.

Gray, H. B., and Solomon, E. I. (1981). Electronic structures of blue copper centers in proteins. *In* "Copper Proteins" (T. G. Spiro, ed.), pp. 1–39. Wiley (Interscience), New York.

Griffith, O. H., and Yost, P. C. (1976). Lipid spin labels in biological membranes. *In* "Spin Labelling Theory and Applications" (L. J. Berliner, ed.), Vol. I, pp. 453–523. Academic Press, New York.

Groh, S. E., Nagahisa, A., Tan, S. L., and Orme-Johnson, W. H. (1983). Electron spin echo modulation demonstrates P-450$_{scc}$ complexation. *J. Am. Chem. Soc.* **105**, 7445–7446.

Gross, A., and Sizer, I. (1959). The oxidation of tyramine, tyrosine and related compounds by peroxidase. *J. Biol. Chem.* **234**, 1611–1615.

Gutteridge, S., and Bray, R. C. (1980). Oxygen-17 splitting of the very rapid molybdenum (V) e.p.r. signal from xanthine oxidase. *Biochem. J.* **189**, 615–623.

Gutteridge, S., Tanner, S. J., and Bray, R. C. (1978). The molybdenum centre of native xanthine oxidase. *Biochem. J.* **175**, 869–878.

Haffner, P. H., and Coleman, J. E. (1973a). High spin and low spin forms of Co(II) carbonic anhydrase. *J. Biol. Chem.* **248**, 6630–6636.

Haffner, P. H., and Coleman, J. E. (1973b). Cu(II)-carbon bonding in cyanide complexes of copper enzymes. *J. Biol. Chem.* **248**, 6626–6629.

Hamilton, J. A., and Blakley, R. L. (1969). ESR studies of ribonucleotide reduction catalysed by the ribonucleotide reductase of *Lactobacillus leichmannia*. *Biochim. Biophys. Acta* **184**, 224–226.

Hamilton, J. A., Tamao, Y., Blakley, R. L., and Coffman, R. E. (1972). ESR studies on cobalamin-dependent ribonucleotide reduction. *Biochemistry* **11**, 4696–4705.

Harbour, J. R., and Bolton, J. R. (1975). Superoxide formation in spinach chloroplasts: electron spin resonance detection by spin trapping. *Biochem. Biophys. Res. Commun.* **64**, 803–807.

Harbour, J. R., Chow, V., and Bolton, J. R. (1974). An electron spin resonance study of the spin adducts of OH and HO_2 radicals with nitrones in the ultraviolet photolysis of aqueous hydrogen peroxide solutions. *Can. J. Chem.* **52**, 3549–3553.

Harris, D. C., Gray, G. A., and Aisen, P. (1974). ^{13}C Nuclear magnetic resonance study of the spatial relation of the metal- and anion-binding sites of human transferrin. *J. Biol. Chem.* **249**, 5261–5264.

Hauser, H., Gains, N., Semenza, G., and Spiess, M. (1982). Orientation and motion of spin-labels in rabbit small intestinal brush border vesicle membranes. *Biochemistry* **21**, 5621–5628.

Hoffman, B. M., Roberts, J. E., and Orme-Johnson, B. H. (1982a). ^{95}Mo and ^{1}H ENDOR spectroscopy of the nitrogenase MoFe protein. *J. Am. Chem. Soc.* **104**, 860–862.

Hoffman, B. M., Venters, R. A., Roberts, J. E., Nelson, M., and Orme-Johnson, W. H. (1982b). ^{57}Fe ENDOR of the nitrogenase MoFe protein. *J. Am. Chem. Soc.* **104**, 4711–4712.

Höhn, M., and Hüttermann, J. (1982). ^{1}H electron–nuclear double resonance of cobalt hemoglobin. *J. Biol. Chem.* **257**, 10554–10557.

Hori, H., Ikeda-Saito, M., and Yonetani, T. (1982). Ligand orientation of oxyproto- and oxymesocobalt porphyrin-substituted myoglobin by single crystal EPR spectroscopy. *J. Biol. Chem.* **257**, 3636–3642.

Hsia, J. C., and Piette, L. H. (1969). Spin-labeling as a general method in studying antibody active site. *Arch. Biochem. Biophys.* **129**, 296–307.

Hunt, M. H., Mackay, A. L., and Edmonds, D. T. (1975). Nuclear quadrupole resonance of nitrogen-14 in imidazole and related compounds. *Chem. Phys. Lett.* **34,** 473–475.

Johnson, M. E., and Hyde, J. S. (1981). 35 GHz (Q-band) saturation transfer electron paramagnetic resonance studies of rotational diffusion. *Biochemistry* **20,** 2875–2880.

Johnson, M. E., Lee, L., and Fung, W.-M. (1982). Models for slow anisotropic rotational diffusion in saturation transfer electron paramagnetic resonance at 9 and 35 GHz. *Biochemistry* **21,** 4459–4467.

Josephy, P. D., Eling, T., and Mason. R. P. (1982). The horseradish peroxidase-catalyzed oxidation of 3,5,3′,5′-tetramethylbenzidine. *J. Biol. Chem.* **257,** 3669–3675.

Kalyanaraman, B., Mason, R. P., Tainer, B., and Eling, T. E. (1982). The free radical formed during the hydroperoxide-mediated deactivation of ram seminal vesicles is hemoprotein-derived. *J. Biol. Chem.* **257,** 4764–4768.

Keana, J. F. W., and Dinerstein, R. J. (1971). A new highly anisotropic dinitroxide ketone spin label. A sensitive probe for membrane structure. *J. Am. Chem Soc.* **93,** 2808–2810.

Keana, J. F. W., Acarregui, M. J., and Boyle, S. L. M. (1982). 2,2-Disubstituted-4,4-dimethyl imidazolidinyl-3-oxy nitroxides: Indicators of aqueous acidity through variation of A_N with pH. *J. Am. Chem. Soc.* **104,** 827–830.

Klimes, N., Lassmann, G., and Ebert, B. (1980). Time-resolved EPR spectroscopy. Stopped-flow EPR apparatus for biological application. *J. Magn. Reson.* **37,** 53–59.

Kosman. D. J. (1972). Electron paramagnetic resonance probing of macromolecules: a comparison of structure/function relationships in chymotrypsinogen, α-chymotrypsin, and anhydrochymotrypsin. *J. Mol. Biol.* **67,** 247–264.

Kosman, D. J., Peisach, J., and Mims, W. B. (1980). Pulsed electron paramagnetic resonance studies of the copper(II) in galactose oxidase. *Biochemistry* **19,** 1304–1308.

Kosman, D. J., Mims, W. B., and Peisach, J. Unpublished observations.

Lendzian, F., Lubitz, W., Scheer, H. Bubenzer, C., and Möbius, K. (1981). *In vivo* liquid solution ENDOR and TRIPLE resonance of bacterial photosynthetic reaction centers of *Rhodopseudomonas sphaeroides* R-26. *J. Am. Chem. Soc.* **103,** 4635–4637.

Lieberman, R. A., Sands, R. H., and Fee, J. A. (1982). A study of the electron paramagnetic resonance properties of single monoclinic crystals of bovine superoxide dismutase. *J. Biol. Chem.* **257,** 336–344.

Likhtenshtein, G. I. (1976). "Spin Labeling Methods in Molecular Biology", p. 10. Wiley (Interscience), New York.

Lowe, D. J., Lynden-Bell, R. M., and Bray, R. C. (1972). Spin–spin interaction between molybdenum and the iron-sulfur systems of xanthine oxidase and its relevance to the enzymic mechanism. *Biochem. J.* **130,** 239–251.

Luckhurst, G. R. (1976). Biradicals as spin probes. *In* "Spin labeling: Theory and Applications" (S. J. Berliner, ed.), Vol. I, pp. 133–181. Academic Press, New York.

McConnell, H. M., and McFarland, B. G. (1970). Physics and chemistry of spin labels. *Q. Rev. Biophys.* **3,** 91–136.

Malthouse, J. P. G., George, G. N., Lowe, D. J., and Bray, R. C. (1981). Coupling of [^{33}S] Sulfur to molybdenum (V) in different reduced forms of xanthine oxidase. *Biochem. J.* **199,** 629–637.

Markham, G. D. (1981). Spatial proximity of two divalent metal ions at the active site of *S*-adenosyl methionine synthetase. *J. Biol. Chem.* **256,** 1903–1909.

Marsh, D. (1981). Electron spin resonance: Spin labels. *In* "Membrane Spectroscopy" (E. Grell, ed.), pp. 51–142. Springer-Verlag, New York.

Marwedel, B. J., Kosman, D. J. Bereman, R. D., and Kurland, R. J. (1981). Magnetic resonance studies of cyanide and fluoride binding to galactose oxidase copper (II): Evidence for two exogenous ligand sites. *J. Am. Chem. Soc.* **103,** 2842–2847.

Meier, P., Blume, A., Ohmes, E., Neugebauer, F. A., and Kothe, G. (1982). Structure and dynamics of phospholipid membranes: An electron spin resonance study employing biradical probes. *Biochemistry* **21,** 526–534.

Melnyk, D., and Ettinger, M. J. (1984). Effects of $H_2^{17}O$ on the spectrum of the copper enzyme, galactose oxidase. *J. Am. Chem. Soc.,* in press.

Mims. W. B. (1974). Measurement of the linear electric field effect in EPR using the spin echo method. *Rev. Sci. Instrum.* **45,** 1583–1591.

Mims, W. B., and Peisach J. (1976a). Assignment of a ligand in stellacyanin by a pulsed electron paramagnetic resonance method. *Biochemistry* **15,** 3863–3869.

Mims, W. B., and Peisach, J. (1976b). The linear electric field effect for low spin feric heme compounds. *J. Chem. Phys.* **64,** 1074–1091.

Mims, W. B., and Peisach, J. (1978). The nuclear modulation effect in electron spin echoes for complexes of Cu^{2+} and imidazole with ^{14}N and ^{15}N. *J. Chem. Phys.* **69,** 4921–2930.

Mims, W. B., and Peisach, J. (1979a). Measurement of ^{14}N superhyperfine frequencies in stellacyanin by an electron spin echo method. *J. Biol. Chem.* **254,** 4321–4323.

Mims, W. B., and Peisach, J. (1979b). Pulsed EPR studies of metalloproteins, "Biological Applications of Magnetic Resonance" (R. G. Shulman, ed.) pp. 221–269. Academic Press, New York.

Mims. W. B., and Peisach, J. (1981). Electron spin echo spectroscopy and the study of metalloproteins. *Biol. Magn. Reson.* **3,** 213–263.

Mims, W. B., Peisach, J., Shaw, R. W., and Beinert, H. (1980). Electron spin echo studies of cytochrome *c* oxidase. *J. Biol. Chem.* **255,** 6843–6846.

Mock, D. M., Bruno, G. V., Griffin, B. W., and Peterson, J. A. (1982). Low temperature EPR spectroscopic characterization of the interaction of cytochrome P-450 cam with a spin label analog of metyrapone. *J. Biol. Chem.* **57,** 5372–5379.

Mondoví, B., Graziani, M. T., Mims, W. B., Oltzik, R., and Peisach, J. (1977). Pulsed electron paramagnetic resonance studies of types I and II copper of *Rhus vernicifera* laccase and porcine ceruloplasmin. *Biochemistry* **16,** 4198–4202.

Mottley, C., Mason, R. P., Chignell, C. F., Sivarajah, K., and Eling, T. E. (1982). The formation of sulfur trioxide radical anion during the prostaglandin hydroperoxidase-catalyzed oxidation of bisulfate (hydrated sulfur dioxide). *J. Biol. Chem.* **257,** 5050–5055.

Moura, J. J. G., Moura, I., Huynh, B. H., Krüger, H. -J-, Teixeira, M., DuVarnery, R. C., DerVartanian, D. V., Xavier, A. V. Peck, H. D., Jr., and LeGall, J. (1982). Unambiguous identification of the nickel EPR signal in ^{61}Ni-enriched *Desulfovibrio gigas* hydrogenase. *Biochem. Biophys. Res. Commun.* **108,** 1388–1393.

Mulks, C. F., Kirste, B., and VanWilligen, H. (1982). ENDOR study of VO^{2+} –imidazole complexes in frozen aqueous solution. *J. Am. Chem. Soc.* **104,** 5906–5911.

Oberley, L. W., and Buettner, G. R. (1979). The production of hydroxyl radical by bleomycin and iron (II). *FEBS Lett.* **97,** 47–49.

Ochiai, E. -I. (1977). "Bioinorganic Chemistry—An Introduction," Ch. 13. Allyn and Bacon, Boston, MA.

Olson, J. S., Ballou, D., Palmer, G., and Massey, V. (1974a). The mechanism of action of xanthine oxidase. *J. Biol. Chem.* **249,** 4363–4382.

Olson, J. S., Ballou, D. P., Palmer, G., and Massey, V. (1974b). The reaction of xanthine oxidase with molecular oxygen. *J. Biol. Chem.* **249,** 4350–4362.

Orme-Johnson, W. H., and Beinert, H (1969). Anaerobic reductive titrations with solid diluted sodium dithionite in an apparatus suitable for EPR spectroscopy. *Anal. Biochem.* **32,** 425–435.

Orme-Johnson, W. H., Beinert, H., and Blakley, R. L. (1974). Cobamides and ribonucleotide reduction. *J. Biol. Chem.* **249,** 2338–2343.

Padlau, E. A., Davis, D. R., Rudikoff, S., and Potter, M. (1976). Structural basis for the specificity of phosphorylcholine-binding immunoglobulins. *Immunochemistry* **13**, 945–949.

Palmer, G., and Beinert, H. (1964). An experimental evolution of the Bray rapid freezing technique. *In* "Rapid Mixing and Sampling Techniques in Biochemistry," pp. 205–213. Academic Press, New York.

Pariyaduth, N., Newton, W. E., and Stiefel, E. I. (1976). Monomeric Mo(V) complexes showing ^1H, ^2H, and ^{14}N shfs in the EPR spectra. Implications for molybdenum enzymes. *J. Am. Chem. Soc.* **98**, 5388–5390.

Peisach, J., and Blumberg, W. B. (1974). Structural implications derived from the analysis of electron paramagnetic resonance spectra of natural and artificial copper proteins. *Arch. Biochem. Biophys.* **165**, 691–708.

Peisach, J., and Mims, W. B. (1973). Linear electric field-induced shifts in electron paramagnetic resonance: A new method for study of ligands of cytochrome P-450. *Proc. Natl. Acad. Sci. USA* **70**, 2979–2982.

Peisach, J., Mims, W. B., and Davis, J. L. (1979). Studies of the electron-nuclear coupling between Fe(III) and ^{14}N in cytochrome P-450 and in a series of low spin heme compounds. *J. Biol. Chem.* **254**, 12379–12389.

Peters, T., Jr., and Blumenstock, F. A. (1967). Copper binding properties of bovine serum albumin and its amino-terminal peptide fragment. *J. Biol. Chem.* **242**, 1574–1578.

Piette, L. (1964). Continuous flow methods adopted for EPR apparatus. *In* "Rapid Mixing and Sampling Techniques in Biochemistry," pp. 131–134. Academic Press, New York.

Roberts, J. E., Hoffman, B. M., Rutter, R., and Hager, L. P. (1981). ^{17}O ENDOR of horseradish peroxidase compound I. *J. Am. Chem. Soc.* **103**, 7654–7656.

Sahlin, M., Gräslund, A., Ehrenberg, A. and Sjöberg, B. M. (1982). Structure of the tyrosyl radical in bacteriophage T4-induced ribonucleotide reductase. *J. Biol. Chem.* **257**, 366–369.

Sando, G. N., Blakley, R. L., Hogenkamp, H. P. C., and Hoffmann, P. J. (1975). Studies on the mechanism of adenosylcobalamin-dependent ribonucleotide reduction by the use of analogs of the coenzyme. *J. Biol. Chem.* **250**, 8774–8779.

Scandella, C. J., Devaux, P., and McConnell, H. M. (1972). Rapid lateral diffusion of phospholipids in rabbit sarcoplamic reticulum. *Proc. Natl. Acad. Sci. USA* **69**, 2056–2060.

Schepler, K. L., Dunham, W. R., Sands, R. H., Fee, J. A., and Ables, R. H. (1975). A physical explanation of the EPR spectrum observed during catalysis by enzymes utilizing coenzyme B_{12}. *Biochim. Biophys. Acta* **397**, 510–518.

Scholes, C. P., Lapidot, A., Mascarenhas, R., Inubushi, T., Isaacson, R. A., and Feher, G. (1982). Electron nuclear double resonance (ENDOR) from heme and histidine nitrogens in single crystals of aquometmyoglobin. *J. Am. Chem. Soc.* **104**, 2724—2735.

Sealy, R. C., Swartz, H. M., and Olive, P. L. (1978). Electron spin resonance spin-trapping. Detection of superoxide formation during aerobic microsomal reduction of nitro-compounds. *Biochem. Biophys. Res. Commun.* **82**, 680–684.

Sjöberg, B. M., Reichard, P., Gräslund, A., and Ehrenberg, A. (1977). Nature of the free radical in ribonucleotide reductase from *Escherichia coli*. *J. Biol. Chem.* **252**, 536–541.

Speckhard, D. C., Wu, F. Y-H., and Wu, C-W. (1977). Role of the intrinsic metal in RNA polymerase from *Escherichia coli*. In vivo substitution of tightly bound zinc with cobalt. *Biochemistry* **16**, 5228–5234.

Spence, J. T., Barber, M. J., and Siegel, L. M. (1982). Determination of the stoichiometry of electron uptake and the midpoint reduction potentials of milk xanthine oxidase at 25°C by microcoulometry. *Biochemistry* **21**, 1656–1661.

Stevens, T. H., Martin, C. T., Wang, H., Brudvig, G. W., Scholes, C. P., and Chan. S. I. (1982). The nature of the Cu_A in cytochrome c oxidase. *J. Biol. Chem.* **257,** 12106–12113.

Tainer, J. A., Getzoff, E. D., Richardson, J. S., and Richardson, D. C. (1983). Structure and mechanism of copper, zinc superoxide dismutase. *Nature* **306,** 284–287.

Tanner, S. J., Bray, R. C., and Bergmann, F. (1970). ^{13}C hyperfine splitting of some molybdenum electron-paramagnetic-resonance signals from xanthine oxidase. *Biochem. Soc. Trans.* **6,** 1328–1330.

Taylor, J. S., and Coleman, J. E. (1972). Nitrogen ligands at the active site of alkaline phosphatase. *Proc. Natl. Acad. Sci. USA* **69,** 859–862.

Taylor, J. S., and Coleman, J. R. (1973). Electron spin resonance of ^{63}Cu and ^{65}Cu carbonic anhydrase. *J. Biol. Chem.* **248,** 749–755.

Taylor, M. G., and Smith, I. C. P. (1981). Reliability of nitroxide spin probes in reporting membrane properties: A comparison of nitroxide and deuterium-labeled steriods. *Biochemistry* **20,** 5252–5255.

Tero-Kubata, S., Ikegami, Y., Kurokawa, T., Sadaki, R., Sugioka, K. and Nakano, M. (1982). Generation of free radicals and initiation of radical reactions in nitrones–Fe^{2+}–phosphate buffer systems. *Biochim. Biophys. Acta* **108,** 1025–1031.

Thomas, D. D., Dalton, L. R., and Hyde, J. J. (1976). Rotational diffusion studied by passage saturation transfer electron paramagnetic resonance. *J. Chem. Phys.* **65,** 3006–3024.

Träuble, H., and Sackmann, E. (1972). Studies of the crystalline-liquid crystalline phase transition of lipid model membranes. III. Structure of a steroid–lecithin system below and above the lipid-phase transition. *J. Am. Chem. Soc.* **94,** 4499–4510.

Twining, S. S., Sealy, R. C., and Glick, D. M. (1981). Preparation and activation of a spin-labeled pepsinogen. *Biochemistry* **20,** 1267–1272.

Valinsky, J. E., Ables, R. H., and Mildvan, A. S. (1974a). ESR studies with dioldehyrase II. *J. Biol. Chem.* **249,** 2751–2755.

Valinsky, J. E., Ables, R. H., and Fee, J. A. (1974b). ESR studies on diol dehyrase. III. *J. Am. Chem. Soc.* **96,** 4709–4710.

VanCamp, H. L., Sands, R. H., and Fee, J. A. (1982). An examination of the cyanide derivative of bovine superoxide dismutase with electron-nuclear double resonance. *Biochim. Biophys. Acta* **704,** 75–89.

Villafranca, J. J., Ash, D. E., and Wedler, F. C. (1976). Manganese (II) and substrate interaction with unadenylated glutamine synthetase (*Escherichia coli* W). *Biochemistry* **15,** 544–553.

Webb, M. R., Ash, D. E., Leyh, T. S., Trentham, D. R., and Reed, G. H. (1982). Electron paramagnetic resonance studies of Mn(II) complexes with myosin subfragment 1 and oxygen 17-labeled ligands. *J. Biol. Chem.* **257,** 3068–3072.

Zweier, J. L. (1980). Electron paramagnetic resonance studies of the binding of copper to conalbumin. *J. Biol. Chem.* **255,** 2782–2789.

Zweier, J. L., and Aisen, P. C. (1977). Studies of transferrin with use of Cu^{2+} as an electron paramagnetic resonance spectroscopic probe. *J. Biol. Chem.* **252,** 6090–6096.

Zweier, J., Aisen, P., Peisach, J., and Mims, W. B. (1979). Pulsed electron paramagnetic resonance studies of the copper complexes of transferrin. *J. Biol. Chem.* **254,** 3512–3515.

Zweier, J. L., Peisach, J., and Mims, W. B. (1982). Electron spin echo studies of the copper complexes of conalbumin. *J. Biol. Chem.* **257,** 10314–10316.

3

Mössbauer Spectroscopy

D. P. E. DICKSON and C. E. JOHNSON

Department of Physics
University of Liverpool
Liverpool, England

I.	INTRODUCTION	246
II.	PRINCIPLES OF MÖSSBAUER SPECTROSCOPY	249
	A. *The Mössbauer Effect*	249
	B. *Mössbauer Spectroscopy*	250
	C. *The Chemical Shift*	251
	D. *Quadrupole Splitting*	253
	E. *Magnetic Splitting*	254
III.	EXPERIMENTAL TECHNIQUES	257
	A. *Spectrometers*	257
	B. *Sample Requirements*	259
	C. *Computer Analysis*	263
IV.	APPLICATIONS TO ISOLATED BIOMOLECULES	265
	A. *Heme Proteins*	265
	B. *Iron–Sulfur Proteins*	266
	C. *Iron-Transport Proteins*	270
	D. *Iron-Storage Proteins*	272
	E. *Iodine Compounds*	273
	F. *Vitamin B_{12}*	274
V.	APPLICATIONS TO MOLECULAR SYSTEMS	274
	A. *Nitrogenase*	274
	B. *Photosynthetic Reaction Centers*	276
	C. *Cytochrome P-450 Ferredoxin Enzyme System*	279
	D. *Hydrogenase*	279
	E. *Oxygenase*	280
VI.	APPLICATIONS TO TISSUE	282
	A. *Lung Samples*	282
	B. *Blood Samples*	283
	C. *Bone Samples*	284
VII.	APPLICATIONS TO WHOLE OR PART ORGANISMS: UPTAKE AND METABOLISM	284
	A. *In Vitro Experiments*	285
	B. *In Vivo Experiments*	286
VIII.	MEASUREMENT OF VIBRATION AND MOVEMENT	288
	A. *Protein Dynamics*	288

      B. Applications in Auditory Physiology   289
      C. Motion of Complete Organisms   290
IX. CONCLUSIONS   290
    REFERENCES   291

I. Introduction

The technique of Mössbauer spectroscopy is based on the Mössbauer effect, which is the phenomenon of the resonant absorption of nuclear gamma rays by nuclei embedded in a solid. Because of the extremely high precision of this resonant absorption it can be used to investigate the very small changes in the nuclear energy levels which result from the hyperfine interactions between the Mössbauer nucleus and its electronic environment. The Mössbauer nucleus in a solid therefore acts as a probe of the chemical state of the atom and molecule in which the nucleus is situated.

Because of certain requirements which must be met if the Mössbauer effect is to occur, Mössbauer spectroscopy can be applied to only certain isotopes, of which ^{57}Fe is the most suitable and is used in the majority of Mössbauer spectroscopic studies. This isotope constitutes only 2% of natural iron. There are also many other Mössbauer isotopes, a few of which may be important in biology.

Mössbauer spectroscopy is completely specific for the type of nucleus being studied; if one is studying ^{57}Fe, no other nucleus will contribute to the spectrum. It is also a very local probe, which can give detailed information about the chemistry in the immediate vicinity of the Mössbauer nucleus. Both of these attributes make the technique particularly appropriate for certain types of problems, such as that of a biological system where the Mössbauer atom has a crucial role, e.g., in the case of the iron atom in hemoglobin. When using Mössbauer spectroscopy to investigate the form of iron in a particular material, one would expect to obtain information on whether the iron is ferrous or ferric, on the nature and arrangement of the ligands, on the spin of the iron atoms and whether they are magnetically ordered, and on whether all the iron atoms are equivalent. Frequently this information can be more readily arrived at by combining the data from Mössbauer spectroscopy with data from other physicochemical techniques, such as electron paramagnetic resonance (EPR) and magnetic susceptibility.

Each different chemical environment of the Mössbauer nucleus within a sample gives rise to a distinct contribution to the Mössbauer spectrum. Thus Mössbauer spectroscopy can be used for analysis, to identify the number of different chemical forms of the Mössbauer atom within a sample and to determine the nature of these chemical forms by comparison of the spectra with those of known materials or by direct interpretation of the spectral parameters.

The many possible areas for the application of Mössbauer spectroscopy in biological research can best be considered in terms of the level of complexity of the system under investigation, from small, isolated biomolecules to complete organisms *in vivo*. Frequently, from the knowledge of relatively simple materials it becomes possible to identify and monitor molecules within a more complex system.

Biological Mössbauer studies have up to now mainly been on the simpler types of biomolecules isolated from all types of organisms. Simple in this context means that while they may have a relative molecular mass of perhaps tens of thousands, they contain only a small number of atoms of the Mössbauer isotope. These isolated biomolecules fall into several groups; the studying of various members of a group reveals a pattern of behavior that is very helpful in understanding the group as a whole. Also in this category comes work on model compounds, which are organic molecules, simpler than the biological molecules but with similar active centers. It is hoped that understanding the model compounds will help in understanding the biomolecules which they emulate. Most of the work on isolated biomolecules has involved the ^{57}Fe Mössbauer isotope, but there have also been investigations of iodine compounds using ^{127}I and ^{129}I. Other isotopes which may have applications to certain biomolecules are ^{67}Zn and ^{133}Cs (substituted for Na$^+$ and K$^+$) and ^{151}Eu (substituted for Ca^{2+}). The work on isolated biomolecules has reached the stage where a complete theoretical understanding of the Mössbauer spectra, with computer fitting, is possible. The computer analysis frequently confirms the more qualitative interpretation obtained by inspection of the spectra and supports the use of qualitative ideas to interpret the spectra of more complex systems. The present understanding of the Mössbauer spectra of isolated biomolecules has essentially provided a set of fingerprints with which to compare the spectra of more complex biological samples.

The information on the various types of known biomolecules can be important in identifying the presence of new types of biological compounds, when the observed spectra cannot be assigned to any of the known types. Mössbauer spectroscopy has provided important evidence in identifying a novel type of ferredoxin and a new iron-storage material.

With the basic information that exists on the simpler molecules, it becomes possible to interpret the Mössbauer spectra of enzymes consisting of complex molecules or molecular systems. These may contain a number of separate molecules or cofactors, all required for biological activity. Such systems frequently contain the Mössbauer isotope in a number of chemical forms, which give rise to multiple components in the spectra. Thus it may be possible to monitor the changes which occur in the different components during biological processes.

A higher level of complexity is that of tissue, and in this area Mössbauer spectroscopy may be able to make a contribution to problems of physiological and medical interest. There has been only a limited amount of work of this type in the past, but it is likely to become more important in the future. So far samples of lung, blood, and bone have been investigated. In the case of one study of blood diseases, some important pathological factors have been identified.

Another physiologically relevant application of Mössbauer spectroscopy is to monitor the uptake and use of a Mössbauer isotope by an organism, e.g., by comparing spectra from the growth medium and from the organism after different periods of growth. Such experiments have normally been performed using material extracted from the organism and examined under *in vitro* conditions, which are frequently very different from those of the living organism, since high concentrations and low temperatures can frequently be necessary to obtain good spectra. It is possible however, under the right conditions, to obtain spectra from complete microorganisms *in vivo,* and this could be very useful in experiments on uptake and metabolism.

Another rather different use of the Mössbauer effect lies in its sensitivity to motion of the Mössbauer nucleus. Any vibration of the Mössbauer atom within a molecule will affect the Mössbauer recoil-free fraction, and information on this vibration can be obtained from measurements of the recoil-free fraction. Thus the Mössbauer source itself can be used as a probe to study the frequency response of auditory mechanisms and the macroscopic motion of complete organisms. The sensitivity to motion of the Mössbauer nucleus leads to line broadening in the spectra obtained from labeled molecules moving through a membrane, and information on this movement can be deduced from the measured linewidth.

It is perhaps appropriate to mention at this stage that the technique imposes rather stringent sample requirements (see Section III.B) and that the limitations on what can be achieved arise more frequently as a result of limitations on what is possible in terms of sample preparation rather than from any lack of ingenuity on the part of the biologists or Mössbauer spectroscopists. Despite these problems, Mössbauer spectroscopy is moving into new areas in which, although sample preparation may be difficult, the possibilities of assisting in the understanding of biological processes will surely make the effort worthwhile.

The next two sections consider the theoretical and experimental aspects of Mössbauer spectroscopy, with particular emphasis on those aspects which are important for biological applications. The subsequent sections cover the various areas of application as outlined here. The intention is to describe the whole range of possible applications, with emphasis on the type of information that can be obtained, without giving an exhaustive review of all previous biological applications of Mössbauer spectroscopy. Because it is intended to highlight future possibilities and potential as well as past achievements, the space devoted to different areas of work does not necessarily reflect the amount of work which has been done in these areas.

II. Principles of Mössbauer Spectroscopy

A. The Mössbauer Effect

Certain nuclei have low-lying excited states which decay to the ground state with the emission of gamma rays. ^{57}Fe is one such nucleus, and the decay scheme from its radioactive parent, ^{57}Co, is shown in Fig. 1. The natural linewidth is related to the mean lifetime of the excited state by the Heisenberg uncertainty principle. In the case of the 14.4 keV state in ^{57}Fe, this gives a linewidth of the order of 10^{-8} eV (2.5 MHz).

When a free nucleus emits a gamma ray there is an accompanying recoil of the nucleus. The energy of the emitted gamma ray is therefore less than the energy difference between the two nuclear levels by an amount equal to the recoil energy. Similarly, for resonant absorption to occur the gamma ray energy must be greater than the energy difference between the two nuclear levels by the amount of the required recoil energy. Thus for free nuclei, the recoil energy prevents resonant absorption of gamma rays under normal circumstances.

Mössbauer's discovery was that a nucleus embedded in a solid can sometimes emit its gamma ray with negligible energy loss to recoil. Although the nucleus is bound to the crystal lattice by elastic forces and its recoil can

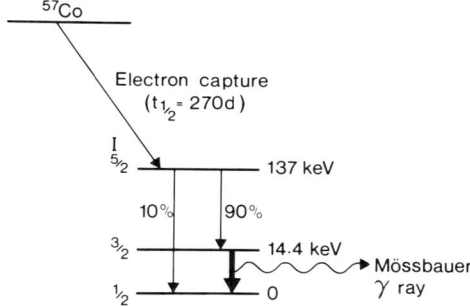

Fig. 1. Nuclear decay scheme of ^{57}Co showing the Mössbauer emission transition.

produce phonons, there exists a fraction of emissions in which no phonon is produced. The effective recoil mass is then that of the entire macroscopic solid, and under these conditions the recoil energy loss is negligible. Similarly, if the absorbing nucleus is bound in a solid, a fraction of absorptions will be recoil free. The probability of a recoil-free event (emission or absorption) depends on certain properties of the solid and the temperature, as well as on the energy and mean lifetime of the nuclear excited state. For ^{57}Fe in the majority of iron compounds this probability is large, particularly below room temperature.

The process of recoil-free resonant absorption of nuclear gamma radiation is the Mössbauer effect.

B. Mössbauer Spectroscopy

Mössbauer spectroscopy is the study of the energy dependence of the resonant absorption. The Mössbauer effect enables the extremely narrow natural linewidth of the excited state to be utilized in an absorption experiment. This linewidth is of the order of the energy of the hyperfine interactions between the nucleus and its surrounding electrons. Mössbauer spectroscopy can therefore be used to investigate these interactions. The experimental measurement provides the energy dependence of the absorption of the gamma radiation, and from this spectrum the nuclear energy-level scheme for the ground and excited states can be deduced. This is in turn interpreted to give information on the electronic arrangement in the vicinity of the Mössbauer nucleus.

A typical ^{57}Fe Mössbauer spectrometer consists of a source of 14.4-keV gamma radiation, usually radioactive ^{57}Co in a suitable matrix, an absorber containing ^{57}Fe (which is not radioactive) in the solid material under investigation, and a gamma-ray detection system. The energy of the incident gamma-ray beam is modulated by moving the source relative to the absorber. This changes the energy as a consequence of the relativistic first-order Doppler effect. The Mössbauer spectrum is generally plotted as counts (or relative absorption) against the velocity of the source relative to the absorber, which constitutes the energy axis. A full discussion of the experimental aspects of Mössbauer spectroscopy, particularly as they affect measurements of biological materials, is given in Section III.

If there were no hyperfine interactions in both source and absorber, the Mössbauer spectrum would consist of a single absorption line at zero velocity. Because of the hyperfine interactions between the nucleus and its environment, Mössbauer spectra are always more complex than this. These interactions fall into three main categories, which are considered in the next three sections.

C. The Chemical Shift

The chemical (or isomer) shift in the position of the nuclear energy levels arises from the electric-monopole interaction between the nucleus and the electronic charge density at the nucleus. The effect depends on the difference in the nuclear radii of the ground and excited states and the difference between the s electron densities at the nucleus for Mössbauer atoms in different materials.

A simple treatment by De Benedetti *et al.* (1961) and Walker *et al.* (1961) assumes that the nucleus is a uniformly charged sphere and that the electronic charge density is uniform over nuclear dimensions. Using these assumptions, a consideration of the difference in the potential energy for a point nucleus and a spherical nucleus leads to an expression for the chemical shift, δ:

$$\delta = \tfrac{2}{5}\pi Ze^2[|\psi_a(0)|^2 - |\psi_s(0)|^2](R_{ex}^2 - R_{gd}^2),$$

where Z is the atomic number, R_{gd} and R_{ex} the nuclear radii of the ground and excited states, and $|\psi_s(0)|^2$ and $|\psi_a(0)|^2$ the electron densities at the nucleus in the source and the absorber, respectively. Fig. 2a shows a diagram of the energy levels of ^{57}Fe and the resulting Mössbauer spectrum for the case where the chemical shift is the only hyperfine interaction operating. When additional hyperfine interactions are present, the chemical shift sets the position of the center of gravity of the whole Mössbauer spectrum.

The chemical shift is clearly not an absolute quantity, as it represents the difference between the electric-monopole interactions in the source and in

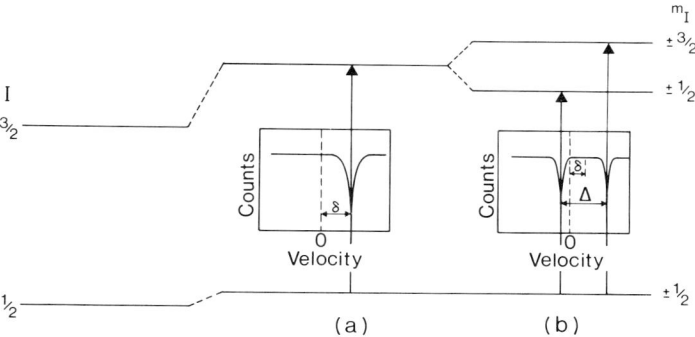

Fig. 2. The effects on the nuclear energy levels of ^{57}Fe of the electric-monopole interaction (a) and the electric-quadrupole interaction (b). The Mössbauer absorption transitions and the resulting spectra, which exhibit the chemical shift δ and the quadrupole splitting Δ, are also shown.

the absorber. Therefore, to permit comparisons, chemical shifts are generally expressed relative to a standard which determines the zero of the velocity axis of the spectrum. Thermal vibration of the nuclei also shift the gamma-ray energy because of the relativistic second-order Doppler effect. As this effect is temperature dependent, the temperature of both source and absorber should be considered when comparing chemical shifts. It is now generally accepted for ^{57}Fe work that chemical shifts should be quoted with respect to metallic iron at room temperature.

The expression for the chemical shift contains a nuclear term, relating to the difference in nuclear radii between the ground and the excited states, and a chemical term concerning the s electron density at the nucleus, which is intimately related to the chemical state of the atom. In Mössbauer spectroscopy the nuclear term is usually known from previous measurements, and chemical information is obtained from a determination of the chemical shift. For ^{57}Fe the nuclear term is negative, because the excited-state nucleus is smaller than the ground-state nucleus.

An important example of information derived from chemical shift measurements concerns the valence state of iron. At first sight it might appear strange that ferrous and ferric iron should exhibit a different chemical shift when their electronic configurations differ by only one d electron, which cannot itself contribute to the charge density at the nucleus. However the $3s$ electrons spend some of their time further from the nucleus than the $3d$ electrons; therefore the size of the $3s$ orbitals (and hence the charge density at the nucleus) depends on the screening of the nuclear charge by the $3d$ electrons. Thus the extra $3d$ electron in a ferrous iron atom reduces the attractive electrostatic potential and allows the $3s$ orbitals to expand. This reduces the charge density at the nucleus and hence increases the chemical shift.

Although the above picture is adequate in ionic iron compounds, other factors also affect the chemical shift. In covalent iron compounds, bond formation involves the mixing of the $4s$ wave functions into lower occupied levels, resulting in a direct contribution to the electron density at the nucleus. There may also be mixing of partly empty $3d$ orbitals with filled ligand orbitals resulting in an increased $3d$ electron density, increased screening, and hence a reduced contribution to the electron density at the nucleus. The first of these effects usually dominates; therefore increased covalency leads to a decrease in the chemical shift. Thus, in covalent ferrous and ferric compounds, the chemical shifts tend to be smaller and less different.

For typical ionic iron salts, the iron atom is in a high-spin state, and the chemical shifts for ferrous and ferric are around 1.5 and 0.5 mm sec^{-1} respectively, being affected also by the coordination of the iron atom. Covalency will have the effect of reducing these values somewhat. In the case of low-spin iron, the chemical shifts are reduced to around 0.2 mm sec^{-1} for both ferrous and ferric. Although molecular orbital calculations may be helpful, it is in general difficult to calculate chemical shifts from first principles, particularly to a sufficient accuracy for meaningful comparison with experimental values. However, chemical shifts are extremely useful as a semiempirical parameter, especially in the study of a series of related materials.

D. Quadrupole Splitting

The excited state of the ^{57}Fe nucleus, which has a nuclear spin $I = \frac{3}{2}$, has a positive electric-quadrupole moment, which means that the nuclear charge distribution is elongated along the spin axis. In an asymmetric electrostatic field, such as that arising from the ligands, the energy of the nucleus is dependent on the orientation of the nuclear spin. This electric quadrupole interaction splits the $I = \frac{3}{2}$ excited state of ^{57}Fe into two doubly degenerate states corresponding to the quantized components of the nuclear spin $m_I = \pm\frac{1}{2}$ and $m_I = \pm\frac{3}{2}$. The nuclear ground state with $I = \frac{1}{2}$ has no quadrupole moment and remains unsplit. The effect of the electric quadrupole interaction on the nuclear energy levels of ^{57}Fe and the resulting Mössbauer spectrum are shown in Fig. 2b. The spectrum with two absorption lines is known as a quadrupole-split doublet, the separation between the lines being the quadrupole splitting, Δ. The quadrupole splitting provides a measure of the charge distribution around the nucleus, whereas the chemical shift provides a measure of the charge density at the nucleus.

The electrostatic field may be quantified by the electric-field gradient. This is a tensor quantity, but it can be reduced to diagonal form in a suitable coordinate system. It can then be completely specified by two independent components: V_{zz}, the second derivative of the electrostatic potential in the z direction, and η, the asymmetry parameter, defined by $\eta = (V_{xx} - V_{yy})/V_{zz}$.

The coordinate system is usually chosen so that $V_{zz} \geq V_{xx} \geq V_{yy}$, and hence $0 \leq \eta \leq 1$. If the nuclear electric-quadrupole moment is expressed as eQ and the principal component of the electric-field gradient (V_{zz}) is denoted by eq, where e is the electronic charge, then the Hamiltonian of the electric-quadrupole interaction is given by

$$\mathcal{H}_Q = \frac{e^2qQ}{4I(2I-1)}[3I_z^2 - I(I+1) + \eta(I_x^2 - I_y^2)].$$

This equation has the eigenvalues

$$E_Q = \frac{e^2qQ}{4I(2I-1)}[3m_I^2 - I(I+1)](1 + \tfrac{1}{3}\eta^2)^{1/2}.$$

The energy separation of the $m_I = \pm\frac{1}{2}$ and $m_I = \pm\frac{3}{2}$ excited state levels of ^{57}Fe, and hence the quadrupole splitting, is given by

$$\Delta = \tfrac{1}{2}e^2qQ(1 + \tfrac{1}{3}\eta^2)^{1/2}.$$

There are essentially two contributions to the electric-field gradient: the valence contribution from electrons in outer unfilled shells of the atom in question and the lattice contribution from electrons in neighboring atoms. In simple systems, calculations can be used to obtain an estimate of the electric-field gradient and hence a value for the quadrupole splitting. By comparing this with the measured value a choice may be made between various possible arrangements of the electrons and atoms around the Mössbauer nucleus.

The quadrupole splitting observed in ^{57}Fe depends on the valence and spin state of the iron atom. High-spin ferric iron has five $3d$ electrons distributed singly through the five $3d$ orbitals, which gives a filled subshell and therefore a spherically symmetrical charge distribution. The component of the quadrupole splitting from the valence electron contribution to the electric-field gradient should therefore be very small. High-spin ferrous iron contains an extra $3d$ electron in addition to the filled subshell. This gives rise to an asymmetry in the charge distribution and hence to a large contribution to the electric-field gradient. Because this extra electron can populate a number of orbitals of closely similar energy, the quadrupole splitting in high-spin ferrous compounds can vary considerably with temperature.

High-spin ferrous iron almost always has a characteristically large quadrupole splitting of greater than 2 mm sec^{-1}. High-spin ferric iron usually has a rather small quadrupole splitting of less than 1 mm sec^{-1}. The low-spin configurations exhibit a range of quadrupole splittings. Apart from high-spin ferrous, the quadrupole splitting cannot be used on its own to distinguish the different states of the iron atom. However, much valuable information can be obtained from the comparison of the quadrupole splittings in related materials.

E. Magnetic Splitting

In the presence of a magnetic field, the nuclear-energy levels are split by the magnetic-dipole interaction, and the resulting Mössbauer spectra can be considerably more complex than in the case of purely electric interactions. The Hamiltonian which describes the interaction between the nucleus and a magnetic field is

$$\mathcal{H}_M = -g_n\mu_n \mathbf{I} \cdot \mathbf{B},$$

where μ_n is the nuclear magneton, g_n the nuclear gyromagnetic ratio, **I** the nuclear spin operator, and **B** the magnetic field at the nucleus. This Hamiltonian has the eigenvalues $E_M = -g_n\mu_n B m_I$. Therefore in the case of ^{57}Fe the magnetic interaction splits the $I = \frac{1}{2}$ ground state into two levels and the $I = \frac{3}{2}$ excited state into four equally spaced levels. For magnetic-dipole transitions, the selection rules are $\Delta m_I = 0, \pm 1$, so that six transitions are allowed. The energy-level diagram in this situation and the resulting Mössbauer spectrum for a system with no electric-quadrupole interaction are shown in Fig. 3.

The transition probabilities for the six transitions, and hence the intensities of the lines in the Mössbauer spectrum, depend on the angle between the direction of the gamma-ray propagation and the direction of the magnetic field. Observation of the intensity of the various lines in a magnetically split Mössbauer spectrum can therefore give considerable information on the magnetic structure of the material.

The magnetic splitting of the nuclear levels, and hence the splitting observed in the Mössbauer spectrum, depends on the size of the magnetic field at the nucleus. This is made up of the internal magnetic field produced by any externally applied magnetic field, together with the magnetic hyperfine field from the atom's own electrons. This depends on the spin and valence

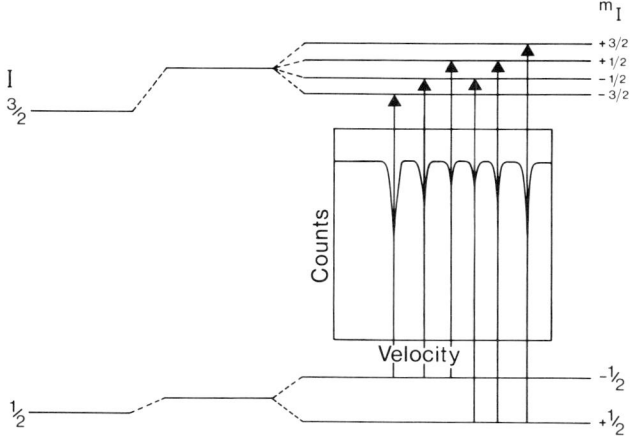

Fig. 3. The effect of a magnetic dipole interaction on the nuclear energy levels of ^{57}Fe, showing the Mössbauer absorption transitions and the resulting spectrum. It is assumed that there is no electric-quadrupole interaction. The splitting of the lines in the Mössbauer spectrum is proportional to the magnetic field at the nucleus.

state of the iron atoms and contains orbital and dipolar terms as well as the Fermi contact term. The latter arises from the direct effect of s electron spin density at the nucleus, which results from the modification of the s electron orbitals by the spin of unpaired $3d$ electrons. In low-spin ferrous compounds the hyperfine field owing to the atom's own electrons is zero, whereas in some high-spin ferric compounds it can be over 60 T. Where the magnetic field at the nucleus is large, the major contribution is generally the Fermi contact term, which is larger than the other hyperfine terms and can also be considerably greater than the largest available externally applied fields.

An important feature of the magnetic hyperfine field is that only unpaired electrons can contribute to it, whether directly or indirectly. In principle therefore, the situation is much less complex than with the quadrupole splitting or the chemical shift, in which contributions may come from all the electrons, both within the atom and in the lattice. Because of this, the calculations of the different contributions to the hyperfine field which have been carried out for a number of systems show good agreement with the experimental values.

In general, the magnetic splitting is modified by the presence of a quadrupole splitting. The resulting Mössbauer spectrum depends on the relative size of the two effects and on the orientation of the magnetic field at the nucleus with respect to the principal axis of the electric-field gradient.

Magnetic splitting is observed when an external magnetic field is applied or in the case of a magnetically ordered material. It can also be observed in the case of a paramagnetic material which has unpaired electron spin on the Mössbauer atom, if the electron spins change their direction very slowly. If the electron spins flip rapidly, no magnetic hyperfine splitting is seen in the Mössbauer spectrum, which then looks like that of a diamagnetic substance. In the intermediate case, the Mössbauer lines broaden, and the magnetic hyperfine splitting cannot be resolved. Therefore, for magnetic hyperfine splitting to be observed from a paramagnetic material, the electron-spin relaxation time must be long.

The electron-spin relaxation time results from the spin–lattice relaxation time and the spin–spin relaxation time. Spin–lattice relaxation takes place via the electron orbit and therefore depends on the orbital magnetic moment. For a ferric ion, the orbital magnetic moment is zero, and the spin–lattice relaxation time is long. For a ferrous ion, there is considerable orbital magnetic moment, and the spin–lattice relaxation time is usually very short. Because of lattice effects, the spin–lattice relaxation time increases with decreasing temperature. The spin–spin relaxation time is dependent on the concentration of spins in the sample. Therefore, to increase the electron-spin relaxation time so that magnetic hyperfine splitting is seen in the Mössbauer spectrum of a paramagnetic material, it is necessary to cool the specimen to low temperatures and to use magnetically dilute samples.

III. Experimental Techniques

Section II explained how the extremely narrow linewidth of the resonant absorption of gamma rays, which results from the Mössbauer effect, leads to a spectroscopy which measures the hyperfine interactions between the nucleus and its electronic environment. The experimental techniques used for Mössbauer spectroscopy will now be discussed. Particular emphasis will be given to the special requirements of work with biological materials, which usually have a low concentration of the Mössbauer isotope.

A. Spectrometers

The Mössbauer spectrometer consists of a radioactive source of Mössbauer gamma rays, a means of modulating the energy of these gamma rays, an absorber of the material under investigation, a counter for the gamma rays, and a means of storing the counts corresponding to a particular energy modulation.

The source contains the Mössbauer isotope in a radioactive excited state which decays to the ground state with the emission of the Mössbauer gamma ray. In the case of ^{57}Fe Mössbauer spectroscopy, the source used is ^{57}Co, which has the decay scheme shown in Fig. 1. The ^{57}Co is in a matrix material chosen to eliminate any hyperfine interactions which would split or broaden the emission line while giving as high a recoil-free fraction as possible, and which allows the high activity needed for investigating samples which are dilute in iron. A suitable source is provided by ^{57}Co diffused into a diamagnetic metal such as palladium or rhodium. The source is in the form of a very thin foil to prevent reabsorption of the gamma rays. A typical source activity for biological work is 100 mCi (3.7×10^9 Bq). Because the half-life of ^{57}Co is 270 days, source replacement represents an important element in the running costs of this technique.

A novel feature of Mössbauer spectroscopy is the method used to modulate the energy of the emitted gamma ray. This is done by moving the source relative to the absorber and hence Doppler shifting the gamma-ray energy. The source velocities used are very small (of the order of millimeters per second in the case of ^{57}Fe), but they produce shifts in the gamma-ray energy which are comparable with the energy of the hyperfine interactions and greater than the natural linewidth of the resonant absorption (about 0.2 mm sec^{-1} for ^{57}Fe). The motion of the source is produced by an electromechanical transducer similar to a loudspeaker, which is fed with a suitable signal from a waveform generator and amplifier. The signal used is usually designed to give a constant acceleration to the source, so that it performs a motion in which it spends equal times at each velocity from a negative value through zero to a positive value. Following this, there is a fast flyback, and the cycle is repeated. The repetition rate is typically of the order of 10 Hz, and the Mössbauer spectrum is accumulated over a large number of cycles. Clearly, the accuracy of the motion of the source is a crucial factor in this spectroscopic technique, and the transducer incorporates a means of monitoring the velocity of the source, which provides a signal used in a feedback network to correct the source motion.

The sample under investigation should consist of a solid containing the Mössbauer isotope, and it should cover the whole of the gamma-ray beam in the solid angle between the source and the window of the counter.

The gamma-ray counting system used for Mössbauer spectroscopy consists of a detector with associated high-voltage power supplies and amplifiers. The detector used depends on the energy of the Mössbauer gamma ray and on the resolution required to separate this gamma ray from x rays or other gamma rays of similar energy which may be present. Argon–methane-filled proportional counters give good resolution for low-energy gamma rays such as the 14.4-keV gamma ray from ^{57}Fe. They can also cope with the high counting rates which occur when working with the strong sources required by samples weak in iron. For other applications, scintillation (e.g., NaI crystal) or solid-state (e.g., lithium-drifted silicon) detectors may be more suitable. After the pulses from the detector have been amplified, they are passed through a single-channel analyzer to select only those pulses arising from Mössbauer gamma rays. This selection is facilitated by using a pulse-height analyzer.

The counts from the single-channel analyzer are stored in a series of memory channels (typically 256 or 512). Each of these corresponds to a particular velocity of the source relative to the absorber, this correspondence being achieved by synchronizing the sequential opening of the memory channels with the velocity waveform applied to the source. Traditionally, the storage functions were performed by a multichannel analyzer operating in the multiscaler mode, but now equivalent microprocessor-based systems are being used. The resulting spectrum consists of number of counts versus channel number (equivalent to velocity of source relative to absorber). A means of displaying the spectrum and outputting the data for analysis is usually incorporated.

To accumulate a spectrum with an adequate signal-to-noise ratio, many velocity sweeps are required, and typical counting times are of the order of hours, although in the case of weak biological samples it may be necessary to count for many days or even several weeks.

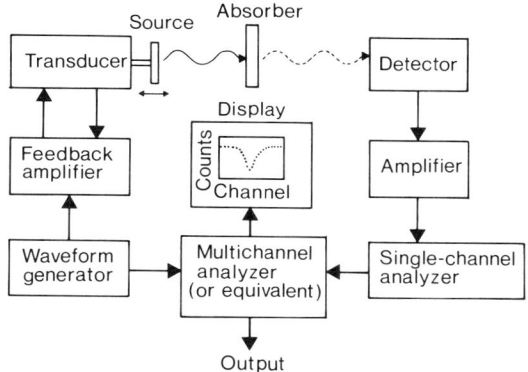

Fig. 4. Schematic representation of a typical Mössbauer spectrometer.

A block diagram indicating the essential elements of a typical Mössbauer spectrometer is shown in Fig. 4. Spectrometers can be assembled using commercially available units or can be purchased complete.

For reasons discussed in the previous and following sections, it is useful to be able to make measurements over a range of temperatures and in large applied magnetic fields. Most spectrometers incorporate arrangements for mounting the sample holder in a cryostat where, by using suitable cryogenic liquids such as liquid nitrogen and liquid helium, temperatures from as low as 0.01 K upwards can be obtained. Large magnetic fields can be obtained by incorporating a superconducting solenoid in a liquid-helium cryostat.

B. Sample Requirements

For the Mossbauer effect to be observed, the first requirement is that the Mössbauer nuclei should be in a solid environment. Thus the samples used for Mössbauer spectroscopy are normally solids. For biomolecules, the measurements are most frequently made on frozen aqueous solutions, as this is usually the simplest solid form which is stable and easily obtainable. Some measurements have been made on concentrated proteins which have been precipitated and separated in solid form by a high-speed centrifuge. In a few instances it has been possible to make measurements on single-crystal samples, which enables useful extra information to be obtained from the analysis. In the case of more complex biological materials such as membranes, tissue, or part or whole organisms, the material can be either frozen directly or freeze-dried to produce a suitable sample. Clearly, in measurements on samples prepared as described here the material is being investigated under *in vitro* conditions, which may differ considerably from those prevailing in the living organism, and this must be borne in mind in any interpretation of the data.

The requirement for a solid sample can sometimes be slightly relaxed. If the nucleus is in a quasi-solid environment, so that any motion within the time scale of the Mössbauer absorption process is negligible, then a Mössbauer spectrum can still be obtained. The necessary quasi-solid environment can be realized by a large molecule in a viscous liquid, conditions which may well occur in a living organism. Thus it can be possible, under the right conditions, to obtain spectra from complete organisms *in vivo*. This could be very useful in experiments on uptake and metabolism and also in measuring the diffusion rates of various molecules in the organism.

To obtain an adequate Mössbauer spectrum, the sample must contain a certain minimum quantity of the Mössbauer isotope. This requirement imposes the biggest constraint on what can be achieved in biological applications of Mössbauer spectroscopy. To understand why the concentration problem is particularly acute, it is necessary to consider the signal-to-noise ratio of the spectrum in some detail.

Mössbauer spectroscopy involves a radioactive decay process. For such a process Poisson statistics apply, and therefore the statistical scatter (standard deviation) is equal to the square root of the number of counts. This statistical scatter provides the noise in the spectrum, and thus to decrease the percentage noise by a certain factor it is necessary to increase the number of counts by the square of that factor. The resulting problem can best be illustrated by an example. If an absorber contains a sufficient amount of the Mössbauer isotope to give a Mössbauer spectrum with 1% absorption, then a spectrum with 10^6 counts per channel will give a signal-to-noise ratio of 10, which is about the lowest value possible for any sort of reliable analysis of the spectrum. If the concentration of the Mössbauer isotope is a factor of 10 less, then the absorption in the Mössbauer spectrum will be reduced by the same factor to 0.1%, but the number of counts per channel required to yield a signal-to-noise ratio of 10 increases to 10^8. For ^{57}Fe Mössbauer spectroscopy, and with strong sources, 10^6 counts per channel corresponds to runs of the order of hours, while 10^8 counts per channel corresponds to runs of the order of weeks. A further reduction in concentration by a factor of 10 would lead to runs taking several years! For many areas of investigation, particularly those involving spectra with multiple components or those requiring complicated computer analysis, a signal-to-noise ratio of substantially more than 10 is necessary.

A number of factors determine the absorption intensity observed in the Mössbauer spectrum. Of these the ones of interest in the present context are the recoil-free fraction (i.e., the fraction of absorption events which take place without recoil and therefore exhibit the Mössbauer effect) and the number of Mössbauer nuclei per unit area.

The recoil-free fraction depends on how tightly the Mössbauer nucleus is bound into the solid and also on the temperature. Biomolecules usually involve rather weak binding, but at low temperatures (e.g., 100 K and below) the recoil-free fraction may be quite adequate. In the case of iron proteins it is typically around 0.7. The recoil-free fraction in biomolecules sometimes falls off very rapidly as the temperature rises above 200 K, and, although the samples may still be solid, no Mössbauer spectrum is observed.

The number of Mössbauer nuclei per unit area depends on the thickness and area of the sample, as well as on the concentration of the Mössbauer isotope within the sample. The thickness is limited by the nonresonant absorption of the Mössbauer gamma rays, which is discussed later, and the area is limited by the geometry of the spectrometer. A typical sample would be 0.5 cm thick and have a cross-sectional area of 2 cm^2, giving a volume of 1 cm^3, and would be contained in a holder similar to that shown in Fig. 5.

The concentration of the Mössbauer isotope depends on both the concentration of the Mössbauer element and the isotopic abundance of the Mössbauer nuclide. In the case of ^{57}Fe this isotopic abundance is 2% and thus natural iron contains 2% of the Mössbauer isotope. This percentage is sufficient to give good spectra in the case of samples of iron salts, alloys, and minerals which are rich in iron, but may not be sufficient in the case of biological materials in which the iron is a minor (albeit important) constituent. It is therefore frequently necessary to enrich the samples with ^{57}Fe. This can be done in a number of ways. In the case of certain proteins it is possible to remove the iron-containing moiety by suitable treatment and then reconstitute the protein using iron enriched in ^{57}Fe. The problems with this technique lie in the possibility of altering the protein and in the fact that many proteins are too unstable for such a procedure to be feasible. Alternatively, the organism can be grown on a medium containing iron enriched in ^{57}Fe and from which other adventitious sources of iron have been removed; then the required material can be extracted. This method is more reliable and more generally applicable than chemical exchange, but it is limited by the amount of ^{57}Fe which may be required. (Iron enriched to over 90% in ^{57}Fe is readily available but costs of the order of $10 per milligram). In the case of large organisms or in medical research, there may be situations where injecting solutions containing ^{57}Fe can provide a suitable means of incor-

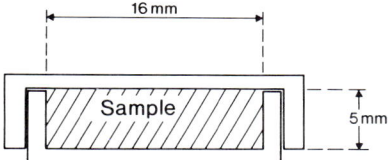

Fig. 5. Cross-sectional view of a typical cylindrical sample holder made from nylon.

poration. The ideas discussed here for the case of ^{57}Fe Mössbauer spectroscopy also apply to work using other Mössbauer isotopes in which some method of isotopic enrichment may be necessary.

The concentration of the Mössbauer element required to produce a sample giving an adequate spectrum is dependent on the particular Mössbauer isotope involved and also on the nature of the spectrum. For ^{57}Fe Mössbauer spectroscopy, the optimum concentration is of the order of several milligrams of natural iron per square centimeter of sample, corresponding to a total of about 10 mg of natural iron (0.2 mg of ^{57}Fe) in the typical sample described here. As biological samples are unlikely to contain this much iron, it is perhaps more appropriate to consider the minimum amount of iron required. This will depend on the spectral shape, but for a typical quadrupole-split doublet, as obtained from many biological materials in the absence of magnetic interactions, of the order of 100 μg of natural iron (2 μg ^{57}Fe) in the sample will give a spectrum with an absorption of around 0.1%, which is about the minimum practical value. Higher concentrations than this will, of course, greatly facilitate both taking the spectrum and its subsequent analysis.

To achieve the required quantity of the Mössbauer isotope in the sample it is often necessary to have the material under investigation in a more highly concentrated form than that required for biological assays or other physicochemical techniques. This must be taken into consideration when comparing the data from different measurements.

Mössbauer spectroscopy is sensitive to all the forms of the Mössbauer isotope in the sample. Although this makes the technique very useful, it imposes rather special constraints on the purity of the sample, which should contain no forms of the Mössbauer isotope other than those under investigation. Particular note should be given to this point when comparing the data from Mössbauer spectroscopy with those from complementary techniques, such as EPR, which are sensitive to only certain forms of the element being studied.

The gamma rays passing through the sample may be absorbed nonresonantly by all the atoms in the sample, which leads to a reduction in intensity. This process is strongly dependent on the atomic number of the absorbing atom, and it is therefore desirable to have as few heavy atoms in the sample as possible. The carbon, hydrogen, and oxygen atoms present in large numbers in biological materials have low atomic numbers and do not produce any problems, but atoms with higher atomic numbers, either in the biological material itself or in any associated reagents, do produce substantial attenuation of the gamma-ray beam and should be avoided. For this reason, the Mössbauer sample holders are made of nylon, PTFE, or other similar materials. Too high a concentration of the Mössbauer isotope leads to high nonresonant absorption, as noted, and also to high resonant absorption. A high resonant absorption necessitates the application of thickness corrections to the spectral analysis, but this is unlikely ever to be necessary in the case of biological samples.

C. Computer Analysis

Computer analysis of the Mössbauer spectra is a well-developed aspect of the technique for two reasons. One is that the raw data of the spectrum is already digitized and is therefore in a convenient form for computer processing. The other and more fundamental reason is that the origin and form of the spectral lines are well understood, which makes fitting to a theoretical model appropriate. This is not the case in many other spectroscopic techniques, such as optical spectroscopy.

The details of the interactions which determine the form of the Mössbauer spectrum were discussed in Section II. Each chemical form of the Mössbauer isotope gives rise to a separate component in the spectrum which, in the case of ^{57}Fe, can be a singlet, a doublet, a sextet, or a more complex form resulting from a combination of hyperfine interactions.

In its simplest form, the computer analysis involves finding the parameters of one or more Lorentzian lines which provide the best possible fit to the experimental spectrum. The Lorentzian lines can be treated individually or fitted as doublet or sextet combinations. This analysis provides the Mössbauer parameters (i.e., chemical shift, quadrupole splitting, magnetic field, linewidth, etc.) of the various spectral components, as well as their relative intensities.

For more complicated situations, the analysis may involve fitting the experimental spectrum to a spectrum calculated from a Hamiltonian which includes a complete description of the hyperfine interactions. From this Hamiltonian the nuclear-energy levels are obtained, and from these the theoretical spectrum can be derived. The fitting involves varying the parameters of the original Hamiltonian until the best fit to the experimental spectrum is obtained. In principle, this technique provides a completely general solution, but for practical reasons it is usually necessary to impose constraints appropriate to the particular type of material under investigation. This type of analysis is often used when measurements are made in an externally applied magnetic field. Details of one such computer program were given by Lang and Dale (1974).

In general, the computer fitting involves varying the parameters of some function, which can be either a set of Lorentzian lines or a more complicated theoretical spectrum and can also include a geometrical baseline correction, to minimize the difference between the function and the experimental spectrum. This minimization is usually performed in terms of a quantity known as chi squared, which is the sum of the squares of the deviation of the experimental spectrum from the function at each channel, normalized to take into account the statistical error. This procedure is therefore often known as least-squares fitting. Theoretically, a good fit should yield a value of chi squared equal to the number of channels minus the number of variable parameters in the fit. The number of variable parameters involved in the fit can be as few as three, in the case of a simple spectrum, or as many as twenty or more in the case of a complex spectrum with multiple components. It is obviously desirable to achieve a satisfactory fit with as few variable parameters as possible.

The values of chi squared obtained in Mössbauer spectroscopy are frequently higher than might be expected theoretically, because the function used is not absolutely correct. This may arise in biological materials for a number of reasons. If the environment of the Mössbauer nuclei is not uniquely defined, this will lead to a spread in certain parameters which can give distortion of the line shape. Another possibility is that there may be impurities present in the sample. There are also instrumental and geometrical factors which may contribute to a higher value of chi squared.

A much greater problem is that, with the relatively noisy spectra often obtained from biological samples weak in the Mössbauer isotope, an apparently satisfactory fit can be obtained too readily if the value of chi squared is used as the sole criterion of the quality of the fit. This arises because a low value of chi squared can result from a large statistical error on the experimental spectrum as well as from a small difference between the experimental and calculated spectra. For this reason other goodness-of-fit parameters have been suggested (Ruby, 1973). A related problem is that the minimization of chi squared gives the best fit within the constraints of the model used and the starting parameters of the fit. Thus the fitting does not necessarily produce the unique best fit which the experimenter is seeking. It is therefore very important to try various models and starting parameters to check the uniqueness of any fit. This problem is particularly acute in the case of data with a poor signal-to-noise ratio, for which there is likely to be large uncertainties in the values of the fitted parameters and considerable doubt about the uniqueness of the fit.

For these reasons, it is necessary to exercise caution when applying computer analysis to Mössbauer spectra. However, used wisely and with the insight which results from experience, it can provide a very powerful tool for extracting information from the data.

The work on isolated biomolecules has now reached the stage where in many cases a complete theoretical understanding of the Mössbauer spectra with computer fitting is possible. This computer analysis frequently confirms the more qualitative interpretation obtained by inspection of the spectrum and supports the use of qualitative ideas to interpret the spectra of more complex systems.

IV. Applications to Isolated Biomolecules

Most Mössbauer measurements on isolated biomolecules have used ^{57}Fe, the most common Mössbauer isotope. Biological molecules containing iron fall into well-defined groups in which the iron usually occurs in a distinctive environment. The main groups of biological molecules that contain iron at their active centers are the heme proteins, the iron–sulfur proteins, the iron-transport proteins, and the iron-storage proteins. In addition, certain biological iodine compounds and also vitamin B_{12} have been investigated by Mössbauer spectroscopy.

A. Heme Proteins

Heme proteins all contain iron in a fixed and characteristic environment known as the heme group. This is a relatively small and stable planar unit in which the iron is coordinated to four nitrogen ligands in the plane. There can be two further ligands above and below the plane, and in the heme proteins one of the nonplanar ligands is attached via a nitrogen atom to a chain of amino acids to make up the protein molecule, whereas the sixth ligand can be attached to a variety of groups.

Hemoglobin, the heme protein in red blood cells, is important in the reversible binding of oxygen. Changing the group attached to the sixth ligand changes the state of the iron atom. In oxyhemoglobin the iron atom is low-spin ferrous and lies in the heme plane, whereas in the deoxyhemoglobin it is high-spin ferrous and lies out of the heme plane. In methemoglobin, where the sixth ligand is a water molecule, the iron is high-spin ferric, whereas in hemoglobin cyanide it is low-spin ferric.

Reviews of the Mössbauer work on heme proteins were given by Lang (1970) and Spartalian and Lang (1980). At high temperatures (195 or 77 K), the spectrum usually consists of a quadrupole-split doublet, and the spin and valence state can be indicated by the chemical shift and the quadrupole splitting (see Table I). Lang and Marshall (1966) showed that the magnetic hyperfine splitting observed at low temperatures provides a powerful method of confirming and characterizing the state of the iron atom. In oxyhemoglobin and in hemoglobin carbon monoxide, the iron is low-spin

Table I Mössbauer Chemical Shifts and Quadrupole Splittings of Typical Hemoglobin Compounds[a]

Compound	State of the iron atom	Chemical shift	Quadrupole splitting
Oxyhemoglobin	Low-spin ferrous	0.20	1.89
Deoxyhemoglobin	High-spin ferrous	0.90	2.40
Hemoglobin cyanide	Low-spin ferric	0.17	1.39
Methemoglobin	High-spin ferric	0.20	2.00

[a] At 195 K, quoted in millimeters per second. (From Lang and Marshall, 1966).

ferrous and shows no magnetic hyperfine splitting even in the presence of a large applied magnetic field. In deoxyhemoglobin, the iron is high-spin ferrous, and the Mössbauer spectrum shows magnetic hyperfine splitting in applied fields. In hemoglobin fluoride and methemoglobin, the high-spin ferric atom gives symmetric six-line magnetic patterns at low temperatures in applied fields, with hyperfine fields of about 50 T. The low-spin ferric atom in hemoglobin cyanide, azide, and hydroxide gives rise to a Mössbauer spectrum showing complex, asymmetric magnetic splitting at low temperatures.

The quadrupole splitting of oxyhemoglobin is large and has an anomalously large temperature dependence, indicating that there are some unusual aspects about the state of the iron. Considerable interest has been shown in making theoretical calculations of the quadrupole splitting data. It is hoped that this may help in the selection of models for the configuration of the oxygen molecule relative to the heme plane and for the electronic states of the iron atom and the oxygen molecule. The Mössbauer data have been shown to be compatible with the widely accepted model for oxyhemoglobin in which the oxygen molecule lies out from the heme plane and at an angle to it (Kirchner and Loew, 1977).

B. Iron–Sulfur Proteins

The iron–sulfur proteins are a group of molecules that are found in plants, animals, and bacteria involved in oxidative electron-transfer processes in many different kinds of functions. Unlike the heme proteins, the iron in the iron–sulfur proteins is not bound in a stable, basic structural unit. Instead, it is held rather loosely in the chain of amino acids via sulfur atoms. Any attempts to remove the iron chemically are likely to cause the whole molecule to break up. Hence spectroscopic methods are especially important for studying these proteins, particularly as the structures of only a few have been determined by x-ray crystallography because of the difficulties of producing single crystals.

The structural information available on the iron–sulfur proteins indicates that although the various molecules may be different in size, structure, and function, they all contain iron atoms in a similar environment, with the iron atom at the center of four sulfur atoms that form an approximate tetrahedron. The simplest of these molecules are the rubredoxins, since they contain only one iron atom per molecule. Next in order of complexity are the plant-type ferredoxins, which contain two iron atoms per molecule. More complex still are the bacterial ferredoxins and high-potential iron proteins (HiPIPs) in which the iron is in cubane units containing four iron atoms and four labile sulfur atoms, or in a planar unit of three iron atoms and three labile sulfur atoms (Cammack, 1980). A review of the Mössbauer work on iron–sulfur proteins was given by Cammack *et al.* (1977).

Mössbauer effect measurements of the one-iron protein rubredoxin in the oxidized and reduced states have been made to find the chemical shifts, quadrupole splittings, and magnetic hyperfine splittings for Fe^{3+} and Fe^{2+} in this tetrahedral sulfur coordination. This effectively calibrates these quantities by allowing for the effects of covalency of the iron atoms in this environment. This information can then be used in the interpretation of the data on the proteins with two and more iron atoms.

The chemical shift is a very useful parameter for characterizing the formal valences of the iron atoms in these proteins. In Table II, the mean chemical shifts in the Mössbauer spectra of some iron–sulfur proteins are listed, together with the assignment of the formal valence states of the iron atoms. However, within the formal valence assignments there are differing degrees of valence-electron localization in the different proteins.

In the proteins with two-iron centers, there are two Fe^{3+} atoms in the oxidized state and one Fe^{3+} and one Fe^{2+} atom in the reduced state. The magnetic moments of the two iron atoms have been shown to be antiferromagnetically coupled, and the magnetically split Mössbauer spectra observed at low temperatures are very different from what is usually observed with a single iron atom. This is a situation unlike that found in inorganic complexes of iron, and it is not easy by any other method to observe directly the antiferromagnetic coupling within the pair of iron atoms.

Proteins with four-iron active centers are of two kinds: those with a negative redox potential such as the four-iron and eight-iron bacterial ferredoxins, and those with a positive redox potential like the HiPIP from *Chromatium*. In all these, the valence electrons are delocalized to varying degrees over the iron atoms. The iron atoms are again antiferromagnetically coupled in a way similar to that found in the proteins with two-iron

Table II Mössbauer Chemical Shifts of the Iron–Sulfur Proteins[a]

Protein	Chemical shift	Average valences	Formal valences
Fe^{3+} in rubredoxin[b]	~0.25	—	—
Fe^{3+} in adrenal ferredoxin[c]	0.26	—	—
Fe^{3+} in spinach ferredoxin[d]	0.26	—	—
Oxidized *Chromatium* HiPIP[e]	0.33	$4Fe^{2.75+}$	$3Fe^{3+}, 1Fe^{2+}$
Reduced *Chromatium* HiPIP[e]	0.42	$4Fe^{2.5+}$	$2Fe^{3+}, 2Fe^{2+}$
Oxidized *B. stearothermophilus* ferredoxin[f]	0.43	$4Fe^{2.5+}$	$2Fe^{3+}, 2Fe^{2+}$
Oxidized *C. pasteurianum* ferredoxin[g]	0.43	$4Fe^{2.5+}$	$2Fe^{3+}, 2Fe^{2+}$
Oxidized *Chromatium* ferredoxin[h]	0.41	$4Fe^{2.5+}$	$2Fe^{3+}, 2Fe^{2+}$
Reduced *B. stearothermophilus* ferredoxin[f]	0.49, 0.59	$2Fe^{2.5+}, 2Fe^{2+}$	$1Fe^{3+}, 3Fe^{2+}$
Superreduced *Chromatium* HiPIP[i]	0.59	$4Fe^{2.25+}$	$1Fe^{3+}, 3Fe^{2+}$
Reduced *C. pasteurianum* ferredoxin[g]	0.57	$4Fe^{2.25+}$	$1Fe^{3+}, 3Fe^{2+}$
Reduced *Chromatium* ferredoxin[h]	0.54	$4Fe^{2.25+}$	$1Fe^{3+}, 3Fe^{2+}$
Fe^{2+} in rubredoxin[b]	0.65	—	—
Fe^{2+} in spinach ferredoxin[d]	0.60	—	—

[a] At 77 K, quoted in millimeters per second. (Adapted from Cammack, et al., 1977).
[b] From Rao et al. (1972).
[c] From Cammack et al. (1971).
[d] From Rao et al. (1971). 195 K value adjusted for second-order Doppler shift.
[e] From Middleton et al. (1980).
[f] From Middleton et al. (1978).
[g] From Thompson et al. (1974).
[h] From Cammack et al. (1977).
[i] From Dickson and Cammack (1974).

centers, but each Mössbauer spectrum is different and can be used to characterize these proteins.

Eight-iron ferredoxins such as those from *C. pasteurianum* and *Chromatium* contain two four-iron centers per molecule. Their Mössbauer spectra are similar to those of the four-iron ferredoxins, but there is an additional spin–spin interaction between the two centers, which leads to certain characteristic differences.

The oxidized four- and eight-iron ferredoxins and reduced HiPIP are in equivalent redox states. The averaged chemical shifts observed in these proteins indicate that their centers contain essentially equivalent iron atoms. These are antiferromagnetically coupled and therefore give a ground state with a total spin of zero. This gives no magnetic hyperfine splitting, and thus no enhancement of an externally applied field is found in the Mössbauer spectra. The resulting spectra, which are strongly characteristic, are shown in Fig. 6.

For many years it was accepted that the centers found in iron–sulfur proteins contained either one, two, or four iron atoms. However, there is considerable evidence for a group of iron–sulfur proteins containing centers with three iron atoms. Mössbauer spectroscopy has provided important data leading to this conclusion. The tetrameric form of the *Desulfovibrio gigas* ferredoxin has been investigated by Huynh *et al.* (1980). The Mössbauer spectrum of the oxidized ferredoxin (Fig. 7a) at 77 K shows essentially a single quadrupole-split doublet, indicating that at this temperature all the iron atoms are equivalent. The Mössbauer parameters are close to those of oxidized rubredoxin, with a chemical shift of 0.27 mm sec^{-1}, characteristic of Fe^{3+} in the tetrahedral sulfur coordination found in iron–sulfur proteins. At 4.2 K the reduced protein shows a spectrum (Fig. 7b) consisting of two

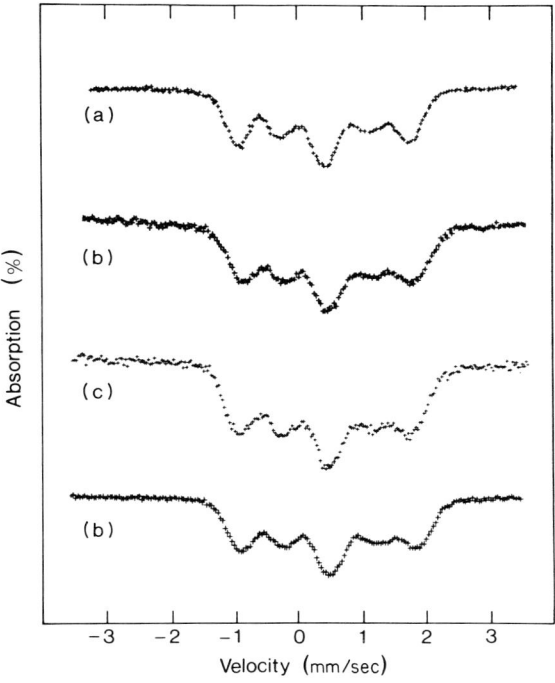

Fig. 6. Mössbauer spectra of proteins with 4Fe-4S centers in the native state, taken at 4.2 K in a perpendicular applied field of 6 T: (a), reduced *Chromatium* HiPIP; (b), oxidized *Bacillus stearothermophilus* ferredoxin; (c), oxidized *Clostridium pasteurianum* ferredoxin; (d), oxidized *Chromatium* ferredoxin. (From Cammack *et al.*, 1977.)

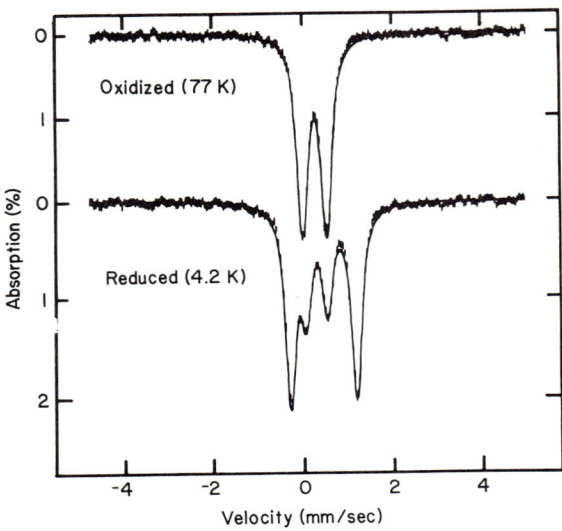

Fig. 7. Mössbauer spectra of ferredoxin II from *Desulfovibrio gigas*. (From Huynh et al., 1980.)

quadrupole-split doublet components with an intensity ratio of 2:1. In the presence of applied magnetic fields, the oxidized ferredoxin gives a spectrum which requires three equal-intensity components to fit it, whereas the spectrum of the reduced ferredoxin requires two components with an intensity ratio of 2:1 to fit it. The authors conclude from these data that the center contains three iron atoms. There is similar evidence from the computer analysis of Mössbauer spectra that the low redox potential center of the ferredoxin from *Azotobacter vinelandii* also contains a center of the three-iron type (Emptage et al., 1980). There is as yet insufficient Mössbauer data on the proteins with three-iron centers to be able to specify completely their characteristic behavior as has been done with the other iron–sulfur proteins.

Mössbauer spectroscopy has been able to contribute to our knowledge of the iron–sulfur proteins by identifying the chemical state of the iron atoms, by providing information on the localization of the $3d$ electrons on the iron atoms in the different redox states, by demonstrating and elucidating the antiferromagnetic coupling between the iron atoms, and by enabling the different sorts of centers to be characterized and identified in new proteins.

C. Iron-Transport Proteins

Iron-transport proteins are molecules which strongly bind iron and enable iron to be taken up from the environment and incorporated into a cell. In primitive organisms, these molecules are called siderochromes and have small relative molecular masses of about 1000. In higher organisms (mammals and birds), iron is transported by transferrin, conalbumin, and lactoferrin, which resemble the siderochromes in many ways but which have much larger relative molecular masses of about 50,000. The iron in these proteins is normally high-spin ferric. The soluble transport proteins which occur in the extracellular fluid frequently contain two metal-binding sites per molecule. Mössbauer studies of iron-transport proteins have been described by Oosterhuis and Spartalian (1976).

Most of the work done so far has been on iron-transport proteins from higher organisms, such as human transferrin and conalbumin from hen egg whites. The ferric ions are well separated from each other, and therefore the Mössbauer spectra show paramagnetic hyperfine splitting at low temperatures when the spin-relaxation times are long, with characteristic high-spin ferric hyperfine fields of around 50–60 T. Since the iron atoms are paramagnetic, they may be aligned by an external magnetic field. In general, the effect of the field is to sharpen the individual lines in the spectrum and reduce the overall splitting by reducing the magnetic field at the nuclei. This behavior is illustrated in Fig. 8 for human transferrin.

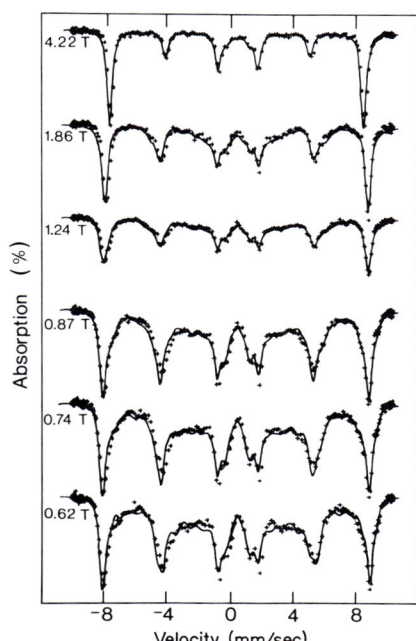

Fig. 8. Mössbauer spectra of 100% iron-saturated human transferrin at 4.2 K in parallel applied magnetic fields. (From Tsang et al., 1976.)

D. Iron-Storage Proteins

The iron-storage proteins are large molecules that contain appreciable concentrations of iron, up to perhaps 20% by weight. The most widespread is ferritin, which occurs in mammalian tissue as well as in plants and fungi. The protein molecules are approximately spherical, with a diameter of typically 12 nm. They consist of an inner core of around 7 nm in diameter, which contains the iron in the form of a small particle of ferric oxyhydroxide, surrounded by a protein envelope. The application of Mössbauer spectroscopy to these proteins has been reviewed by Oosterhuis and Spartalian (1976).

The Mössbauer spectra show the characteristic superparamagnetic behavior of a finely divided antiferromagnetic material. At low temperatures, a simple six-line pattern characteristic of an ordered magnetic material is observed, with a hyperfine field of about 50 T. Unlike a bulk antiferromagnet, where the hyperfine field disappears at the Néel temperature T_N, the magnetically split spectrum in a superparamagnet transforms to a doublet at temperatures below T_N when fluctuations of the spin direction of each particle become sufficiently rapid to smear out the magnetic interaction. The temperature range over which this occurs depends on the mean size and the distribution of sizes of the particles. Mössbauer spectra for horse ferritin at different temperatures are shown in Fig. 9. From spectra of this sort an estimate of the particle sizes can in principle be made. Spartalian *et al.* (1975) found that the iron cores in the ferritin from the fungus *Phycomyces blakesleanii* appear to be smaller than those in horse ferritin.

A series of Mössbauer investigations (Bauminger *et al.,* 1979a, 1980; Dickson and Rottem, 1979) have demonstrated the presence of another type of iron-storage material in bacteria. The Mössbauer spectra show evidence of a magnetic ordering transition at low temperatures, indicating that the iron must be in a fairly concentrated form, as would be expected in an iron-storage compound. However, the spectra are significantly different from those of ferritin, and in particular do not show the characteristic superparamagnetic behavior observed in that protein. Thus, although electron micrographs show a structure very similar to that of ferritin, the Mössbauer data indicate that the iron-containing core must be different.

In some novel experiments on magnetotactic bacteria it has been established that these microorganisms use magnetite single-domain particles for orientation purposes. Mössbauer spectroscopy has been useful in identifying the form of iron, which is found in a relatively highly concentrated state, and can therefore be regarded as being a type of storage iron (Frankel *et al.,* 1979).

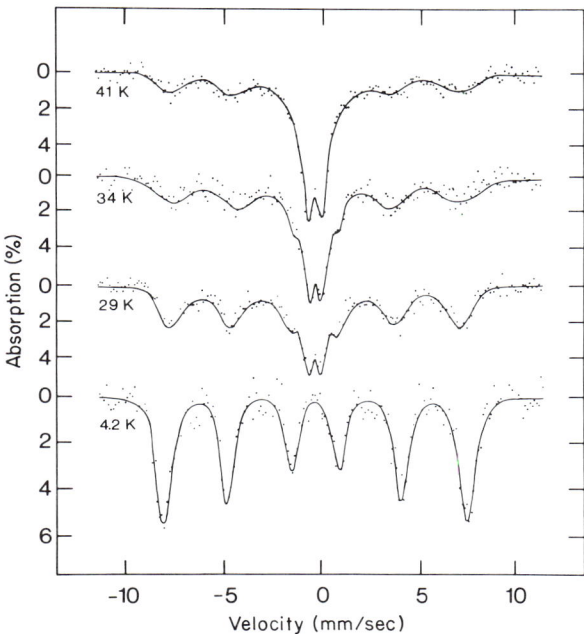

Fig. 9. Mössbauer spectra of horse ferritin at various temperatures. (From Boas and Window, 1966.)

E. Iodine Compounds

Iodine compounds and hormones play an important part in certain physiological processes. They can be investigated with either ^{129}I or ^{127}I Mössbauer spectroscopy. The latter technique is usually preferable because it utilizes the naturally occurring isotope, although ^{129}I Mössbauer spectroscopy gives much better resolution.

Groves et al. (1973) used ^{129}I Mössbauer spectroscopy to study iodine bonding in some iodine-containing hormones. They made measurements on thyroxine, which is synthesized by the thyroid gland and contains four iodine atoms, and on di-iodo-tyrosine, which is believed to be a precursor for thyroxine and contains two iodine atoms. Identical Mössbauer spectra were found for the two compounds, showing that the iodine sites in each compound could not be distinguished by Mössbauer spectroscopy.

Oberley et al. (1974) and Oberley and Erhardt (1975) observed significant Mössbauer absorption for these and some related thyroid hormones using the ^{127}I isotope. The results obtained are consistent with previous ^{129}I Mössbauer studies.

Haffner et al. (1976) investigated tobacco mosaic virus using ^{129}I. The spectra of tobacco mosaic virus with certain amino acids iodinated were compared with the spectra of di-iodo-tyrosine, and the authors interpreted the results in relation to the bonding which exists within these compounds.

F. Vitamin B_{12}

Vitamin B_{12} and its derivates (cobalamins) contain cobalt in an environment somewhat similar to that of iron in the heme proteins. If ^{57}Co is incorporated into the molecule, Mössbauer gamma rays are emitted from an ^{57}Fe atom occupying a cobalt site in the molecule, and the resulting Mössbauer emission spectra can be measured against a single-line absorber.

Nath et al. (1968) demonstrated that the Mössbauer spectra are not drastically affected by the aftereffects associated with the radioactive decay. They found that the Mössbauer parameters of cobalamins are similar to those of iron compounds with similar structures.

In another study Inoue and Nath (1977) obtained emission Mössbauer spectra from vitamin B_{12} and related complexes in three redox states. The differences in chemical shift and quadrupole splitting observed in the spectra are discussed in terms of the electronic structure and bonding of the ^{57}Fe atom on the cobalt site. These spectra constitute fingerprints for the various forms of vitamin B_{12}, which could be useful in identifying the intermediates in reactions involving this vitamin.

V. Applications to Molecular Systems

The biological reactions which take place in living cells are catalyzed by enzymes which can contain a number of proteins that function together, each needing the others for the reaction to occur. Mössbauer spectroscopy can be used, in suitable circumstances, both to investigate the various components of the system and to study their interrelationship to obtain information on their biological function. Mössbauer spectroscopy has been applied to a number of enzymes and enzyme systems in which iron is involved.

A. Nitrogenase

Nitrogenase is the enzyme system which catalyzes the fixing of atmospheric nitrogen to form ammonia. It consists of two proteins, an iron protein and a molybdenum–iron protein. Mössbauer measurements were made on the nitrogenases from *Klebsiella pneumoniae* (Kelly and Lang, 1970; Smith and Lang, 1974; Smith *et al.*, 1980) and *Azotobacter vinelandii* (Münck *et al.*, 1975; Huynh *et al.*, 1979). There are close similarities between the spectra obtained by the two groups, which is encouraging in view of the difficulties involved in purification and characterization.

The data indicate that the iron protein is very similar to a ferredoxin with a four-iron center, and most of the work has involved elucidating the nature of the molybdenum–iron protein. Measurements such as EXAFS indicate that this protein contains a novel type of center in which there are molybdenum, iron, and labile sulfur atoms. The Mössbauer spectroscopy of the whole protein is complicated by the fact that in addition to the molybdenum–iron center it appears to contain iron in three other forms, although one of these makes only a minor contribution and may be an impurity. Figure 10 is the Mössbauer spectrum of the molybdenum–iron protein at 30 K, showing the three main components.

One of the Mössbauer spectral components shows a change on reduction. This component has been associated with the EPR signals which have been observed from two centers per molecule, each of which undergoes a one-

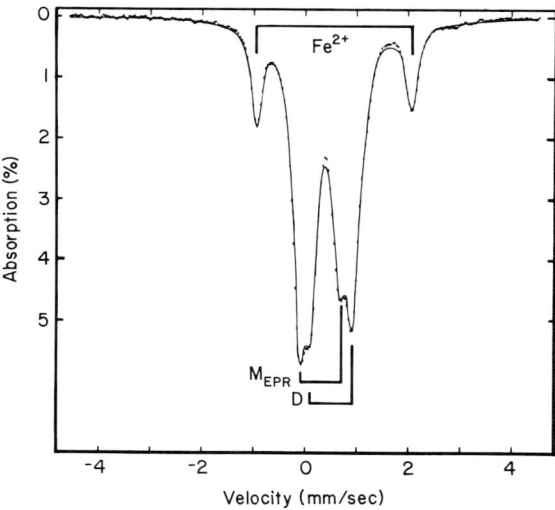

Fig. 10. Mössbauer spectra of the molybdenum–iron protein of the nitrogenase form *Azotobacter vinelandii* at 30 K, showing the three main components: M_{EPR}, the component associated with the EPR signal; D, the main diamagnetic component; and Fe^{2+}, the high-spin ferrous component. (From Münck *et al.*, 1975.)

electron reduction. Mössbauer data obtained in large applied magnetic fields, taken together with the EPR data, suggest that each of these centers has a total spin of 3/2 in the nonreduced form. Mössbauer spectra obtained from the cofactor of the molybdenum-iron protein which contains the molybdenum-iron centers indicate that it is these centers which give both the EPR signal and the Mössbauer signal, which changes on reduction. Another component of the Mössbauer spectra has an intensity corresponding to around four high-spin ferrous iron atoms, somewhat similar to those of reduced rubredoxin. In addition, there is a diamagnetic component corresponding to nearly half the iron atoms and apparently arising from low-spin ferrous ions or antiferromagnetically coupled high-spin ferrous ions.

While the Mössbauer measurements can give a determination of the relative number of iron atoms associated with the various types of centers, the absolute number depends on the total number of iron atoms. Data indicate a value of between five and eight iron atoms and one molybdenum atom in each of the two cofactor centers in a protein containing 33 ± 5 iron atoms and two molybdenum atoms. To decide on the possible nature of the molybdenum-iron centers, the stoichiometry plays an important role, but so far the errors are too large for an unequivocal assignment.

Smith *et al.* (1980) used Mössbauer spectroscopy to monitor the redox properties of the molybdenum-iron protein from *K. pneumoniae*. They observed two redox processes, one at -216 mV, associated with the iron-molybdenum centers, and the other at -340 mV, associated with the low-spin ferrous component. Following a consideration of the structural implications of their own and other published data, these authors conclude that it is not yet possible to assign the two ferrous species unambiguously to iron-sulfur centers.

Another approach to understanding the nature of the centers in nitrogenases which contain iron, molybdenum, and labile sulfur atoms has been to try to synthesize clusters containing these atoms and compare their Mössbauer spectra and other physicochemical properties with those of the enzyme. Tieckelmann *et al.* (1980) and Coucouvanis *et al.* (1980) investigated compounds containing simple dimetallic FeS_2Mo clusters, and Wolff *et al.* (1978) and Christou *et al.* (1979) studied compounds containing $MoFe_3S_4$ clusters which are thought to be closer to those found in the enzyme.

B. Photosynthetic Reaction Centers

The basic process of photosynthesis, which involves the absorption of light followed by the photooxidation of a chlorophyll molecule, takes place in a special system known as a reaction center. Photosynthetic organisms fall into two groups: the photosynthetic bacteria, which use reduced sulfur compounds or organic substrates as electron donor; and the cyanobacteria (blue-green algae), algae, and higher plants, which use water as electron donor. In the reaction centers of the latter group of organisms there are Photosystem I and Photosystem II, which transfer an electron in sequence from water to NADP. Reaction centers are fequently associated with insoluble membrane components of the cell, and hence it can be difficult to isolate and analyze the component molecules in the conventional way. As reaction centers contain iron, Mössbauer spectroscopy is an appropriate technique for their investigation.

Debrunner *et al.* (1975) investigated reaction centers from the photosynthetic bacterium *Rhodopseudomonas spheroides* in both the native and reduced states. At 77 K the spectra of both forms are closely similar and are indicative of a high-spin ferrous state. At low temperatures there are some small differences between the magnetic hyperfine interactions in the two states. These data indicate that the iron is probably not the primary electron acceptor of the reaction center, although the authors suggest that the differ-

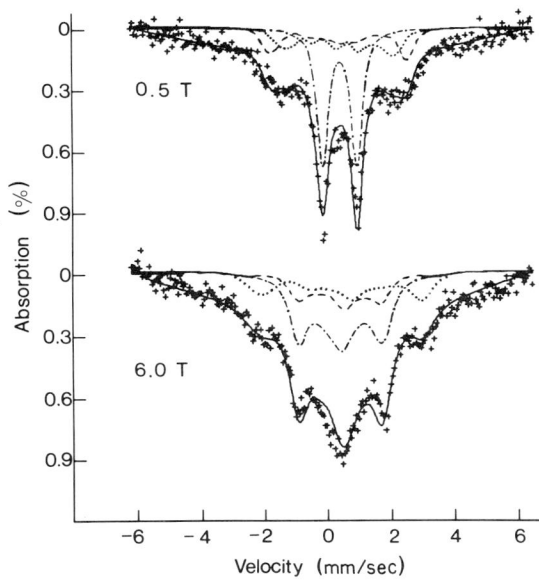

Fig. 11. Mössbauer spectra of reduced Photosystem-I fractions from *Chlorogloea fritschii* at 4.2 K in magnetic fields applied parallel to the direction of the gamma-ray beam and computer fitted to one oxidized (– · –) and two reduced (· · ·, – – –) components. (From Evans *et al.,* 1979.)

ences in the magnetic hyperfine interactions might result from a free-radical coupling to the iron atom.

In a Mössbauer investigation of membrane preparations of the cyanobacteria *Chlorogloea fritschii* and *Anacystis nidulans,* Evans *et al.* (1977) observed a ferredoxinlike component which subsequent work (Evans *et al.,* 1979) showed to be associated with Photosystem-I reaction centers. Measurements on photochemically reduced samples show spectral components very similar to those of reduced four-iron ferredoxins, which is consistent with the EPR signals with *g* values of below 2 obtained from reduced samples. Spectra were computer simulated on the assumption that they consist of two components, one corresponding to an oxidized four-iron center and the other corresponding to a reduced four-iron center. The agreement of the simulation with the experimental data (see Fig. 11) confirms the validity of this model. The EPR data indicate the presence of three types of center within the Photosystem I of cyanobacteria, and Mössbauer measurements

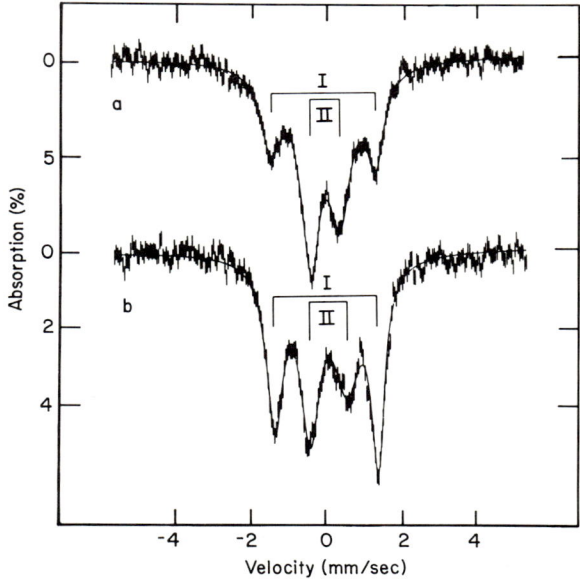

Fig. 12. Mössbauer spectra of oxidized ^{57}Fe-cytochrome *P*-450 in the presence of a saturating concentration of camphor: (a), with no additions; (b), with the addition of oxidized ^{56}Fe-ferredoxin. The two fitted doublet components represent low-spin (I) and high-spin (II) ferric configurations. The spectra were obtained at approximately 195 K. (From Sharrock, 1973.)

(Evans *et al.*, 1981) are consistent with all three centers being similar to the four-iron centers found in ferredoxins.

C. Cytochrome *P*-450 – Ferredoxin Enzyme System

The camphor hydroxylation system of *Pseudomonas putida* involves NADH and a flavoprotein, as well as two proteins which contain iron: cytochrome *P*-450 and a two-iron ferredoxin (putidaredoxin). The ferredoxin and the cytochrome interact at two points in the enzymatic cycle, and various data indicate that they form a well-defined complex.

Mössbauer spectroscopy can give information on the cytochrome and the ferredoxin, and in addition, by selectively enriching the two proteins with ^{56}Fe and ^{57}Fe, the effect of complex formation can be studied (Sharrock, 1973). Figure 12 shows the changes in the Mössbauer spectrum of ^{57}Fe-cytochrome *P*-450 which occur in the presence of ^{56}Fe-ferredoxin. The spectra show two doublets arising from iron atoms in high-spin and low-spin ferric states. With the ferredoxin present (Fig. 12b), the low-spin fraction is considerably larger than with the ferredoxin absent. The mechanism of this partial spin transition is not yet clear. The corresponding experiment using ^{56}Fe-cytochrome *P*-450 and ^{57}Fe-ferredoxin indicates that the oxidized ferredoxin is not noticeably affected by complex formation with the cytochrome.

D. Hydrogenase

Hydrogenases are enzymes containing iron–sulfur centers which catalyze the terminal steps of hydrogen evolution and utilization in microorganisms. There is considerable interest in hydrogenases, since, coupled with chloroplasts and ferredoxins, they can be used to catalyze the photoproduction of hydrogen from water, a process that is ideal for the conversion of solar energy to storable hydrogen fuel. Because of limitations on this process which result from the instability of the hydrogenase, it is desirable to identify the nature and mode of action of the iron-containing active centers to be able to produce more stable synthetic analogs.

Mössbauer spectroscopy can clearly provide information which is relevant to the study of hydrogenases, but the problems of producing samples of sufficient concentration and purity have limited its use. Bell *et al.* (1984) obtained Mössbauer spectra of native (oxidized) and dithionite-reduced samples of the hydrogenase from *Desulfovibrio desulfuricans* (Norway). The data indicate that the iron centers in this enzyme are of the four-iron type. This can be clearly seen in Fig. 13, which shows the spectra of the

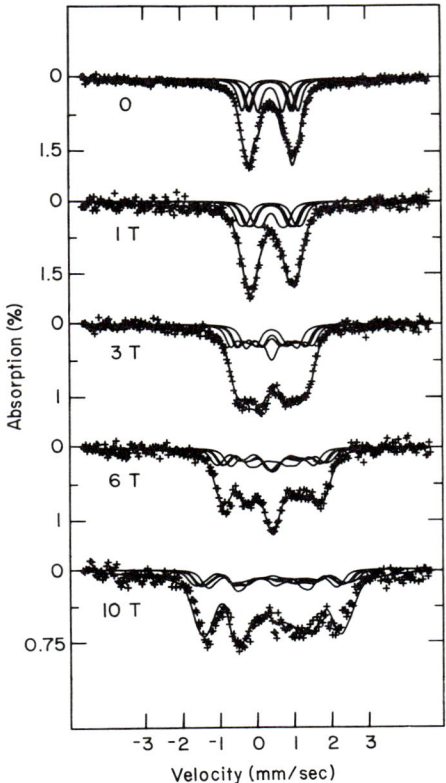

Fig. 13. Mössbauer spectra of oxidized *Desulfovibrio desulfuricans* (Norway) hydrogenase at 4.2 K in various perpendicular magnetic fields. The computer fits correspond to the spectra of a four-iron ferredoxin under the same conditions. (From Bell *et al.*, 1984.)

oxidized enzyme obtained in various magnetic fields, computer fitted using the known parameters of a four-iron ferredoxin.

E. Oxygenase

Oxygenases are enzymes involved in the insertion of one (monooxygenases) or two (dioxygenases) atoms of oxygen into an aromatic ring. Many oxygenases are known to contain iron, but the reaction mechanism and the role of this iron is as yet poorly understood. A number of Mössbauer studies of oxygenases have provided useful information in these areas.

Mössbauer measurements on the terminal dioxygenase protein of the benzene dioxygenase system from *Pseudomonas putida* (Geary and Dickson, 1981) show that it contains two-iron centers similar to those of the two-iron plant-type ferredoxins. In the oxidized form, the spectrum (Fig. 14) indicates that the two iron atoms within the center are high-spin ferric but with more inequivalence between the iron atoms than in the two-iron ferredoxins. The spectrum of the reduced form (Fig. 14) shows that the extra reducing electron is localized on one of the iron atoms, which becomes high-spin ferrous. The activation of oxygen by the enzyme requires the presence of free iron in addition to the iron–sulfur redox centers, which is a novel feature of this enzyme. The involvement of iron atoms of different types in the catalysis makes Mössbauer spectroscopy using samples partly enriched with ^{57}Fe particularly suitable for further investigation of this enzyme.

The multienzyme system 4-methoxybenzoate-*O*-demethylase from *Pseudomonas putida* contains the monooxygenase putidamonooxin, which has

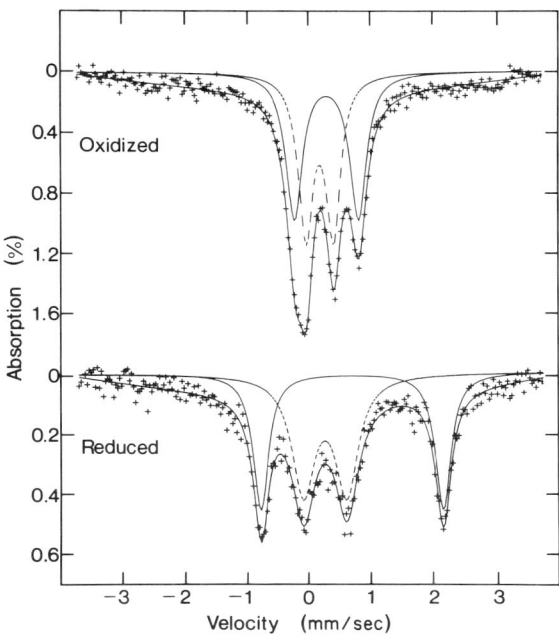

Fig. 14. Mössbauer spectra of the terminal dioxygenase protein of benzene dioxygenase from *Pseudomonas putida* at 77 K. (From Geary and Dickson, 1981).

iron–sulfur centers. The Mössbauer spectra of putidamonooxin in the oxidized and reduced state are similar to those of the dioxygenase previously discussed and indicate that it also contains two-iron centers (Bill et al., 1980). This enzyme also requires an additional iron for activity, and measurements indicate that one free iron atom is linked to each two-iron center.

An investigation of the enzyme protocatechuate 3,4-dioxygenase from *Pseudomonas aeruginosa* was reported by Münck et al. (1976) and Zimmerman et al. (1978). The spectra indicate that the iron is high-spin ferric in the native state of the enzyme and high-spin ferrous in the reduced state. The Mössbauer parameters suggest that the iron is in a novel environment.

VI. Applications to Tissue

A. Lung Samples

^{57}Fe Mössbauer spectroscopy can be used to monitor the iron content of lung samples. The iron in the lung may result from occupational exposure or from pathological factors.

Rush et al. (1975) made Mössbauer measurements on lyophilized samples of lung tissue taken from rats experimentally exposed to arc-weld fume. Spectra were also obtained from the weld fume itself. The parameters of the spectra and their temperature dependence are indicative of Fe_2O_3 in the form of superparamagnetic small particles. The average particle size is estimated to be about 18 nm in the fume and appreciably less in the lung.

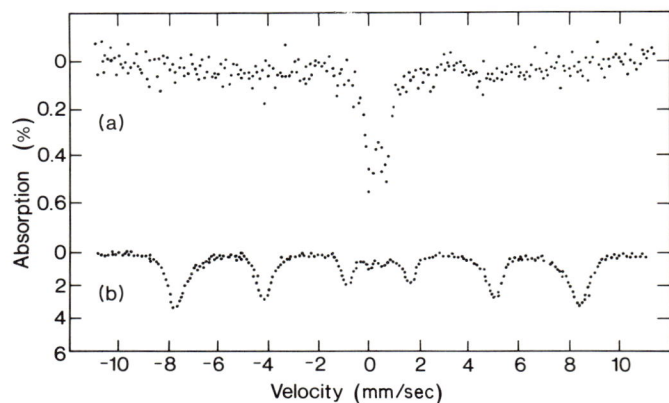

Fig. 15. Mössbauer spectra of human dried lung material at 4.2 K: (a) from a healthy control and (b) from a hemosiderosis victim. (From Johnson, 1971.)

3. MÖSSBAUER SPECTROSCOPY   283

Certain diseases lead to the build-up of iron in the lung. Johnson (1971) showed spectra from dried samples of human lung material from both a healthy subject and a hemosiderosis victim (Fig. 15). The healthy lung showed a weak signal corresponding to the presence of hemoglobin, whereas the diseased lung showed a strong absorption with magnetic splitting, corresponding to the presence of an iron-storage material like ferritin or hemosiderin.

In a comprehensive study of various human subjects, Guest (1976, 1978) obtained Mössbauer spectra showing components resulting both from exposure to iron-containing aerosols and from a metabolic buildup of iron. Information on the relative amounts of the iron compounds in the lung was obtained and correlated with the occupational and medical history of the subjects.

B. Blood Samples

There have been many Mössbauer studies on the properties of hemoglobin in its various forms (see Section IV.A) but relatively few investigations have been made on blood samples rather than on isolated protein, and there has been comparatively little work with a medical objective.

Kellershohn *et al.* (1976) used Mössbauer spectroscopy to study the effects of x radiation and heat treatment on arterial red blood cells. Before treatment, the spectra are those of oxyhemoglobin. Varying dosages of x radiation produce additional iron species corresponding to deoxyhemoglobin and two as yet uncharacterized high-spin ferric compounds. Heating produces the two high-spin ferric compounds but without any deoxyhemoglobin.

Mössbauer spectra have been obtained from frozen whole blood, from red blood cells of patients with different forms of thalassemia and other blood diseases, and from normal, healthy adults (Bauminger *et al.*, 1979b). All the spectra contain two components which correspond to oxyhemoglobin and deoxyhemoglobin. In the spectra of blood samples from patients with thalassemia and certain of the other blood diseases (Fig. 16), there is also another component with an intensity which is variable between the different samples and with parameters essentially identical to those obtained from the isolated iron-storage proteins, ferritin and hemosiderin. The amounts of iron in the ferritinlike form were determined to be comparable with those in the form of hemoglobin and were particularly large in reticulocytes (young red blood cells). The diseases in which the ferritinlike iron was observed are those having an abnormality of hemoglobin structure and composition which leads to intracellular denaturation of the hemoglobin, with presumably a deposition of the excess iron in the ferritinlike form. The observed differ-

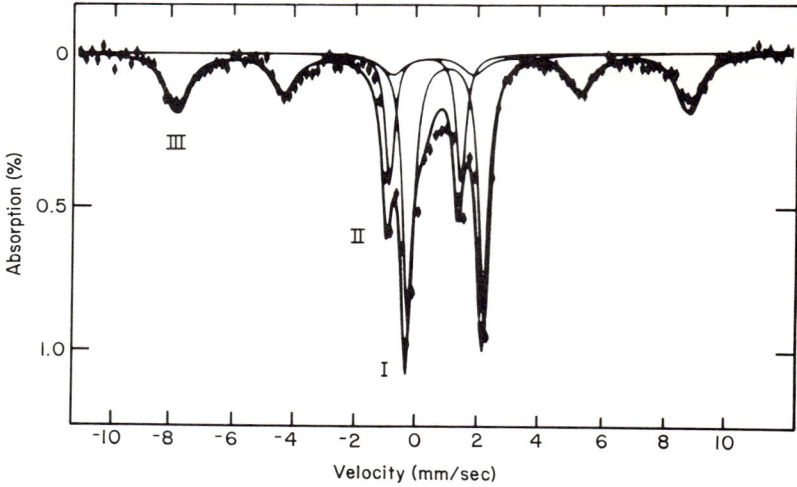

Fig. 16. Mössbauer spectrum of thalassemic red-blood cells at 4.1 K. Subspectrum I corresponds to deoxyhemoglobin, subspectrum II corresponds to oxyhemoglobin, and subspectrum III corresponds to the ferritinlike compound. (From Bauminger *et al.*, 1979b)

ences in the quantities of ferritinlike iron in the reticulocytes and the mature red blood cells could be an important factor in understanding these diseases.

C. Bone Samples

Although there is no Mössbauer isotope of calcium, Mössbauer spectroscopy of bone samples could be carried out using the Mössbauer isotope ^{151}Eu, since this element is isoelectronic with calcium. No measurements of this type have yet been performed, but Marshall (1968) obtained ^{133}Cs Mössbauer spectra using ^{133}Ba (the parent isotope of ^{133}Cs) fixed onto bone powder as the source. Kellershohn (1979) demonstrated the feasibility of monitoring the uptake of rare earths by using ^{161}Tb fixed onto bone as the source for ^{161}Dy Mössbauer spectroscopy.

VII. Applications to Whole or Part Organisms: Uptake and Metabolism

By obtaining Mössbauer spectra of whole or part organisms it is possible to identify the presence of various chemical compounds which contain the

Mössbauer nuclide. In addition, one can in principle identify the location of the various biomolecules within the organism. In its most advanced form, this work can involve obtaining spectra from whole living organisms *in vivo.*

To monitor the way an organism uses iron, or any other metabolically important Mössbauer isotope, the organism can be grown under varying conditions of nutrient medium, concentration, temperature, etc., and then samples can be prepared and spectra taken and analyzed to give information on the different components and their relative proportions.

A. *In Vitro* Experiments

In their work on the fungus *Phycomyces blakesleanii,* Spartalian *et al.* (1975) obtained Mössbauer spectra from different parts of the fungus and from the growth medium. They observed that iron in the same chemical state as in the growth medium is present in all parts of the fungus but in diminishing proportions as one looks closer to the top. This can be interpreted to give information on the digestion of the iron as it rises in the fungus and on the role of the various parts in nutrient absorption.

Evans *et al.* (1977) investigated membrane preparations from the cyanobacteria *C. fritschii* and *A. nidulans* by means of Mössbauer spectroscopy. The Mössbauer spectra of the membranes are composed of two components with similar relative areas. One component is a quadrupole-split doublet, which is interpreted as resulting from the photosynthetic reaction centers, while the other component shows a six-line magnetically split pattern at 4.2 K, somewhat similar to that in iron transport or storage proteins. To investigate the effects on the two components of different growth conditions, Mössbauer spectra were obtained from membranes of *A. nidulans* grown under three different regimes of iron nutrition. The organism grew well when the iron concentration was reduced, with the 10% (relative to the normal conditions) growth showing no significantly reduced mean generation times, whereas on 1% iron the mean generation times were reduced by less than a factor of two. Figure 17 shows the 4.2 K spectra of membranes from *A. nidulans* grown under these different conditions. It can be seen that there is a large reduction in the magnetic component (I) in the 10 and 1% growths, and in the most iron-deficient growth conditions it virtually disappears. The authors interpret these measurements as confirming that the six-line magnetically split Mössbauer component can be associated with a form of iron concerned with the process of iron transport or storage in the membrane. This work is an example of the possibilities of using Mössbauer spectroscopy to monitor the uptake of a certain element by an organism and to indicate which chemical forms of this element are more essential for healthy metabolism.

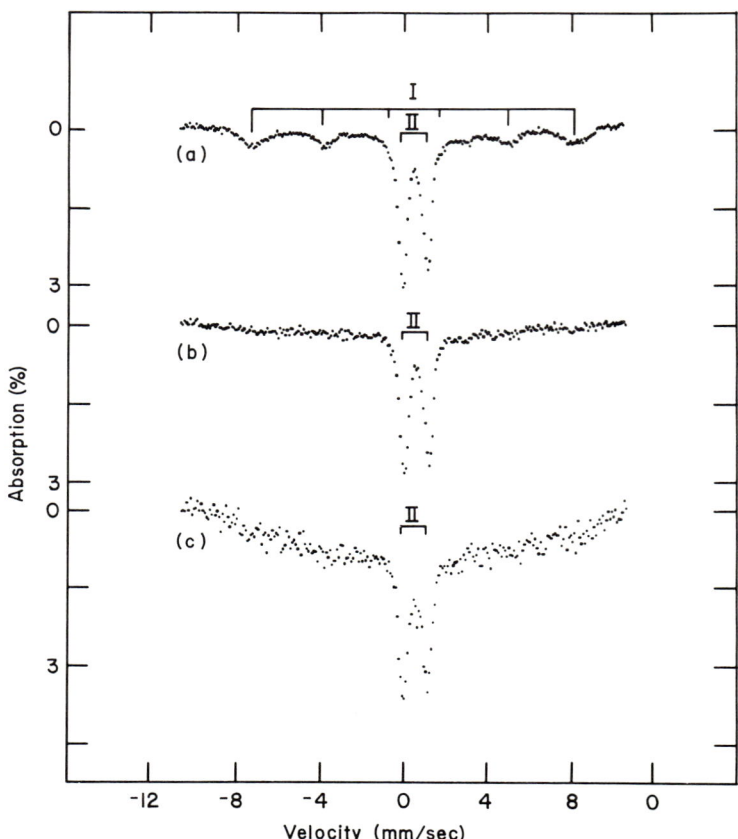

Fig. 17. Mössbauer spectra of *Anacystis nidulans* membranes at 4.2 K after growth with three different concentrations of iron in the medium: (a) normal, (b) 10% of normal, and (c) 1% of normal. The lowest spectrum is plotted with a different absorption scale, giving an exaggerated background curvature which is a purely geometric factor and should not be confused with absorption. (From Evans *et al.*, 1977.)

B. *In Vivo* Experiments

The vast majority of Mössbauer measurements of biological materials take place under conditions far removed from those of a living organism. Although much very useful information can be obtained from measurements made under these conditions, it might be helpful to obtain spectra under *in vivo* conditions to compare with the other measurements. In addition, *in vivo* experiments can enable a particular metabolic process to be monitored as it happens.

Obviously, experiments of this nature are extremely difficult, but the two reported so far (Giberman *et al.*, 1974; Bauminger *et al.*, 1976), although essentially exploratory, showed that *in vivo* experiments are both possible and give novel and exciting information.

Giberman *et al.* (1974) obtained Mössbauer spectra from ^{57}Co-enterochelin in *Escherichia coli* cells, including measurements at 3°C under living conditions. Enterochelin is a chelating agent associated with the transport of iron into this bacterium. The experiments were carried out using bacteria containing the ^{57}Co-enterochelin complex as the source of Mössbauer gamma rays. A small absorption was obtained in the first 24 hours after preparation, but no absorption in the subsequent 72 hours. The authors associate the absorption with certain special sites in the freshly prepared ("live") bacteria. In discussing the origin of these special sites, which have a

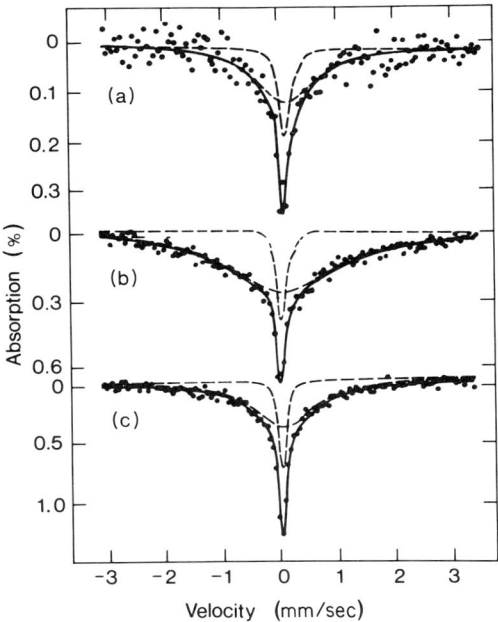

Fig. 18. Mössbauer spectra of *Escherichia coli* cells at 3°C: (a) strain K12 with ~1 mg ^{57}Fe/g of cells, (b) strain K12 with ~500 μg ^{57}Fe/g of cells, and (c) strain H7 with ~60 μg ^{57}Fe/g of cells. The solid lines (———) are computer fits to the spectra using the two Lorentzian broadened components indicated by the dashed lines (– – –). (From Bauminger *et al.*, 1976.)

quasi-solid environment and give recoil-free emission of gamma rays, they considered it likely that the special sites are in the membrane. On this assumption the broadening of the Mössbauer line at 3°C can be ascribed to the diffusion of the ^{57}Co through the membrane. Although the preliminary nature of these measurements is stressed, they do show that spectra can be obtained under *in vivo* conditions and can be interpreted to give information of real biological significance.

A subsequent study involving Mössbauer spectroscopy of whole bacterial cells, including measurements under *in vivo* conditions, was reported by Bauminger *et al.* (1976). In this work, samples of *E. coli* and *Halobacterium* were grown on an ^{57}Fe-enriched medium and used as the absorber in a Mössbauer spectrometer. One particular motivation in these experiments was to identify any membrane-bound iron sites.

Mössbauer spectra of unfrozen *E. coli* cells were obtained at 3°C and are shown in Fig. 18. These spectra are remarkable in that they exhibit appreciable recoil-free effects. They also show very large linewidths, indicative of motional broadening, with two components, one relatively narrow and the other much wider. The different growth conditions do not affect these basic features, although the exact parameters and the intensity ratio of the two components do differ in the various cases. The spectra were fitted to two Lorentzian lines of different widths and intensities. The authors consider that the broadening may be due either to the motion of iron-containing proteins in the membrane or in solution within the cell, or to the motion of the iron atoms within the proteins.

VIII. Measurement of Vibration and Movement

Mössbauer spectroscopy is usually used to study hyperfine interactions to obtain information on the nature of the atoms and molecules containing the Mössbauer nuclide. Other features of the Mössbauer effect are the recoil-free fraction, which determines the spectral intensity, and line broadening and line shifting which result from motion of the source or absorber. These factors enable information to be obtained on movement of the Mössbauer atom within a molecule, motion of the molecule within a system, and motion of the whole system.

A. Protein Dynamics

Measurements of the temperature dependence of the recoil-free fraction can give information on the dynamics of the Mössbauer atom within a molecule. Dwivedi et al. (1979) made such measurements on frozen solutions of oxymyoglobin, deoxymyoglobin, carboxymyoglobin, and rubredoxin. They observed that the molecular dynamics of these systems differ considerably from those of ordinary solids. In another study of a similar type (Belonogova et al., 1979; Prusakov et al., 1979), Mössbauer spectroscopy was used to investigate the effects on the recoil-free fraction and linewidth of the interaction of water with various proteins. Shaitan and Rubin (1980) interpreted the abnormally low Mössbauer recoil-free fraction observed for ^{57}Fe in certain proteins between 200 and 250 K in terms of gamma-ray emission and absorption by an oscillator in a highly viscous medium. Cohen et al. (1981) observed a broadened component in the Mössbauer spectra of ferritin crystals above 265 K, which is interpreted in terms of the dynamics of the iron-containing core undergoing bounded diffusive motion within a cage.

B. Applications in Auditory Physiology

Problems relating to pitch perception and frequency discrimination are a major area of interest in auditory physiology and have been related to the movement of the cochlear partition. Johnstone and Boyle (1967) reported measurements of the vibration of the basilar membrane in the first turn of the guinea pig cochlea, *in vivo,* using the Mössbauer effect. These measurements were the first of their kind to be carried out on a living animal and used the relatively low levels of acoustic stimulation made possible by the sensitivity of the technique. The measurements were made with very small (0.3 μg) sources of ^{57}Co in stainless steel surgically mounted on the basilar membrane and the stapes (to monitor the input). An absorber of ^{57}Fe-enriched stainless steel was used so that, with no acoustic stimulation, the source and absorber were stationary and absorption was at a maximum. When the source moved in response to a stimulus of known frequency, the count rate increased by an amount that could be interpreted to give information on the amplitude of the vibration. Rhode (1971) used the same technique, refined to yield both phase and amplitude information, to investigate the response of the basilar membrane of the squirrel monkey. Rhode and Robles (1974) carried out further experiments, including the observation of postmortem changes and the transient response to clicks, in addition to the steady-state response *in vivo* investigated previously. A central question is whether the Mössbauer source follows the motion of the basilar membrane in a faithful manner, or somehow interacts with the cochlear mechanism. Because of the discrepancy between the various investigations concerning the linearity of the basilar membrane response, Johnstone and Yates (1974) made a further investigation using guinea pigs, and a fuller study of the transient response of the basilar membrane of the squirrel monkey was made by Robles et al. (1976).

C. Motion of Complete Organisms

Bonchev et al. (1968) performed an extremely novel series of experiments in which the motion of ants was investigated using Mössbauer spectroscopy. The ants used, *Formica pratensis,* had SnO_2 powder glued onto their abdomens, and 50 were packed into a holder to form the absorber for ^{119}Sn Mössbauer spectroscopy. The influence of temperature and air supply on the width of the resonance line was investigated, with the natural linewidth being measured by killing the ants.

The linewidth was observed to increase from under 2 mm sec^{-1} for dead ants to a value for live ants of about 3 mm sec^{-1} at 5°C and approximately 5 mm sec^{-1} at 25°C. It was found that dead ants gave the same linewidth as that of live ants extrapolated to 0°C. Spectra obtained for ants at the same temperature, with and without an inflow of air, showed that there is a slight increase in linewidth when the flow is reduced. This is attributed to an increased frequency of the breathing cycle.

This work suggests many possibilities whereby the laws governing the movement of individual small animals could be tested and changes in the average speed could be readily monitored.

IX. Conclusions

Mössbauer spectroscopy has a valuable role in certain areas of biological research and can often provide information which could not be obtained by any other method. It has already shown itself to be a powerful and detailed probe for the study of iron-containing biological molecules and is being used more widely in the study of biochemical reactions and in investigations of more complex biological systems. The latter area includes work of physiological and medical significance; studies of this type may become increasingly important in the future.

The main difficulties to be overcome for this technique to find even more applications in biology concern the sample preparation. The constraints resulting from the requirement for a solid sample containing a relatively large amount of the Mössbauer isotope have imposed certain limitations on what has been achieved so far. However, with ingenuity these problems can be overcome, and Mössbauer spectroscopy will become an increasingly important member of the group of physicochemical techniques used in the study of biological materials and processes.

References

Bauminger, E. R., Cohen, S. G., Giberman, E., Nowik, I., Ofer, S., Yariv, J., Werber, M. M., and Mevarech, M. (1976). *J. Phys.* **37**, C6-227.
Bauminger, E. R., Cohen, S. G., Dickson, D. P. E., Levy, A., Ofer, S., and Yariv, J. (1979a). *J. Phys.* **40**, C2-523.
Bauminger, E. R., Cohen, S. G., Ofer, S., and Rachmilewitz, E. A. (1979b). *Proc. Natl. Acad. Sci.* **76**, 939.
Bauminger, E. R., Cohen, S. G., Dickson, D. P. E., Levy, A., Ofer, S., and Yariv, J. (1980). *Biochim. Biophys. Acta* **623**, 237.
Bell, S. H., Dickson, D. P. E., Cammack, R., Hall, D. O., and Rao, K. K. (1984). *Biochem. J.*, forthcoming.
Belonogova, O. V., Frolov, E. H., Illyustrov, N. W., and Likhtenshtein, G. I. (1979). *Mol. Biol. Moscow* **13**, 567.
Bill, E., Bernhardt, R. H., Marathe, V. R., and Trautwein, A. (1980). *J. Phys.* **41**, C1-485.
Boas, J. F., and Window, B. (1966). *Aust. J. Phys.* **19**, 573.
Bonchev, T., Vassilev, T., Sapundzhiev, T., and Evtimov, K. (1968). *Nature London* **217**, 96.
Cammack, R. (1980). *Nature* **286**, 442.
Cammack, R., Rao, K. K., Hall, D. O., and Johnson, C. E. (1971). *Biochem. Biophys. Res. Commun.* **125**, 849.
Cammack, R., Dickson, D. P. E., and Johnson, C. E. (1977). *In* "Iron-Sulfur Proteins" (W. Lovenberg, ed.), Vol. III. Academic Press, New York.
Christou, G., Garner, C. D., King, T. J., Johnson, C. E., and Rush, J. D. (1979). *J. Chem. Soc. Chem. Commun.*, 503.
Cohen, S. G., Bauminger, E. R., Nowik, I., Ofer, S., and Yariv, J. (1981). *Phys. Rev. Lett.* **46**, 1244.
Coucouvanis, D., Baenziger, N. C., Simhon, E. D., Stremple, P., Swenson, D., Simopoulos, A., Kostikas, V., Petrouleas, V., and Papaefthymiou, V. (1980). *J. Am. Chem. Soc.* **102**, 1730.
De Benedetti, S., Lang, G., and Ingalls, R. (1961). *Phys. Rev. Lett.* **6**, 60.
Debrunner, P. J., Schulz, C. E., Feher, G., and Okamura, M. Y. (1975). *Biophys. J.* **15**, 226a.
Dickson, D. P. E., and Cammack, R. (1974). *Biochem. J.* **143**, 763.
Dickson, D. P. E., and Rottem, S. (1979). *Eur. J. Biochem.* **101**, 291.
Dwivedi, A., Pederson, T., and Debrunner, P. G. (1979). *J. Phys.* **40**, C2-531.
Emptage, M. H., Kent, T. A., Huynh, B. H., Rawlings, J., Orme-Johnson, W. H., and Münck, E. (1980). *J. Biol. Chem.* **255**, 1793.
Evans, E. H., Carr, N. G., Rush, J. D., and Johnson, C. E. (1977). *Biochem. J.* **166**, 547.
Evans, E. H., Rush, J. D., Johnson, C. E., and Evans, M. C. W. (1979). *Biochem. J.* **182**, 861.
Evans, E. H., Dickson, D. P. E., Johnson, C. E., Rush, J. D., and Evans, M. C. W. (1981). *Eur. J. Biochem.* **118**, 81.
Frankel, R. B., Blakemore, R. P., and Wolfe, R. S. (1979). *Science* **203**, 1355.
Geary, P. J., and Dickson, D. P. E. (1981). *Biochem. J.* **195**, 199.
Giberman, E., Yariv, Y., Kalb, A. J., Bauminger, E. R., Cohen, S. G., Froindlich, D., and Ofer, S. (1974). *J. Phys.* **35**, C6-371.
Groves, J. L., Potasek, M. J., and De Pasquali, G. (1973). *Phys. Lett.* **42A**, 493.
Guest, L. (1976). *Ann. Occup. Hyg.* **19**, 49.
Guest, L. (1978). *Ann. Occup. Hyg.* **21**, 151.
Haffner, H., Andle, A., Appel, H., Buche, G., Holmes, K. C., and Morris, S. (1976). *J. Phys.* **37**, C6-223.
Huynh, B. H., Münck, E., and Orme-Johnson, W. H. (1979). *Biochim. Biophys. Acta* **579**, 192.

Huynh, B. H., Moura, J. G., Moura, I., Kent, T. A., LeGall, J., Xavier, A. V., and Münck, E. (1980). *J. Biol. Chem.* **255**, 3242.
Inoue, K., and Nath, A. (1977). *Bioinorg. Chem.* **7**, 159.
Johnson, C. E. (1971). *Phys. Today* **24**, 35.
Johnstone, B. M., and Boyle, A. J. F. (1967). *Science* **158**, 389.
Johnstone, B. M., and Yates, G. K. (1974). *J. Acoust. Soc. Am.* **55**, 584.
Kellershohn, C., Rimbert, J. N., Chevalier, A., and Hubert, C. (1976). *J. Phys.* **37**, C6-185.
Kellershohn, C., Rimbert, J. N., Fortier, D., and Maziere, M. (1979). *J. Phys.* **40**, C2-505.
Kelly, M., and Lang, G. (1970). *Biochim. Biophys. Acta* **233**, 86.
Kirchner, R. F., and Loew, G. H. (1977). *J. Am. Chem. Soc.* **99**, 4639.
Lang, G. (1970). *Q. Rev. Biophys.* **3**, 1.
Lang, G., and Dale, B. W. (1974). *Nucl. Instrum. Methods* **116**, 567.
Lang, G., and Marshall, W. (1966). *Proc. Phys. Soc.* **87**, 3.
Marshall, J. H. (1968). *Phys. Med. Biol.* **13**, 15.
Middleton, P., Dickson, D. P. E., Johnson, C. E., and Rush, J. D. (1978). *Eur. J. Biochem.* **88**, 135.
Middleton, P., Dickson, D. P. E., Johnson, C. E., and Rush, J. D. (1980). *Eur. J. Biochem.* **104**, 289.
Münck, E., Rhodes, H., Orme-Johnson, W. H., Davis, L. C., Brill, W. J., and Shah, V. K. (1975). *Biochim. Biophys. Acta* **400**, 32.
Münck, E., Zimmermann, R., Que, L., Lipscomb, J. D., and Orme-Johnson, W. H. (1976). *J. Phys.* **37**, C6-203.
Nath, A., Harpold, M., Klein, M. P., and Kundig, W. (1968). *Chem. Phys. Lett.* **2**, 471.
Oberley, L. W., and Erhardt, J. C. (1975). *J. Chem. Phys.* **63**, 2329.
Oberley, L. W., Herskowitz, V., and Erhardt, J. C. (1974). *Phys. Lett.* **50A**, 77.
Oosterhuis, W. T., and Spartalian, K. (1976). *In* "Applications of Mössbauer Spectroscopy" (R. L. Cohen, ed.), Vol. I, p. 141. Academic Press, New York.
Prusakov, V. E., Belonogova, O. V., Frolov, E. V., Stukan, R. A., Gol'danskii, V. I., Berg, A. I., and May, L. (1979). *Mol. Biol. Moscow,* **13**, 443.
Rao, K. K., Cammack, R., Hall, D. O., and Johnson, C. E. (1971). *Biochem. J.* **122**, 257.
Rao, K. K., Evans, M. C. W., Cammack, R., Hall, D. O., Thompson, C. L., Jackson, P. J., and Johnson, C. E. (1972). *Biochem. J.* **129**, 1063.
Rhode, W. S. (1971). *J. Acoust. Soc. Am.* **49**, 1218.
Rhode, W. S., and Robles, L. (1974). *J. Acoust. Soc. Am.* **55**, 588.
Robles, L., Rhode, W. S., and Geisler, C. D. (1976). *J. Acoust. Soc. Am.* **59**, 926.
Ruby, S. L. (1973). *In* "Mössbauer Effect Methodology" (I. J. Gruverman and C. W. Seidel, eds.), Vol. III, p. 263. Plenum, New York.
Rush, J. D., Dickson, D. P. E., Johnson, C. E., Hewitt, P. J., and Lam, H. F. (1975). *Phys. Med. Biol.* **20**, 128.
Shaitan, K. V., and Rubin, A. B. (1980). *Mol. Biol. Moscow* **14**, 1323.
Sharrock, M. P. (1973). Thesis, University of Illinois, Urbana, Illinois.
Smith, B. E., and Lang, G. (1974). *Biochem. J.* **137**, 169.
Smith, B. E., O'Donnell, M. J., Lang, G., and Spartalian, K. (1980). *Biochem. J.* **191**, 449.
Spartalian, K., and Lang, G. (1980). *In* "Applications of Mössbauer Spectroscopy" (R. L. Cohen, ed.), Vol. II, p. 249. Academic Press, New York.
Spartalian, K., Oosterhuis, W. T., and Smarra, N. (1975). *Biochim. Biophys. Acta* **399**, 203.
Thompson, C. L., Johnson, C. E., Dickson, D. P. E., Cammack, R., Hall, D. O., Weser, U., and Rao, K. K. (1974). *Biochem. J.* **139**, 97.
Tieckelmann, R. H., Silvis, H. C., Kent, T. A., Huynh, B. H., Waszczak, J. V., Teo, B. K., and Averill, B. A. (1980). *J. Am. Chem. Soc.* **102**, 5550.

Tsang, C. P., Bogner, L., and Boyle, A. J. F. (1976). *J. Chem. Phys.* **65,** 4584.
Walker, L. R., Wertheim, G. K., and Jaccarino, V. (1961). *Phys. Rev. Lett.* **6,** 98.
Wolff, T. E., Berg, J. M., Warrick, C., Hodgson, K. O., Holm, R. H., and Frankel, R. B. (1978). *J. Am. Chem. Soc.* **100,** 4630.
Zimmermann, R., Huynh, B. H., Münck, E., and Lipscomb, J. D. (1978). *J. Chem. Phys.* **69,** 5463.

4

X-Ray Absorption Spectroscopy

ROBERT A. SCOTT

School of Chemical Sciences
University of Illinois
Urbana, Illinois

I.	INTRODUCTION	295
II.	THEORY	297
	A. *The X-Ray Absorption Spectrum*	298
	B. *Theoretical EXAFS Expression*	299
	C. *Information from X-Ray Absorption Edges*	306
III.	EXPERIMENTAL CONSIDERATIONS	308
	A. *The X-Ray Absorption Spectrometer*	308
	B. *Data Acquisition*	318
	C. *Data Reduction and Analysis*	332
IV.	EXAFS AS A STRUCTURAL TECHNIQUE IN BIOLOGY	344
	A. *Limitations*	344
	B. *Specific Applications*	346
V.	CONCLUSION	358
	REFERENCES	358

I. Introduction

The technique of x-ray absorption spectroscopy (XAS) makes use of the fact that the energy dependence of the x-ray absorption coefficient (i.e., the x-ray absorption spectrum) of a material exhibits features characteristic of the particular atoms which make up the material. For each type of atom, these features consist of a series of discontinuities (absorption edges) which occur at energies that reflect the electronic structure of that atom. In addition, if the atom is surrounded by a regular array of other atoms (e.g., in a crystalline matrix, or a homogeneous chemical compound), the absorption coefficient in the region of the spectrum just above (to higher energy of) each absorption edge will exhibit oscillatory behavior, referred to as extended x-ray absorption fine structure (EXAFS). This spectral region contains structural information about the makeup of the "atomic neighborhood" (of the absorbing atom). What has come to be known as the EXAFS technique involves acquisition and analysis of data in the EXAFS spectral region to gain information concerning the local structural environment of the absorbing atom. It is the purpose of this chapter to introduce the reader to the EXAFS technique and to discuss its application to biological systems.

The presence of element-specific x-ray absorption edges was demonstrated as early as the 1920s (Kossel, 1920; Siegbahn, 1925). In the early 1930s, Kronig (1931, 1932) observed the fine structure just beyond the absorption edge, which is now known as EXAFS. However, the structural content of the EXAFS remained unrecognized until the early 1970s when Sayers, Lytle, and Stern presented a derivation of an expression for the oscillatory nature of EXAFS based on single-scattering theory (Sayers, 1971; Sayers *et al.*, 1970, 1971, 1972, 1974; Stern, 1974; Stern *et al.*, 1975). Even with this theoretical advance, x-ray absorption data could be conveniently measured only for strong absorbers (e.g., Cu foil) because of the inherently low intensity of the x-ray sources (conventional x-ray tubes) available. (To record an x-ray absorption spectrum with a conventional copper target x-ray tube, one must make use of the broadband portion of the source spectrum —the *Bremsstrahlung* radiation—which is approximately two orders of magnitude less intense than the monochromatic Cu K_α emission line.)

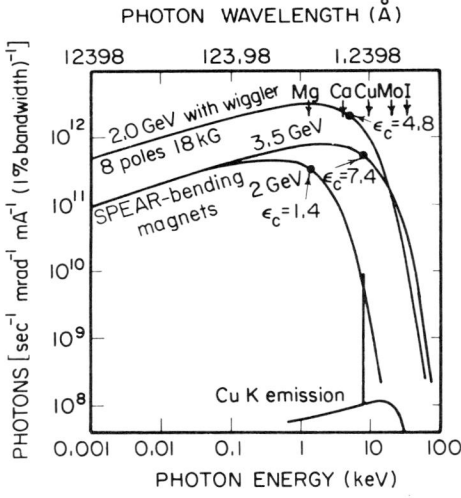

Fig. 1. Comparison of spectral distributions for a synchrotron radiation (SR) source (the SPEAR ring at Stanford) and a conventional copper x-ray tube. The critical energy (ϵ_c) of the SR spectral distribution is indicated for each set of running conditions. Note the increase in both flux and ϵ_c with the wiggler. (Two 8-pole wigglers are currently in use at SPEAR.) (From Cramer and Hodgson, 1979.)

With the advent of synchrotron radiation as a source, XAS came into its own. As shown in Fig. 1, the spectral distribution of synchrotron radiation (in this case, from the Stanford facility) is extremely broadband and exhibits average intensities which surpass the *Bremsstrahlung* background of a conventional copper target x-ray tube by 4–5 orders of magnitude. In 1974, the Stanford Synchrotron Radiation Project (SSRP, now known as the Stanford Synchrotron Radiation Laboratory, SSRL) was established to make use of the synchrotron radiation generated by the electrons circulating in the Stanford Positron Electron Accelerating Ring (SPEAR). SPEAR was designed to allow observation of subatomic particles created during collisions between electrons and positrons in counter-rotating bunches, and so the initial use of synchrotron radiation as an intense x-ray source was termed "parasitic." It was not until 1980 that SPEAR converted 50% of its operation to the "dedicated" production of synchrotron radiation. In the dedicated mode, only electrons are circulated in SPEAR, and the running parameters are optimized for production of synchrotron radiation.

Since 1974, other synchrotron and storage-ring facilities have been used as synchrotron radiation sources. Most operated in the parasitic mode. There are now storage rings built for the sole purpose of generating synchrotron radiation. This is an obvious reflection of the promising future of x-ray absorption spectroscopy and EXAFS as structural techniques.

II. Theory

Since the initial theoretical work of Sayers, Lytle, and Stern, several good discussions of EXAFS theory have appeared (Ashley and Doniach, 1975; Lee and Pendry, 1975; Lee and Beni, 1977; Lee, 1981; Pendry, 1981) which cover (to one extent or another) the theoretical derivations given in this section. In addition, many good reviews of EXAFS theory and application have appeared (Chan and Gamble, 1978; Eisenberger and Kincaid, 1978; Shulman *et al.*, 1978b; Cramer and Hodgson, 1979; Kincaid and Shulman, 1980; Teo, 1980; Eisenberger, 1981; Lee *et al.*, 1981; Teo, 1981a). Because the purpose of this section is to give the reader a physical (as opposed to a mathematical) view of the EXAFS phenomenon, the reader is referred to these other reviews and theoretical works for some of the mathematical details.

Recognition of the intimate relationship between the structural information obtained from EXAFS and the electronic information available in the structure on and near the absorption edge has prompted the inclusion of a discussion (Section II.C) of absorption edges and some of the recent theoretical work involved in their interpretation.

A. The X-Ray Absorption Spectrum

To understand the origin of the x-ray absorption spectrum, one should begin with the theoretical basis of the photoelectric effect. Light incident on a conducting plate only results in the emission of photoelectrons (measured by a current across two charged conducting plates) if it has a frequency higher than some threshold value. This threshold energy (frequency) is just equal to the ionization potential of a bound electron. The ionization of this bound electron gives rise to the observed photoelectron. If, instead of measuring emitted photoelectrons, one measures the absorption of light, the threshold energy would show up as a sharp (virtually discontinuous) increase (an "edge") in the absorption coefficient. If the ionized electron is a core electron (e.g., $1s$, $2s$, $2p$), the threshold energy (for most of the useful elements) falls in the x-ray region of the spectrum, giving rise to x-ray absorption edges.

A typical atomic energy-level diagram is shown in Fig. 2. Above it is shown the corresponding x-ray absorption spectrum with the pertinent edges labeled. The K edge results from ionization of a $1s$ electron, while the L edges result from ionization of $n = 2$ ($2s$, $2p$) electrons. In biological applications of XAS, we will typically be interested in edges above ~ 4000 electron volts (eV) owing to the strong absorption below this energy by protein, water, and air. This limits the observable elements to Ca and above (for K edges) or Sb and above (for L edges). By far the most common biological application occurs for metal sites in metalloproteins or metalloenzymes. Effectively, one can examine by XAS any element above K (by either K or L edges). Absorption-edge energies for all elements have been tabulated by Bearden and Burr (1967).

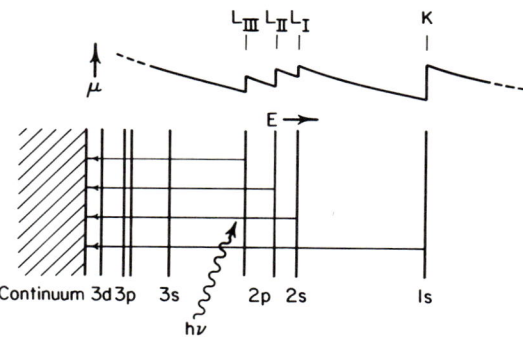

Fig. 2. Energy-level diagram for a typical x-ray absorbing atom. The lower portion indicates the possible ionizations that give rise to the K, L_I, L_{II}, and L_{III} absorption edges indicated in the spectrum in the upper portion. The spectrum is plotted as absorption coefficient (μ) versus photon energy ($E = h\nu$).

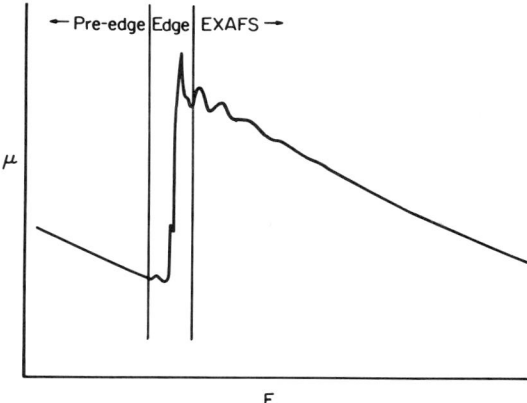

Fig. 3. Schematic x-ray absorption spectrum indicating the energy regions of interest. The divisions between regions are vague with the edge region consisting of any features before the major inflection and any features after the inflection that are not part of the analyzable EXAFS region.

A more detailed schematic of a typical x-ray absorption spectrum is shown in Fig. 3. The entire spectrum can be broken down into three regions. (It should be noted that the end points of these regions are ill-defined.) The pre-edge region consists of background absorption from all the atoms in the sample which have lower atomic numbers than the absorbing atom and from the L edges of the absorbing atom (assuming the K edge is being examined). The absorption coefficient in the pre-edge region is thus smoothly, monotonically decreasing. The edge region consists of any features caused by bound-state transitions (from $1s$ to levels just below the continuum) as well as the actual edge and features just beyond (within ~ 50 eV of the inflection). Owing to complications in the simple single-scattering theory that is used to derive the EXAFS expression, the usable EXAFS region is often assumed to begin $\sim 50-100$ eV past the edge inflection. The damped oscillations which constitute the EXAFS can extend up to ~ 1000 eV beyond the edge, although in many biological applications this range is shortened as a result of poor signal-to-noise ratio or the presence of another absorption edge.

B. Theoretical EXAFS Expression

Absorption of an x-ray photon with energy in the EXAFS region of the absorbing atom results in the production of a photoelectron (the ionized $1s$

electron for a K absorption). It is convenient to consider this photoelectron as a wave with a de Broglie wavelength given by

$$\lambda = h/m_e v, \tag{1}$$

where h is Planck's constant, m_e the electron mass, and v the velocity of the photoelectron. The photoelectron velocity is a result of the conversion of the excess energy from the absorbed photon into kinetic energy of the photoelectron:

$$(E - E_0) = \tfrac{1}{2} m_e v^2, \tag{2}$$

where E is the photon energy and E_0 the energy needed to ionize the electron (the threshold energy). The wave vector k of the photoelectron is related to the inverse of the de Broglie wavelength:

$$k = 2\pi/\lambda \tag{3}$$

which, on substitution from Eqs. (1), (2), becomes

$$k = [(2m_e/\hbar^2)(E - E_0)]^{1/2}, \tag{4}$$

or

$$k = [0.262449(E - E_0)]^{1/2}.$$

In the latter expression, E and E_0 have units of electron volts (eV) and k has units of inverse angstroms (Å^{-1}). The EXAFS is most conveniently expressed as a function of k.

The phenomenon responsible for the EXAFS oscillations is the elastic backscattering of this photoelectron from nearby electron density surrounding the neighboring atoms. (In the case of a free atom or a monoatomic gas, no EXAFS is observed.) To derive the form of the absorption coefficient, we must consider the initial and final states between which the photon is promoting a transition. The initial state (for a K-absorption edge) is just the $1s$ ground state centered on the absorbing atom. The final state is some product of the ionized atom ground state and a combination of both the outgoing and backscattered photoelectron waves. We are particularly interested in the photon-energy dependence of the absorption coefficient, and, since the ionized ground-state term of the final-state wave function is not expected to be significantly energy dependent, we will concentrate on the combination of the outgoing and backscattered photoelectron waves. The situation is schematically illustrated in Fig. 4. Here, the outgoing and backscattered photoelectron waves are shown at two different photon energies $(E_1 < E_2)$. For absorption of a photon with energy E_1, the situation is shown in Fig. 4a. At this energy the wavelength of the outgoing and backscattered photoelectron waves is such that the product of the two waves and

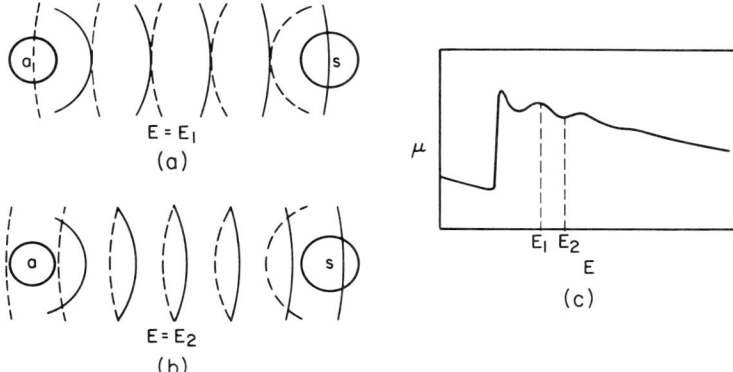

Fig. 4. Schematic diagram of the scattering processes that give rise to the final-state wave function used to calculate the EXAFS quantity. (a) Representation of the outgoing and backscattered waves at energy E_1 giving rise to a maximum in the absorption coefficient [in both (a) and (b), the solid (———) arcs represent the "peaks" of the outgoing wave from absorber a, and the dashed (– – –) arcs represent the backscattered wave from scatterer s]; (b) representation of the waves at energy E_2 giving rise to a minimum in the absorption coefficient; (c) the resulting spectrum indicating the value of μ at E_1 and E_2.

the initial state (a 1s wave function centered at the absorber) is maximal. The square of this product is proportional to the absorption coefficient.† In other words, the backscattered photoelectron "arrives back" at the absorbing atom *in phase* with the outgoing photoelectron.

Absorption of a photon with energy E_2 (higher than E_1) gives rise to the situation depicted in Fig. 4b. At this energy, the wavelength of the outgoing and backscattered photoelectron waves is such that the magnitude of the backscattered wave at the absorber is a minimum. Thus, the product of the two waves and the initial state is also a minimum. The backscattered photoelectron arrives back at the absorbing atom 180° *out of phase* with the outgoing photoelectron. The response of the absorption coefficient is shown in Fig. 4c. Thus, scanning the photon energy through the EXAFS region causes the absorption coefficient to trace out an interference pattern (between the outgoing and backscattered photoelectron waves). This predicts that each neighboring atom will contribute a periodic oscillation (e.g., a sine wave) to the x-ray absorption coefficient. Of course, this is oversimplified, but the basic conclusion is correct.

† The absorption coefficient is, of course, actually proportional to the square of the electric-dipole-moment matrix element [see Eq. (6) in the text], but for this simple physical representation, the product of the wave functions is adequate.

Mathematically, the EXAFS (χ) is defined as the oscillatory portion of the absorption coefficient (above the edge) normalized by the absorption coefficient one would observe for the free atom (μ_0):

$$\chi \equiv (\mu - \mu_s)/\mu_0. \tag{5}$$

Here, μ_s is the smooth background which, when subtracted from the overall absorption coefficient μ, leaves the oscillatory portion. We will discuss this in detail in conjunction with data analysis procedures. Quantum mechanically, the absorption coefficient is proportional to the square of the electric-dipole-moment matrix element:

$$\mu \propto |\mathbf{M}|^2 = |\langle f|\mathbf{r}|i\rangle|^2, \tag{6}$$

where $\langle f|$ is the final-state wave function and $|i\rangle$ the initial-state wave function ($1s$ for a K-absorption edge). To evaluate the matrix element, we must write down $\langle f|$. As already noted, this wave function is a combination of the outgoing and backscattered photoelectron waves:

$$\langle f| = \langle f^o| + \langle f^{bs}|. \tag{7}$$

We are most interested in the behavior of the backscattered wave. Without going into the details, this wave can be described by the following function:

$$\langle f^{bs}| = i \frac{\exp(ikR_{as} + i\delta_1)}{kR_{as}} f_s(\pi)$$

$$\times \frac{\exp(ikR_{as})}{R_{as}} \exp(-ik\mathbf{R}_{as} \cdot \mathbf{r}) \quad (r > R_0), \tag{8}$$

$$\langle f^{bs}| = i \frac{\exp(2ikR_{as} + 2i\delta_1)}{2kR_{as}^2} f_s(\pi)\psi_0 \quad (r < R_0).$$

Outside the atomic radius R_0 of the absorbing atom, the final-state wave function looks like the first function; inside R_0, it looks like the second (where ψ_0 is the wave function in the absence of a scatterer). In these expressions, R_{as} is the distance between the absorber (a) and the scatterer (s); δ_1, some function describing the phase shift that the backscattered photoelectron wave undergoes upon reentering the potential of the absorbing atom; $f_s(\theta)$, a scattering function which describes the amplitude and phase of the photoelectron wave scattered from the neighboring atom at an angle θ. Only the portion that is backscattered ($\theta = \pi$) is important here.

Using these backscattered photoelectron wave functions to calculate the dipole-moment matrix element allows us to derive the following expression for the EXAFS:

4. X-RAY ABSORPTION SPECTROSCOPY

$$\chi(k) = \frac{\Delta|\mathbf{M}|^2}{|\mathbf{M}|^2} = -\frac{|f_s(\pi,k)|}{kR_{as}^2} \sin(2kR_{as} + 2\delta_1 + \alpha_s). \tag{9}$$

Here, we have broken up $f_s(\pi)$ into a magnitude ($|f_s(\pi, k)|$) and a phase (α_s). The term α_s can be considered as the phase shift that the photoelectron wave undergoes owing to the potential of the scatterer. Also, it is recognized that the scattering function (amplitude and phase) may depend on k (i.e., on the kinetic energy of the photoelectron). Examination of Eq. (9) reveals that the EXAFS oscillations are not only dependent on the distance between the absorber and scatterer but also on the phase shifts that the photoelectron undergoes as it passes out of the absorber potential δ_1, through the scatterer potential α_s, and back into the absorber potential δ_1.

Equation (9) was derived for a fixed absorber and scatterer. To adequately predict the observed effects, vibrational motion of the two atoms must be taken into account. This leads to multiplication of the simple expression in Eq. (9) by the Debye–Waller factor $\exp(-2\sigma_{as}^2 k^2)$, where σ_{as}^2 is the relative mean-square displacement of the two atoms from their equilibrium separation. Actually, σ_{as}^2 is a combination of two effects:

$$\sigma_{as}^2 = \sigma_{stat}^2 + \sigma_{vib}^2, \tag{10}$$

where σ_{stat}^2 represents the contribution from static disorder in the interatomic distances which make up a "shell," and σ_{vib}^2 represents the dynamic contribution resulting from vibrational motion of the two atoms. For a simple system with m atoms at a distance R_m and n atoms at a distance R_n, it can be shown (Teo, 1980; Teo, et al., 1979) that

$$\sigma_{stat} \simeq [(mn)^{1/2}/(m+n)]|R_m - R_n|. \tag{11}$$

The vibrational contribution can be calculated, assuming harmonic motion, if the vibrational frequency v of the bond-stretching vibration is known (Teo, 1980):

$$\sigma_{vib}^2 = (h/8\pi^2 \mu v) \coth(hv/2kT), \tag{12}$$

where μ is the reduced mass, k Boltzmann's constant, and T the temperature. In systems which exhibit large anharmonicity in the vibrational motion or abnormally high disorder in bond lengths, the simple EXAFS theory must be improved upon (Hayes et al., 1978; Eisenberger and Brown, 1979).

Including the Debye–Waller factor, the EXAFS contribution from each scatterer is predicted to be a damped sine wave. The total EXAFS is just a sum of damped sine waves from all the atoms neighboring the absorber:

$$\chi(k) = \sum_s \frac{N_s |f_s(\pi,k)|}{kR_{as}^2} \exp(-2\sigma_{as}^2 k^2) \sin[2kR_{as} + \alpha_{as}(k)]. \tag{13}$$

This holds for N_s equivalent scatterers at a distance R_{as} from the absorber, and the summation is over all the scatterers, s. In addition, all the phase shifts have been consolidated into a total phase shift:

$$\alpha_{as}(k) = 2\delta_1 + \alpha_s - \pi. \tag{14}$$

(The π is included as a matter of convention so that the EXAFS amplitude will be positive.)

It is important to recognize the simplifications that have been incorporated into the derivation of the EXAFS expression [Eq. (13)]. First of all, it is assumed that we are dealing only with "neighboring" atoms (e.g., the first coordination sphere of a metal complex). If "next-nearest neighbors" are involved, consideration will have to be given to additional effects. These include multiple scattering (the photoelectron scattering from more than one atom before "returning" to the absorber) and inelastic losses of the photoelectron. The latter effect gives rise to another factor of the form, $\exp(-R_{as}/\lambda)$, where λ is some mean free path for the photoelectron. Multiple scattering effects include additional phase shifts of the photoelectron wave and a complicated relationship between the frequency of the sine wave and the "path of travel" of the photoelectron (rather than an individual absorber–scatterer distance). Figure 5 illustrates a hypothetical multiple scattering situation. In this case, the amplitude of the photoelectron wave which arrives at s_2 is proportional to the amount scattered from s_1 at an angle θ_1 (i.e., $|f_{s1}(\theta_1,k)|$), whereas the amplitude of the photoelectron that scatters back to the absorbing atom is proportional to $|f_{s1}(\theta_1,k)| \times |f_{s2}(\theta_2,k)|$. The term α_{s2} represents the additional phase shift of the scattered photoelectron

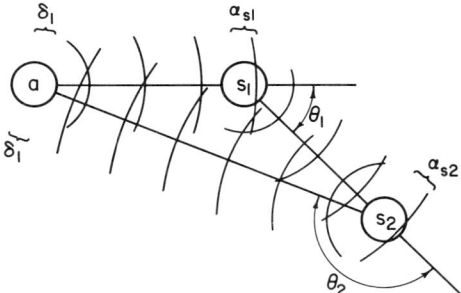

Fig. 5. Schematic diagram of one possible arrangement of atoms giving rise to multiple scattering. The photoelectron wave peaks (indicated by solid arcs) trace out one possible multiple scattering pathway. In this case, the observed scattering by s_2 of a photoelectron from the absorber a is affected by the "intervening" atom s_1; δ_1, α_{s1}, and α_{s2} are the phase shifts undergone by the photoelectron wave owing to the potentials of the appropriate atoms.

wave resulting from the presence of the second scatterer, s_2. In the situation with $\theta_1 \to 0$ and $\theta_2 \to \pi$ (three collinear atoms), one must take into account forward scattering ($\theta = 0$) from s_1 for both the outgoing *and* the backscattered photoelectron wave in calculating the EXAFS effect from s_2. The collinear arrangement leads to actual intensification of the effective scattering from s_2 (above that expected in the absence of s_1), as well as a shift in the effective frequency of the EXAFS oscillations. This is sometimes referred to as the focusing effect (Teo, 1981b). A few workers have attempted to come to grips with the multiple scattering problem (Beni *et al.,* 1976; Stern *et al.,* 1980; Teo, 1981b; Boland *et al.,* 1982).

It should be clear, from the discussion so far, what types of information are available from an EXAFS spectrum. In each component sine wave, there are basically three observables: frequency, phase, and amplitude. The frequency of the sine wave is directly related to the absorber–scatterer distance R_{as}. However, since the phase shift $\alpha_{as}(k)$ is k dependent, the distance is really dependent on both frequency and phase. (This is easily demonstrated by expanding $\alpha_{as}(k)$ as a Taylor series in k. The term in this series which is first order in k is really a frequency, and the distance thus depends on the value of the coefficient of this term.) The phase of the sine wave allows one to calculate $\alpha_{as}(k)$ (for a given R_{as}), and $\alpha_{as}(k)$ is indicative of the type of scatterer (atoms of different atomic number show different phase shifts). The amplitude of the sine wave is made up of several contributions: the scattering power of the neighboring atom [measured by $|f_s(\pi,k)|$], the Debye–Waller factor, and the number of atoms in the shell (i.e., the number of equivalent atoms at R_{as}). To extract information concerning a particular contribution, one must know the other contributions. It proves to be difficult in practice to achieve this with high accuracy.

The general approach to analyzing an EXAFS spectrum goes as follows:

(1) The individual sine wave components are isolated for study, if possible.

(2) For a particular scatterer, one uses known phase-shift [$\alpha_{as}(k)$] and amplitude [$|f_s(\pi, k)|$] values (obtained either from theory or empirically determined from compounds of known structure) to fit a specific sine wave component (shell).

(3) The fit of the theoretical EXAFS expression to the measured data is optimized by varying the number of scatterers N_s and the absorber–scatterer distance R_{as}.

(4) If the fit is unsatisfactory, the type of scattering atom may have been chosen incorrectly — a different atom can be used.

Depending on the specific approach, the Debye–Waller factor can either be assumed constant or also varied to optimize the fit. EXAFS data analysis procedures will be discussed in more detail in Section III.C.2.

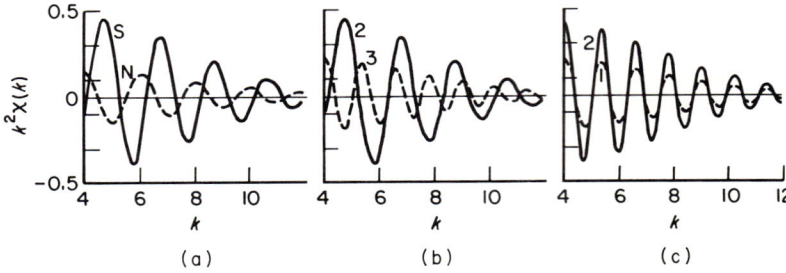

Fig. 6. Comparison of simulated Cu EXAFS patterns showing the predicted effects of distance, coordination number, and type of scatterer. (a) Comparison between one nitrogen (N) scatter and one sulfur (S) scatterer (both at a distance of 2 Å). Note the larger amplitude for S and the very different phases. (b) Comparison between one S at 2 Å and one S at 3 Å. Note both the lower amplitude and higher frequency of the wave owing to S at the longer distance. (c) Comparison between one S and two S (both at 3 Å). All of these simulations were done using parameterized phase, amplitude, and Debye–Waller functions empirically determined from structurally characterized Cu(II) complexes.

Figure 6 illustrates the effect of each physical quantity (type of scatterer, R_{as} and N_s) on the shape of the EXAFS curves. The spectra displayed in Fig. 6 are simulated using phase-shift and amplitude parameters measured from structurally characterized copper complexes. It should be noted that the plots are $k^2\chi(k)$ versus k, the k^2 weighting of $\chi(k)$ enhancing the high-k portion of the spectra. [Some researchers prefer to plot $k^3\chi(k)$ versus k.]

C. Information from X-Ray Absorption Edges

EXAFS studies of metal complexes or metal-containing active sites of proteins and enzymes yield only a radial distribution function of atoms surrounding the absorbing metal atom. In particular, EXAFS yields no information regarding the chemical (electronic) nature of the metal environment and no information regarding the geometrical arrangement of ligand atoms about the metal. In contrast, the position and shape of the x-ray absorption edge does give information regarding these other aspects of the metal-coordination environment. The study of x-ray absorption edges has been variously referred to as AES (absorption edge spectroscopy), AEFS (absorption edge fine structure), and XANES (x-ray absorption near edge structure).

For a particular metal, the position of the K-absorption edge depends on the oxidation state of the metal. More precisely, it depends on the effective charge density at the metal (Sarode *et al.*, 1979; Manthiram *et al.*, 1980) — a more meaningful quantity than the formal oxidation state. The edge position shifts up in energy as the oxidation state becomes more positive. This is to be expected, since K absorption is just ionization of a core ($1s$) electron. The amount of the edge shift is usually a few electron volts or less per oxidation state (depending on the ligand environment). This shift in edge position is useful for defining oxidation states in comparisons between the same (or similar) metal complexes at different levels of oxidation. However, care should be used in making comparisons between complexes with very different ligand environments, since the absolute position of an absorption edge of a metal in a particular oxidation state is definitely not a constant.

The dependence of the structure (shape) of an absorption edge on the geometry of the metal coordination sphere is less well studied. In many cases, the presence or absence of a bound-state transition just before the main edge can allow arguments to be made (based on selection rules) regarding the general symmetry of the metal site. For example, Cu(II) complexes often exhibit a $1s \rightarrow 3d$ bound-state transition at ~ 8979 eV. This transition is usually observed to be weaker (although not completely absent) in centrosymmetric complexes (since, in these cases, the $1s \rightarrow 3d$ transition is strictly electric-dipole forbidden†). Other resolvable structure observed on absorption edges of Cu(II) complexes seems to be indicative of the geometry of the site, although more study is needed to determine the relationships involved.

Some effort has gone into theoretical calculations aimed at simulating x-ray absorption edges (Kutzler *et al.*, 1980). The calculational approach consists of an SCF-X_α multiple-scattered-wave (MSW) formalism. The metal site (metal atom plus immediately surrounding atoms) is modeled by a muffin-tin (MT) potential, and the x-ray absorption cross section is calculated (as a function of photon energy) by using initial and final-state wave functions that are solutions to the Schrödinger equation that includes the iterated self-consistent potential. In several highly symmetric small molecules (CrO_4^{2-}, MoO_4^{2-}, MoS_4^{2-}), the calculation was able to simulate energy differences between edge features with reasonable accuracy. One interesting finding is that much of the structure on the absorption edge can be accounted for by "continuum resonances" rather than bound-state transitions (Kutzler *et al.*, 1980).

The study of x-ray absorption edges is in its infancy, but the time is rapidly approaching when we will be able to extract as much information from the structure of edges as from the EXAFS portion of the spectrum.

---

† In centrosymmetric copper(II) complexes, this transition seems to have mainly electric-quadrupole origins (Hahn *et al.*, 1982).

III. Experimental Considerations

Of equal importance with the theoretical developments (discussed in Section II.B) in the renaissance of XAS as a powerful structural technique was the advent of the technical capabilities to perform the experiments. Foremost among these was the use of synchrotron radiation as the x-ray source. Certainly, the use of XAS techniques with biologically important systems is dependent on this advance. In this section, we will discuss the general components of an "XAS spectrometer" (whether it be a laboratory facility or a beam line at a synchrotron source), the techniques and considerations required in the acquisition of XAS data, and the analysis of EXAFS data. Particular emphasis is placed on application of these methods to XAS studies of biological systems.

A. The X-Ray Absorption Spectrometer

Figure 7 diagrammatically illustrates the general equipment components necessary to perform XAS measurements. These can be divided into three main requirements: a source of x-rays, some means of separating individual frequency components from the x-radiation (one type of monochromator is shown in Fig. 7, although dispersive elements may also be used), and detection equipment. In addition, of course, a means of control of spectrometer operation and data acquisition is required. In the case of XAS, a (dedicated) computer plays this role.

1. Source

For most biological applications involving edge and EXAFS studies on first and second-row transition metals, the source requirement is for photons with energies between ~ 7 and 25 keV. Other necessary features are a reasonably flat spectrum (i.e., "white" radiation), high intensity, and high collimation (so that reasonably small samples can be used). Three types of x-ray sources are available which produce x-rays in the desired energy range: conventional sealed-tube x-ray sources, rotating anodes, and synchrotron radiation. Other x-ray sources such as cold-cathode x-ray tubes (Azoulay, 1980) or laser-produced plasmas (Mallozzi *et al.*, 1979, 1980) produce x-rays which are too soft for these applications.

Conventional x-ray tubes (e.g., a copper target source) produce a spectral distribution consisting of a *Bremsstrahlung* background with characteristic emission lines (e.g., copper K_α, K_β emissions). To scan the x-ray photon energy, the *Bremsstrahlung* background must be used, and typical photon fluxes (through a 0.1-mrad^2 aperture, for example) are $\sim 10^4$ photons sec^{-1}

4. X-RAY ABSORPTION SPECTROSCOPY 309

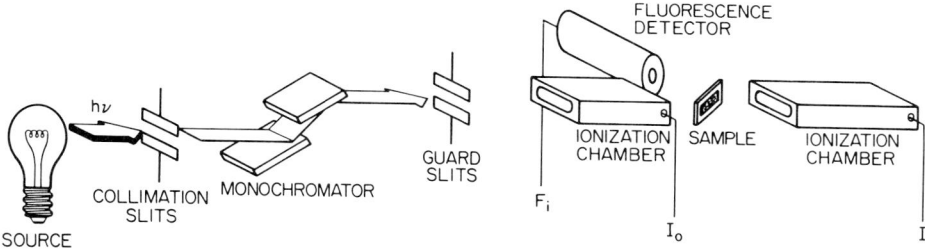

Fig. 7. Diagrammatic illustration of a general XAS spectrometer. The x-ray source can be a conventional x-ray tube, a rotating anode, or synchrotron radiation. The monochromator shown is that described in Fig. 8b. The detection system can measure either transmission ($\mu x = \ln(I_0/I)$) or fluorescence excitation ($\mu x \propto \Sigma_i F_i/I_0$) spectra. One possible position for a fluorescence detector is indicated. See the text for further details.

eV^{-1} at 10 keV. In rotating anode x-ray tubes, the target is rotated (to increase the effective target surface area), allowing higher photon fluxes to be obtained (approximately one to two orders of magnitude higher than a similar conventional x-ray tube). The source spectrum still contains features characteristic of the target material. Aside from the relatively low photon flux, these "laboratory"-type x-ray sources share the disadvantage of having distinctly "nonflat" spectra in particular energy regions. This can be partially overcome by choosing the target material wisely. Their main advantages are the relatively low cost and good long-term stability. However, their low intensities have precluded their use for XAS studies on dilute metalloprotein systems.

Synchrotron radiation first became available as a source for XAS studies at the Stanford Synchrotron Radiation Laboratory (SSRL)† in 1974. Since that time, many more storage rings and synchrotrons have become available for XAS work, and some facilities [notably the National Synchrotron Light Source at Brookhaven (Hastings, 1981)] are in use as dedicated producers of synchrotron radiation. Table I lists the facilities currently available for synchrotron radiation research. Several good reviews are available for the reader who requires familiarization with the properties of synchrotron radiation (Leigh and Rosenbaum, 1976; Lindau and Winick, 1976; Chan *et al.*, 1978; Watson and Perlman, 1978). As can be seen in Fig. 1, under properly chosen operating conditions the spectral distribution of synchrotron radiation can be reasonably flat in the spectral range of interest. It also displays extremely high brightness resulting from high photon fluxes and high colli-

† For a recent discussion of available facilities at SSRL, the reader is referred to two reviews by Bienenstock (1980, 1981).

Table I Synchrotron Radiation Sources and Their Characteristics by Country (as of August, 1981)[a]

Machine (SR Fac.[b])	Location	Type	Electron energy (GeV)	Current (mA)	Radius[c] (m)	Photon critical energy[c] (keV)	SR use[d]
PEP	Stanford, CA (USA)	Storage ring	15	50	165.5	45.2	Fac. planned
			12	45	(23.6)	(163)	(w/17-kG wiggler)
CESR (CHESS)	Ithaca, NY (USA)	Storage ring	8	50	32.5	35.5	Parasitic
SPEAR (SSRL)	Stanford, CA (USA)	Storage ring	4	50	12.7	11.1	50% dedicated
			3	100	(5.5)	(10.8)	(w/18-kG wiggler)
NSLS	Brookhaven Natl. Lab., NY (USA)	Storage ring (x-ray)	2.5	500	6.88	5.0	Dedicated [1982]
					(1.67)	(20.5)	(w/50-kG wiggler)
ALADDIN	Stoughton, WI (USA)	Storage ring	1.0	500	2.08	1.07	Dedicated [1981]
NSLS	Brookhaven Natl. Lab., NY (USA)	Storage ring (VUV)	0.7	500	1.90	0.40	Dedicated [1981]
SURF II	Washington, D.C. (USA)	Storage ring	0.25	25	0.84	0.041	Dedicated
TANTALUS I	Stoughton, WI (USA)	Storage ring	0.24	200	0.64	0.048	Dedicated
DCI	Orsay (France)	Storage ring	1.8	500	4.0	3.63	Partly dedicated
ACO	Orsay (France)	Storage ring	0.54	100	1.1	0.32	Dedicated
PETRA	Hamburg (Germany)	Storage ring	15	50	192	39.0	Possible future use
DESY	Hamburg (Germany)	Synchrotron	7.5	10–30	31.7	29.5	Dedicated
DORIS	Hamburg (Germany)	Storage ring	5	50	12.1	22.9	30% dedicated
			2.5	300	12.1	2.9	
BONN I	Bonn (Germany)	Synchrotron	2.5	30	7.6	4.6	
BESSY	West Berlin (Germany)	Storage ring	0.8	500	1.83	0.62	Dedicated [1982]
BONN II	Bonn (Germany)	Synchrotron	0.5	30	1.7	0.16	
PTB	Braunschweig (Germany)	Storage ring	0.14	150	0.46	0.013	Dedicated

Name	Location	Type					Status
ADONE	Frascati (Italy)	Storage ring	1.5	60	5.0 (2.8)	1.5 (2.7)	Partly dedicated (w/18-kG wiggler)
PHOTON FACTORY	Tsukuba (Japan)	Storage ring	2.5	500	8.33 (1.67)	4.16 (20.5)	Dedicated [1982] (w/50-kG wiggler)
INS-ES	Tokyo (Japan)	Synchrotron	1.3	30	4.0	1.22	
ETL	Tsukuba (Japan)	Storage ring	0.66	100	2	0.32	Dedicated
UVSOR	Okazaki (Japan)	Storage ring	0.60	500	2.2	0.22	Dedicated [1983]
SOR	Tokyo (Japan)	Storage ring	0.40	250	1.1	0.13	Dedicated
LUSY	Lund (Sweden)	Synchrotron	1.2	40	3.6	1.06	
MAX	Lund (Sweden)	Storage ring	0.50	100	1.20	0.23	Dedicated
SRS	Daresbury (UK)	Storage ring	2.0	500	5.55 (1.33)	3.2 (13.3)	Dedicated (w/50-kG wiggler)
VEPP-4	Novosibirsk (USSR)	Storage ring	7	10	16.5 (18.6)	46.1 (10.9)	(w/8-kG wiggler)
ARUS	Erevan (USSR)	Synchrotron	4.5		24.6	8.22	
VEPP-3	Novosibirsk (USSR)	Storage ring	2.25	1.5 100	6.15 (2.14)	4.2 (11.8)	Partly dedicated (w/35-kG wiggler)
SIRIUS	Tomsk (USSR)	Synchrotron	1.36	15	4.23	1.32	
PAKHRA	Moscow (USSR)	Synchrotron	1.3	300	4.0	1.22	
FIAN C-60	Moscow (USSR)	Synchrotron	0.68	10	1.6	0.44	
VEPP-2M	Novosibirsk (USSR)	Storage ring	0.67	100	1.22	0.54	Partly dedicated
KURCHATOV	Moscow (USSR)	Storage ring	0.45		1.0	0.21	Dedicated [?]
N-100	Karkhov (USSR)	Storage ring	0.10	25	0.50	0.004	

[a] The author acknowledges Herman Winick of SSRL for compilation of this list.
[b] Name of synchrotron radiation facility (if different from machine name).
[c] Numbers in parentheses indicate effective values for use with wigglers.
[d] Dates in square brackets indicate expected completion dates (as of August, 1981) for facilities under construction.

mation. For a 0.1-mrad² aperture, fluxes can easily be $\geq 10^{10} - 10^{11}$ photons sec⁻¹ eV⁻¹ at 10 keV. With the incorporation of insertion devices like wigglers, these fluxes can be increased by as much as two to three orders of magnitude again. It is possible with the conventional x-ray sources to collect photons over a wider angle and focus them onto a small sample, thus increasing the photon flux by as much as three orders of magnitude (Knapp et al., 1978). As we will see in Section III.A.2, this is achieved at the expense of resolution.

2. Monochromator

In recording XAS data, the normal procedure involves collecting data points at discrete photon energies, one at a time. This procedure requires monochromatization of the white source radiation. In general, this is accomplished by taking advantage of Bragg reflection from a certain diffraction plane of a single crystal:

$$n\lambda = 2d \sin \theta, \qquad (15)$$

where n is the order of the reflection, λ the wavelength of the diffracted ray, d the d spacing of the crystal lattice planes, and θ the angle between the Bragg planes and the incident ray. For wavelengths in the proper energy region (~4–25 keV) and reasonable incident angles θ, the optimum d spacing is ~1.5–3.5 Å (for use of a first-order reflection, $n = 1$). Typical crystal planes used for x-ray monochromators include Si[111], Si[220], Si[400], Ge[111], Ge[220], Ge[311], etc. Silicon proves to be a good choice, since large single-crystal specimens of silicon are available for cutting and polishing.

Flat-crystal monochromators make use of crystals polished parallel (or nearly parallel) to a particular set of Bragg planes. Several different arrangements have been tried (at one time or another). These are illustrated in Fig. 8a–d. The simplest monochromator is a single flat crystal which can be rotated during a scan to different incident angles θ. This arrangement proves to be clumsy, since the experiment (sample and detectors) has to track the 2θ reflected beam during the scan. Figure 8b shows a conventional two-crystal monochromator (the type used at SSRL, for example). Here, the incident beam is Bragg reflected from two flat crystals in succession, resulting in a reflected ray parallel to the incident ray. The two crystals are generally mounted on a goniometer (at some fixed intercrystal separation), and the goniometer is turned (by computer-controlled stepping motors) to effect a scan. One coincidence of the fixed crystal separation is that the vertical position of the beam changes during the scan. This can rather easily be compensated by having the computer raise or lower the experimental apparatus to track the beam (relatively slight movements are required). The

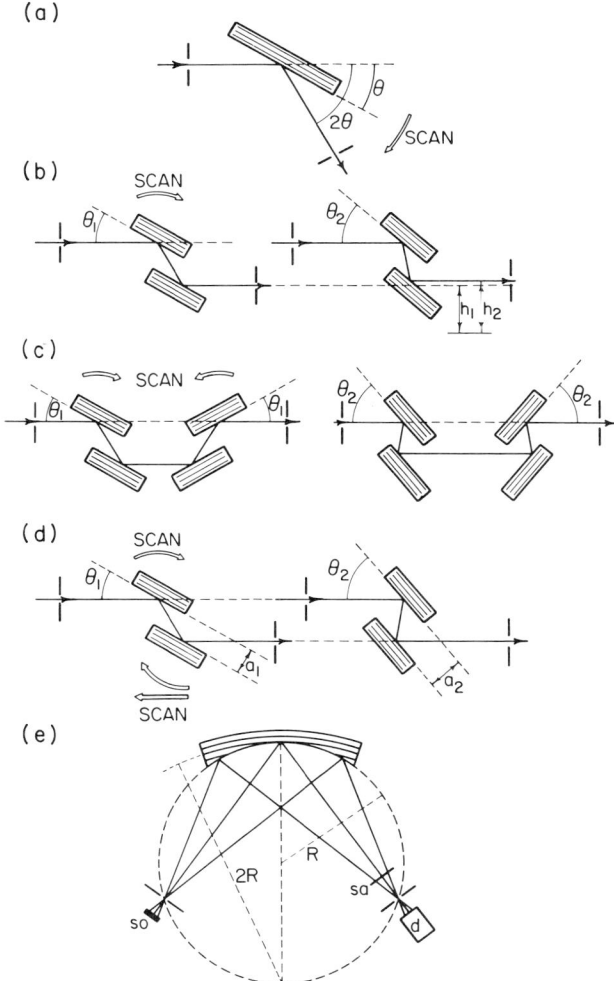

Fig. 8. Various designs for crystal x-ray monochromators that make use of Bragg reflection. Each diagram illustrates the diffracting crystal(s) in cross section (the lines indicate the Bragg planes) and typical ray tracings. (a) A simple flat crystal monochromator; (b) the common double-crystal (fixed separation) monochromator; (c) a double-double crystal (fixed separation, fixed in/fixed out) monochromator; (d) the Golovchenko double-crystal (variable separation, fixed in/fixed out) monochromator; (e) a bent-crystal monochromator using the Johansson geometry (d = detector, sa = sample, so = source).

advantage of the fixed-separation double-crystal arrangement is its simplicity of construction and operation.

Two possible methods to avoid having to track the reflected beam with the experiment are illustrated in Fig. 8c,d. The first is a double-double crystal monochromator which rotates two pairs of fixed-separation crystals in tandem. The disadvantages of this arrangement are the use of twice as many crystals and the relative complexity of alignment and operation. Figure 8d shows a double-crystal monochromator with variable crystal separation (the Golovchenko design). In this monochromator, both crystals are rotated, maintaining parallel alignment (these two motions can be mechanically connected), while one crystal (the lower one in Fig. 8d) is also translated (along an air bearing) by a computed amount parallel to the ray-propagation direction. This allows the second crystal to intercept the ray reflected from the first crystal at the same vertical position, regardless of the angle. The latter two crystal arrangements are often referred to as "fixed-in, fixed-out" monochromators. The main disadvantage to such monochromators is usually the relative complexity of their operation.

One example of an arrangement for a focusing (bent-crystal) monochromator is shown in Fig. 8e. This arrangement has been used (Knapp et al., 1978; Cohen et al., 1980; Haensel, 1980; Khalid et al., 1982) with laboratory-type x-ray sources to increase the total photon flux at the sample by approximately three orders of magnitude (Knapp et al., 1978). In this monochromator, one bent crystal is used, and the source and detector collimation are then required to be located on points equidistant from the crystal on the circumference of the Rowland circle (radius R). The arrangement shown in Fig. 8e with the crystal bent to a radius of $2R$ and polished to a radius of R is called the Johansson geometry (whereas the Johann geometry entails bending to $2R$ only). For use of this arrangement, the spectrometer is set up so that, as the crystal is rotated by θ, the sample and detector are rotated around the same pivot by 2θ.

In studies on biological systems (as well as other applications) where edge structure is of interest in helping characterize the metal site, one must be concerned with the intensity–resolution trade-off involved in the monochromator arrangement to be selected. Since edge structure involves features which may be separated by only a few electron volts, one must have resolution at least as good to see such structure. For flat-crystal monochromators, the energy resolution is given by

$$\Delta E = E \cot \theta \, \Delta \theta, \tag{16}$$

convolved with the rocking curve of the crystal reflection being used. In Eq. (16), ΔE is the absolute energy resolution (bandwidth), θ the Bragg angle at energy E, and $\Delta \theta$ the spread in θ for rays incident on the first crystal. For synchrotron radiation, the resolution is thus determined by the vertical opening angle of the entrance aperture (collimation slits, see Fig. 7) of the monochromator. For a typical high-resolution XAS scan at SSRL, the opening aperture might be ~1 mm at 20 m from a source point 1 mm high. This gives $\Delta\theta \simeq 0.1$ mrad. For a Bragg reflection with a sufficiently narrow rocking curve (e.g., Si[220]), $\theta = 18.84°$ at 10 keV, and Eq. (16) gives $\Delta E \lesssim 3$ eV (at 8 keV, this decreases to $\lesssim 2$ eV). This compares with a reported resolution for a bent-crystal, focusing monochromator of ~10–15 eV at 10 keV (Knapp *et al.*, 1978). This poor resolution for the bent-crystal arrangement results from the relatively broad rocking curve of the bent crystal and the fact that the source image is not ideal (i.e., it is not a perfect line source). Resolution this poor is unacceptable for examining edge structure. However, since EXAFS features are usually separated by at least 30–50 eV, an EXAFS spectrum can be recorded (without significant distortion) with the focusing arrangement. The resolution–intensity trade-off for synchrotron radiation comes about because any increase in the opening angle of the source aperture to gain intensity causes contamination with rays propagating at slightly different angles, yielding a decrease in resolution.

One disadvantage of using a Bragg reflection for monochromatization is the presence of n in the Bragg relationship [Eq. (15)]. The crystals not only diffract in first order, but also in second order ($n = 2$), third order ($n = 3$), etc. These harmonics occur at twice, three times, etc., the fundamental frequency. Thus, if the Bragg angle of a monochromator is set to give 10-keV photons in first order, it will also reflect 20-keV photons along precisely the same path. This contamination by harmonics can lead to distortions in the data (this is discussed in more detail in Section III.B.3), and some methods are available for discriminating against these higher-order beams.

The most direct method for harmonic discrimination is adjustment of the source spectrum to decrease the relative amount of higher-energy photons being emitted. For laboratory-type sources, one can just turn down the power on the x-ray tube, whereas for synchrotron radiation one must choose to run at lower electron energy (shifting the source spectrum to lower critical energy), or turn down a wiggler magnet field (if possible). In any case, some sacrifice in intensity of the fundamental is necessary. Another method is to choose a crystal for which the harmonic reflection is forbidden. For example, for Si[111], the even-order harmonics (e.g., Si[222], Si[444]) are nearly forbidden. This method is not always available for a particular experiment. For flat double-crystal monochromators, one can take advantage of the relatively narrow rocking curve of the harmonic reflection (compared to the fundamental). These monochromators are usually equipped with some means of remotely rotating one crystal with respect to the other (tuning).

This allows precise alignment of the Bragg planes, maximizing the intensity throughput. However, owing to the nature of the rocking curves, if the tune is set slightly off maximum (i.e., the monochromator is detuned slightly), the ratio of harmonic to fundamental decreases dramatically with only a small loss in total fundamental intensity. For synchrotron radiation applications, detuning is a relatively common practice. At CHESS (Cornell High Energy Synchrotron Source), a monochromator has been designed with automatic feedback control to maintain a precise amount of detuning (Mills and Pollock, 1980).

A slightly more drastic approach is to condition the source spectrum before the light reaches the monochromator by the use of some kind of x-ray-reflecting optics. At x-ray energies, total external reflection from a material occurs only at glancing incident angles (below some critical angle) (Gamble, 1980a). This critical angle depends on photon energy; the higher the energy, the smaller the critical angle. Thus, with proper choice of material and incident angle, one can design an x-ray mirror which will reflect the fundamental but absorb the harmonic. One can use a flat mirror, assuming that a material with adequate optical flatness can be found (float glass has been used for this purpose). Platinum- (or gold-) coated quartz has been used for x-ray optics in focusing applications (Gamble, 1980a). Although focusing mirrors are designed to gather a much wider horizontal acceptance of the source beam and focus it onto the sample, a secondary result is harmonic discrimination. (It should be noted here that one is dealing with another intensity–resolution trade-off with focusing mirrors, since horizontal divergence is translated into vertical *con*vergence after reflection from the mirror.)

With monochromators, an XAS spectrum is recorded as discrete data points collected successively. This takes a significant amount of time and is not applicable for short-lived species or dynamic studies. Some effort has gone into the development of the dispersive XAS technique (Gamble, 1980b; Matsushita, 1980), which uses a bent crystal to disperse synchrotron radiation onto some type of position-sensitive detector (PSD) (Gamble *et al.*, 1979; Matsushita, 1980; Sano *et al.*, 1980; Gamble, 1980b). Figure 9 shows a typical arrangement for a dispersive spectrometer. In this ray tracing, each ray diffracted from the crystal has a different energy. The rays are first focused (through the sample), then dispersed onto x-ray film or a PSD. With proper choice of crystal and Rowland circle radius, one can obtain an entire EXAFS spectrum in one "exposure" with a resolution of 2–3 eV (Matsushita, 1980). It is hoped that this will eventually be useful for the study of dynamic chemical or biochemical processes (perhaps on the millisecond time scale).

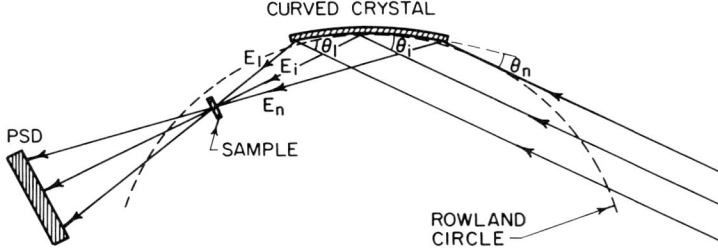

Fig. 9. Dispersive XAS spectrometer with a curved crystal (in the Johann geometry). Ray tracings are shown for reflection of three different x-ray energies (E_1, E_i, E_n). The x-rays are focused through the sample then dispersed onto a position-sensitive detector (PSD). The linear displacement of each ray on the detector surface depends on the energy of the radiation in that ray.

3. Detection Equipment

The absorption coefficient μ of a sample of thickness x can be measured using the relationship

$$I = I_0 \exp(-\mu x), \tag{17}$$

where I_0 is the intensity of x-rays incident on the sample and I the intensity transmitted. Thus, the detection equipment must be able to measure I_0 and either I or some quantity proportional to it. The most direct measurement technique is the transmission technique (measurement of both I_0 and I directly). Indirect techniques involve measurement of processes that are a result of absorption of a photon. One can either detect x-ray fluorescence (e.g., K_α emission for K-edge absorption) or secondary electrons generated by the primary absorption event. Of the three measurement methods, the fluorescence technique is most useful in biological applications.

The detectors that are available for measurement of x-ray intensities include both pulse (photon-counting) detectors and dc (integrating) detectors (Stern, 1980). Measurement of I_0 requires a reasonably transparent detector, so that the beam still strikes the sample. This is usually accomplished with a gas-ionization chamber (or, less commonly, a gas proportional counter). The ionization chamber is a dc detector filled with an appropriate gas that can be ionized by the x-radiation. The small current generated across the plates by the ionizing x-rays is measured and is proportional to the photon flux through the chamber. The gas is chosen so as to optimize the fraction of the radiation absorbed in measuring I_0 (usually about 20–30%). For transmission XAS, the detector measuring I should be chosen to absorb all the remaining photons (for highest measurement efficiency). This is usually accomplished with a larger ionization chamber. For some applications, it proves useful to allow a small amount of radiation to pass through the I detector (e.g., for alignment purposes or for use in a simultaneous calibration measurement).

For detection of fluorescent photons (in measurement of a fluorescence excitation spectrum), it is almost always advantageous to use pulse detectors (because of the lower count rates). Also, pulse detectors allow some amount of energy discrimination of the photons detected. This is useful for fluorescence detection, since the fluorescent photons are at lower energy than the Compton- and elastically scattered photons which make up the background. (This subject is treated in more detail in the discussion on data acquisition.) Pulse detectors include gas proportional counters, scintillation counters, gas scintillation counters, and solid-state detectors. Several considerations go into the choice of fluorescence detection equipment for a particular problem. These include: maximum count rates, energy resolution, and ability to cover a large solid angle. One can sometimes circumvent a low-energy resolution problem by use of appropriate filters. Several different fluorescence-detection systems have been constructed, and each has been shown to have advantages and disadvantages. A discussion of these systems will be given in Section III.B.2.

B. Data Acquisition

1. Transmission

The most common experimental setup for collection of transmission XAS data consists of a monochromator (usually of the double flat-crystal design), an I_0 ionization chamber, the sample, and an I ionization chamber (see Fig. 7). The ionization chambers are filled with an inert gas (e.g., He, N_2, Ar, or a mixture) which is chosen so that I_0 absorbs a certain fraction of the total photon flux. From statistical considerations, the optimum fraction of absorption by I_0 for a sample of total absorption coefficient μ_t (in cm^{-1}) and thickness x (in cm) is given by (Stern, 1980)

$$f = [1 + \exp(\mu_t x/2)]^{-1}. \tag{18}$$

For a sample made up of an absorber A (with absorption coefficient μ_A) and various other constituents represented by i (giving a background absorption coefficient $\mu_B = \Sigma \mu_i (i \neq A)$, it is possible to calculate the signal-to-noise ratio (S/N) for a transmission experiment (Shulman *et al.*, 1978b; Lee *et al.*, 1981; Teo, 1981a):

$$(S/N)_{tr} = \mu_A x \exp(-\mu_t x/2)(I_0^{1/2} \Delta \mu_A/\mu_A), \tag{19}$$

where $\Delta\mu_A/\mu_A$ is the EXAFS signal and

$$\mu_t = \mu_A + \sum_{i \neq A} \mu_i. \tag{20}$$

Setting $\partial(S/N)_{tr}/\partial x$ to 0 in Eq. (19) gives the optimum sample thickness:

$$\mu_t x_{opt} = 2,$$

or, if we include the noise in the I_0 measurement:

$$\mu_t x_{opt} = 2.56, \tag{21}$$

giving the optimal fraction of photon flux absorbed by I_0 as ~20–25% (Stern, 1980; Teo, 1981a). This fractional absorption is obtained by using a low-atomic-number (low-Z) gas for lower photon energies and a high-Z gas for higher photon energies. It is convenient to use N_2 gas for photon energies between ~7 and 12 keV and Ar gas for higher energies.

Of course, the S/N calculated by Eq. (19) is a maximum value—other experimental effects can serve to worsen the actual S/N observed. Such effects include beam instability, incorrect sample positioning, and heterogeneous sample thickness (e.g., pinholes, etc.). It is necessary to be extremely careful about preparing solid samples, making sure the material is homogeneously ground and (usually) pressed so that these artifacts are not important. Even with these precautions, other factors such as harmonic contamination of the beam and monochromator "glitches" can cause severe distortions of the data. Some of these practical aspects of data collection will be discussed in Section III.B.3.

In most XAS measurements, the data is collected at discrete energies, selected by stepping-motor positioning of the monochromator crystals. It is necessary to collect data with variable intervals between successive points, since often much smaller steps are used across the edge. In the EXAFS region, both variable energy intervals and variable integration times are often used to compensate for the eventual k^n weighting of the processed EXAFS data. Thus, the energy intervals can be selected to step fixed intervals in k space, and the integration times should be considerably longer in the high-k region to give constant S/N throughout the EXAFS region after multiplication by k^n. In practice, the total desired scan time precludes perfect compensation, so that most reported EXAFS spectra still exhibit more noise in the high-k region.

It is necessary in general to calibrate the monochromator by performing a preliminary (high-resolution) scan of an appropriate calibrant. Often, a thin metal foil of the element being examined is used for this purpose, and the exact energy of the inflection point of the metal edge (Bearden and Burr, 1967) is used as a single-point calibration. The calibration procedure has

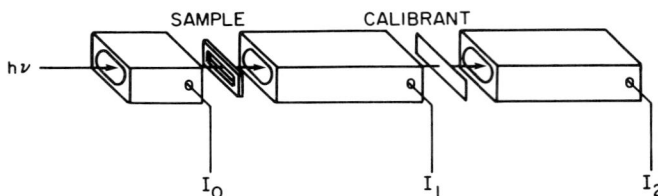

Fig. 10. Detector arrangement for measurement of internal calibration. The sample absorption is given by $\mu x = \ln(I_0/I_1)$ and the small proportion of radiation that goes through the I_1 detector is used to measure the calibrant absorption, $\mu x = \ln(I_1/I_2)$.

conventionally been performed before and after sample data collection to monitor any drift in the monochromator mechanism. A better method is to use an "internal" calibration procedure that makes use of a third ionization chamber detector, as shown in Fig. 10. With this arrangement, three measurements are recorded: I_0, I_1, and I_2. A small amount of radiation is not absorbed by I_1 but passes through a calibrant and into I_2. While the transmission of the sample is being recorded $[\mu_t x = \ln(I_0/I_1)]$, the calibration is also being recorded $[\mu_c x_c = \ln(I_1/I_2)]$. In this manner, each scan of the sample is internally calibrated simultaneously with data acquisition. This can be done for fluorescence-excitation data as well, providing that at least some radiation passes *through* the sample.

2. Fluorescence

The transmission technique using gas-ionization chambers provides the best S/N for samples which are relatively concentrated in absorber. For dilute samples (as in most biological applications), fluorescence detection is the preferred method of data collection. In practice, for first-row transition metals the lower limit of concentration for which transmission XAS is feasible is probably ~ 10 mM (Shulman *et al.*, 1978b). (For higher-Z transition metals this estimate may be somewhat high.)

The fluorescence XAS technique takes advantage of the fact that a certain fraction of the atoms from which a K shell ($1s$) electron has been ionized (by absorption of an x-ray photon) relax by emitting a fluorescent photon (see Fig. 11a). The fluorescent yield is a monotonically increasing function of atomic number Z. For example, K_α fluorescent yields for Fe and Mo are 0.347 and 0.764, respectively (Teo, 1981a). The data collection proceeds in the same manner as for transmission XAS, with the monochromator scanned through the energy range containing the sample absorption edge and EXAFS. However, instead of monitoring the transmitted intensity, the number of K_α photons emitted are measured (usually by photon-counting

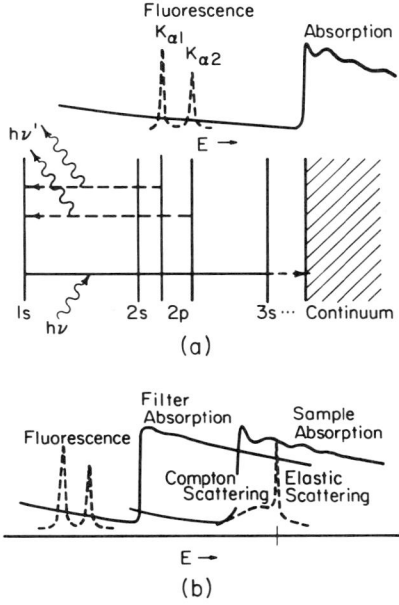

Fig. 11. (a) Energy-level diagram of a typical absorbing atom indicating K_α fluorescence. The absorption and fluorescence spectra of the sample are shown above the energy-level diagram. Measurement of the number of K_α photons fluoresced as a function of incident photon energy constitutes a fluorescence excitation spectrum. (b) The relationship of the filter absorption to the sample absorption, fluorescence, and scatter. The elastic and Compton scattering are shown for the particular incident photon energy marked on the abscissa. The filter is chosen so that it has large absorption in the energy region of the scattered photons, but small absorption in the energy region of the fluorescent photons. This is typically an elemental filter with atomic number one less than the absorbing atom ("Z-1" filter).

techniques). This quantity is directly proportional to the number of photons absorbed and thus contains the same information. I_0 is still measured with a partially transparent ionization chamber, and the signal of interest is F/I_0 (where F is the measured fluorescence signal). The advantage of the fluorescence technique for dilute (or thin) samples comes about because of the difference in energy of the photons making up the signal (at the K_α emission energy) and making up the background (the elastic and Compton scattered photons at higher energies). This is illustrated in Fig. 11b. The fluorescence technique gains sensitivity by making use of various scatter-rejection schemes (i.e., being able to energy-discriminate against scattered photons). This can be accomplished either by using detector systems with high-energy resolution or by choosing an appropriate low-pass filter which selectively absorbs the higher-energy scattered photons (Fig. 11b). For study of an element with atomic number Z, an appropriate filter can be made from a thin foil or deposited layer of the element with atomic number $Z-1$.

Four basic types of detectors have been used for fluorescence-detected XAS work: scintillation counters (Shulman et al., 1978a; Stern and Heald, 1979; Cramer and Scott, 1981; Phillips, 1981), solid-state detectors (Jaklevic et al., 1977), gas-filled detectors (either ionization chambers or proportional counters) (Stern, 1980), and the so-called "barrel monochromator" or crystal analyzer (Hastings et al., 1979; Marcus et al., 1980). Scintillation counters use a material that can convert photon-produced photoelectrons into electronic excitation energy, which is discharged by emission of a visible or near-ultraviolet photon. A photomultiplier tube behind the scintillator detects this fluorescence. The most commonly used scintillators are NaI(Tl) or certain plastics. The advantage of scintillation counters is that they can reach fairly high count rates, and they are inexpensive enough that arrays of detectors can be built (Cramer and Scott, 1981; Phillips, 1981) to cover a large solid angle (as high as $\sim 20\%$ of 4π) around the sample. The maximum count rate available to any photon-counting detector is determined by its characteristic dead time τ. This is defined as the minimum time (Δt) which must separate adjacent pulses for each of them to be detectable. If $\Delta t < \tau$, the two pulses are counted as only one. Therefore, the observed count rate N' is related to the actual count rate N by

$$N' = N(1 - N\tau). \quad (22)$$

Equation (22) holds up to $N\tau \simeq 0.4$ (Stern, 1980). Thus, *with* appropriate dead time corrections, a particular detector will yield reliable count rates up to N_{max}, where

$$N_{max} \simeq 0.4/\tau. \quad (23)$$

For NaI(Tl) scintillation counters, the dead time is ~ 1 μsec, and $N_{max} \simeq 4 \times 10^5$ counts/sec (cps). With plastic scintillators, the dead time can be about an order of magnitude shorter (Stern, 1980), yielding another factor of 10 in maximum count rate. The main disadvantage of scintillation counters is the low-energy resolution (no better than $\sim 50\%$ at 7 keV), which is not sufficient to separate K_α emission from scattering. Therefore, scintillation detector systems are generally used with appropriately designed filter assemblies (Stern and Heald, 1979; Cramer and Scott, 1981).

Solid-state detectors use semiconductor materials (Si or Ge) as ionization detectors. In some cases, to reduce the dark current of the semiconducting material, impurities are added (Li is most common), and the detector is operated at liquid-nitrogen temperature. These detectors (most commonly SiLi or intrinsic Ge) have very good energy resolution (perhaps ~ 100 eV at 7 keV), but they can only operate at a maximum of ~ 20,000 cps (Stern and Heald, 1979). They also tend to be expensive, and this (plus the liquid-nitrogen dewars) makes it prohibitive to build an array to collect a significant solid angle of fluorescence.

The barrel monochromator (Hastings et al., 1979; Marcus, et al., 1980) achieves its high-energy resolution by using Bragg reflection from graphite or LiF crystals to monochromatize the emitted (and scattered) radiation and using a Rowland circle geometry to focus a cone of rays from a point source (the sample) to a point on the detector (a solid-state detector was used in both reported systems). The crystals are formed (as a mosaic) into a concave "barrel" geometry to accomplish the focusing with monochromatization of the proper photon-energy range (e.g., the Fe K_α emission line). The main advantage of this system is its high-energy resolution, although this seems to be outweighed by the disadvantage of small subtended solid angle [~ 1.5% of 4π (Stern and Heald, 1979)] and the fact that a new crystal arrangement must be developed for use at each energy (e.g., one for Fe, one for Cu, etc.).

The optimal arrangement for fluorescence-detection studies on low-Z metals (e.g., Fe, Cu) has the sample at 45° to both the incident radiation and the fluorescence detector, with the detector in the horizontal plane (Sandstrom and Fine, 1980) (see Fig. 7). This not only allows the fluorescence detector to see an optimal projection of sample surface area, but also is the geometry which gives the minimum amount of background scatter. With an array of detectors, one packs the detectors around this particular orientation to collect as much solid angle as possible (Cramer and Scott, 1981). For XAS studies of higher-Z metals (e.g., Mo), one can use a sample with greater path length (up to ~ 1 cm of H_2O at 20 keV), still allowing transmission of some radiation for internal calibrations, yet allowing more surface area of the sample to be "viewed" by the fluorescence detectors.

Calculation of S/N for the fluorescence-detection technique is more complicated than for the transmission technique. It is dependent on the method and extent of scatter rejection. Assuming a thick, dilute sample (usually a good assumption for biological samples) and an imperfect scatter-rejection scheme, one can derive the following expression for the S/N of fluorescence data (Teo, 1981a):

$$(S/N)_\text{fl} = \left\{ \frac{I_0(\Omega/4\pi)\phi\mu_A}{(1+K)[\mu_t(E) + \mu_t(E_f)]} \right\}^{1/2} \Delta\mu_A/\mu_A, \qquad (24)$$

where $\Omega/4\pi$ is the relative solid angle subtended by the detector(s) (as a fraction of 4π), ϕ the fluorescent yield, K the ratio of background scatter to fluorescence, $\mu_t(E)$ the total absorption coefficient of the sample at the elastic scattering energy E, and $\mu_t(E_f)$ the total absorption coefficient of the sample at the fluorescence energy E_f. The other quantities have been defined previously. Calculations show that S/N similar to that for the transmission technique is obtained at absorber concentrations which are lower by approximately two orders of magnitude in the fluorescence experiment (Jaklevic et al., 1977; Teo, 1981a) (assuming the sample is dilute). If filters are used as the scatter-rejection scheme, Eq. (24) has to be modified. With filters, one must take into account the absorption coefficient of the filter for scatter, $\mu_f(E)$, and for fluorescent photons, $\mu_f(E_f)$ (which is nonzero), as well as considering the filter fluorescence generated by absorption of scattered photons by the filter. These filter-fluorescent photons also make it to the detector and are counted. In addition, depending on the filter thickness, some of the filter fluorescence is reabsorbed by the filter and does not make it to the detector. Defining the absorption coefficient of the filter for its own fluorescence as $\mu_f(E_{ff})$, one can calculate the nominal *improvement* factor Q of S/N over the S/N expected in the absence of any scatter rejection by the following expression (Cramer and Scott, 1981):

$$Q = \left(\frac{(1+K)\exp[-\mu_f(E_f)t]}{1 + K\{\exp[-\mu_f(E)t] + \phi(\Omega/4\pi)\alpha\}\exp[\mu_f(E_f)t]} \right)^{1/2}, \quad (25)$$

where

$$\alpha = \left[\frac{\mu_f(E)}{\mu_f(E) - \mu_f(E_{ff})} \right] \{\exp[-\mu_f(E_{ff})t] - \exp[-\mu_f(E)t]\}$$

and t is the thickness of the filter. This expression has been used (Cramer and Scott, 1981) to indicate that at low values of K (i.e., higher concentrations), the S/N improvement is insensitive to the filtering scheme used, and filters do slightly better than, for example, the barrel monochromator (Hastings et al., 1979). At high values of K (past $K \simeq 10$), the potential S/N improvement from filtering is significantly degraded by the presence of filter fluorescence. Thus, for the filtering scheme to compete with other scatter-rejection techniques at very low concentrations, one must be careful about limiting the amount of filter fluorescence which reaches the detector. This can be accomplished through the use of a set of Soller slits (Stern and Heald, 1979).

In general, the same (dedicated) minicomputer which controls the XAS experiment also takes care of data acquisition. Data from ionization chambers start out as currents and go through a current preamplifier and a voltage-to-frequency converter to be recorded as counts per second (cps). For scintillation counters, the individual pulses must be amplified (by fast delay-line amplifiers, for example) and then go through single-channel analyzers (which allow discrimination against harmonic scatter) before being recorded as cps. For applications that use multiple detector arrays, the outputs of all the detectors can be summed in an analog fashion to give one signal (assuming the computer interface can handle the overall count rate), but for statistical reasons it is best to collect data from each channel (detector) separately and use an appropriate weighting scheme (Scott et al., 1981).

Each channel will have a different S/N (depending on the specific electronic components and on the geometrical location of the detector), and, to avoid overemphasizing the data from a channel with large scatter-to-fluorescence ratio K, one should use a weighting scheme based on estimated S/N. The S/N for each channel can be estimated before data collection is begun by examining the actual data to get an estimate of the background and fluorescence counts and then assuming that photon statistics are obeyed. Figure 12 shows a hypothetical spectrum [of $(F/I_0)_i$ versus energy] which might be observed for the ith fluorescence channel. The signal of interest (e.g., the EXAFS) is assumed proportional to the size of the edge jump (measured as cps and normalized to I_0), and the noise is estimated (by photon statistics) as proportional to the square root of the total count rate (again normalized to I_0). For the ith channel, one need make only two measurements (see Fig. 12): a count rate B_i at an energy E_B just before the edge of interest and a count rate A_i at an energy E_A just after the edge of interest. Then the signal S_i and noise N_i for the ith channel are estimated as

$$N_i \propto A_i^{1/2}, \qquad (26)$$

$$S_i \propto (A_i - B_i) \equiv \Delta_i. \qquad (27)$$

The proper weighting factor W_i is proportional to the square of the S/N but also must include normalization of the data in the ith channel to unit edge

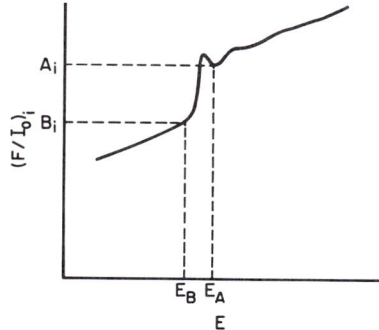

Fig. 12. Measurements used for calculation of fluorescence weighting factors. The proper weighting factor can be calculated to be proportional to the relative edge jump ($\Delta_i' = (A_i - B_i)/A_i$), where A_i is the value of F/I_0 at an energy E_A after the edge and B_i is the value of F/I_0 at an energy E_B before the edge. See text for further details.

jump (this is so that channels with equal S/N contribute equally to the weighted average edge jump):

$$W_i \propto (S_i/N_i)^2/\Delta_i = \Delta_i/A_i \equiv \Delta_i'. \tag{28}$$

Thus, the proper weighting factors are just proportional to the relative edge jumps Δ_i' in each channel. The properly weighted fluorescence $\langle F \rangle$ is then calculated from the fluorescence in each channel F_i by

$$\langle F \rangle = \sum_i W_i F_i \Big/ \sum_i W_i. \tag{29}$$

With the aid of the computer, the proper weighting may be done at the time of data acquisition. (This saves storage space, since only one number is stored per energy.) However, in some cases, fluorescence detectors at different locations respond differently to beam or monochromator instabilities (see following). Since it is difficult to know the extent of such behavior at data acquisition, it may be advantageous to postpone the proper averaging until the individual channels can be examined in detail.

3. *Experimental Difficulties*

It seems appropriate to mention here some of the effects that can (and do) serve to distort the XAS data acquired by either of the techniques previously described. Included in this discussion are effects owing to leakage radiation in the beam (usually harmonic content), "glitches" in the monochromator, nonlinearity of detector response, and inappropriate design of sample preparation (including the "thickness effect"). The distortions of the XAS data resulting from such effects (except for glitches) are almost always manifested as reductions in the amplitude of the EXAFS from its true value. Glitches are defined as localized anomalies (usually sharp spikes or dips) in the ratioed data [i.e., $\ln(I_0/I)$ or F/I_0]. In the ideal experiment, the monochromatized beam would be purely the fundamental Bragg energy and the detectors would be perfectly linear over their entire dynamic range. If this ideal situation were attainable, there would be no thickness effect and no glitches in the ratioed data. In the real experiment, things are never quite ideal, and since the EXAFS constitutes a small change on top of a large background (especially at high k), the measurements are susceptible to small nonidealities.

We must first recognize the origin of the nonideal behavior of the radiation that is used in the typical XAS experiment. Leakage radiation usually refers to the presence in the radiation beam of energy components other than the desired fundamental energy. The sample absorption characteristics are usually very different for these other energy components. Commonly, the harmonic reflection makes up the other component and it passes through the sample with negligible absorption (hence the name leakage radiation). The effects of leakage radiation can be mimicked by leakage of the desired (fundamental) radiation through the sample. This might occur if the sample were not homogeneously prepared (e.g., having pinholes or thin spots). If it were possible to use detectors that could discriminate against leakage radiation, it would have no effect (at least in the transmission experiment). Of course, in the case of pinholes, this is impossible, since the leakage radiation is identical to the transmitted radiation. But, even in the case of harmonic content, ionization chambers have very poor energy resolution, and the harmonic will always make some contribution to the measured intensity.

Glitches in the radiation spectrum used for an XAS experiment are generated by the crystals used in the monochromator. Since any single crystal has numerous diffraction planes at many different orientations, as the crystal angle is scanned, the crystal may pass through an angle that satisfies the Bragg condition for another set of planes yielding diffraction of the *same wavelength* of radiation which satisfies the fundamental Bragg condition. Thus, at some particular energy the intensity of radiation in the desired fundamental beam will be decreased, since some of this radiation has been diffracted by the "spurious" reflection at some other angle (thus not contributing to the reflected fundamental beam). Because of this, a monochromator glitch always appears as a (sharp) dip in the I_0 intensity. The glitch does not contribute equally to the fundamental and harmonic (since it would be fortuitous for the primary and spurious reflections to have identical relative reflectivity for their harmonics). Thus we have another situation wherein a change in the fundamental energy component is accompanied by a smaller (or no) change in the other energy component. (For leakage radiation, the change in the fundamental occurred because of absorption by the sample — at the edge, for instance.) In the ideal situation, the glitch would contribute equally (on a relative basis) to the incident and transmitted intensity measurements and would not be present in the ratioed data (i.e., it would ratio out). But, if the detectors are nonlinear *or* leakage radiation is present, the glitch does not ratio out. In synchrotron radiation applications, similar nonratioing effects may occur for other beam anomalies which occur because of sudden changes in beam characteristics (e.g., brightness, position, etc.).

It is useful to examine in detail the effects that this nonideal behavior can cause. Other discussions of some of these ideas can be found, especially in reference to the thickness effect (Heald and Stern, 1977; Pease *et al.*, 1979; Stern *et al.*, 1979, 1980; Stern and Kim, 1981). Three test cases come to mind: a transmission measurement in the presence of leakage radiation, a transmission measurement with nonlinear detectors, and a fluorescence measurement with nonlinear fluorescence detectors. In each case, we will try to determine the differences between the ideal and the actual measured (nonideal) data, represented by the absorption coefficient μ. In effect, we are looking at some *change* in a signal and calculating by how much we underestimate this change because of nonideal behavior. Thus, the analysis will apply both to EXAFS amplitude reductions *and* the nonratioing of beam anomalies (e.g., glitches).

In the case of leakage radiation in a transmission experiment, the assumption is made that the leakage radiation contributes an additive component (α) to the I_0 signal and that this leaks through the sample unaltered. Thus, if the ideal measurement of incident and transmitted intensities are I_0 and I, respectively, the nonideal measurement gives

$$I_{0m} = I_0 + \alpha,$$
$$I_m = I + \alpha = I_0 \exp(-\mu x) + \alpha. \tag{30}$$

The subscript m denotes the measured quantity, x the thickness of the sample, and μ the ideal (true) absorption coefficient. We desire to calculate the measured (nonideal) absorption coefficient μ_m:

$$\exp(\mu_m x) = \frac{I_{0m}}{I_m} = \frac{I_0 + \alpha}{I_0 \exp(-\mu x) + \alpha}, \tag{31}$$

$$\mu_m x = \ln \left[\frac{1 + (\alpha/I_0)}{\exp(-\mu x) + (\alpha/I_0)} \right]. \tag{32}$$

In Fig. 13a, μ_m/μ (the factor by which the measured absorption coefficient is in error) is plotted against α/I_0 (the fraction of the I_0 measurement contributed by the leakage radiation). Each curve in Fig. 13 represents a different value of μx (the product of the true absorption coefficient and the true thickness). As expected, the higher the amount of leakage radiation, the larger the error in the measured absorption coefficient. But also, the thicker (or more concentrated) the sample, the larger the error in μ_m (at any particular α). This is because less fundamental radiation is transmitted through a thicker sample, but the leakage remains the same. Thus, the relative amount of leakage in I is higher for a thicker sample. This is the so-called thickness effect (Stern and Kim, 1981). It is present due to the nonratioing of I_0 and I because of the additive leakage α. The importance of this effect is clear: for the statistically optimum concentration ($\mu x \simeq 2.5$), a 1% contribution from leakage radiation gives an error in μ of ~5%. For thicker samples, the error is even larger.

Even if leakage radiation is completely absent, similar nonratioing effects can be caused by detector nonlinearities (although, if ionization chambers are used, this is not as much of a problem, since they are extremely linear

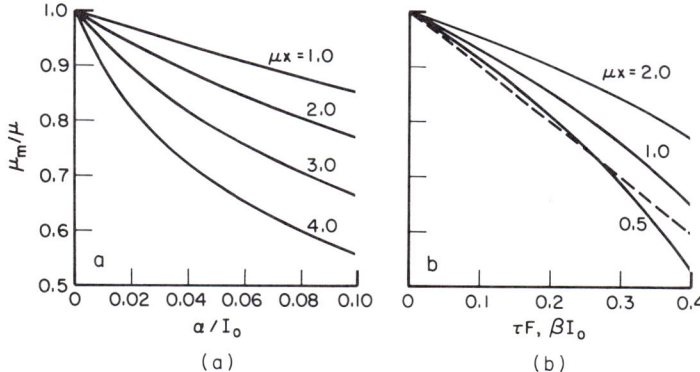

Fig. 13. The effect of leakage radiation (a) and detector nonlinearities (b) on the measurement of μ, the x-ray absorption coefficient. (a) Plot of the relative size of the measured absorption coefficient μ_m compared with the true absorption coefficient μ as a function of the fractional contribution of leakage radiation α to the incident beam I_0 and as a function of sample thickness (concentration) μx. (b) Plot of μ_m/μ as a function of nonlinearity in transmission detectors (measured by βI_0, see text) and as a function of μx (———); (– – –) is μ_m/μ as a function of nonlinearity in fluorescence detectors (measured by τF, see text).

throughout a wide dynamic range). Assuming that any nonlinearities show up as a quadratic decrease in the measured intensities:

$$I_{0m} = I_0 - \beta I_0^2,$$
$$I_m = I - \beta I^2 = I_0 \exp(-\mu x) - \beta I_0^2 \exp(-2\mu x). \tag{33}$$

Then μ_m can be calculated using Eq. (34):

$$\mu_m x = \ln\{(1 - \beta I_0)/[\exp(-\mu x) - \beta I_0 \exp(-2\mu x)]\}. \tag{34}$$

Figure 13b plots μ_m/μ versus βI_0 for different values of μx. Similar effects are seen, except that nonlinearities have a *relatively* larger effect for *dilute* samples. This is because the transmitted intensity is higher for dilute samples, and the nonlinearity is worse at higher intensities. In the usual transmission EXAFS experiment, nonlinearity effects are expected to be negligible, since the conventional ionization chambers are extremely linear even at high count rates. However, if one were using a gas proportional counter (with $\beta \simeq 10^{-7}$ sec) at count rates of 10^6 cps, one would realize an error in μ_m of $\sim 5\%$ for $\mu x = 2.0$.

In pulse detectors, nonlinearities result from dead time effects, and this will come into play in fluorescence experiments. If I_0 is measured with an ionization chamber, it can be assumed perfectly linear compared to measurement of F (the fluorescence–scatter signal). The measured F_m will be given by

$$F_m = F(1 - \tau F), \tag{35}$$

where τ is the effective dead time. [Note that this has the same form as Eq. (33).] The relationship to μ is simpler for fluorescence:

$$\mu x = F/I_0, \tag{36}$$

$$\mu_m x = \mu x(1 - \tau F), \tag{37}$$

and the dashed line in Fig. 13b shows the error associated with this effect (independent of sample concentration). The effect is relatively serious, giving an error in μ_m of 10% for scintillation counters ($\tau \simeq 10^{-6}$ sec) operating at 100,000 cps. This calculation assumes that no dead-time correction is made and points out the usefulness of such a correction (assuming knowledge of the effective dead time of the fluorescence detectors being used). It should be noted that any dead-time correction must be applied to data from individual detectors before calculating any weighted average, since the correction depends on *actual* count rates.

The thickness effect per se does not apply to a fluorescence experiment, but the presence of "leakage" radiation in the form of harmonic content does have some effect. One must take into account the ability of the harmonic radiation to generate a fluorescent photon that will contribute to the detected fluorescence signal. The catch is that the fluorescent yield is energy dependent and, in particular, $\phi(E) \neq \phi(2E)$, where E is the energy of the fundamental (the harmonic is at $2E$). For this case, we have

$$I_{0m} = I_0 + \alpha, \tag{38}$$

$$F/I_0 = \phi(E)\mu x, \tag{39}$$

$$F_m = I_0\phi(E)\mu x + \alpha\phi(2E)\mu x, \tag{40}$$

and, using Eqs. (38), (40) to calculate μ_m:

$$\mu_m x = \mu x \left\{ \frac{1 + (\alpha/I_0)[\phi(2E)]/[\phi(E)]}{1 + \alpha/I_0} \right\}. \tag{41}$$

Thus, if the fluorescent yields were the same there would be no distortion of μ_m. If $\phi(2E) \ll \phi(E)$, Eq. (41) simplifies to

$$\mu_m x = \mu x(1 + \alpha/I_0)^{-1}. \tag{42}$$

Some partial solutions to the related problems of amplitude distortions and nonratioing of glitches can be identified. For transmission work, one should be very careful to prepare the sample homogeneously and at low enough concentration to avoid the thickness effect. One should also do as much as possible to remove harmonic content from the radiation. As discussed in Section III.A.2, crystal detuning or use of x-ray mirrors are possibilities. Of course, if a focusing mirror is used, the disadvantage of low resolution has to be weighed against the advantage of having no harmonic. In addition, one should always take care to operate detectors in a linear portion of their response curve. This is especially critical for scintillation and solid-state detectors which have relatively long dead times. Finally, dead-time corrections should be applied if at all possible.

Before proceeding with a discussion of data analysis, it seems appropriate to mention here one more advantage to the use of fluorescence techniques for dilute samples. Since the appearance of glitches and other beam anomalies in the ratioed data is dependent on nonideal ratioing, and the relative size

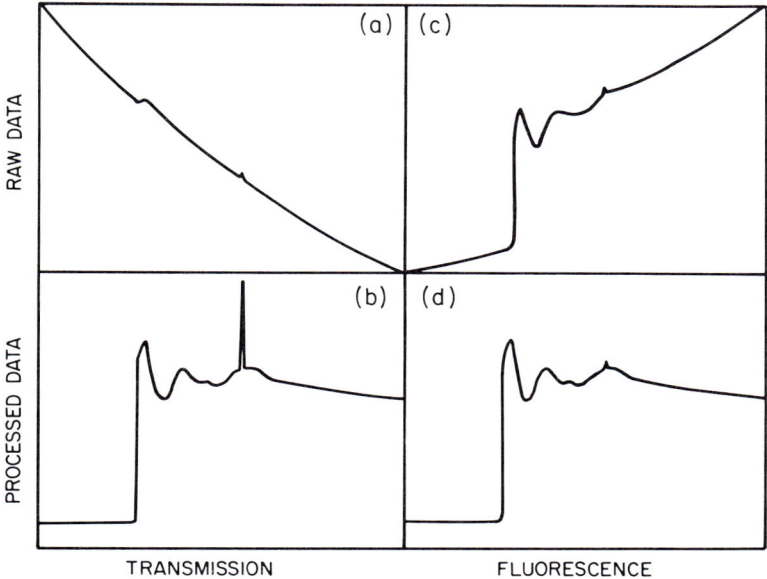

Fig. 14. Comparison of the effect of a monochromator glitch on processed data in the transmission and fluorescence modes. (a) and (b) are the unprocessed and processed (background-subtracted) data, respectively, for the transmission mode; (c) and (d) are the same for the fluorescence mode. Since the glitch size is approximately proportional to the total measured quantity (signal *plus* background) and since the fluorescence technique allows background rejection, yielding a much higher signal-to-background ratio, the glitch has relatively little effect on the fluorescence data.

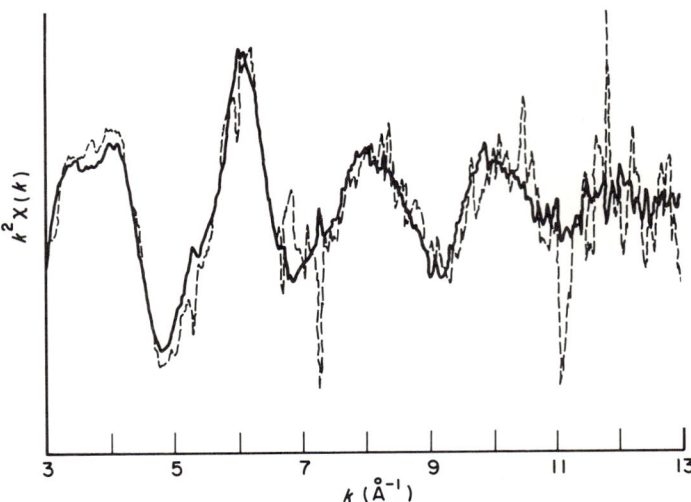

Fig. 15. Comparison of Cu EXAFS data on the copper protein, hemocyanin, collected by both transmission (---) and fluorescence (——). Note the effective suppression of glitches in the fluorescence data compared with the transmission data, as well as the improved S/N.

of these features is generally directly related to the total size of the ratioed quantity, one would expect that the scatter (background) rejection schemes used in the fluorescence technique will also reduce the size of these anomalies *relative* to the EXAFS amplitude. This of course is a direct result of the relatively large signal-to-background ratio in fluorescence data as compared with transmission data on the same dilute sample. Figures 14 and 15 give a graphic illustration of this phenomenon. Figure 14a,b shows hypothetical transmission data on a dilute sample before and after processing (see following). Since the glitch size is proportional to total measured quantity, and the edge is relatively small compared to the background in transmission, the glitch is large relative to the edge. For fluorescence (Fig. 14c,d) the edge is relatively large compared to the glitch. Figure 15 shows an example of this effect (this is Cu EXAFS data on hemocyanin at ~10 mM concentration). In this case, the glitches are almost completely suppressed in the fluorescence data.

C. Data Reduction and Analysis

Unlike most other spectroscopic techniques, it is difficult if not impossible to interpret a raw XAS spectrum in terms of structural parameters. Without a certain amount of data reduction and analysis, the data are virtually useless (especially in the case of data on dilute samples). Thus, it is not surprising that proper interpretation of EXAFS data is critically dependent on careful data analysis.

EXAFS data-analysis procedures can be naturally divided into two parts. The first involves extraction of the EXAFS portion of the data from the absorption edge and background. The second consists of simulating (or fitting) the EXAFS data using a theoretical EXAFS expression to gain information about structural parameters (e.g., distances, coordination numbers) of the metal site being examined. For this discussion, the first part (extraction of the EXAFS) will be referred to as data reduction and the second part as data analysis. Data reduction procedures are well developed and reasonably standardized, whereas data analysis procedures are not yet standardized, with different procedures being used by different researchers.

1. Data Reduction

Since biological applications are of primary interest here, we will concentrate on data reduction procedures that are useful for data collected by the fluorescence-excitation technique. However, virtually the same procedures are used for transmission XAS data (except for differences in detail), so the discussion is of general applicability.

The raw XAS data is generally recorded as a function of an artificial independent variable ("motor steps") which is directly proportional to the Bragg angle of the monochromator crystal. Thus, the first step of data reduction involves a calibration procedure to convert the data to a function of energy. This is accomplished by use of a reference material as a calibrant. This calibrant (often a metal foil) has a known energy associated with its edge position and allows a single-point calibration of the energy scale. The edge spectrum of the calibrant is recorded either separately from (e.g., before and after) the sample ("external" calibration) or simultaneously with the sample ("internal" calibration), as discussed earlier. In general, the first inflection point of the calibrant edge is used as the calibration point, and this inflection point is found by numerically calculating the second derivative of the edge spectrum and searching for down-going zero crossings. The energy for the ith data point $E(i)$ in the sample spectrum can then be calculated using

$$E(i) = hc/2d \sin \theta(i), \qquad (43)$$

$$\theta(i) = \theta_c + [S(i) - S_c]/\delta, \qquad (44)$$

$$\theta_c = \sin^{-1}(hc/2dE_c), \qquad (45)$$

where hc is Planck's constant times the speed of light, d the monochromator crystal d spacing, $S(i)$ the motor-step value for the ith data point, S_c the motor-step value at the calibration point, E_c the energy at the calibration point, and δ the number of motor steps per degree of monochromator-crystal rotation.

Once the raw data is placed on an energy scale, it is necessary to perform a background subtraction procedure. However, for data on dilute samples (whether fluorescence or transmission), it is typical to have many separate scans that need to be eventually averaged into one resultant data set. Assuming that the backgrounds for all the scans are reasonably similar, it is usually convenient to form the averaged data set before subtracting the background. For fluorescence data, the background (which is represented by the data before the edge) is owing to the Compton and elastic scattering portion of the detected radiation. This scatter background increases with increasing energy (see Fig. 16). In a transmission experiment, the background results from absorption by other atoms and by the L edges of the atom of interest and obeys the Victoreen equation,† decreasing with increasing energy. In either case, the best method of determining the shape of this background through the EXAFS energy region is by measuring an experimental background. This is a fairly simple matter for fluorescence data, since one can use a sample of the solvent (e.g., buffer for a protein solution) to mimic the scattering curve of the sample. (In a transmission experiment, it is often difficult to assemble a reference with the appropriate absorbers, and thus modeling the background with some polynomial or with a Victoreen is usually preferred.) Of course, to avoid introducing noise from the experimental background data, the background is fitted with a smooth curve (usually a second or third-order polynomial), which is then used to subtract from the data. An example of this procedure is shown in Fig. 16. In the absence of an experimental background, the data before the edge can be fitted with a polynomial in an effort to model the scatter background. The main difficulty with this procedure is that extrapolation of such a fit through the EXAFS region often does a poor job of modeling the background. An alternative method, which we have used with some success, is performing a polynomial fit through the EXAFS region (above the edge) and then subtracting a constant value from this fit to set the value of the data just before the edge to zero (after subtraction).

From this discussion, the selection of the scatter-background curve to use for subtraction may seem rather arbitrary. It is important to realize that the

---

† Coefficients for the Victoreen equation ($C\lambda^3 - D\lambda^4$) for all elemental absorption edges can be found in "International Tables for X-Ray Crystallography," vol. III (C. H. MacGillavry and G. D. Rieck, eds.), pp. 171–173. Kynoch, Birmingham (1968).

4. X-RAY ABSORPTION SPECTROSCOPY

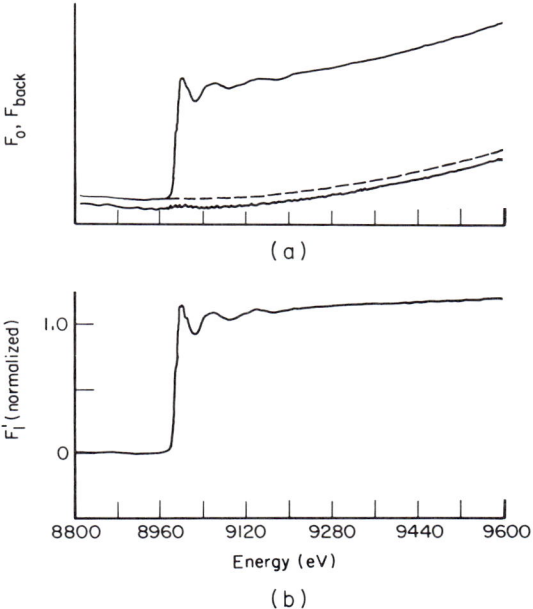

Fig. 16. Example of background subtraction procedures. The data used as an example in Fig. 16–19 are Cu EXAFS data of resting state (fully oxidized) cytochrome c oxidase collected by fluorescence excitation at SSRL with an array of 20 NaI(Tl) scintillation detectors. (a) The solid line (———) spectra are the raw data from the sample (upper spectrum) and the buffer (lower spectrum, offset downward by an arbitrary amount). The dashed (- - -) spectrum represents the fit background (by second-order polynomial least squares fit to the buffer spectrum over the energy range 8800–9600 eV). (b) Subtraction of the fit background from the raw data yields this background-subtracted spectrum, which has also been normalized to a unit edge jump. [The labels on the ordinate correspond to the symbols used in Eqs. (46)–(54) in the text.]

exact nature of this curve is not critical to the data reduction. In principle, the raw data (F_0) can be written as

$$F_0 = \gamma\mu + F_{scat}, \qquad (46)$$

where γ is some proportionality constant (related to concentration, fluorescence yield, sample thickness, etc.) and F_{scat} represents the scatter background. Ideally, the procedure described here would find a perfect model of

F_{scat} (as a function of energy), and subtraction would give the signal (F_1):

$$F_1 \equiv (F_0 - F_{\text{scat}}) = \gamma\mu. \tag{47}$$

Ultimately, we want to extract the EXAFS (χ), which is theoretically represented by

$$\chi \equiv (\mu - \mu_0)/\mu_0, \tag{48}$$

where μ_0 is the absorption expected for a free atom. Ideally, we know γ, and the next step in data reduction would involve subtracting and normalizing F_1 by the function $\gamma\mu_0$:

$$(\chi=)F_2 = (F_1 - \gamma\mu_0)/\gamma\mu_0. \tag{49}$$

Unfortunately, in the real world, we can never perfectly model F_{scat}, and we do not know γ. The procedure already described yields some imperfect approximation to F_{scat} (defined as F_{back}), and subtraction yields

$$F_1' \equiv (F_0 - F_{\text{back}}) = \gamma\mu + \Delta F, \tag{50}$$

where $\Delta F = F_{\text{scat}} - F_{\text{back}}$. In this nonideal case, we fail to completely remove the scatter background, and the next step, which should involve subtraction of the scaled atomic absorption $\gamma\mu_0$, will not work. However, the appropriate quantity to subtract can be found by simply fitting the EXAFS region to a smooth curve, μ_s. This is generally accomplished by the use of a polynomial spline. Typically, the EXAFS region is broken up into two or three spline regions, and each region is fitted with a polynomial (of third order, for example). The spline criterion is that the polynomials meet at the region limits with equal value and equal slope. This calculation of μ_s is found to be generally useful for both transmission and fluorescence data. Care must be exercised so that μ_s does not track any EXAFS oscillations (thus eliminating them), and that there is not any low-frequency curvature left in the EXAFS data after subtraction of μ_s. The first point is usually not a problem as long as the polynomial order used in the spline is low enough, and the second point causes a problem only when μ_s does not fit the "background" curvature in the EXAFS region. This is the reason for the F_{back} subtraction; it gives an F_1' [Eq. (50)] which is reasonably flat (except for the EXAFS oscillations) and usually allows the spline to fit the overall curvature more precisely.

The assumption in the preceding discussion is that the spline fit μ_s takes care of both the atomic falloff and the residual scatter,

$$\mu_s = (\gamma\mu_0 + \Delta F), \tag{51}$$

so that using Eq. (50) gives

$$F_1' - \mu_s = \gamma(\mu - \mu_0). \tag{52}$$

All that remains is to normalize the data to the atomic falloff, and the result will be the EXAFS. The proper normalization function (N) is the atomic falloff (modeled by the Victoreen equation), scaled so that the atomic absorption has an edge-jump identical to the F_1' data:

$$N = (\gamma \mu_0), \tag{53}$$

$$\chi = (F_1' - \mu_s)/N. \tag{54}$$

An example of the spline subtraction and Victoreen normalization is shown in Fig. 17.

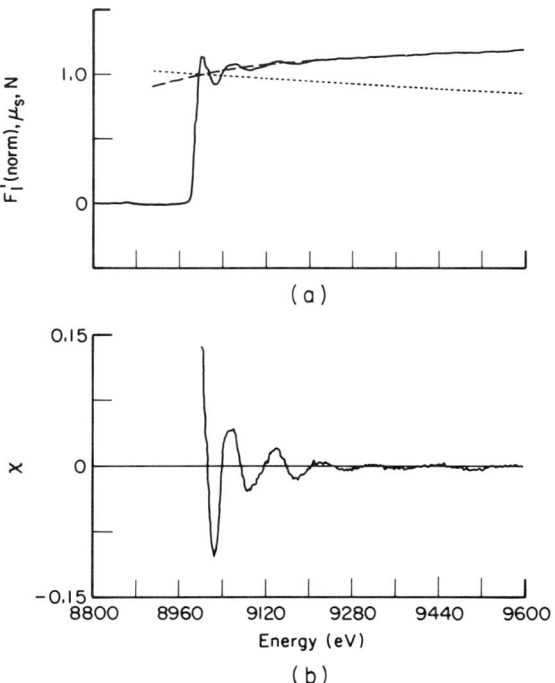

Fig. 17. Example of extraction of EXAFS from the background-subtracted spectrum. The data is the same as in Fig. 16. (a) The solid line (———) is the background-subtracted data calculated in Fig. 16b, the dashed line (– – –) is the spline fit (μ_s), and the dotted line ($\cdot \cdot \cdot$) is the Victoreen approximation to the atomic falloff N. The spline fit is calculated as a cubic spline over the range 9025–9650 eV with a spline point at 9400 eV. (b) Subtraction of the spline from the background-subtracted data followed by division by the Victoreen yields the EXAFS as shown here. The spectrum starts at 9000 eV, which was chosen as the energy at which $k = 0$.

In general, χ is treated as a function of the photoelectron wave vector k, rather than energy, and to better visualize the high-k oscillations, the $\chi(k)$ data are weighted by k^n ($n = 2,3$) for display. Figure 18 compares representations of $k^n\chi(k)$ for $n = 0,2,3$. Care should be taken to determine the quantity being plotted before making comparisons between literature spectra.

Calculation of k requires a knowledge of the threshold energy E_0 [see Eq. (4)]. The choice of E_0 for a particular element is not trivial. First, because there may be absorption processes other than $1s$ ionization occurring in the K-edge region, the position of the edge inflection for a particular sample does not necessarily reflect the threshold energy. Second, since the actual thresh-

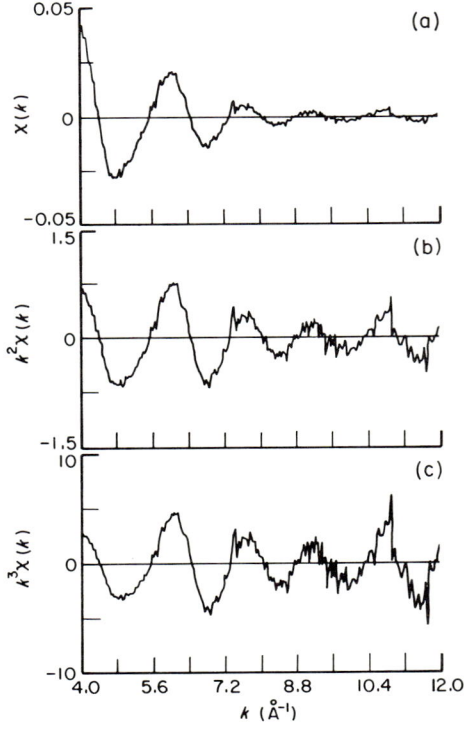

Fig. 18. Relationship between various representations of $\chi(k)$. The data are the same as used in Fig. 16 and 17. (a) $\chi(k)$ versus k (compare this with Fig. 17b and notice how quickly the EXAFS oscillations decay). (b) $k^2\chi(k)$ versus k (this is a commonly used plot to visualize the high-k oscillations). (c) $k^3\chi(k)$ versus k (k^3 weighting is conventionally used for Fourier transformation and weighting of fits).

old energy (and the inflection energy) is somewhat dependent on metal oxidation state and chemical environment, it is not true that all samples of a particular element (e.g., all Cu EXAFS) exhibit identical E_0 values. Thus, both the assumption that E_0 is a fixed energy away from the measured inflection and the assumption that E_0 is a fixed value for each element do not hold. The extent by which they are bad assumptions is not large (perhaps estimating E_0 with an error of 5–10 eV at the most), but they introduce artifactual phase shifts into the EXAFS data (especially at low k, where small energy differences are most important) and can be responsible for small errors in calculated distances. The way this problem is handled depends on the data-analysis technique adopted and is discussed further in Section III.C.2.

Examining the extracted EXAFS data, we find another difference between XAS and other spectroscopic techniques. With most other techniques, the data in frequency space (or k space for EXAFS) is readily interpretable (in terms of transitions or resonances, for example). However, for EXAFS data, we must alter our reference frame. $\chi(k)$ [as manifested in Eq. (13)] is a sum of damped sine waves, and one must think in terms of the frequencies, amplitudes, and phases of these sine-wave components. A useful visual aid for this purpose is the Fourier transform (FT) technique. Generally speaking, the FT of a sum of sine waves gives a series of delta functions in the inverse space, one at the frequency of each sine wave. If the sine waves were in the time (sec) domain, the FT would be in the frequency (sec^{-1}) domain. The FT of $\chi(k)$ (in the k domain, Å^{-1}) is in the real-space domain (Å). Since $\chi(k)$ is a sum of damped, phase-shifted sine waves, the FT shows peaks (broadened from delta functions) at the frequencies of the sine waves ($2R + \alpha$) where α is some phase shift directly related to $\alpha_{as}(k)$ in Eq. (13). Thus, the FT of $\chi(k)$ can be thought of as a radial distribution function about the absorbing atom. It is conventional to perform the FT so that the 2 in the frequency is factored out, and one then observes FT peaks at $R + \alpha'$. It is also conventional to weight $\chi(k)$ by k^n ($n = 3$, usually) before performing the FT. The FT of the $\chi(k)$ data in Fig. 18 is shown in Fig. 19. Although procedures have been developed for determining structures based on the fitting of Fourier transforms of EXAFS (Gurman and Pendry, 1976; Hayes et al., 1976), it is fairly standard to merely use the FT as a starting point to visualize what shells of atoms may lie at specific distances, but to perform the actual fits on the $\chi(k)$ data.

Figure 19 also illustrates another significant use of the FT. Since the individual sine-wave components of $\chi(k)$ are separated into peaks in the FT, one can use a Fourier filtering technique to isolate the $\chi(k)$ contribution from one shell of scatterers. Convolving the FT with a window that includes one FT peak and backtransforming to k space generates the EXAFS component which gave rise to that FT peak. Using this filtering technique, one can examine the characteristics of the EXAFS wave of each shell of scatterers separately to determine structural parameters for that shell. It should be noted that such filtering techniques can lead to distortion of the backtransformed data because the window cannot perfectly separate one FT peak from adjacent ones. Also, square window functions can introduce distortions due to discontinuous data truncation. It is often helpful to use a window with half-Gaussian wings, as shown in Fig. 19. Fourier filtering is also used to remove high-frequency noise from $\chi(k)$ data. This is accomplished by using a wide window (perhaps from 0.5 to 5.0 Å) for backtransformation of the FT data to k space. This generates filtered $\chi(k)$ data, which is the form of data often used in fits. Although such filtered $\chi(k)$ data appears to be noise free, it is important to recognize that a typical noise spectrum is

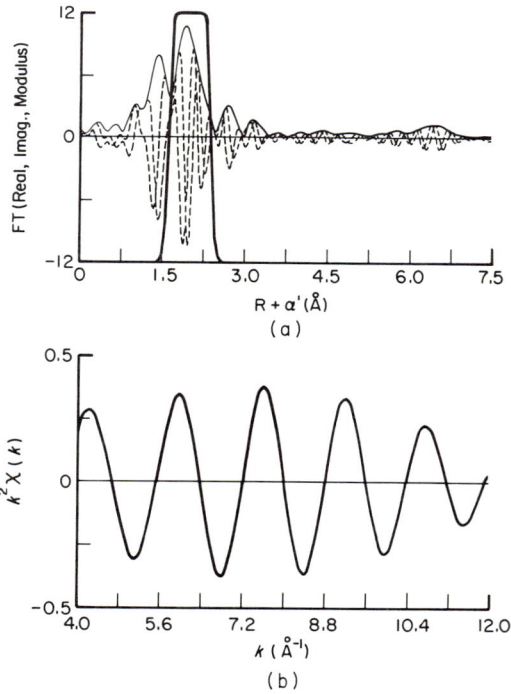

Fig. 19. Example of Fourier transformation and the Fourier filtering techniques. (a) The k^3-weighted Fourier transform (FT) of the EXAFS data of Fig. 18. The solid line (——) is the FT magnitude and the dashed lines (– – –) are the imaginary and real components. The bold line (——) indicates the filter window used to produce the Fourier-filtered data in (b).

fairly white, and low-frequency noise that falls within the filter window is still present in the filtered data.

2. Data Analysis

The actual analysis of EXAFS data in terms of shells of atoms at particular distances is performed, in general, by hypothesizing a "model" structure and calculating, using the EXAFS expression [Eq. (13)], what the spectrum of this hypothetical structure would be (simulation). The hypothetical structure is then optimized (minimizing the difference between the simulated and actual spectrum) by varying the parameters of the individual shells, and some type of goodness-of-fit criterion is used to determine the adequacy of the structural model.

Each shell of the hypothetical model consists of N_s atoms (of one particular element), all at a distance R_{as} from the absorbing atom. For example, in a heme protein, the absorber is Fe, and one shell might be the porphyrin nitrogens: $N_s = 4$, $R_{as} \simeq 2$ Å. To calculate a simulated spectrum, one must know the nature of the scattering expected for a particular scatterer type. In other words, for each shell, some knowledge of the scattering amplitude, $|f_s(\pi,k)|$, and phase shift, $\alpha_{as}(k)$, is required. (The Debye–Waller factor provides some special problems which will be discussed later.) Once the amplitude and phase functions of a particular shell are known, they can be plugged into the EXAFS expression for calculation of a simulated spectrum. The fit to the actual data is then accomplished by varying R_{as} and N_s (and in some cases, σ_{as}) in an optimization algorithm. If it is found that the optimized fit is still unsatisfactory, one can redefine the scatterer type for that shell (by plugging in different amplitude and phase functions) and reoptimize R_{as} and N_s.

This is a completely general procedure for fitting $\chi(k)$ data [actually $k^n\chi(k)$ data are used in the fit, typically $n = 3$]. There exist two specific methods for accomplishing this. They differ in the manner in which the amplitude and phase functions are determined for a particular scatterer type. One method [the theoretical method (Teo and Lee, 1979)] is to calculate (from first principles) the scattering amplitude and phase expected for a particular scatterer. The alternative is the empirical approach (Cramer and Hodgson, 1979; Cramer et al., 1978a), which uses phases and amplitudes measured from (model) compounds that are structurally characterized. Both methods have been used on a wide variety of chemical and biological systems with what seems to be comparable success. It will suffice for this discussion to mention briefly the details of each method (the reader is referred to the cited references for more details).

The theoretical approach (Teo and Lee, 1979) makes use of electron–

atom scattering theory (Lee and Beni, 1977) to calculate the amplitude $F(k)$ of backscattering from a particular scatterer, and the phase shift undergone by the scattered wave leaving (or reentering) a particular absorber $\phi_a(k)$ and scattering from a particular scatterer $\phi_b(k)$. The total phase shift for the wave from a particular absorber–scatterer pair is then given (for K and L_I edges) by

$$\phi_{ab}(k) = \phi_a(k) + \phi_b(k) - \pi. \tag{55}$$

The values of $F(k)$, $\phi_a(k)$, and $\phi_b(k)$ are tabulated (for most elements) from $k = 4.0 - 15.0$ Å^{-1} (Teo and Lee, 1979), and they are used in optimization of fits to experimental EXAFS data. In these fits, since the calculated phase shifts are dependent on E_0 and (as already discussed) there is no good way of accurately measuring E_0 in a particular compound, the value of E_0 is allowed to vary in the optimization procedure. In addition, the calculated amplitudes can be in error by some scale factor (perhaps as large as a factor of two), so that they have to be scaled in a semiempirical way by examining structurally characterized compounds. The Debye–Waller factor is also varied as part of the optimization procedure. The calculated amplitudes and phase shifts can be used in their numerical form or as parameterized functions in the fitting procedure. The amplitudes are generally parameterized (Teo et al., 1977) as

$$F(k) = A/[1 - B^2(k - C)^2], \tag{56}$$

and the phase shifts are usually parameterized as

$$\phi_a(k) = a_0 + a_1 k + a_2 k^2 + a_3/k^3,$$
$$\phi_b(k) = b_0 + b_1 k + b_2 k^2 + b_3/k^3 \tag{57}$$

Detailed studies of structurally characterized systems indicate that use of these fitting procedures yields (for single-shell systems) a distance accuracy of ~ 0.01 Å, accuracy in the Debye–Waller σ_{as} of $\sim 10\%$, and accuracy in the coordination number of $\sim 20\%$ (Teo and Lee, 1979). In multishell systems, these numbers will be somewhat degraded, but distance determinations are still expected to be accurate to ~ 0.03 Å.

The empirical data analysis procedures (Cramer et al., 1978a) involve parameterization of the amplitude and phase contributions to the EXAFS expression and use of EXAFS data on structurally characterized compounds to establish the parameters for a certain absorber–scatterer pair. The amplitude portion of the EXAFS expression (including the Debye–Waller factor) is parameterized as the completely empirical function:

$$A(k) = c_0 \exp(-c_1 k^2)/k^{c_2}, \tag{58}$$

4. X-RAY ABSORPTION SPECTROSCOPY   343

and the phase shift is parameterized as

$$\alpha_{as}(k) = a_0 + a_1 k + a_2 k^2, \tag{59}$$

or, in some cases

$$\alpha_{as}(k) = a_0 + a_1 k + a_{-1} k^{-1}. \tag{60}$$

Fits are first performed to single-shell EXAFS data (perhaps one shell filtered from a FT) of structurally characterized model compounds in which all the a_i and c_i parameters are optimized (with N_s and R_{as} known). Once the a_i and c_i have been determined in this way for a particular absorber–scatterer pair, this defines the amplitude and phase functions to be used in unknown fits in which N_s and R_{as} are optimized. Two assumptions are inherent in this procedure: for a particular absorber, E_0 is selected as some rather arbitrary value, and it is assumed not to change; and for a particular absorber–scatterer pair, the Debye–Waller factor is constant. Even given these rather crude assumptions, the accuracy in distances reported for the empirical method is ~0.015 Å, and the accuracy in coordination numbers ~20–25%, which compares favorably with the results using the theoretical phases and amplitudes. The strategy behind assuming E_0 constant is simply that there is no good way of predicting how it would change from one compound to another, and including it as a variable parameter just introduces another degree of freedom and perhaps parameter-correlation problems. There is some sentiment (Peisach *et al.*, 1982) that this assumption (fixed E_0) may introduce small systematic errors in calculated distances, but this has yet to be verified.

In the empirical methods, the Debye–Waller factor is assumed to be constant for a particular absorber–scatterer pair. Although this is clearly not the case, inclusion of it as a variable parameter is not satisfactory, since it will be highly correlated with the overall amplitude-scale factor (i.e., N_s). The Debye–Waller factor has proven particularly difficult to measure independently, and little is known about its dependence on bond length, metal-oxidation state, etc. Because it is so highly correlated with other amplitude parameters, it is virtually impossible to use EXAFS of structurally characterized compounds to measure the Debye–Waller factor. However, in simple cases it may be possible to extract differences in σ_{as}^2 ($\Delta\sigma_{as}^2$) between compounds of the same absorber–scatterer pair by examining ratios of EXAFS amplitudes (Lee *et al.*, 1981). If we assume that we can write the EXAFS amplitude as

$$A(k) = (N_s/kR_{as}^2)|f_s(\pi,k)| \exp(-2\sigma_{as}^2 k^2), \tag{61}$$

then, for a model m and an unknown compound u, we can form the ratio of the empirically measured amplitudes $A_m(k)/A_u(k)$, and taking the natural logarithm yields

$$\ln[A_m(k)/A_u(k)] = \ln(N_m R_u^2/N_u R_m^2) + 2k^2(\sigma_u^2 - \sigma_m^2). \quad (62)$$

If we assume that σ_m^2 is known (or if we are only looking for the difference, $\Delta\sigma^2 = \sigma_u^2 - \sigma_m^2$), and that N_m, R_m are known and R_u has been determined from the standard data analysis procedures, then a plot of the left-hand term versus k^2 will yield values for N_u and σ_u^2 (or $\Delta\sigma^2$). This, coupled with calculations of σ_{vib}^2 and σ_{stat}^2 contributions as described earlier [see Eqs. (11), (12)] may lead eventually to a better handling and understanding of the Debye–Waller factor.

IV. EXAFS as a Structural Technique in Biology

Up to this point, the discussion of the XAS technique has been completely general. However, from this discussion it should be apparent how important EXAFS is as a structural technique applied to metal atoms in biological systems. First of all, EXAFS is the only technique that can yield direct information about (local) structural characteristics without the need for samples with long-range order. X-ray crystallography of proteins gives structural information but, of course, only on proteins which can be grown as single crystals. In biological XAS applications, the sample is typically a solution or frozen solution of a metalloprotein, -enzyme, etc. Secondly, EXAFS is a *localized* technique, giving information only about the immediate environment of the metal without "interference" from the atoms making up the general protein backbone. Although this may be interpreted as a limitation, for most systems of interest this is precisely the spatial region of the metalloprotein which is of importance (i.e., the active site). Third, EXAFS is a *selective* technique—it gives information about only one element without significant interference from other elements (e.g., other metal atoms). Thus, one can selectively study the structural environment of one metal atom in the presence of another.

A. Limitations

One must also recognize the limitations of the technique, to put it to effective use. One general limitation inherent in the technique is the lack of any geometrical information—only a radial distribution function is measured. Of course, as already discussed, in the future x-ray absorption-edge measurements may give some clues regarding geometry. Also, if one has access to an oriented sample, polarization measurements can give this type of information. Another inherent limitation of EXAFS is the inability to determine scatterer type with great precision. Thus, for ligand atoms which differ in atomic number (Z) by one, the amplitude and phase functions are sufficiently similar that it is difficult to distinguish the scattering patterns. In biological applications, this means that C, N, and O are difficult to differentiate as ligands.

Some limitations are specific for biological applications of EXAFS. The most obvious of these is the requirement of high concentrations of samples for adequate S/N. This was one of the primary reasons for development of fluorescence detection techniques. The current state-of-the-art instrumentation puts an operational lower limit on concentration of ~ 1 mM (at ~ 10 keV) and ~ 0.2 mM (at ~ 20 keV) for fluorescence data collection using $\sim 10^{15}$ photons total irradiation (a typical dedicated run at SSRL with ~ 10 hours of data collection). For an enzyme with one metal atom per 100,000 molecular weight, this represents concentrations of 100 and 20 mg ml^{-1}, respectively. Standard beam sizes ($\sim 0.2 \times 1.5$ cm^2 cross section) require sample volumes of ~ 0.1 and 0.5 ml, respectively (taking into account the longer pathlengths desirable at the higher photon energy). These numbers probably represent ultimate lower limits with two- to five-fold higher concentrations desirable for good S/N.

Another type of problem specific to biological applications is heterogeneity. This may apply to heterogeneity of metal sites (e.g., binding of extraneous metal atoms to nonspecific sites or more than one binding site for stoichiometric metal atoms) or heterogeneity of metal-ligand distances (usually referred to as static disorder). In the former case, one must take care to remove all extraneous metals or, if there is more than one distinct binding site, use chemical means to selectively perturb the structure of one of the metal sites (to identify their individual contributions to the EXAFS). In the latter case (static disorder), there may be two types of occurrences: low symmetry of the metal site contributing to a spread of distances to otherwise identical ligand atoms; or actual microheterogeneity of the "same" metal sites among different protein molecules (this may also encompass a certain dynamic contribution from protein-chain motion). The latter is difficult to document, but an important observation is that high-resolution edge spectra of metalloproteins tend to exhibit significantly broader features than inorganic complexes of the same metal recorded at the same resolution (compare with Fig. 25).

With the high flux of x-rays used for biological XAS experiments, one must also be concerned with the possibility of radiation damage resulting from the creation of radicals in aqueous solution by the ionizing radiation. Such radiation-produced radicals could cause either protein denaturation or undergo redox reactions with a metalloprotein active site. In either case, the XAS data would correspond to a species that is different from the one placed in the beam. Performing XAS work at low temperature seems to be the best defense against radiation-induced sample damage, but one should always be aware of the possibility and use appropriate tests to assure the integrity of the XAS sample before, during, and after the experiment.

Even with these limitations in mind, it is clear that XAS techniques have played and will continue to play an important role in structural studies of metals in biological systems. The remainder of this chapter will be devoted to a discussion of several examples of applications of XAS to biological systems. These will illustrate some of the strengths and some of the limitations of the technique for solving structural problems in biology.

B. Specific Applications

The following discussion is not meant to be a comprehensive survey of XAS work on biological systems. It is meant to give examples of the use of the technique for solving biological problems. It is a fairly safe bet that XAS work has been *attempted* on at least one biological example of every naturally occurring metal (including nickel, tungsten, selenium, and others). This is not to say that all the relevant questions have been answered (or even asked). Some systems are much better studied than others, but on the average, researchers have barely scratched the surface of the biological XAS field.

1. Iron Proteins

The Fe K-absorption edge falls at ~ 7.1 keV, an energy that is easily accessible using most synchrotron radiation sources. This, in combination with the widespread occurrence of iron in biological systems, is responsible for the large amount of XAS work done on iron proteins. Examples of each of the classes of iron proteins (heme proteins, iron–sulfur (Fe—S) proteins, and other nonheme iron proteins) have been studied. We will look at the Fe—S proteins first.

The simplest Fe—S protein consists of the crystallographically characterized rubredoxin (from several different bacterial sources), [Rd], that contains a monomeric iron-active site ligated to the sulfur atoms of four cysteine residues in a nearly tetrahedral environment. This is a prime example of a single-shell environment, and the simplicity of the EXAFS made it a good test case for the use of the technique on metalloproteins. In addition, the availability of synthetic analogs which are very good structural models for the [Rd] active site allows comparisons of the protein with low molecular-

Table II EXAFS Results on Fe—S Proteins, Models

Analog/Protein[a]	Cluster type	EXAFS R(Fe—S)[b] (Å)	EXAFS R(Fe—Fe)[b] (Å)	Ref.	Crystallography R(Fe—S)[b] (Å)	Crystallography R(Fe—Fe)[b] (Å)	Ref.
[Fe(S₂-o-xyl)₂]⁻	[Rd]³⁺	2.279(13)		c,d,e	2.267(2)		f
[Fe(S₂-o-xyl)₂]²⁻	[Rd]²⁺	2.340(14)		c,d,e	2.356(5)		f
Rdox(solid)	[Rd]³⁺	2.265(13)		c,d,e	2.29(3)		g
Rdox(solid)	[Rd]³⁺	2.30(4)		h			
Rdox(solution)	[Rd]³⁺	2.256(16)		c,d,e			
Rdred(solution)	[Rd]²⁺	2.32(2)		c,d,e			
[Fe₂S₂(S₂-o-xyl)₂]²⁻	[2Fe—2S]²⁺	2.234(15)	2.704(23)	c	2.257(2)	2.698(1)	i
Fdox(solid)	[2Fe—2S]²⁺	2.227(15)	2.696(47)	c			
Fdox(solution)	[2Fe—2S]²⁺	2.233(44)	2.726(80)	c			
Fdred(solution)	[2Fe—2S]⁺	2.241(56)	2.762(96)	c			
[Fe₄S₄(S-benzyl)₄]²⁻	[4Fe—4S]²⁺	2.270(13)	2.717(24)	c	2.286(2)	2.747(2)	j
HiPIPox(solution)	[4Fe—4S]³⁺	2.262(13)	2.705(26)	c	2.24(5)	2.73(4)	k
HiPIPred(solution)	[4Fe—4S]²⁺	2.251(13)	2.659(50)	c	2.30(7)	2.81(5)	k
bact. Fdox(solution)	[4Fe—4S]²⁺	2.249(16)	2.727(35)	c			
bact. Fdred(solution)	[4Fe—4S]⁺	2.262(14)	2.744(32)	c			

[a] Rd = *Peptococcus aerogenes* (ref. c,d,e) or *Clostridium pasteurianum* (ref. h) rubredoxins; Fd = rhubarb ferredoxin; HiPIP = *Chromatium vinosum* [4Fe—4S] ferredoxin (high potential iron protein); bact. Fd = *C. pasteurianum* 2[4Fe—4S] ferredoxin.
[b] Estimated standard deviation in parentheses.
[c] Teo et al., 1979.
[d] Shulman et al., 1978a.
[e] Shulman et al., 1975.
[f] Lane et al., 1977.
[g] Watenpaugh et al., 1980.
[h] Sayers et al., 1976.
[i] Mayerle et al., 1975.
[j] Averill et al., 1973.
[k] Jensen, 1978.

weight compounds with more precise independent (crystallographic) determinations of bond lengths. Table II shows a comparison of crystallographically determined distances and distances determined by EXAFS for models and proteins of the Fe—S type. As expected, the EXAFS of both rubredoxin and the mononuclear analogs exhibit a single wave (see Fig. 20), and the FT shows a single peak owing to Fe—S scattering. The Fe—S distances for the [Rd]-type site agree within 0.02 Å (for the models) or 0.03 Å (for the protein) with the crystallographically determined distances. In most of these measurements (Shulman et al., 1975, 1978a; Teo et al., 1979), the coordination number was fixed as 4, and the Debye–Waller σ_{as} varied to match the EXAFS amplitude (using theoretical phase and amplitude functions). For the models, the measured σ_{as} compared favorably with the value calculated assuming a typical $\sigma_{vib} \simeq 0.045$ Å, the σ_{stat} calculated from Eq. (11), and the known bond-length disorder (from the crystal structures) (Teo et al., 1979).

The absorption edge of rubredoxin has been examined by several workers (Sayers et al., 1976; Shulman et al., 1975, 1976) and is shown in Fig. 20b. The most interesting feature is the relatively intense pre-edge absorption which has been assigned to the Fe $1s \rightarrow 3d$ transition. The intensity of this transition is directly affected by the site symmetry of the Fe, with only a very weak transition occurring in octahedral iron complexes. However, in the nearly tetrahedral environment of the rubredoxin site, the lack of a center of

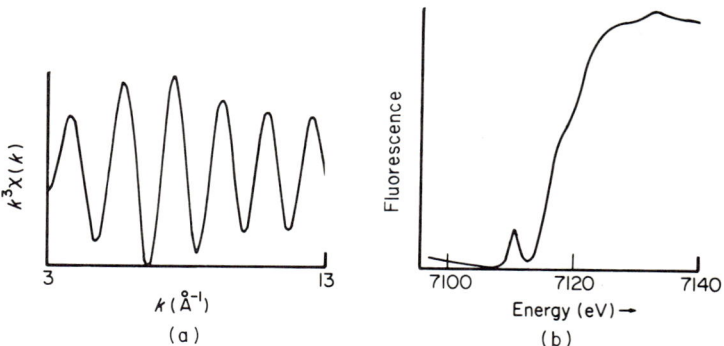

Fig. 20. Fe EXAFS and edge spectra of the oxidized form of the Fe—S protein, rubredoxin. (a) The simplicity of the EXAFS oscillations indicates the presence of only one major bond distance (the Fe—S bond). (b) The relatively large pre-edge peak is the enhanced Fe $1s \rightarrow 3d$ transition owing to the noncentrosymmetric (approximately tetrahedral) crystal field. [(a) From Shulman et al., 1978. (b) From Shulman et al., 1976. © 1978 Academic Press, Inc. (London) Ltd.]

symmetry allows extensive $p-d$ mixing to lend intensity to this otherwise forbidden transition. Thus, this pre-edge feature can be used to predict gross features of the symmetry of the iron sites in proteins of unknown structure.

For the plant-type ferredoxins (with [2Fe—2S] clusters) and the bacterial ferredoxins (with [4Fe—4S] clusters), the EXAFS exhibits a beat pattern indicating the presence of at least two sine-wave components. An example (*Chromatium* HiPIP) is shown in Fig. 21 in which the FT exhibits the expected two peaks. The first peak consists of scattering from Fe—S interactions, and the second is from Fe—Fe scattering. Although the crystallographic results show that the Fe—S* (S* is acid-labile sulfide) distances are actually ~0.04 Å longer than the Fe—S(Cys) distances, they cannot be resolved in the FT. This points out the difference between accuracy and resolution. Although calculated bond distances from EXAFS are expected to be accurate to ~0.02 Å, EXAFS is unable to resolve shells that are different by ≤ 0.10 Å (or perhaps more). The shells are then observed as one shell with a large σ_{stat} (static disorder). Bond distances determined by EXAFS and crystallography are compared in Table II for some [2Fe—2S] and [4Fe—4S] proteins and synthetic analogs.

A significant amount of XAS work has also been done on heme proteins. As with the Fe—S proteins, this work has often benefited from the existence of structurally characterized synthetic iron porphyrins with known axial ligands that are good structural models of the active sites. In the case of a heme group, one knows that the first coordination sphere contains four porphyrin nitrogens. The question usually centers around identification of the axial ligands, and determination of bond lengths and their dependence

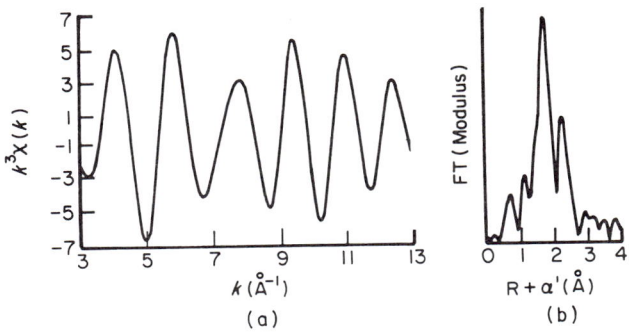

Fig. 21. Fe EXAFS and Fourier transform (FT) of the oxidized form of *Chromatium vinosum* high-potential iron protein (HiPIP). (a) EXAFS spectrum showing the beat pattern owing to the presence of two different frequencies (Fe—S and Fe—Fe). (b) The FT clearly indicates the separate Fe—S and Fe—Fe EXAFS frequencies in the data. (From Teo *et al.*, 1979.)

on oxidation state. The porphyrin ligand is relatively unique in that it presents an extensive array of rigidly fixed scattering atoms past the first coordination sphere, and in model systems FT peaks are often observed for these outer shells (Cramer *et al.*, 1978d) (see Fig. 22). The presence of all these shells makes the EXAFS rather complicated, and often it proves difficult to separate the axial ligand scattering from the porphyrin contribution.

EXAFS studies have been performed on hemoglobin solutions (Kincaid *et al.*, 1975; Eisenberger *et al.*, 1978) to determine that the average iron-to-porphyrin nitrogen distance changed from 1.98 to 2.06 Å on going from oxy- to deoxyhemoglobin (Eisenberger *et al.*, 1978). In this work, the O_2-binding complex Fe(TpivPP) (N-MeIm) was used as a structural model for the EXAFS interpretation. Eisenberger *et al.* (1978) used triangulation to determine that the Fe could be no more than 0.2 Å out of the plane of the porphyrin nitrogens, which contradicted the Perutz hypothesis of hemoglobin cooperativity (Perutz, 1970). EXAFS studies of several other structurally characterized metalloporphyrins has demonstrated that the precision of the EXAFS distances is probably not sufficient to calculate an Fe out-of-plane distance by triangulation (Perutz *et al.*, 1982).

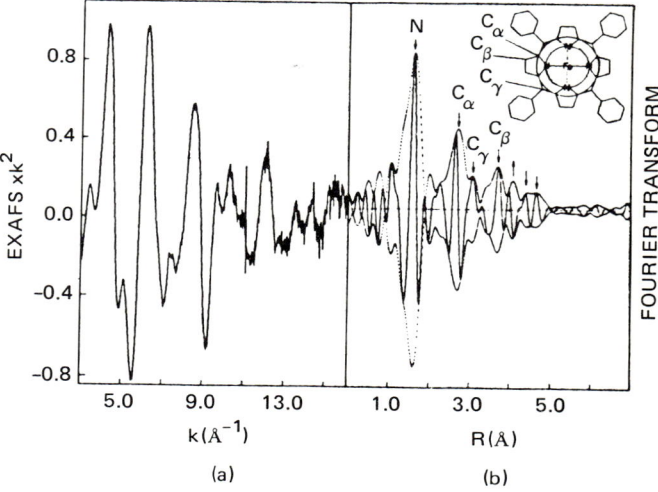

Fig. 22. Fe EXAFS and Fourier transform (FT) of Fe(II)(TPP) (TPP = tetraphenylporphyrin). (a) EXAFS spectrum exhibiting a complicated combination of sine waves. (b) The FT separates out the frequency components indicating that the different shells of the porphyrin (pyrrole nitrogens, α-carbons, meso-carbons, and β-carbons) all contribute to the EXAFS. (From Cramer *et. al.*, 1978d.)

Other heme proteins that have been examined by XAS include cytochrome *P*-450 and chloroperoxidase (Cramer *et al.*, 1978d), cytochromes *c* (Labhardt and Yuen, 1979; Korszun *et al.*, 1982), and cytochrome *c* oxidase (Hu *et al.*, 1977a, 1977b, Powers *et al.*, 1979, 1981; Chance *et al.*, 1980; Scott *et al.*, 1981; Scott, 1982). The EXAFS work on cytochrome *P*-450 and chloroperoxidase identified a sulfur as one of the axial ligands (Cramer *et al.*, 1978d), the most direct evidence yet for the existence of the postulated thiolate ligand in these enzymes. The most recent EXAFS work on a series of high-potential cytochromes *c* indicated no significant differences in iron-axial ligand bond length among the series, although significant variation in reduction potential is seen (Korszun *et al.*, 1982). The Fe—N(His) bond length was always 1.97–1.99 Å, and the FeS(Met) bond length was always 2.29–2.32 Å. This information was used to rule out bond length variation as a controlling influence on heme reduction potential.

XAS studies of both the heme Fe and Cu components of beef heart cytochrome *c* oxidase have been performed by several groups. With this multisite enzyme, we run into another problem. The enzyme contains two distinct types of heme Fe (heme *a* and heme a_3) and two distinct types of Cu (Cu_A, the EPR-visible copper, and Cu_B). Thus, both Fe and Cu EXAFS exhibit a superimposition of contributions from each site. Thus an "average" structure is seen, and, to isolate individual site contributions, we must selectively structurally perturb one site without disturbing the other. For the copper sites, this can be done by selective reduction (to the mixed valence state) of one copper, thereby altering its local structure (Scott, 1982). For the hemes, very little structural change accompanies reduction, so that the only option left is to assume a structure for one heme and assign the remaining EXAFS to the other. This may be possible with cytochrome *c* oxidase (Powers *et al.*, 1981), since other evidence suggests that heme *a* is six-coordinate with two histidine ligands, but one must be aware of the assumption when interpreting results for the remaining heme a_3.

Most of the useful information regarding the active site structures in cytochrome *c* oxidase has come from the Cu EXAFS. The most direct evidence for the presence of sulfur in the copper coordination sphere comes from such data (Scott *et al.*, 1981). Work on various oxidation states has resulted in the assignment of most (if not all) of the sulfur ligands to Cu_A (Scott, 1982). It is instructive to go through the argument that gives this conclusion. Figure 23 shows the transforms of the Cu EXAFS of cytochrome *c* oxidase in three oxidation states: resting state (Cu_A^{2+}, Cu_B^{2+}), mixed valence-formate (Cu_A^+, Cu_B^{2+}), and fully reduced (Cu_A^+, Cu_B^+). The two main FT peaks at lower and higher *R* in each case result from Cu—N and Cu—S scattering, respectively. Using empirical amplitude and phase functions, the fits give the effective coordination numbers and distances

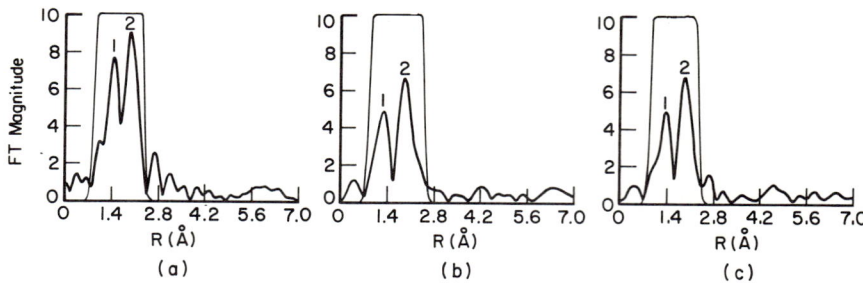

Fig. 23. Fourier transform (FT) of Cu EXAFS of various oxidation state derivatives of cytochrome c oxidase (CCO). (a) FT of resting state (fully oxidized) CCO showing the results of curve fitting of the Fourier-filtered EXAFS indicated by the filter window around the two major peaks: 1 is 1.8 N(O) at 1.96 Å; 2, 1.6 S at 2.26 Å. (b) FT of fully reduced CCO with similar curve fitting results: 1 is 1.2 N(O) at 1.95 Å; 2, 1.4 S at 2.32 Å. (c) FT and curve fitting results for the mixed-valence–formate derivative of CCO: 1 is 1.3 N(O) at 1.96 Å; 2, 1.4 S at 2.32 Å. See the text for further details.

indicated in Fig. 23. It is clear that the major change occurs in the Cu—S distance, and it occurs when Cu_A is reduced (i.e., the resting-state data are different from the mixed valence-formate data). This leads directly to the interpretation that all the redox-sensitive sulfur ligands reside at the Cu_A site. What is important to note is that this conclusion is completely independent of the validity of any fitting procedure, since it does not depend on absolute bond distances and coordination numbers. This is a good example of the proper experiments allowing conclusions to be made based solely on the data, *not* on the results of fits of the data.

A combined study of both Fe and Cu EXAFS data on cytochrome c oxidase has resulted in the postulation of a structure for the binuclear heme a_3–Cu_B site in the resting enzyme (Powers *et al.*, 1981). The structure features a thiol bridge between the Fe of heme a_3 and Cu_B. It will remain for future work to determine whether this structure can be verified.

Among the nonheme iron proteins (other than Fe—S) that have been studied by XAS are ferritin (Heald *et al.*, 1979; Theil *et al.*, 1979) and hemerythrin (Co *et al.*, 1982, pers. comm.). Some attention is also being paid to the iron dioxygenases (Que, 1983, pers. comm.). The studies on azidomethemerythrin are important in that they have identified the presence of a μ-oxo bridge between the two Fe atoms of the binuclear active site with an Fe—O distance of 1.76 Å in the presence of several other Fe—O interactions at ~2.0 Å. This was possible even though the two Fe—O shells were not resolved in the FT. Also, these workers have identified Fe—Fe scattering that seems to proceed by a multiple scattering pathway through

2. Copper Proteins

The copper K-absorption edge occurs at ~9.0 keV. Compared with iron, it is relatively easy to study copper proteins by XAS. This is found to be generally the case as one moves to higher atomic number. First of all, the fluorescent yield increases with atomic number so that fluorescence detection becomes more efficient. Also, at higher photon energies there are fewer photons lost through absorption by window materials, sample cells, etc., so that more of the available photons can be used.

Whether it is due to these reasons or due to the relatively varied coordination environments available to biological copper, a considerable amount of effort has been spent on XAS of copper proteins. Some of the initial EXAFS studies of metalloproteins were performed on the common blue (type 1) copper proteins, azurin (Tullius et al., 1978), plastocyanin (Scott et al., 1982; Tullius and Hodgson, 1982, pers. comm.), and stellacyanin (Peisach et al., 1982; Tullius and Hodgson, 1982, pers. comm.). The initial work on azurin (in the oxidized state) indicated the presence of a short (~2.1 Å) Cu—S interaction and two Cu—N interactions at ~2.0 Å. The subsequent crystallographic study of oxidized plastocyanin (Colman et al., 1978) confirmed the presence of a short Cu—S(Cys) bond and indicated the presence of two histidine ligands at the proper distance and a methionine with the thioether sulfur at a long distance (~2.9 Å) from the Cu. Additional work (Tullius and Hodgson, 1982, pers. comm.) on both oxidized and reduced blue copper proteins show that the reduced site is "expanded" with two Cu—N interactions at 1.97 Å (oxidized) and 2.05 Å (reduced), and one Cu—S interaction at 2.11 Å (oxidized) and 2.21 Å (reduced).

One of the more interesting features of this work is the failure to observe the Cu—S(Met) interaction in any of these proteins in either oxidation state. Recent polarized XAS studies on single crystals of Cu(II)-plastocyanin were designed to maximize the contribution of the Cu—S(Met) scattering to the EXAFS by proper orientation of the crystal in the polarized synchrotron radiation beam; but even with this selectivity, no scattering from the methionine sulfur was observed (Scott et al., 1982). This is probably caused by a much larger Debye–Waller σ_{as} damping out this contribution, which is to be interpreted in terms of a very weak Cu(II)—S(Met) bond.

Two separate groups have been involved in examination of Cu XAS of hemocyanin. The results of these studies have been used to propose structures for the dimeric active site in the oxy- and deoxy- states. Hodgson and coworkers (Co and Hodgson, 1981; Co et al., 1981a) propose the structures shown in Fig. 24a, and Spiro and coworkers (Brown *et al.*, 1980) propose the structures in Fig. 24b. In both cases, X represents some endogenous bridge containing a low-Z bridging atom (perhaps tyrosinate ?). The main difference between the hypothesized structures (aside from coordination number differences) is the predicted Cu—Cu distance in the deoxy- derivative. Spiro and coworkers interpret one of the higher-R peaks in the deoxy- FT as a Cu—Cu peak (Brown *et al.*, 1980), whereas Hodgson and coworkers attribute all the "second-shell" peaks to outer atoms of imidazole ligands (Co and Hodgson, 1981). They are able to fit the deoxy- EXAFS data reasonably well with two imidazoles per copper using an imidazole group-fitting technique (Co *et al.*, 1981b). This technique treats the entire imidazole ligand as a group and includes the outer scattering shells with the ligated nitrogen by constraining the phase functions and amplitude functions for the outer shells to track the phase functions and amplitude functions of the ligated nitrogen. The technique has been tested out on $Cu(imid)_4^{2+}$ model compounds as well as blue copper proteins and has been shown to work well in accounting for the total scattering from all shells of an imidazole ligand.

Copper K-absorption edge spectra can be used quite successfully in determining copper oxidation state, as demonstrated with oxy- [Cu(II)] and

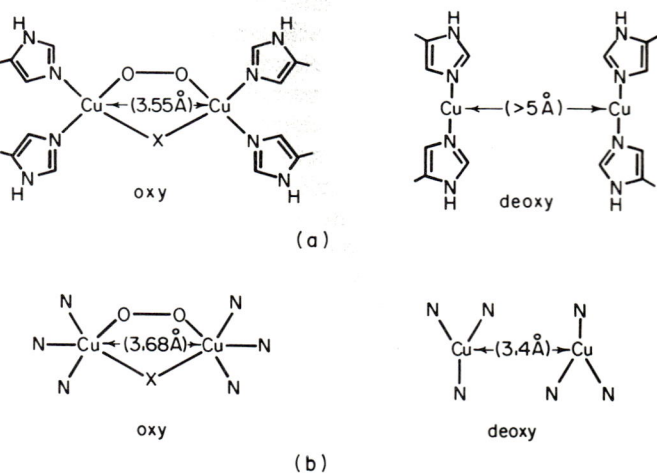

Fig. 24. Proposed structures (based on independent EXAFS studies) for the hemocyanin active site in both oxy- and deoxy- derivatives. (a) Structural models based on EXAFS work by Co *et al.* (1981a) and Co and Hodgson (1981). The identification of the imidazole ligands resulted from use of an imidazole group-fitting procedure. (b) Models based on EXAFS work by Brown *et al.* (1980). The major difference between the two proposed models is the Cu—Cu distance in the deoxy- derivative.

4. X-RAY ABSORPTION SPECTROSCOPY

deoxy- [Cu(I)] hemocyanin. Figure 25 shows a comparison of these two edges with those of aqueous solutions of Cu(II)-imidazole and Cu(I)-imidazole. The Cu(I) edge is characteristically ~3–5 eV to lower energy and exhibits a transition at 8983–8984 eV unique to Cu(I). These features have been used to identify the copper oxidation states in the type 2-depleted (T2D) and the H_2O_2-treated T2D derivatives of *Rhus vernicifera* laccase (LuBien *et al.*, 1981). This study was a rather elegant demonstration by XAS that the absence of the 330-nm band in T2D laccase is a result of reduction rather than loss of the type 3 copper site. The edges of these two laccase derivatives are also shown in Fig. 25.

Other XAS work on copper proteins includes edge studies of the Cu and

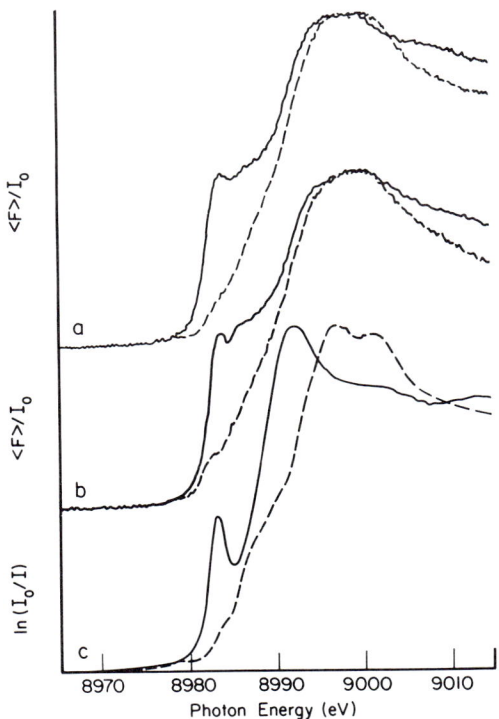

Fig. 25. Comparison of Cu(I) and Cu(II) *K*-edges in copper proteins and copper complexes. Curve a, copper *K*-absorption edges for type-2-depleted (———) and H_2O_2-treated type-2-depleted *Rhus* laccase (– – –); curve b, Cu *K*-edges for oxy- (– – –) and deoxy-hemocyanin (———); curve c, Cu *K*-edges for aqueous solutions of $Cu(II)(imidazole)_4^{2+}$ (– – –) and $Cu(I)(imidazole)_4^+$ (———). In each case, the Cu(I) species shows an edge with inflection at lower energy and a feature at ca. 8984 eF (courtesy M.S.Co).

Zn sites of bovine erythrocyte superoxide dismutase (Blumberg et al., 1978), and edge studies on cobalt substituted into Zn and Cu sites in several metalloproteins (Desideri et al, 1981). (Copper EXAFS of cytochrome c oxidase was discussed in the section on iron proteins.)

3. Molybdenum Enzymes

The main difficulty in performing XAS studies on molybdenum in biological systems is the relatively low photon flux usually available at 20 keV (Mo K edge). With dedicated synchrotron sources running at higher electron energies and currents (or with wigglers), however, this problem is not so severe. The fluorescent yield of Mo is quite high, and 20-keV photons are "hard" enough to pass through most window and sample cell materials without significant absorption loss. In addition, the relative harmonic content is lower than at Fe or Cu, and it is relatively easier to discriminate against, since it is separated from the fundamental by 20 keV. Finally, for most popular monochromator crystals (e.g., Si[220] or Si[400]), the glitches in the Mo region are virtually nonexistent.

The element specificity of the XAS technique has proven extremely useful for examination of some molybdenum enzymes. This is particularly true in the case of nitrogenase, which contains large amounts of iron as well as molybdenum. In the MoFe protein, the Mo : Fe ratio is $\sim 1:16$, whereas in the extracted FeMo cofactor, it is $\sim 1:8$. In both cases, the irons exist in nonidentical environments, making interpretation of Fe EXAFS extremely difficult. However, Mo XAS is specific for the unique Mo site and can yield information regarding the local Mo environment. The Mo EXAFS indicates contributions from both Fe and S scatterers (Cramer et al., 1978b,c), suggesting some kind of sulfur-bridged (Mo, Fe) cluster as the active site (Cramer et al., 1978b; Teo and Averill, 1979).

A significant amount of work has been done on other Mo enzymes, including sulfite oxidase (Berg et al., 1979; Cramer et al., 1979, 1980, 1981), xanthine dehydrogenase (Cramer et al., 1981), and xanthine oxidase (Bordas et al., 1979; Tullius et al., 1979). Major differences are found between the structures of these other enzymes and the Mo site in nitrogenase as determined by XAS. These center around the presence of terminal oxo-(Mo=O) groups in the nonnitrogenase Mo enzymes. A primary indication of the presence of oxo- groups is a shoulder on the Mo K-absorption edge which can be ascribed to a bound-state transition to molecular orbitals associated with the Mo=O moiety. Figure 26 compares Mo K edges for sulfite oxidase, xanthine dehydrogenase, and xanthine oxidase with nitrogenase in its dye-oxidized (active) form and its air-oxidized (inactive) form. It is clear from this comparison that the other enzymes differ from nitrogen-

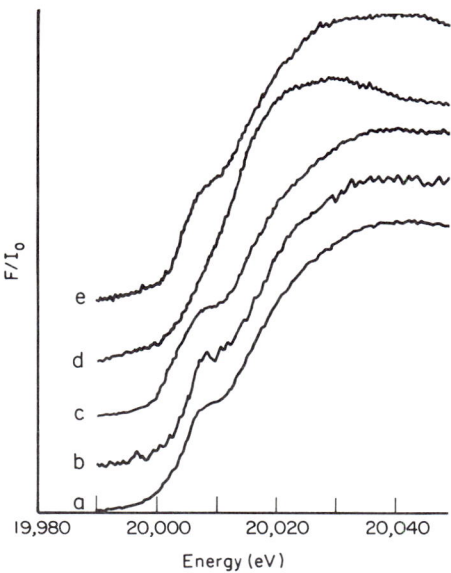

Fig. 26. Comparison of Mo K-edges of several molybdenum-containing enzymes in their oxidized [Mo(VI)] states. Curve a, chicken-liver sulfite oxidase; curve b, chicken-liver xanthine dehydrogenase; curve c, milk xanthine oxidase; curve d, MoFe protein of *Azotobacter vinelandii* nitrogenase, dye oxidized; curve e, nitrogenase MoFe protein, air oxidized. The shoulders on the edges indicate the presence of terminal oxo (Mo = O) groups in all but the dye-oxidized (active) nitrogenase sample.

ase by the presence of terminal oxo- groups and that aerobic inactivation of nitrogenase occurs with formation of Mo=O groups.

The presence of pre-edge shoulders (or peaks) like that exhibited by Mo=O is often indicative of oxo- groups for other metals as well. Several examples including CrO_4^{2-}, MoO_4^{2-}, VO^{2+}, and others, all show resolved pre-edge features ascribable to the presence of oxo- groups.

4. Other Metals

As examples of the types of applications XAS can have in a wide variety of metal-associated biological systems, a few of the other types of metals examined are listed here. The reader is referred to the primary references cited for further information on specific examples.

Calcium EXAFS has been used to study Ca^{2+} binding to phospholipids and several Ca^{2+}-activated or Ca^{2+}-modulated proteins (Powers *et al.*, 1978), as well as the structures of Ca^{2+} environments in bone minerals (Miller *et al.*,

1981). Vanadium edge and EXAFS spectra have been used to identify the form of the vanadium complex concentrated by ascidians in vanadocytes (Tullius et al., 1980). Zinc EXAFS has been used to identify sulfur-containing ligands in the Zn enzyme aspartate transcarbamylase (Phillips et al., 1982). Gold edges and EXAFS are being used to examine the structure and oxidation states of gold complexes in use as antiarthritic drugs (Mazid et al., 1980; Elder, 1982, pers. comm.).

V. Conclusion

The primary purpose of this chapter was to provide both the novice and the occasional user of x-ray absorption spectroscopy with instructions and information regarding theoretical and, more important, practical aspects of the application of XAS in solving biological problems. It is hoped that the first part of the chapter was able to accomplish this goal. The specific applications discussed in the last part were selected to indicate the range of problems for which XAS can be of use, as well as to stress the fact that the utility of the technique is just *beginning* to bear fruit. This is best reflected in the use of edges to identify coordination geometries, a field that will certainly make great strides in the near future.

References

Ashley, C. A., and Doniach, S. (1975). *Phys. Rev. B* **11**, 1279–1288.
Averill, B. A., Herskovitz, T., Holm, R. H., and Ibers, J. A. (1973). *J. Am. Chem. Soc.* **95**, 3523–3534.
Azoulay, J. (1980). *In* "Laboratory EXAFS Facilities—1980," *AIP Conf. Proc. No. 64* (E. A. Stern, ed.), pp. 93–95. American Institute of Physics, New York.
Bearden, J. A., and Burr, A. F. (1967). *Rev. Mod. Phys.* **39**, 125–142.
Beni, G., Lee, P. A., and Platzman, P. M. (1976). *Phys. Rev. B* **13**, 5170–5178.
Berg, J. M., Hodgson, K. O., Cramer, S. P., Corbin, J. L., Elsberry, A., Pariyadath, N., and Stiefel, E. I. (1979). *J. Am. Chem. Soc.* **101**, 2774–2776.
Bienenstock, A. (1980). *Nucl. Instrum. Methods* **172**, 13–20.
Bienenstock, A. (1981). *In* "EXAFS Spectroscopy: Techniques and Applications" (B.-K. Teo and D. C. Joy, eds.), pp. 185–196. Plenum, New York.
Blumberg, W. E., Peisach, J., Eisenberger, P., and Fee, J. A. (1978). *Biochemistry* **17**, 1842–1846.
Boland, J. J., Crane, S. E., and Baldeschwieler, J. D. (1982). *J. Chem. Phys.* **77**, 142–153.
Bordas, J., Bray, R. C., Garner, C. D., Gutteridge, S., and Hasnain, S. S. (1979). *J. Inorg. Biochem.* **11**, 181–186.
Brown, J. M., Powers, L., Kincaid, B., Larrabee, J. A., and Spiro, T. G. (1980). *J. Am. Chem. Soc.* **102**, 4210–4216.
Chan, S. I., and Gamble, R. C. (1978). *Methods Enzymol.* **54**, 323–345.

Chan, S. I., Hu, V. W., and Gamble, R. C. (1978). *J. Mol. Struct.* **45**, 239–266.
Chance, B., Angiolillo, P., Yang, E. K., and Powers, L. (1980). *FEBS Lett.* **112**, 178–182.
Co, M. S., and Hodgson, K. O. (1981). *J. Am. Chem. Soc.* **103**, 3200–3201.
Co, M. S., Hodgson, K. O., Eccles, T. K., and Lontie, R. (1981a). *J. Am. Chem. Soc.* **103**, 984–986.
Co, M. S., Scott, R. A., and Hodgson, K. O. (1981b). *J. Am. Chem. Soc.* **103**, 986–988.
Cohen, G. G., Fischer, D. A., Colbert, J., and Shevchik, N. J. (1980). *Rev. Sci. Instrum.* **51**, 273–277.
Colman, P. M., Freeman, H. C., Guss, J. M., Murata, M., Norris, V. A., Ramshaw, J. A. M., and Venkatappa, M. P. (1978). *Nature* **272**, 319–324.
Cramer, S. P., and Hodgson, K. O. (1979). *In* "Progress in Inorganic Chemistry" (S. J. Lippard, ed.), Vol. 25, pp. 1–39. Wiley, New York.
Cramer, S. P., and Scott, R. A. (1981). *Rev. Sci. Instrum.* **52**, 395–399.
Cramer, S. P., Hodgson, K. O., Stiefel, E. I., and Newton, W. E. (1978a). *J. Am. Chem. Soc.* **100**, 2748–2761.
Cramer, S. P., Hodgson, K. O., Gillum, W. O., and Mortensen, L. E. (1978b). *J. Am. Chem. Soc.* **100**, 3398–3407.
Cramer, S. P., Gillum, W. O., Hodgson, K. O., Mortensen, L. E., Stiefel, E. I., Chisnell, J. R., Brill, W. J., and Shah, V. K. (1978c). *J. Am. Chem. Soc.* **100**, 3814–3819.
Cramer, S. P., Dawson, J. H., Hodgson, K. O., and Hager, L. P. (1978d). *J. Am. Chem. Soc.* **100**, 7282–7290.
Cramer, S. P., Gray, H. B., Rajagopalan, K. V. (1979). *J. Am. Chem. Soc.* **101**, 2772–2774.
Cramer, S. P., Gray, H. B., Scott, N. S., Barber, M., and Rajagopalan, K. V. (1980). *In* "Molybdenum Chemistry of Biological Significance" (W. E. Newton and S. Otsuka, eds.), pp. 157–168. Plenum, New York.
Cramer, S. P., Wahl, R., and Rajagopalan, K. V. (1981). *J. Am. Chem. Soc.* **103**, 7721–7727.
Desideri, A., Comin, F., Morpurgo, L., Cocco, D., Calabrese, L., Mondovi, B., Maret, W., and Rotilio, G. (1981). *Biochim. Biophys. Acta* **661**, 312–315.
Eisenberger, P. (1981). *Hyperfine Interact.* **10**, 915–929.
Eisenberger, P., and Brown, G. S. (1979). *Solid State Commun.* **29**, 481–484.
Eisenberger, P., and Kincaid, B. M. (1978). *Science* **200**, 1441–1447.
Eisenberger, P., Shulman, R. G., Kincaid, B. M., Brown, G. S., and Ogawa, S. (1978). *Nature* **274**, 30–34.
Gamble, R. C. (1980a). *In* "Laboratory EXAFS Facilities—1980, "*AIP Conf. Proc. No. 64* (E. A. Stern, ed.), pp. 113–116. American Institute of Physics, New York.
Gamble, R. C. (1980b). *AIP Conf. Proc. No. 64* (E. A. Stern, ed.), pp. 123–126.
Gamble, R. C., Baldeschwieler, J. D., and Giffin, C. E. (1979). *Rev. Sci. Instrum.* **50**, 1416–1420.
Gurman, S. J., and Pendry, J. B. (1976). *Solid State Commun.* **20**, 287–290.
Haensel, R. (1980). *In* "Laboratory EXAFS Facilities—1980," *AIP Conf. Proc. No. 64* (E. A. Stern, ed.), pp. 73–82. American Institute of Physics, New York.
Hahn, J. E., Scott, R. A., Hodgson, K. O., Doniach, S., Desjardins, S., and Solomon, E. I. (1982). *Chem. Phys. Lett.* **88**, 595–598.
Hastings, J. B. (1981). *In* "EXAFS Spectroscopy: Techniques and Applications" (B.-K. Teo and D. C. Joy, eds.), pp. 205–211. Plenum, New York.
Hastings, J. B., Eisenberger, P., Lengeler, B., and Perlman, M. L. (1979). *Phys. Rev. Lett.* **43**, 1807–1810.
Hayes, T. M., Sen, P. N., and Hunter, S. H. (1976). *In* "Structure and Excitations of Amorphous Solids," *AIP Conf. Proc. No. 31* (G. Lucovsky and F. L. Galeener, eds.), pp. 166–170. American Institute of Physics, New York.

Hayes, T. M., Boyce, J. B., and Beeby, J. L. (1978). *J. Phys. C* **11**, 2931–2937.
Heald, S. M., and Stern, E. A. (1977). *Phys. Rev. B* **16**, 5549–5559.
Heald, S. M., Stern, E. A., Bunker, B., Holt, E. M., and Holt, S. L. (1979). *J. Am. Chem. Soc.* **101**, 67–73.
Hu, V. W., Chan, S. I., and Brown, G. S. (1977a). *Proc. Natl. Acad. Sci. USA* **74**, 3821–3825.
Hu, V. W., Chan, S. I., and Brown, G. S. (1977b). *FEBS Lett.* **84**, 287–290.
Jaklevic, J., Kirby, J. A., Klein, M. P., Robertson, A. S., Brown, G. S., and Eisenberger, P. (1977). *Solid State Commun.* **23**, 679–682.
Jensen, L. H. (1978). *In* "Energy Conservation in Biological Membranes" (G. Schäfer and M. Klingenberg, eds.), pp. 74–83. Springer-Verlag, Berlin.
Khalid, S., Emrich, R., Dujari, R., Shultz, J., and Katzer, J. R. (1982). *Rev. Sci. Instrum.* **53**, 22–33.
Kincaid, B. M., and Shulman, R. G. (1980). *In* "Advances in Inorganic Biochemistry" (G. L. Eichhorn and L. G. Marzilli, eds.), Vol. 2, pp. 303–309. Elsevier, Amsterdam.
Kincaid, B. M., Eisenberger, P., Hodgson, K. O., and Doniach, S. (1975). *Proc. Natl. Acad. Sci. USA* **72**, 2340–2342.
Knapp, G. S., Chen, H., and Klippert, T. E. (1978). *Rev. Sci. Instrum.* **49**, 1658–1666.
Korszun, Z. R., Moffat, K., Frank, K., and Cusanovich, M. A. (1982). *Biochemistry* **21**, 2253–2258.
Kossel, W. (1920). *Z. Phys.* **1**, 119–134.
Kronig, R. deL. (1931). *Z. Phys.* **70**, 317–323.
Kronig, R. deL. (1932). *Z. Phys.* **75**, 191. 468.
Kutzler, F. W., Natoli, C. R., Misemer, D. K., Doniach, S., and Hodgson, K. O. (1980). *J. Chem. Phys.* **73**, 3274–3288.
Labhardt, A., and Yuen, C. (1979). *Nature* **277**, 150–151.
Lane, R. W., Ibers, J. A., Frankel, R. B., Papaefthymiou, G. C., and Holm, R. H. (1977). *J. Am. Chem. Soc.* **99**, 84–98.
Lee, P. A. (1981). *In* "EXAFS Spectroscopy: Techniques and Applications" (B.-K. Teo and D. C. Joy, eds.), pp. 5–11. Plenum, New York.
Lee, P. A., and Beni, G. (1977). *Phys. Rev. B* **15**, 2862–2883.
Lee, P. A., and Pendry, J. B. (1975). *Phys. Rev. B* **11**, 2795–2811.
Lee, P. A., Citrin, P. H., Eisenberger, P., and Kincaid, B. M. (1981). *Rev. Mod. Phys.* **53**, 769–806.
Leigh, J. B., and Rosenbaum, G. (1976). *Annu. Rev. Biophys. Bioeng.* **5**, 239–270.
Lindau, I., and Winick, H. (1976). *In* "Scientific and Industrial Applications of Small Accelerators—Fourth Conference," pp. 215–224. IEEE, New York.
LuBien, C. D., Winkler, M. E., Thamann, T. J., Scott, R. A., Co, M. S., Hodgson, K. O., and Solomon, E. I. (1981). *J. Am. Chem. Soc.* **103**, 7014–7016.
Mallozzi, P. J., Schwerzel, R. E., Epstein, H. M., and Campbell, B. E. (1979). *Science* **206**, 353–355.
Mallozzi, P. J., Schwerzel, R. E., and Epstein, H. M. (1980). *In* "Laboratory EXAFS Facilities—1980," *AIP Conf. Proc. No. 64* (E. A. Stern, ed.), pp. 96–98. American Institute of Physics, New York.
Manthiram, A., Sarode, P. R., Madhusudan, W. H., Gopalakrishnan, J., and Rao, C. N. R. (1980). *J. Phys. Chem.* **84**, 2200–2203.
Marcus, M., Powers, L. S., Storm, A. R., Kincaid, B. M., and Chance, B. (1980). *Rev. Sci. Instrum.* **51**, 1023–1029.
Matsushita, T. (1980). *In* "Laboratory EXAFS Facilities—1980," *AIP Conf. Proc. No. 64* (E. A. Stern, ed.), pp. 109–110. American Institute of Physics, New York.

Mayerle, J. J., Denmark, S. E., DePamphilis, B. V., Ibers, J. A., and Holm, R. H. (1975). *J. Am. Chem. Soc.* **97**, 1032–1045.
Mazid, M. A., Razi, M. T., Sadler, P. J., Greaves, G. N., Gurman, S. J., Koch, M. H. J., and Phillips, J. C. (1980). *J. Chem. Soc. Chem. Commun.,* 1261–1263.
Miller, R. M., Hukins, D. W. L., Hasnain, S. S., and Lagarde, P. (1981). *Biochem. Biophys. Res. Commun.* **99**, 102–106.
Mills, D., and Pollock, V. (1980). *Rev. Sci. Instrum.* **51**, 1664–1668.
Pease, D. M., Azaroff, L. V., Vaccaro, C. K., and Hines, W. A. (1979). *Phys. Rev. B* **19**, 1576–1581.
Peisach, J., Powers, L., Blumberg, W. E., and Chance, B. (1982). *Biophys. J.* **38**, 277–285.
Pendry, J. B. (1981). *Daresbury Lab. Rep.* **17**, 5–12.
Perutz, M. F. (1970). *Nature* **228**, 726–739.
Perutz, M. F., Hasnain, S. S., Duke, P. J., Sessler, J. L., and Hahn, J. E. (1982). *Nature* **295**, 535–538.
Phillips, J. C. (1981). *J. Phys. E* **14**, 1425–1428.
Phillips, J. C., Bordas, J., Foote, A. M., Koch, M. H. J., and Moody, M. F. (1982). *Biochemistry* **21**, 830–834.
Powers, L., Eisenberger, P., and Stamatoff, J. (1978). *Ann. N.Y. Acad. Sci.* **307**, 113–123.
Powers, L., Blumberg, W. E., Chance, B., Barlow, C. H., Leigh, J. S., Smith, J., Yonetani, T., Vik, S., and Peisach, J. (1979). *Biochim. Biophys. Acta* **546**, 520–538.
Powers, L., Chance, B., Ching, Y., and Angiolillo, P. (1981). *Biophys. J.* **34**, 465–498.
Sandstrom, D. R., and Fine, J. M. (1980). *In* "Laboratory EXAFS Facilities—1980," *AIP Conf. Proc. No. 64* (E. A. Stern, ed.), pp. 127–128. American Institute of Physics, New York.
Sano, M., Taniguchi, K., and Yamatera, H. (1980). *Chem. Lett.,* 1285–1286.
Sarode, P. R., Ramasesha, S., Madhusudan, W. H., and Rao, C. N. R. (1979). *J. Phys. C* **12**, 2439–2445.
Sayers, D. E. (1971). Ph.D. Dissertation, University of Washington.
Sayers, D. E., Lytle, F. W., and Stern, E. A. (1970). *In* "Advances in X-Ray Analysis," Vol. 13, pp. 248–271. Plenum, New York.
Sayers, D. E., Lytle, F. W., and Stern, E. A. (1971). *Phys. Rev. Lett.* **27**, 1204–1207.
Sayers, D. E., Lytle, F. W., and Stern, E. A. (1972). *J. Non-Cryst. Solids* **8-10**, 401–407.
Sayers, D. E., Lytle, F. W., and Stern, E. A. (1974). *In* "Amorphous and Liquid Semiconductors," *Proc. 5th Int. Conf. Amorph. Liq. Semicond.* (J. Stuke and W. Brenig, eds.), pp. 403–412. Taylor and Francis, London.
Sayers, D. E., Stern, E. A., and Herriott, J. R. (1976). *J. Chem. Phys.* **64**, 427–428.
Scott, R. A. (1982). *In* "The Biological Chemistry of Iron" (H. B. Dunford, D. H. Dolphin, K. N. Raymond, and L. C. Sieker, eds.), pp. 475–484. Reidel, Boston.
Scott, R. A., Cramer, S. P., Shaw, R. W., Beinert, H., and Gray, H. B. (1981). *Proc. Natl. Acad. Sci. USA* **78**, 664–667.
Scott, R. A., Hahn, J. E., Doniach, S., Freeman, H. C., and Hodgson, K. O. (1982). *J. Am. Chem. Soc.* **104**, 5364–5369.
Shulman, R. G., Eisenberger, P., Blumberg, W. E., and Stombaugh, N. A. (1975). *Proc. Natl. Acad. Sci. USA* **72**, 4003–4007.
Shulman, R. G., Yafet, Y., Eisenberger, P., and Blumberg, W. E. (1976). *Proc. Natl. Acad. Sci USA* **73**, 1384–1388.
Shulman, R. G., Eisenberger, P., Teo, B.-K., Kincaid, B. M., and Brown, G. S. (1978a). *J. Mol. Biol.* **124**, 305–321.
Shulman, R. G., Eisenberger, P., and Kincaid, B. M. (1978b). *Annu. Rev. Biophys. Bioeng.* **7**, 559–578.

Siegbahn, M. (1925). "The Spectroscopy of X-Rays," pp. 131–149. Oxford University Press, London.
Stern, E. A. (1974). *Phys. Rev. B* **10**, 3027–3037.
Stern, E. A. (1980). *In* "Laboratory EXAFS Facilities—1980," *AIP Conf. Proc. No. 64* (E. A. Stern, ed.), pp. 39–50. American Institute of Physics, New York.
Stern, E. A., and Heald, S. M. (1979). *Rev. Sci. Instrum.* **50**, 1579–1582.
Stern, E. A., and Kim, K. (1981). *Phys. Rev. B* **23**, 3781–3787.
Stern, E. A., Sayers, D. E., and Lytle, F. W. (1975). *Phys. Rev. B* **11**, 4836–4846.
Stern, E. A., Heald, S. M., and Bunker, B. A. (1979). *Phys. Rev. Lett.* **42**, 1372–1375.
Stern, E. A., Bunker, B. A., and Heald, S. M. (1980). *Phys. Rev. B* **21**, 5521–5539.
Teo, B.-K. (1980). *Acc. Chem. Res.* **13**, 412–419.
Teo, B.-K. (1981a). *In* "EXAFS Spectroscopy: Techniques and Applications" (B.-K. Teo and D. C. Joy, eds.), pp. 13–58. Plenum, New York.
Teo, B.-K. (1981b). *J. Am. Chem. Soc.* **103**, 3990–4001.
Teo, B.-K., and Averill, B. A. (1979). *Biochem. Biophys. Res. Commun.* **88**, 1454–1461.
Teo, B.-K., and Lee, P. A. (1979). *J. Am. Chem. Soc.* **101**, 2815–2832.
Teo, B.-K., Lee, P. A., Simons, A. L., Eisenberger, P., and Kincaid, B. M. (1977). *J. Am. Chem. Soc.* **99**, 3854–3856.
Teo, B.-K., Shulman, R. G., Brown, G. S., and Meixner, A. E. (1979). *J. Am. Chem. Soc.* **101**, 5624–5631.
Theil, E. C., Sayers, D. E., and Brown, M. A. (1979). *J. Biol. Chem.* **254**, 8132–8134.
Tullius, T. D., Frank, P., and Hodgson, K. O. (1978). *Proc. Natl. Acad. Sci. USA* **75**, 4069–4073.
Tullius, T. D., Kurtz, D. M., Conradson, S. D., and Hodgson, K. O. (1979). *J. Am. Chem. Soc.* **101**, 2776–2779.
Tullius, T. D., Gillum, W. O., Carlson, R. M. K., and Hodgson, K. O. (1980). *J. Am. Chem. Soc.* **102**, 5670–5676.
Watenpaugh, K. D., Sieker, L. C., and Jensen, L. H. (1980). *J. Mol. Biol.* **138**, 615–633.
Watson, R. E., and Perlman, M. L. (1978). *Science* **199**, 1295–1302.

5

Macromolecular Crystallography

Keith Moffat

Section of Biochemistry, Molecular and Cell Biology
Cornell University
Ithaca, New York

I.	INTRODUCTION	364
	A. Nature of the Problem	364
	B. The Crystallographic Approach: Advantages and Limitations	365
	C. Outline	366
II.	CRYSTALLIZATION OF MACROMOLECULES	367
	A. Strategies	367
	B. Apparatus and Techniques	371
III.	X-RAY DIFFRACTION FROM CRYSTALS	374
	A. Introduction: The Electron-Density Equation	374
	B. Crystals and Symmetry	376
	C. Diffraction by Crystals	379
	D. The Patterson Function	386
	E. Conclusion	388
	F. Outline of Experimental Aspects	389
IV.	PHASE DETERMINATION	391
	A. Introduction	391
	B. Isomorphous Replacement	391
	C. Anomalous Scattering	395
	D. Treatment of Errors	399
	E. Molecular Replacement and Noncrystallographic Symmetry	401
V.	HEAVY-ATOM DERIVATIVES	408
	A. Preparation	408
	B. Location: The Difference Patterson Function	409
	C. Refinement of Heavy-Atom Parameters	411
	D. Difference Fouriers	413
VI.	THE ELECTRON-DENSITY MAP	414
	A. Interpretation	414
	B. Refinement	416
	C. Relation to Biochemical Problems	421
VII.	NEW DIRECTIONS	423
	A. Uses of Synchrotron Radiation	423
	B. Partial Structure Resolved Anomalous Scattering	425
	C. Membrane Proteins	426

APPENDIX A. THE BASIC MATHEMATICS OF
           CRYSTALLOGRAPHY   428
APPENDIX B. A SPECIFIC STRATEGY FOR PROTEIN
           CRYSTALLIZATION   430
APPENDIX C. A CRYSTALLOGRAPHIC PAPER PRIMER   431
APPENDIX D. SUGGESTIONS FOR FURTHER READING   432
REFERENCES   433

I. Introduction

A. Nature of the Problem

It is generally accepted that an understanding of the mechanism of action — the function — of biological macromolecules requires elucidation of their three-dimensional arrangement of atoms — their structure. That is, physiological processes such as respiration, motility, hormone recognition, energy transduction, and DNA replication will be explained only by examination of the structures of the macromolecules involved, of the way in which they interact with each other and with small molecules, and of the equilibrium free energies and free energies of activation involved. Needless to say, this is a tall order; no such physiological process is yet understood in molecular detail. However, much of the molecular basis for the transport of oxygen and carbon dioxide by the red blood cell protein, hemoglobin, and the influence on transport of pH, chloride-ion concentration, and 2,3-diphosphoglycerate, has been revealed by structural examination of hemoglobin and related compounds at the atomic level. This example suggests that "molecular physiology" is an attainable goal. Were it not, physical biochemists would be confined to examining macromolecular structure for structure's sake. While macromolecular structures may reveal the principles of folding and have a beauty of their own, molecular sculpture, they are more often viewed as a means to a functional end, molecular architecture.

X-ray crystallography remains the single technique capable of providing information about the complete atomic structure of a macromolecule. Electron microscopy, in principle a rival technique, has been limited by sample staining and preservation difficulties to somewhat lower resolution. In the most favorable case of bacteriorhodopsin (Henderson and Unwin, 1975), the application of sophisticated image reconstruction techniques to electron micrographs of unstained, highly ordered, two-dimensional arrays of membrane proteins has revealed the location of α helices. Extension of these techniques to higher resolution, adequate to reveal, for example, the retinal in bacteriorhodopsin, is under way, but is a most challenging problem. Atomic resolution is a more distant goal.

On the other hand, ingenious nuclear magnetic resonance (NMR) techniques are beginning to yield sufficiently numerous and accurate internuclear distances to permit triangulation and hence determination of macromolecular structure in solution (see Chapter 1). These NMR techniques have been applied so far only to small proteins of (crystallographically) known structure, thus seeking to validate the techniques and to demonstrate that no gross structural changes have occurred on crystallization. Whether they will prove adequate to predict three-dimensional structure when applied to macromolecules of unknown crystal structure is a key question, now being tackled.

Early confidence that knowledge of the amino acid sequence of proteins, or the nucleotide sequence of nucleic acids, would lead to successful prediction of atomic structure (as in principle it must) has been severely shaken. Despite the experimental and theoretical efforts of numerous biophysical chemists, no satisfactory macromolecular folding algorithm has been devised. To be most useful, a folding algorithm would have to generate a set of model atomic coordinates which could be refined against restricted experimental data (such as x-ray diffraction amplitudes from crystals of a native macromolecule alone, or a set of internuclear distances from NMR studies) and which would converge to the correct structure. That is, the initial model structure would have to lie in the correct free-energy well containing the global minimum, and not in a well containing a local minimum, from which the global minimum might be inaccessible. Unfortunately, no complete protein or tRNA structure has been successfully predicted in this way (Schulz et al., 1974; Matthews, 1975; Schulz and Schirmer, 1979), although limited features of the real structures, such as α helices in proteins and tertiary base pairing in tRNAs, have been correctly identified. Failure to incorporate the influence of solvent adequately may lie at the heart of this problem; however, recognition of that fact hardly advances a solution.

B. The Crystallographic Approach: Advantages and Limitations

The advantages of macromolecular crystallography may be readily stated. It is (as noted previously) the only technique in practice capable of yielding the complete, three-dimensional structure; it is reasonably objective, in the sense that derivation of an electron density map does not require critical subjective decisions by the crystallographer, which might lead to a totally erroneous structure determination; related structures may be solved more readily, by molecular replacement or difference Fourier techniques (see Sections IV.E and V.D); and molecules in the crystal can usually be demonstrated to be biologically active. That is, chemistry can be conducted in the crystalline state to verify that the structure determined is active and biologically relevant.

The disadvantages must also be recognized. Obtaining crystals suitable for structure determination is challenging (see Section II); the crystal structure is determined under nonphysiological conditions; the accuracy of the atomic coordinates may not be high enough to answer the most urgent chemical and biochemical questions, even after extensive crystallographic refinement (see Section VI.B); the relationship between the atomic structure and the mechanism of action is often subtle (see Section VI.C); and only quasi-static structures can be examined, which are both a space average (over all the molecules in the crystal, typically 10^{15}) and a time average (over the tens of hours required for x-ray data collection).

The last is potentially the most serious disadvantage. Although the final output of a crystallographic analysis is usually a single set of atomic coordinates, it is essential to realize that the structures of macromolecules in the cell, in solution, and in the crystal are continuously varying: side chains, bases, and the polypeptide or polynucleotide backbones are all in motion. This is an automatic consequence of the fact that the tertiary structures of macromolecules are stabilized by very large numbers of weak noncovalent interactions: hydrogen bonds, electrostatic interactions, van der Waals forces, and solvent-related "hydrophobic" forces. The free energy of stabilization of any one of these interactions between individual groups is quite low, comparable with kT. Hence, these interactions are continually being made and broken; the structure is dynamic, fluctuating. The time scale of these fluctuations may range from 10^{-15} sec, characteristic of interatomic vibrations, through 10^{-9} sec, associated with, for example, cooperative motions of α helices or motion of domains with respect to each other, through 10^{-6} sec, more characteristic of changes in quaternary structure. Without these structural fluctuations, biological processes such as catalysis would be impossible. The "brass model mentality," which holds that macromolecules have a *single,* crystallographically determinable structure, is inappropriate; the challenge is to infer the nature of the structural fluctuations — macromolecular dynamics — from examination of the space and time-average crystal structure — macromolecular statics.

Finally, it must be acknowledged that a macromolecular structure determination is still lengthy (measured in the 3–10 person-years range for an "average" macromolecule if all goes well), employs reasonably sophisticated apparatus, and involves extensive computing on at least medium-size computers.

C. Outline

The remainder of this chapter deals with some of the details of macromolecular crystallography, in an order which approximates that followed in a structure determination: crystallization (Section II), phase determination (Section IV), and interpretation and refinement of the electron-density map (Section VI). The theory underlying x-ray crystallography is dealt with briefly (Section III and Appendix A), largely to introduce the terminology, as this theory is treated extensively in numerous textbooks (see Appendix D). Its extension to phase determination in macromolecules (Section IV and V) is also straightforward. Since the subject is still evolving, some important new developments are considered in Section VII. Experimental details are considered only briefly except for the all-important area of crystallization (Section II). Crystallization of macromolecules is often viewed as a black art, in which muttering appropriate incantations at the critical moment plays a major role. While divine or other assistance should not be rejected, the general chemical procedures for crystallization are well understood and easy to conduct quickly. With hope of encouraging more scientists to attempt crystallization of their favorite macromolecule, an explicit experimental protocol, based on the general principles outlined in Section II, is given in Appendix B. Finally, Appendix C poses key questions which should be applied to any crystallographic results. Like all other techniques, macromolecular crystallography is fallible; its limitations, and possible sources of error, should not be ignored.

II. Crystallization of Macromolecules

A. Strategies

The procedures necessary to grow crystals of macromolecules are straightforward, simple to execute, and do not require large amounts of time or complicated apparatus. An extensive and valuable review article and a recent book by McPherson (1976a, 1982) are devoted to the procedures necessary to grow crystals of macromolecules; much of what follows is based on a shorter article, aimed at biologists and other noncrystallographers, by Moffat (1980). Although the procedures themselves are straightforward, the difficulties arise (as we shall see) not in executing them, but in achieving success: reproducible, large, well-ordered crystals suitable for high-resolution x-ray analysis.

Crystallization of macromolecules occurs when their concentration just exceeds the solubility under the given solution conditions; that is, the solution is very slightly supersaturated. Further, the free energy of the macromolecules in the ordered, crystalline state must be lower than in the disordered, amorphous state, to ensure that formation of crystals is thermodynamically favored over amorphous precipitate. All techniques for crystallization are therefore directed towards adjusting the solution conditions so that intermolecular interactions are maximized and insolubility smoothly approached, in a manner which (it is hoped) will favor crystallization over precipitation. As we shall see, numerous variables affect solubility, and one cannot predict exactly what these solution conditions will be, nor even predict with confidence that a certain macromolecule will crystallize under *any* solution conditions. Part of the difficulty stems from a remarkable lack of understanding of the physical chemistry of macromolecular crystal growth. The initial crystallization of each new macromolecule is therefore an independent problem. Knowledge of the *general* principles of crystallization does not provide a *specific* strategy which is guaranteed to work.

Given this degree of uncertainty, it is always necessary to explore a wide range of solution conditions, seeking just those which will yield appropriate crystals—a shotgun rather than a hunting rifle approach. This imposes the first two general requirements: the availability of appreciable amounts of material; and (since there is never enough to satisfy crystallographers) the use of micro techniques to make the best use of the limited supply. In the absence of extraordinary luck, it is probably not worth a serious attempt to crystallize a macromolecule with the ultimate intention of solving its structure unless several milligrams are available. The minimum quantity used in a complete atomic-structure determination thus far is 1.5 mg for a snake-venom toxin, a particularly favorable case. The first requirement may be relaxed if a macromolecule is particularly stable or if it does not precipitate irreversibly, since a small amount can then be recycled through numerous crystallization attempts. Unfortunately, many of the most biochemically interesting molecules are available only in small amounts, and are not noted for their stability over the several weeks it may take for crystallization to occur.

Since the essence of a crystal is order, crystallization is promoted if the molecules in solution are themselves "ordered." That is, they must satisfy a third general requirement, that they be both chemically and structurally uniform and remain so over the period of crystallization. Although certain proteins such as hemoglobin and lysozyme will crystallize from quite impure solutions (and for which crystallization is therefore a form of purification), it is generally found that the higher the purity, the greater the probability of crystallization and the higher the quality of the resultant crystals. If a macromolecule is demonstrated to be chemically heterogeneous by sensitive chromatographic tests such as gel electrophoresis or isoelectric focusing, it will almost always be advantageous to purify it to homogeneity before attempting crystallization. Structural heterogeneity of a chemically homogeneous molecule is harder both to detect and to modify. Some forms of structural heterogeneity, such as the existence of a monomer—dimer or other reversible aggregation equilibrium, do not rule out crystallization, though they generally have a profound effect on the types of crystal obtained and may produce a particularly strong dependence of crystal quality on concentration. Other forms of structural heterogeneity such as the presence of even small amounts of irreversibly denatured molecules may greatly hinder crystallizability. Further difficulties may arise if an otherwise homogeneous molecule is unusually flexible and the various conformers have closely similar solubilities, as is indeed likely.

If crystals are obtained, it must be demonstrated that these are truly macromolecular crystals and not crystals of a small molecule, one of the buffer components. Inadvertent crystallization of buffer components is very common; fortunately, even small buffer crystals can be rapidly distinguished from macromolecular crystals by their x-ray diffraction pattern, by their much greater mechanical stability (macromolecular crystals are very fragile because of the weak intermolecular interactions, and are readily crushed when prodded with a dissecting needle), and by their failure to react with protein or nucleic acid stains. The crystals must also be shown to contain the desired, intact macromolecule; it is embarrassing to discover, perhaps years later, that the beautiful crystals in fact contain only a large fragment of the molecule! The more chemically heterogeneous the crystallization solution, the more grounds for caution. Immunological tests alone are likely to be insufficient; careful biochemical analysis of the initial crystallization solution and of dissolved crystals is called for.

Finally, supposing crystals of the macromolecule are indeed obtained, they have to satisfy three further requirements to be suitable for high-resolution structure determination. First, they must be obtained reproducibly, an obvious requirement which has turned out to be difficult to ensure. As McPherson (1976a) notes, many structure determinations have gone through frustrating periods of crystal shortage, when previously established crystallization conditions obstinately refuse to yield further suitable crystals. This is generally attributed to slight variations in structural homogeneity from preparation to preparation, although variation in other solution conditions, such as trace-metal content of the buffer, is often a contributor. For example, 3 M $(NH_4)_2SO_4$, enzyme grade, may easily contain 20 μM calcium. Second, the crystals must be quite large, with an average dimension not less than 0.3 mm (although this depends somewhat on the shape of the crystal and on how strongly it diffracts x rays); and third, they must be well ordered. These last two requirements can also be hard to satisfy; indeed, the second stage of crystallization is devoted to enlarging the microcrystals, which are the usual first products of crystallization attempts, and maximizing their order. It is usually (though not always) found that the slower the crystal growth, the larger the resultant crystals and the better their order; many crystals take weeks or months to grow.

The variables which affect the solubility of macromolecules, and hence crystallization, are numerous. For the macromolecule the most important variables include its purity, stability, flexibility, and concentration; for the solution, its pH, ionic strength, the nature of the buffer and its concentration, the nature of ions and their concentration, the nature of the precipitant and its concentration, and temperature. Other variables such as light levels, presence or absence of substrates, effectors, or other physiologically relevant small molecules, freedom from or presence of vibration (or even whether or not the crystallographer and his colleagues in the same laboratory are pipe smokers!) can influence crystallization.

Given this degree of complexity, a systematic investigation of the entire crystallization phase diagram is not feasible. For example, suppose one wishes to explore the crystallization of a hormone over the pH range from 4 to 9 in increments of 0.5 pH units (11 values); 4 different buffers at each pH, of 3 different concentrations (12 values); 4 different precipitants such as $(NH_4)_2SO_4$, 2-methyl-2,4-pentane diol, polyethylene glycol-1000, and ethanol, each at 6 concentrations; 4 different counterions such as K^+, Na^+, Ca^{2+}, and Mg^{2+}; 3 trace metals such as Cu, Zn, or Mn; and temperature in the range from 4° to 37°C (say, 4 values). This straightforward scheme produces $11 \times 12 \times 4 \times 6 \times 4 \times 3 \times 4$, or 152,064 separate conditions! If each experiment requires 3 μl of hormone (a minimum) at a concentration of 15 μg/μl (about average), then this scheme uses 6.8 gm of hormone, which is unlikely to be available.

If this were the whole story, then *no* macromolecule would have been crystallized, except by luck. What saves the day is that macromolecules crystallize in many regions of the solubility surface; perhaps 1 in 100 experiments (or 1500 of the 150,000) would yield crystals of some sort, and solution conditions in the vicinity of these initial conditions can then be explored in more detail, seeking large, well-ordered crystals.

Since, as noted, one cannot predict in advance *which* one experiment in one hundred would yield crystals, initial crystallization strategy is subject to two restrictions. The first to explore only those solution conditions such as pH range and temperature where the macromolecule is already known to be stable, and to use only those precipitants known not to denature it irreversibly, for which preliminary solubility information is available. Such preliminary information is usually derived from the purification scheme, which may involve such steps as ammonium sulfate fractionation, precipitation with organic solvents or at low ionic strength, or isoelectric precipitation. The second is to make use of the fact that relatively few precipitants have been employed in the growth of crystals used for high-resolution x-ray analysis. Of those crystals listed by McPherson (1976a), roughly 45% were grown from ammonium sulfate, 20% from solutions of low ionic strength, 15%

from organic solvents such as ethanol, methanol, 2-methyl-2,4-pentanediol, and acetone, 7% from phosphate, and the remainder from several other precipitants such as polyethylene glycols of different ranges of molecular weight (McPherson, 1976b; Moffat, 1980). Furthermore, it is generally not necessary to explore a wide range of macromolecule concentrations in initial experiments; 80% of protein crystals grow in the range from 2 to 40 mg/ml (McPherson, 1976a; Moffat, 1980), and a single initial concentration of, say, 15 mg/ml may be selected.

The number of solution variables to be explored initially can thus be restricted to a few precipitants, at a single macromolecule concentration, at one or two temperatures (say, 4°C and room temperature). Since most macromolecules have a strongly pH-dependent solubility, it is usual to examine pH dependence at an early stage. Other variables such as the buffer concentration and the presence or absence of cofactors, effectors, and ions may be examined later.

One further solution variable demands special mention here: flexibility. Structural homogeneity (or near homogeneity) is essential; macromolecules which are unusually flexible, such as intact immunoglobulins, have proved very difficult to crystallize, and the crystals once obtained are often disordered, probably containing molecules in a wide variety of conformations. Of course, all macromolecules are flexible to some extent, and this flexibility is central to their mechanism of action (Section I.B). It seems very likely that the repeated failure to crystallize some smaller proteins such as the hormone ACTH is due to their intrinsic flexibility. It may be possible to reduce flexibility in several ways. Raising the concentration further will promote aggregation, and the aggregates may be less flexible; such is the case for the glucagon trimer (Sasaki et al., 1975). Certain small molecules such as ions or inhibitors may bind to and constrain the structure. Finally, many examples are known where limited proteolysis, leading to nicking of the sequence or trimming off flexible loops on the surface of the protein, greatly aided crystallization (see Table III of McPherson, 1976a). The most extreme form of proteolysis involves cleavage of the protein into several domains, which may be readily crystallizable separately. Examples are again the immunoglobins (F_{ab} and F_c fragments) and the lambda repressor (Amzel and Poljak, 1979; Pabo et al., 1979).

B. Apparatus and Techniques

Historically, the most widely used crystallization technique has been the batch method, where the solvents are mixed to form the desired final solution conditions, the macromolecule is added, the container sealed and then examined from time to time. This technique is appropriate once optimum crystallization conditions have been established and when large amounts are available, but it is not well suited to preliminary investigations on a micro scale. Two main techniques are in wide use: vapor diffusion and dialysis, both of which can be carried out on micro samples (down to about 3 μl) or scaled up to any larger volume desired.

In vapor diffusion, the chemical composition of a solution is continuously altered by equilibration with a reservoir of different composition, via diffusion through the vapor phase of a volatile component such as water, ethanol, acetone, acetic acid, or ammonium hydroxide, in a closed container. For example, suppose it has been established that a protein is soluble in 35% $(NH_4)_2SO_4$, but precipitates at 40% under specified conditions of temperature and pH. The initial protein solution could then be made up in weakly buffered solution at the desired pH and temperature in 35% $(NH_4)_2SO_4$ and equilibrated with a much larger volume of a solution otherwise identical except for the absence of protein and the presence of 40% $(NH_4)_2SO_4$. Water will then slowly diffuse out of the protein solution and into the reservoir; the concentration of $(NH_4)_2SO_4$ in the protein solution will rise until it, too, reaches 40% saturation (the protein and buffer concentration will also increase slightly, as the volume of the protein solution diminishes). The intent is that the slow rise in $(NH_4)_2SO_4$ concentration and consequent decrease in protein solubility will lead to crystallization. If precipitation occurs instead, then the reservoir solution can be replaced by 35% $(NH_4)_2SO_4$; water will diffuse back into the protein solution and may cause the precipitate to redissolve. Vapor diffusion against slightly lower reservoir concentrations of $(NH_4)_2SO_4$, say, 39% or 38%, can then be attempted. Alternatively, the pH in the protein solution can be slowly raised or lowered by adding NH_4OH or CH_3COOH, respectively, to the reservoir. If control of the final pH is required, the experiment can be conducted the other way around. The pH of a buffered solution is adjusted by the addition of concentrated NH_4OH or CH_3COOH and protein added; as the base or acid diffuses out, the pH returns slowly to the buffer point. Although the end point may be the same, it is often found that the results depend on the direction from which the pH endpoint is approached, and both should be tried. Almost all macromolecules have strongly pH-dependent solubilities, which are usually at a minimum near their isoelectric point. This pH variation technique is, therefore, generally applicable and has proved quite successful (McPherson, 1976a).

In the initial stages, numerous vapor-diffusion experiments are conducted on small volumes of solution, which have to be easy to set up and readily examined. Tissue culture multiwell plates are easy to use. Into each well is placed 1 ml of reservoir solution. Droplets of a solution of the macromolecule ranging from 3–40 μl are placed on microscope cover slips previously coated with dimethyldichlorosilane, which are then inverted over each well to produce small hanging drops. The joint is temporarily sealed with light oil or vacuum grease, and may be readily broken when the reservoir solution is to be changed. This technique is ideal when solutions are to be equilibrated against numerous different reservoir solutions. The use of depression slides for the solutions, placed in clear plastic sandwich boxes which hold the reservoir solution, has been advocated by McPherson (1976a). In each case the droplets are readily observed with a low-power microscope, 10–100×, of at least 2-cm working distance.

The advantage of this experimental approach is that it is readily set up, uses a minimum amount of protein (say 50 μg per experiment at a protein concentration of 15 mg ml^{-1}) and uncomplicated apparatus, and is readily monitored. However, certain solution conditions such as those of very low ionic strength are not readily attained by vapor diffusion; for them, microdialysis is frequently used.

Various types of microdialysis cell, suitable for volumes down to about 3 μl, have been devised. Equilibration can occur across a conventional dialysis membrane or an acrylamide plug. The most common type of apparatus is that proposed by Zeppezauer et al. (1968), where the protein sample is contained in a capillary tube sealed by a dialysis membrane held in place by a Tygon collar. The modification proposed by Weber and Goodkin (1968) to provide greater control of the rate of equilibration across the membrane has also been popular. In my experience, these cells are cumbersome to set up, prone to leaks, and less readily monitored (particularly if using the smallest volumes, when the Tygon collar can obscure the sample) than vapor diffusion. A third cell design, which could be used for either microdialysis or vapor diffusion, has been presented by Lagerkvist et al. (1972).

Other experimental techniques such as sequential extraction, free interface diffusion, concentration dialysis, controlled evaporation, and controlled temperature variation have been employed with great success in crystallizing certain proteins, and may be of more general utility (McPherson, 1976a). It is generally agreed that if a macromolecule can be crystallized using one particular technique, it could probably be crystallized using several others; the technique itself is not critical. Crystallographers have tended to use one favorite technique repeatedly, often for historical reasons rather than because of any overriding scientific preference. For example, hemoglobins have been crystallized almost exclusively by batch techniques; many polypeptide hormones have been crystallized by controlled cooling.

Some general points should be noted. All glassware or other labware should be scrupulously clean and free of residual traces of cleaning compounds. All reagents used for crystallization must be of the highest (or at least of known) purity, and their lot numbers must be noted. Exceptional care must be taken to record seemingly minor experimental details, which may subsequently turn out to be of critical importance for crystallization. To retard growth of microorganisms, solutions should be subjected to ultrafiltration, particularly those used for crystallization itself; a trace of toluene, azide, or pyridine can also be added. Sulfhydryl groups can be protected by mild reducing agents such as dithiothreitol. Temperature control, low light levels, and freedom from vibration may be necessary. Though hard to ensure, frequent disturbance by an over-eager experimenter should also be avoided; many macromolecules have crystallized most readily when the experimenter was on vacation.

A detailed experimental strategy for initial macromolecule-crystallization attempts is given in Appendix B.

III. X-Ray Diffraction from Crystals

A. Introduction: The Electron-Density Equation

X-rays are scattered or diffracted by the electrons in a molecule, as the electrons are accelerated by the electric field of the incident radiation, and reradiate as point sources. The scattering may be either coherent (where there is no change in wavelength and a uniform change in phase of π, 180°) or incoherent (where energy is absorbed by the scatterer, which results in an increase in wavelength and a random change in phase). Crystallography deals with coherent scattering.

Before considering crystals, it is instructive to describe the scattering from a general object whose electron density (electrons per unit volume) as a function of position $\mathbf{r}$ is denoted $\rho(\mathbf{r})$. (See Appendix A for a description of the notation). Let x-rays with wave vector $\mathbf{k}_0$ ($|\mathbf{k}_0| = 1/\lambda$, where λ is the wavelength of the x-rays) fall on the object, and consider the coherent scattered x-rays with wave vector $\mathbf{k}$ (Fig. 1). Then the path difference between x-rays scattered by electrons at the origin O and at the point $\mathbf{r}$ is $\mathbf{r} \cdot (\mathbf{k} - \mathbf{k}_0)$, and hence the phase difference is $2\pi \mathbf{r} \cdot (\mathbf{k} - \mathbf{k}_0)$. We now describe the scattered x-ray wave by a complex expression with real and imaginary parts of the form (Appendix A)

$$\text{amplitude} \times \exp(i \times \text{phase}).$$

The amplitude of the scattered wave is proportional to the number of electrons contained in the volume element $d\mathbf{r}$, and hence the scattered wave is given by

$$\rho(\mathbf{r}) \exp[2\pi i \mathbf{r} \cdot (\mathbf{k} - \mathbf{k}_0)] \, d\mathbf{r}.$$

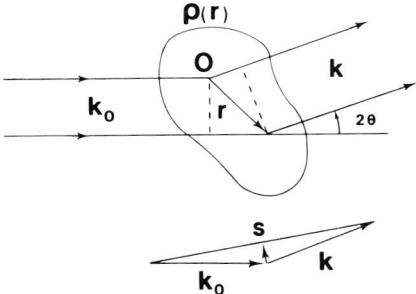

Fig. 1. Scattering of x-rays from a general object $\rho(\mathbf{r})$. Incident x-rays of wave vector $\mathbf{k}_0$ are scattered from a point $\mathbf{r}$ through an angle 2θ, and the outgoing wave vector is $\mathbf{k}$. Since scattering occurs without change in wavelength, $|\mathbf{k}_0| = |\mathbf{k}|$. The scattering vector $\mathbf{s}$ is given by $\mathbf{s} = \mathbf{k} - \mathbf{k}_0$.

To obtain the scattering from the whole object, we must integrate over all volume elements $d\mathbf{r}$ to obtain the scattering factor or structure factor $\mathbf{F}$:

$$\mathbf{F}(\mathbf{k} - \mathbf{k}_0) = \int_{\text{all object}} \rho(\mathbf{r}) \exp[2\pi i \mathbf{r} \cdot (\mathbf{k} - \mathbf{k}_0)] \, d\mathbf{r}.$$

Making the substitution $\mathbf{s} = \mathbf{k} - \mathbf{k}_0$, where $\mathbf{s}$ is known as the scattering vector (Fig. 1), this equation becomes

$$\mathbf{F}(\mathbf{s}) = \int_{\text{all object}} \rho(\mathbf{r}) \exp(2\pi i \mathbf{r} \cdot \mathbf{s}) \, d\mathbf{r}. \tag{1}$$

Equation (1) expresses a mathematical relationship known as a Fourier transform; it states that the structure factor $\mathbf{F}(\mathbf{s})$ is the Fourier transform of the electron-density distribution $\rho(\mathbf{r})$.

Having made the useful identification with a Fourier transform of an equation describing the scattering of x rays by an object, we exploit other mathematical properties of Fourier transforms to derive crystallographically useful results. Four properties are important and are described in Appendix A: inversion, additivity, convolution, and the relation between sections and projections. Applying the inversion property, if $\mathbf{F}(\mathbf{s})$ is the Fourier transform of $\rho(\mathbf{r})$, then $\rho(\mathbf{r})$ is given by the inverse Fourier transform of $\mathbf{F}(\mathbf{s})$:

$$\rho(\mathbf{r}) = A \int_{\text{all s}} \mathbf{F}(\mathbf{s}) \exp(-2\pi i \mathbf{r} \cdot \mathbf{s}) \, d\mathbf{s}, \tag{2}$$

where A is a normalizing factor. Equation (2) is known as the electron-density equation. It shows that if we could determine $\mathbf{F}(\mathbf{s})$ experimentally for all values of **s**, then computation of the inverse Fourier transform of **F(s)** by Eq. (2) would yield the desired electron-density distribution $\rho(\mathbf{r})$ of the object—its structure. Now **F(s)** is a complex quantity, composed of both an amplitude |**F(s)**| and a phase ϕ (Appendix A): $\mathbf{F(s)} = |\mathbf{F(s)}| \exp[i\phi]$. The scattering amplitude or structure amplitude |**F(s)**| can be readily measured experimentally, as it is proportional to the square root of the intensity $I(\mathbf{s})$ of the x-rays scattered in the direction characterized by the scattering vector **s**: $|\mathbf{F(s)}| = a\sqrt{I(\mathbf{s})} = a\sqrt{\mathbf{F(s)F^*(s)}}$, where a is a constant and * denotes the complex conjugate. The intensity, in units such as x-ray photons per unit area per unit time, can be directly measured by devices such as scintillation counters, geiger counters, or photographic film. Unfortunately, there is no direct way of measuring the phase ϕ of the structure factor—the phase is lost in the act of recording the photons. This constitutes the *phase problem,* and its existence makes structure determination by x-ray scattering a nontrivial experiment. As we shall see, much of macromolecular crystallography is devoted to complicated techniques aimed at recovering the phase by indirect means.

B. Crystals and Symmetry

All crystals are composed of a regular, periodically repeated arrangement of objects. In macromolecular crystals, the molecules themselves are the objects, loosely held together by noncovalent, weak intermolecular interactions. A typical crystal contains on the order of 10^{15} molecules. The intervening space is occupied largely by disordered (not periodically repeated) solvent molecules, in equilibrium with the liquid from which the crystals were grown. The fraction of the total volume of the crystal occupied by the solvent is quite variable, from around 30% for a tightly packed crystal up to 80% for a very loosely packed crystal, such as those of several tRNAs. Thus, the crystals contain large liquid-solvent channels through which small molecules (such as substrates) can diffuse, enabling chemistry to be conducted in the crystalline state. Macromolecular crystals can be thought of as resembling a very concentrated solution, with an important solvent component. In this they differ from crystals of small molecules, from which solvent is largely or entirely excluded. Since the intermolecular interactions are neither numerous nor strong, it is generally found that the structure in the crystal does not depend on the crystallographic environment. For example, two molecules of oxidized cytochrome c, differing in their environment and hence in intermolecular interactions, are essentially identical (Takano and Dickerson, 1981). Thus, it is believed that crystal structures do not differ radically from those in solution.

The unit cell of a crystal is the smallest parallelepiped within the crystal which, by translation alone, will build up the entire macroscopic crystal.

Each parallelepiped is identical in shape, content, and orientation to all others. If the same single point is chosen in each parallelepiped, the array of points defines a space lattice which completely describes the repetition characteristics. The axes which make up the edges of the unit cell are denoted **a**, **b**, and **c**. The requirement for the "smallest" parallelepiped leads to a primitive lattice, denoted P, with one lattice point per unit cell. However, it is often convenient to relax this requirement to allow a choice of orthogonal or other conveniently disposed unit-cell axes. In three dimensions, there are fourteen distinct space lattices, the Bravais lattices [see, e.g., Fig. 4.5 of Blundell and Johnson (1976)]. Seven of these fourteen are primitive (the six P lattices and the one rhombohedral R lattice), and correspond to the seven crystal systems: triclinic, monoclinic, orthorhombic, hexagonal, tetragonal, cubic, and rhombohedral. In each crystal system, definite relationships exist between the dimensions of **a**, **b**, and **c** and the interaxis angles of the unit cell (Table I).

The unit cell usually contains internal symmetry elements which relate the contents of different parts of the unit cell to each other. The smallest unique volume within the unit cell is the asymmetric unit, whose volume is an integral fraction of the unit-cell volume. Often the asymmetric unit contains only one molecule. If it contains more than one molecule, the asymmetric unit is said to exhibit noncrystallographic symmetry, which may be exploited in subsequent structure determination (Section IV.E). If it contains less than one molecule, then the molecule itself must possess some symmetry. For example, horse methemoglobin crystallizes with one $\alpha\beta$ dimer in the asymmetric unit, thus conclusively establishing that the $\alpha_2\beta_2$ hemoglobin molecule, a tetramer, is composed of two structurally identical $\alpha\beta$ dimers. Only two types of symmetry operations can be exhibited by crystals of macromolecules: rotation axes (2-, 3-, 4- and 6-fold) and screw axes (2_1, 3_1, 3_2, 4_1, 4_2, 6_1, 6_2, 6_3, 6_4, and 6_5). The notation M_N for a screw operation along, for example, the **a** axis denotes a rotation of $2\pi/M$ radians followed by a translation of (N/M)**a**. (Other symmetry operations such as mirror planes and centers of symmetry found in crystals of small molecules are not allowed for macromolecules, as they would convert, for example, an L-amino acid to a D-amino acid). If we are restricted to rotation and screw operations only, then there are only 65 distinct ways in which these symmetry operations can be arranged in space. That is, there are 65 enantiomorphic space groups. The space group symbols are also listed in Table I, and a complete description of each space group is given in the "International Tables for X-Ray Crystallography" (see Appendix D).

One other type of symmetry is important: the point symmetry or point group. A point group describes the possible arrangements of symmetry operators (for macromolecules, rotations axes), all of which pass through a single point. Point groups are important from a molecular aspect, as they

Table I Crystallographic Relationships

Crystal system	Minimum symmetry	Dimension and angle requirements	Class	Space group symbols
Triclinic	None	None	1	P1
Monoclinic	Twofold axis (parallel to **b**)	**a** and **c** both perpendicular to **b**	2	P2, P2$_1$, C2
Orthorhombic	Three mutually perpendicular twofold axes	**a**, **b**, and **c** mutually perpendicular	222	P222, P2$_1$2$_1$2, P2$_1$2$_1$2$_1$, P222$_1$, C222, C222$_1$, F222, I222, I2$_1$2$_1$2$_1$
Tetragonal	Fourfold axis (parallel to **c**)	**a**, **b**, and **c** mutually perpendicular; |**a**| = |**b**|	4	P4, P4$_1$, P4$_2$, P4$_3$, I4, I4$_1$
			422	P422, P42$_1$2, P4$_1$22, P4$_1$2$_1$2, P4$_2$22, P4$_2$2$_1$2, P4$_3$22, P4$_3$2$_1$2, I422, I4$_1$22
Trigonal	Threefold axis (parallel to **c**)	**a** and **b** both perpendicular to **c** and at 120° to each other; |**a**| = |**b**|	3	P3, P3$_1$, P3$_2$, R3
			32	P312, P321, P3$_1$12, P3$_1$21, P3$_2$12, P3$_2$21, R32
Hexagonal	Sixfold axis (parallel to **c**)	Same as trigonal	6	P6, P6$_5$, P6$_4$, P6$_3$, P6$_2$, P6$_1$
			622	P622, P6$_1$22, P6$_2$22, P6$_3$22, P6$_4$22, P6$_5$22
Cubic	Threefold axes along cube diagonals	**a**, **b**, and **c** mutually perpendicular; |**a**| = |**b**| = |**c**|	23	P23, P2$_1$3, F23, I23, I2$_1$3
			432	P432, P4$_1$32, P4$_2$32, P4$_3$32, F432, F4$_1$32, I432, I4$_1$32

describe the mode of association of identical subunits: the number and type(s) of subunit interfaces, and hence the likely mode of association. They are central to, for example, problems in the structure and self-assembly of regular viruses. There are 14 possible point groups for macromolecules (see Table I of Matthews and Bernhard, 1973).

C. Diffraction by Crystals

We now consider what happens when x-rays fall on a crystal in which $\rho(\mathbf{r})$ in Eq. (1) is a three-dimensionally periodic function of position, with translational periods $\mathbf{a}$, $\mathbf{b}$, and $\mathbf{c}$, the unit cell axes of the crystal lattice. It is easiest to visualize scattering from crystals by considering first a one-dimensionally periodic object such as a diffraction grating (Fig. 2). Diffraction occurs through angles μ such that the path difference between rays from adjacent elements of the grating, $|\mathbf{a}| \sin \mu = \lambda \mathbf{a} \cdot \mathbf{k}$, is an integral number of wavelengths, $h\lambda$. Thus, $\mathbf{a} \cdot \mathbf{k} = h$. More generally, $\mathbf{a} \cdot (\mathbf{k} - \mathbf{k}_0) = h$, or $\mathbf{a} \cdot \mathbf{s} = h$. This is known as a Laue equation, and diffraction is confined to those directions $\mathbf{s}$ which satisfy this equation. Extending this derivation to three-dimensionally periodic objects such as crystals leads to three Laue equations:

$$\mathbf{a} \cdot \mathbf{s} = h; \quad \mathbf{b} \cdot \mathbf{s} = k; \quad \mathbf{c} \cdot \mathbf{s} = l,$$

which must be satisfied simultaneously for diffraction to occur in a direction characterized by the scattering vector $\mathbf{s}$. To satisfy them, consider a new lattice characterized by vectors $\mathbf{a}^*$, $\mathbf{b}^*$ and $\mathbf{c}^*$, defined by $\mathbf{a} \cdot \mathbf{a}^* = \mathbf{b} \cdot \mathbf{b}^* = \mathbf{c} \cdot \mathbf{c}^* = 1; \mathbf{a} \cdot \mathbf{b}^* = \mathbf{a} \cdot \mathbf{c}^* = \mathbf{b} \cdot \mathbf{a}^* = \mathbf{b} \cdot \mathbf{c}^* = \mathbf{c} \cdot \mathbf{a}^* = \mathbf{c} \cdot \mathbf{b}^* = 0$. That is, each new lattice vector is perpendicular to the other two (real) crystal-lattice vectors, and vice versa. For example, $\mathbf{a}^*$ is perpendicular to both $\mathbf{b}$ and $\mathbf{c}$, and $\mathbf{c}^*$ is perpendicular to both $\mathbf{a}$ and $\mathbf{b}$. The dimensions of $\mathbf{a}^*$, $\mathbf{b}^*$, and $\mathbf{c}^*$ are (1/length), causing the new lattice to be denoted the reciprocal lattice, in contrast to the real crystal lattice $\mathbf{a}$, $\mathbf{b}$, and $\mathbf{c}$, whose dimensions are of course (length). Notice also that the larger is $\mathbf{a}$ (or $\mathbf{b}$ or $\mathbf{c}$), the smaller is $\mathbf{a}^*$ (or $\mathbf{b}^*$ or

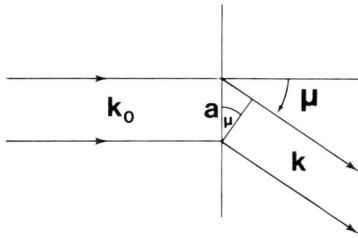

Fig. 2. Scattering by a diffraction grating with slits a distance $\mathbf{a}$ apart. The path difference for rays scattered from adjacent slits is $|\mathbf{a}| \sin \mu$.

c*); there is an inverse relationship of scale between the real lattice and the reciprocal lattice.

Take a point in the reciprocal lattice, $h\mathbf{a}^* + k\mathbf{b}^* + l\mathbf{c}^* = \mathbf{d}^*$, say, and substitute $\mathbf{d}^* = \mathbf{s}$ in the three Laue equations. For example, $\mathbf{a} \cdot \mathbf{s} = \mathbf{a} \cdot (h\mathbf{a}^* + k\mathbf{b}^* + l\mathbf{c}^*) = h\mathbf{a} \cdot \mathbf{a}^* + k\mathbf{a} \cdot \mathbf{b}^* + l\mathbf{a} \cdot \mathbf{c}^* = h$, as is required; similarly for the other two equations. Thus, a formal solution of the Laue equations is $\mathbf{s} = h\mathbf{a}^* + k\mathbf{b}^* + l\mathbf{c}^*$. In words, for diffraction to occur, the scattering vector $\mathbf{s}$ must also be a reciprocal lattice vector $\mathbf{d}^*$. In a very real sense, diffraction space is reciprocal space.

The real lattice vectors $\mathbf{a}$, $\mathbf{b}$, and $\mathbf{c}$ are fixed in the crystal, and hence the reciprocal lattice vectors $\mathbf{a}^*$, $\mathbf{b}^*$, and $\mathbf{c}^*$ are also fixed in the crystal. In general, the continuous variable $\mathbf{s}$ will not coincide with a reciprocal lattice vector, since $\mathbf{s}$ is defined solely with reference to the incident ($\mathbf{k}_0$) and diffracted ($\mathbf{k}$) x-rays. We must reorient the crystal with respect to the incident x-ray beam $\mathbf{k}_0$ in such a way that a reciprocal lattice vector coincides with $\mathbf{s}$. The vector $\mathbf{s}$ was originally defined by $\mathbf{s} = \mathbf{k} - \mathbf{k}_0$, which represents a sphere. Thus, diffraction will occur only when a reciprocal lattice vector $\mathbf{d}^*$ lies on the surface of this sphere, whose radius is $|\mathbf{k}_0| = |\mathbf{k}| = 1/\lambda$ and which is known as the sphere of reflection or Ewald sphere (Fig. 3). Then and only then will $\mathbf{s} = \mathbf{d}^*$.

The entire diffraction pattern from a crystal thus consists of a three-dimensional array of points, the points of the reciprocal lattice. Each point is also known as a spot, or reflection, and is characterized by its coordinates in reciprocal space, h, k, and l. Each plane of the reciprocal lattice will intersect the Ewald sphere in a circle, and for a given crystal orientation only a few reciprocal lattice points will happen to lie on this circle and hence give rise to diffraction (Fig. 3). To obtain diffraction from each and every lattice point, the crystal (and hence the crystal lattice and the reciprocal lattice embedded in it) must be moved in such a way that all lattice points traverse the sphere of reflection at some stage in the motion. One common way of achieving this

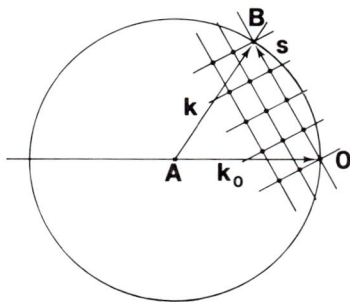

Fig. 3. A small section of a reciprocal lattice, in which the reciprocal lattice point B (where OB = $\mathbf{s}$) lies on the surface of the Ewald sphere.

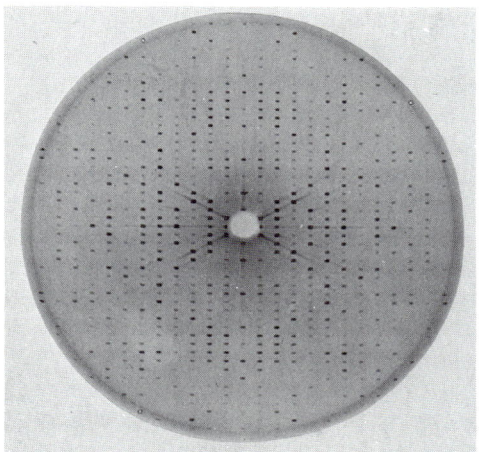

Fig. 4. An example of a precession photograph. The incident x-ray beam is directed perpendicular to the plane of the photograph; the white circle in the center is caused by the shadow of a lead disk which prevents the incident beam from striking the film. The photograph is the $h0l$ plane of a crystal of bovine intestinal calcium-binding protein (Szebenyi *et al.*, 1981), in which the **a*** axis is vertical and the **c*** axis is horizontal.

is to cause the crystal to precess about the incident x-ray beam, and to record the diffraction pattern on a photographic film which also precesses in a coupled manner. This type of pattern is known as a precession photograph. Figure 4 illustrates one plane of the reciprocal lattice and the point-by-point nature of the diffraction pattern. Each dark spot on this photograph represents a different reciprocal lattice point, with different values of h, k, and l. This particular photograph shows an $h0l$ plane, where $k = 0$ for all spots on this film. Note also that the spacing and position of the spots depend only on **a***, **b***, and **c***, and hence on **a**, **b**, and **c**; that is, they depend only on the size and shape of the unit cell and not on the structure of its contents, the macromolecules. As we shall see, the structure gives rise to the variation in intensity of each of the spots, $I(\mathbf{s})$ or $I(h,k,l)$.

There is another very common way, due to W. L. Bragg, of considering the diffraction from crystals. He showed that each diffraction spot could be thought of as arising from "reflection" of x-rays from a family of parallel planes of electrons in the crystal, a definite distance d apart (Fig. 5). Hence he derived Bragg's law:

$$2d \sin \theta = \lambda,$$

where 2θ is the angle through which the x-rays are reflected (or scattered) and

θ is their wavelength. Notice that the scattering vector **s** is perpendicular to the family of planes, and thus each family corresponds to a single diffraction maximum or reflection, $h\,k\,l$. Furthermore, it can be shown that the planes intercept the unit-cell axes at points $\mathbf{a}/h$, $\mathbf{b}/k$, and $\mathbf{c}/l$, which are integer fractions of the unit-cell edges.

The interplanar spacing d governs the detail with which a structure is examined; from Bragg's law, the larger the scattering angle 2θ, the greater the value of $\sin\theta$, the smaller the corresponding value of d, and the finer the details of the structure being probed. The minimum value of d used in a structure determination of a macromolecule is known as the resolution. A high-resolution structure determination has a low value of d, and vice versa (see Table II, Section VI).

We have concentrated so far on effects which govern the positions of the diffraction spots. We now turn to those effects which govern their intensities and demonstrate that the effect of the crystal lattice in which the macromolecules are arranged is to sample the continuous Fourier transform (in which **s** is a continuous variable) of the contents of the unit cell at the reciprocal lattice points. That is, the intensity of scattering $I(\mathbf{s})$ at each of the reciprocal lattice points $h\mathbf{a}^* + k\mathbf{b}^* + l\mathbf{c}^*$ is the (suitably scaled) intensity of the transform of the contents of a single unit cell, measured at that point.

The easiest way to prove this result is to regard a macroscopic crystal as the result of the mathematical operation of convolution (see Appendix A): the crystal represents the convolution of the contents of one unit cell with the real crystal lattice. The mathematical form of a lattice is a function with value 1 at the lattice points and 0 everywhere else, a three-dimensional δ function. A two-dimensional illustration of convolution is shown in Fig. 6. We now use a further property (Appendix A) of Fourier transforms, which states that the transform of the convolution of two functions is the

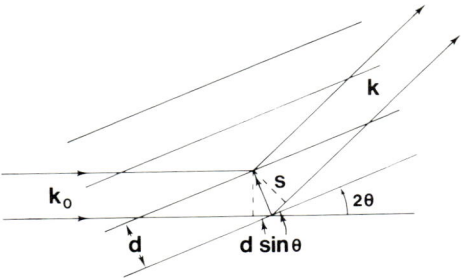

Fig. 5. According to Bragg, diffraction of incident x-rays with wave vector $\mathbf{k}_0$ through an angle 2θ may be thought of as occurring by "reflection" of the x-rays from a family of planes of electron density a distance **d** apart. The scattering vector **s** is perpendicular to this family of planes.

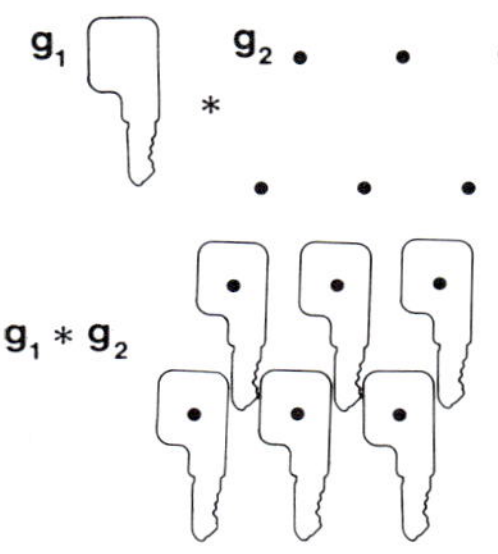

Fig. 6. The mathematical operation of convolution, in a simple two-dimensional example. The convolution of the function $\mathbf{g}_1$, the key, with the function $\mathbf{g}_2$, a lattice, yields the array of keys, resembling a small section of a two-dimensional crystal of keys. The operation of convolution is denoted by *.

product of the transforms of the two functions separately. Since the crystal is the convolution of the contents of a unit cell with the real lattice, the transform (diffraction pattern) of the crystal is given by the product of the transform of the contents and the transform of the real lattice. From Eq. (1), the transform of the contents $\rho(\mathbf{r})$ is $\mathbf{F}(\mathbf{s})$. The transform of the real lattice may be shown to be the reciprocal lattice, again with value 1 at the reciprocal lattice points and 0 everywhere else. Thus, $\mathbf{F}(\mathbf{s})$ is sampled at the reciprocal lattice points, where $\mathbf{s} = h\mathbf{a}^* + k\mathbf{b}^* + l\mathbf{c}^*$ as before; $\mathbf{F}(\mathbf{s})$ is nonzero only at these points.

The electron-density equation, Eq. (2), was originally expressed as an integral over the continuous variable $\mathbf{s}$:

$$\rho(\mathbf{r}) = A \int \mathbf{F}(\mathbf{s}) \exp(-2\pi i \mathbf{r} \cdot \mathbf{s}) \, d\mathbf{s}.$$

If we now define $\mathbf{r}$ to be a point (x,y,z) in the unit cell $\mathbf{r} = x\mathbf{a} + y\mathbf{b} + z\mathbf{c}$, and $\mathbf{s}$ to be a reciprocal lattice point $\mathbf{s} = h\mathbf{a}^* + k\mathbf{b}^* + l\mathbf{c}^*$, then the integral becomes a Fourier summation:

$$\rho(x,y,z) = A \sum_h \sum_k \sum_l F(h,k,l) \exp[-2\pi i(hx + ky + lz)] \qquad (3)$$

over integer values of h, k, and l from $-\infty$ to ∞.

This Fourier summation and the more general Fourier integral from which it was just derived do not depend in any way on the nature of the electron-density distribution $\rho(\mathbf{r})$. If $\rho(\mathbf{r})$ satisfies certain conditions, then its diffraction pattern $\mathbf{F}(\mathbf{s})$ exhibits simplifying features. First, if $\rho(\mathbf{r})$ is everywhere real, with no imaginary component, then $\rho(\mathbf{r}) = \rho^*(\mathbf{r})$, where the * again denotes the complex conjugate. Now

$$\mathbf{F}(\mathbf{s}) = \int \rho(\mathbf{r}) \exp[2\pi i \mathbf{r} \cdot \mathbf{s}] \, d\mathbf{r} \quad \text{from Eq. (1);}$$

$$\therefore \mathbf{F}^*(\mathbf{s}) = \int \rho^*(\mathbf{r}) \exp[-2\pi i \mathbf{r} \cdot \mathbf{s}] \, d\mathbf{r} = \int \rho(\mathbf{r}) \exp[-2\pi i \mathbf{r} \cdot \mathbf{s}] \, d\mathbf{r} = \mathbf{F}(-\mathbf{s}).$$

Since the intensity of diffraction $I(\mathbf{s}) = \mathbf{F}(\mathbf{s})\mathbf{F}^*(\mathbf{s})$, it follows that $I(\mathbf{s}) = I(-\mathbf{s})$. That is, the intensity of diffraction from the front ($\mathbf{s}$) and back ($-\mathbf{s}$) of an object, or from a set of planes in the crystal, is identical; the diffraction pattern has a center of symmetry. This is known as Friedel's law. A corollary is that x-ray scattering from real objects cannot yield the "hand" of the object. For example, the diffraction pattern alone could not reveal whether a crystal of an amino acid contained L-amino acids or D-amino acids. As we shall see in Section IV.C, the electron density is not always real, which leads to deviations from Friedel's law. These deviations may enable the hand to be deduced.

Second, if the object itself possesses a center of symmetry, then $\rho(\mathbf{r}) = \rho(-\mathbf{r})$, and it can similarly be shown that $\mathbf{F}(\mathbf{s}) = \mathbf{F}^*(\mathbf{s})$. That is, the structure factors are real, with phases 0 (+ sign) or π (− sign). No intermediate phase values are possible. In this case, the phase problem reduces to a sign-selection problem. Although macromolecules do not possess a center of symmetry, it frequently happens that their projections (Appendix A) do. For example, all projections down a twofold rotation or screw axis will be centrosymmetric. We now make use of a further property of Fourier transforms (Appendix A) which relates projections to the Fourier transform of the corresponding central section, passing through the origin. In this example, the structure factors for the plane of the reciprocal lattice passing through the origin, normal to the twofold rotation or screw axis, will all have phase 0 or π. The reduction of the phase problem to a sign-selection problem enormously simplifies crystallographic operations such as the location of heavy atoms (Section IV.B).

Third, for all crystal systems except triclinic, the electron density within the unit cell will possess some symmetry, and this is carried over to the diffraction pattern. In general, if the electron density possesses an n-fold rotation axis, then its Fourier transform, the diffraction pattern, will also possess a parallel n-fold rotation axis. For example, if the unit cell contains a twofold rotation axis parallel to $\mathbf{b}$, then all planes of the reciprocal lattice normal to $\mathbf{b}$ will also possess a twofold axis of symmetry. Thus, $\mathbf{F}(h,k,l) = \mathbf{F}(-h,k,-l)$ or, as it is more usually written $\mathbf{F}(hkl) = \mathbf{F}(\bar{h}k\bar{l})$, with the commas omitted and the minus signs above the coordinates. This feature of the diffraction pattern is also illustrated in the precession photograph shown in Fig. 4, where a twofold rotation axis is horizontal. Hence, the intensities in the diffraction pattern on the top are identical to those on the bottom. If the unit cell contains an n-fold screw axis, then the diffraction pattern will again contain a parallel n-fold rotation axis. In addition, the diffraction pattern will contain one other feature which enables rotation and screw axes to be distinguished. If the entire structure containing an n-fold screw axis is projected onto that axis (a linear projection: Appendix A), then the projected structure repeats in one/nth of the unit-cell dimension along that axis, and the reflections along the parallel axis in the diffraction pattern will be n times farther apart. The intervening reflections will have zero intensity, and they are said to be systematically absent. Observation of systematic absences along certain axes in the diffraction pattern is thus diagnostic of screw axes. In Fig. 4, a twofold screw axis is vertical; every other spot is absent along the $\mathbf{a}^*$ axis, vertical. Systematic absences are also present if a nonprimitive unit cell has been chosen (Section III.B), since the diffraction pattern depends only on the primitive unit cell and is unaffected by this choice. The crystallographer's nonprimitive unit cell is always larger than the primitive unit cell (two times larger for A, B, C, and I lattices, four times for an F lattice). Hence, one-half of the reflections are systematically absent if an A, B, C, or I lattice is chosen and three-fourths if an F lattice is chosen. For example, in space group C2, the lattice is chosen to be c-face centered, and all reflections for which $h + k$ is odd are of zero intensity, systematically absent. The systematic absences for each space group, arising both from screw axes and choice of a nonprimitive unit cell, are listed in the "International Tables for X-ray Crystallography" (noted in Appendix D).

The existence of Friedel's law means that the Fourier summation by which the electron density is calculated, Eq. (3), may be simplified. Writing $\mathbf{F}(hkl) = |\mathbf{F}(hkl)| \exp[i\phi_{hkl}]$ and noting that $|\mathbf{F}(hkl)| = |\mathbf{F}(\bar{h}\bar{k}\bar{l})|$ by Friedel's law, and $\phi_{hkl} = -\phi_{(\bar{h}\bar{k}\bar{l})}$, then Eq. (3) reduces to

$$\rho(xyz) = \frac{2}{V} \sum_h \sum_k \sum_l |\mathbf{F}(hkl)| \cos[2\pi(hx + ky + lz) - \phi_{hkl}],$$

where the summations now run from 0 to $+\infty$ and we have replaced the constant A by its correct value $2/V$, where V is the volume of the unit cell. Relations between the structure factors generated by symmetry and systematic absences, such as $|\mathbf{F}(hkl)| = |\mathbf{F}(\bar{h}\bar{k}\bar{l})|$ and $|\mathbf{F}(hkl)| = 0$ for $h + k = 2n + 1$, further simplify the summation.

In an exactly similar way, the Fourier integral which yielded the structure factor $\mathbf{F}(\mathbf{s})$, Eq. (1), may be rewritten for a crystal. Thus, the original equation

$$\mathbf{F}(\mathbf{s}) = \int_{\text{all object}} \rho(\mathbf{r}) \exp[2\pi i \mathbf{r} \cdot \mathbf{s}] \, d\mathbf{r}$$

becomes

$$\mathbf{F}(hkl) = \sum_{j=1}^{N} f_j(hkl) \exp[2\pi i(hx_j + ky_j + lz_j)], \tag{4}$$

where we consider that the electron density in the unit cell is made up of N atoms located at positions (x_j, y_j, z_j) and $f_j(hkl)$ is the atomic scattering factor, the Fourier transform of the electron density of a single atom. Equation (4) is known as the structure-factor equation. It states that, if the atomic positions and type are known, then the x-ray scattering from them can be directly calculated, both in magnitude $|F(hkl)|$ and phase ϕ_{hkl}, yielding $F(hkl)$.

D. The Patterson Function

Since the intensities $I(hkl)$ are the only experimental data which can readily be measured, it is useful to demonstrate the nature of the structural information which can be extracted from the intensities alone, with no knowledge of the phases. This problem was first tackled by A. L. Patterson, who considered the function now named after him, the Fourier transform of the intensities $I(\mathbf{s})$:

$$P(\mathbf{u}) = \int_{-\infty}^{\infty} I(\mathbf{s}) \exp(2\pi i \mathbf{u} \cdot \mathbf{s}) \, d\mathbf{s} \tag{5}$$

by analogy with Eq. (2). Note first that the Patterson function P is both real and centrosymmetric:

$$P(\mathbf{u}) = \int_{0}^{\infty} I(\mathbf{s}) \exp(2\pi i \mathbf{u} \cdot \mathbf{s}) \, d\mathbf{s} + \int_{-\infty}^{0} I(\mathbf{s}) \exp(2\pi i \mathbf{u} \cdot \mathbf{s}) \, d\mathbf{s}$$

$$= \int_{0}^{\infty} I(\mathbf{s}) \exp(2\pi i \mathbf{u} \cdot \mathbf{s}) \, d\mathbf{s} + \int_{0}^{\infty} I(-\mathbf{s}) \exp(-2\pi i \mathbf{u} \cdot \mathbf{s}) \, d\mathbf{s}$$

$$= 2 \int_{0}^{\infty} I(\mathbf{s}) \cos(2\pi \mathbf{u} \cdot \mathbf{s}) \, d\mathbf{s} = P(-\mathbf{u}). \tag{6}$$

Second, since $I(\mathbf{s}) = F(\mathbf{s})F^*(\mathbf{s})$, Eq. (5) may be written as

$$P(\mathbf{u}) = \int_{-\infty}^{\infty} F(\mathbf{s})F^*(\mathbf{s}) \exp(2\pi i \mathbf{u} \cdot \mathbf{s}) \, d\mathbf{s},$$

which shows that $P(\mathbf{u})$ is the Fourier transform of the product of two functions, $F(\mathbf{s})$ and $F^*(\mathbf{s})$. A further property of Fourier transforms (Appendix A) is that the transform of the product of two functions is given by the convolution of the transforms of the two functions separately. Hence, $P(\mathbf{u}) =$ (transform of F) convoluted with (transform of F^*). Now the transform of F is the desired structure $\rho(\mathbf{r})$, by Eq. (2), and it can readily be shown that the transform of F^* is the desired structure inverted through the origin, $\rho(-\mathbf{r})$. Thus

$$P(\mathbf{u}) = \rho(\mathbf{r}) \quad \text{convoluted with} \quad \rho(-\mathbf{r}) = \int_{\text{all } \mathbf{r}} \rho(\mathbf{r})\rho(\mathbf{u} + \mathbf{r}) \, d\mathbf{r},$$

by the definition of convolution (Appendix A). In words, this convolution operation may be stated as follows. Take the electron density $\rho(\mathbf{r})$ of one unit cell with origin O, and superimpose on it the same electron density, but with a different origin O' such that $OO' = \mathbf{u}$. Take the product of the two electron-density functions, repeat for all possible origin pairs in the unit cell $\mathbf{u}$, and sum all the products.

It is easiest to see what this means physically by considering a simple two-dimensional structure with point atoms (Fig. 7). A peak will occur in the Patterson function when the shift vector $\mathbf{u}$ coincides with the vectors $\mathbf{r}_i - \mathbf{r}_j$, which relate two atoms i and j in the structure, and the magnitude of the peak will be proportional to the product of the number of electrons in the atoms $Z_i Z_j$, where Z denotes atomic number. The Patterson function is thus a representation of all possible interatomic vectors. For N atoms, there will be N^2 peaks; but N of these are self-vectors, relating each atom to itself, and hence occur at the origin. Thus, there are $N^2 - N$ peaks at nonorigin positions in the Patterson function.

For small molecules where N is in the range of 10–20, it may be possible to deduce the atomic structure simply by inspection of the Patterson function, or by application of certain more sophisticated image-seeking methods. In practice, as N increases beyond 20 up to the several hundred characteristic of

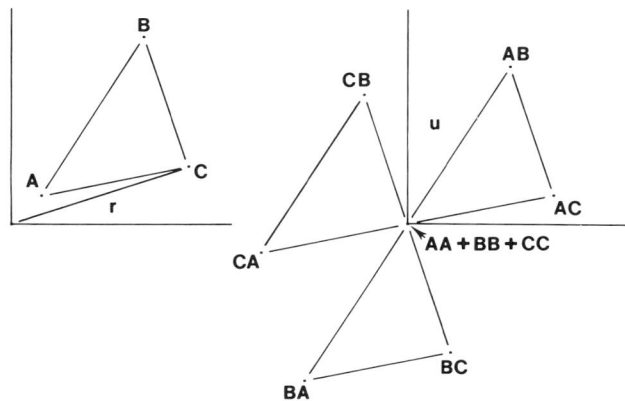

Fig. 7. Illustration of a Patterson function. Given the array of three atoms A, B, and C on the left, the Patterson function on the right depicts all possible interatomic vectors (nine here, of which three fall at the origin). For example, the vector between atoms C and A on the left falls at the point CA on the Patterson function on the right; the vector between atoms A and C falls at the centrosymmetrically related point AC.

macromolecules, the number of peaks becomes so large that they tend to overlap and cannot be distinguished from one another, a drawback accentuated by the fact that the x-ray data are of limited resolution and the atoms are not points. Hence, Patterson functions are of limited utility in macromolecular crystallography, except in the form of difference Patterson functions (Section V.B).

E. Conclusion

We thus see that diffraction from a crystal exists only at distinct points, the points of the reciprocal lattice defined by $\mathbf{a^*}$, $\mathbf{b^*}$, and $\mathbf{c^*}$, which depend only on the real crystal lattice, $\mathbf{a}$, $\mathbf{b}$, and $\mathbf{c}$. The intensity at each lattice point $I(hkl)$ depends on the symmetry contained in the unit cell, on the choice of unit cell, and, most important, on the distribution of electron density within the unit cell. From $\sqrt{I(hkl)} = \sqrt{\mathbf{F}(hkl)\mathbf{F^*}(hkl)}$, we can measure directly $|\mathbf{F}(hkl)|$, but the phase ϕ_{hkl} necessary to reconstruct $\mathbf{F}(hkl) = |\mathbf{F}(hkl)| \times \exp(i\phi_{hkl})$ cannot be measured — the phase problem.

Preliminary investigation of the x-ray diffraction pattern of a newly crystallized macromolecule may be conducted very rapidly, within a week or so, and certain useful results obtained. The location of the spots yields $\mathbf{a^*}$, $\mathbf{b^*}$, and $\mathbf{c^*}$, and $\mathbf{a}$, $\mathbf{b}$, and $\mathbf{c}$ can immediately be derived from them. When coupled with the symmetry of the diffraction pattern and identification of systematic absences, the space group can be selected from the 65 possibilities listed in Table I, and the volume of the unit cell and the asymmetric unit can be deduced. If the molecular weight of the macromolecule is known, then the parameter V_M = (volume of the asymmetric unit)/(molecular weight) can be calculated, in $Å^3$/dalton. Matthews (1976) has tabulated the distribution of V_M values found for macromolecular crystals. A very low value of V_M relative to this distribution suggests that the molecule may possess internal symmetry, so that only a fraction of the molecule (one-half for twofold symmetry, one-third for threefold symmetry, and so forth) lies in the asymmetric unit. That is, a molecular-symmetry axis coincides with a crystallographic-symmetry axis. Conversely, a high value of V_M suggests that either the crystal is very loosely packed and a large fraction of the unit cell volume is solvent, or that the asymmetric unit contains more than one molecule and can exhibit noncrystallographic symmetry. These possibilities can be quickly distinguished by careful measurement of the density of the crystal (Blundell and Johnson, 1976).

Identification of molecular symmetry is often a biochemically valuable result. For example, crystallization of the important regulatory enzyme aspartate transcarbamoylase in several different crystal forms with different space groups (Wiley and Lipscomb, 1968) clearly revealed that the enzyme itself possesses both a twofold and a threefold axis of symmetry, and strongly suggested that the molecule contains six copies of both the catalytic and the regulatory chains. Even from this preliminary x-ray investigation, it was possible to correct extensive and seemingly reliable biochemical work which had suggested a stoichiometry of four copies of each chain.

F. Outline of Experimental Aspects

Two types of x-ray laboratory source are in common use: sealed-tube and rotating-anode x-ray generators. In both, an electron gun is aimed at a water-cooled copper block, and characteristic copper K radiation is emitted. The K_β radiation is preferentially filtered out by a thin nickel foil, so that the radiation which strikes the crystal is largely a mixture of K_{α_1} and K_{α_2} radiation. For practical purposes, the radiation may be regarded as monochromatic with a wavelength of 1.542 Å (closely similar to the length of a C—C single bond).

Unfortunately, most macromolecular crystals are sensitive to x rays and experience radiation damage. Radiation damage leads to a degradation in quality of the diffraction pattern, particularly at high resolution, and finally to the almost complete loss of the diffraction pattern as the molecules are destroyed. For reasons which are unclear, some crystals such as those of the proteolytic enzyme elastase are radiation insensitive, and a crystal can be exposed for several hundred hours without loss of diffraction quality. In contrast, crystals of the closely related enzymes chymotrypsin and trypsin have higher, more typical radiation sensitivity, in which average intensities can fall by 20% after about 50 hours irradiation by a sealed-tube x-ray generator. Some crystals are exquisitely radiation sensitive, which severely hinders x-ray data collection. It appears that radiation damage is diminished at lower temperatures, and that it often contains a time-dependent component as well as a dose-dependent component. Thus, it may be advantageous to use a high-intensity source to minimize exposure times and radiation damage in those cases.

As noted previously (Section III.C), for diffraction to occur, the crystal must be oriented with respect to the incident x-ray beam. Two types of x-ray camera which use film as a recording medium for the diffraction pattern are used, distinguished by the type of motion the crystal undergoes: precession and rotation–oscillation. The former yields an undistorted representation of the reciprocal lattice (see Fig. 4), but at the price of screening out all the diffraction pattern except the desired reciprocal lattice plane with an annular aperture, known as a layer-line screen. In the latter, the crystal is oscillated to and fro about a fixed axis, over an angular range of a few degrees; this yields a distorted representation of segments of several reciprocal lattice planes simultaneously, with no necessity for screening out portions of the lattice. To record all diffraction out to a defined resolution (a complete, three-dimensional data set) with either camera requires that many photographs be taken, perhaps 20–40; exposures of 5–40 hours per photograph are common.

Each of the spots on the films must then be digitized by an optical device known as a densitometer, which measures the optical density point by point across the x-ray film and integrates over each spot, to yield a number proportional to the intensity of the spot, $I(hkl)$. These devices are of the flat-bed or rotating-drum type, under computer control. Recording on film followed by densitometry yields structure amplitudes $|F|$ accurate to about 4%, as judged by the disagreement between the amplitudes of reflections which should be equal, by symmetry. Strong reflections are more accurate than this, and weak ones less so. It is common for a data set to consist of many tens of thousands of spots, or even hundreds of thousands. For example, the structure analysis of tomato bushy-stunt virus crystals is based on densitometry of 198,000 spots to 2.85 Å resolution for the native data set, obtained from 50 oscillation photographs on 25 crystals (Harrison *et al.*, 1978).

For crystals with smaller unit cells (and hence larger values of a^*, b^*, and c^*, and fewer diffraction maxima to a given resolution) it is often desirable to measure the diffraction point by point, using a digital x-ray photon counter. Orientation of the crystal to bring each reflection in turn into diffracting position, on the sphere of reflection, is accomplished by a device called a four-circle diffractometer, so called because it contains three circles in an Eulerian cradle to orient the crystal, and a fourth to orient the counter at the desired scattering angle 2θ. Structure amplitudes can be measured somewhat more accurately with a diffractometer than by film techniques, to within (say) 2–3% on average. Electronic area detectors are being developed (Phizackerley *et al.*, 1980) which aim at combining the area-detection efficiency of film with the accuracy and digital output of diffractometers, but are not yet in routine use.

Given the prodigious quantities of data to be analyzed, it is fortunate that the development of high-speed computers followed closely on the first demonstration of phase determination in macromolecular crystals by isomorphous replacement (Section IV.B). Without computers, macromolecular crystallography would be experimentally unfeasible. Performing a Fourier summation such as Eq. (3) with over 10,000 terms using a hand calculator would be daunting to the most enthusiastic crystallographer—and to his or her graduate students. Computers are also essential for conducting multiparameter least-squares refinement (Section VI.B) and for controlling the display of three-dimensional results on sophisticated black-and-white or color graphics terminals (Feldmann, 1976; Langridge et al., 1981).

IV. Phase Determination

A. Introduction

Since little tertiary structural information can be extracted by Patterson syntheses using the intensities alone (Section III.D), it is essential that phase information be obtained. Although the first diffraction patterns from crystals of macromolecules were obtained in 1934 (Bernal and Crowfoot, 1934), it was 20 years before the first experimental determination of phases was accomplished (by Green, Ingram, and Perutz, 1954). Green, Ingram, and Perutz demonstrated that addition of a heavy atom such as Hg to crystals of hemoglobin produced a readily measurable change in the intensities of diffraction, that the heavy atom could be located in the unit cell by comparison of the intensities alone from the original, native crystal and the Hg derivative, and that the phases could then be determined. The technique of using appropriate heavy atoms for phase determination is known as isomorphous replacement. The term "isomorphous" arises from the necessity that the heavy-atom derivative have the "same form" (that is, exactly the same tertiary structure and crystal lattice) as the native crystal. With one notable exception (Section VII.B), all known macromolecular structures have been determined by isomorphous replacement, either directly or by comparison with earlier structures established by that route. The comparison process is known as molecular replacement (Section IV.E). It also turns out that heavy atoms do not merely scatter x-rays in phase, but they can, under suitable circumstances, resonantly absorb x-rays and produce a definite, $\pi/2$ phase shift. This absorption process is known as anomalous scattering, which has also been exploited to assist in phase determination (Section IV.C).

B. Isomorphous Replacement

The principle behind phase determination by isomorphous replacement is easy to demonstrate. Suppose we have a macromolecule described by the electron-density distribution $\rho_M(\mathbf{r})$, to which we add a heavy atom with electron density $\rho_H(\mathbf{r})$. The addition is intended to leave the original molecule otherwise unaltered, so that the resultant derivative electron density, $\rho_D(\mathbf{r})$, is given by

$$\rho_D(\mathbf{r}) = \rho_M(\mathbf{r}) + \rho_H(\mathbf{r}), \tag{7}$$

where M denotes native macromolecule; H, heavy atom; and D, heavy-atom derivative. Now, the Fourier transform of the sum of two functions is given by the sum of their individual transforms (Appendix A), and hence, on taking the Fourier transform of both sides of Eq. (7), we obtain

$$\mathbf{F}_D(\mathbf{s}) = \mathbf{F}_M(\mathbf{s}) + \mathbf{f}_H(\mathbf{s}) \tag{8}$$

by Eq. (1). Equation (8) is known as the isomorphous replacement equation. If we now regard $\rho_M(\mathbf{r})$ as the electron-density distribution of a macromolecular crystal, then the isomorphous-replacement equation states that, for every reflection $\mathbf{s}$ (where $\mathbf{s} = h\mathbf{a}^* + k\mathbf{b}^* + l\mathbf{c}^*$), the structure factor $\mathbf{F}_D(\mathbf{s})$ which describes x-ray scattering from the crystal containing the heavy atom is given by the sum of the structure factor for the native crystal, $\mathbf{F}_M(\mathbf{s})$, and the structure factor which describes scattering from the heavy atom alone, $\mathbf{f}_H(\mathbf{s})$. All three structure factors are vector quantities: they have both an amplitude, the structure amplitude, and a phase. Since the transform of a crystal is the transform of the contents of the unit cell sampled at the reciprocal lattice points (the transform of the crystal lattice; Section III.C), it is necessary that the reciprocal lattice of the native and derivative crystals be identical. In other words, addition of the heavy atom must leave both the tertiary structure of the molecules, and their packing in the real crystal lattice, unaltered.

Suppose the nature and location of the heavy atom in the unit cell are known; that is, its atomic scattering factor f_j, characteristic of the element, and coordinates x_j, y_j, and z_j are known. Then $\mathbf{f}_H(\mathbf{s})$ can be readily calculated, both in amplitude and phase, by Eq. (4). Furthermore, $|\mathbf{F}_D(\mathbf{s})|$ and $|\mathbf{F}_M(\mathbf{s})|$ can be measured experimentally, from crystals of the derivative and native macromolecule. If the isomorphous replacement equation for a single reflection $\mathbf{s}$ is presented graphically (the Harker construction: Fig. 8), then possible solutions for $\mathbf{F}_M(\mathbf{s}) = \mathbf{F}_D(\mathbf{s}) + (-\mathbf{f}_H(\mathbf{s}))$ are given by the vectors **OA** and **OB**, where the phase circles for the native and derivative structure factors intersect. Thus, the original phase uncertainty has been reduced to a phase ambiguity: the phase of $\mathbf{F}_M(\mathbf{s})$ is either ϕ_A or ϕ_B. Notice that these two possibilities are symmetrically disposed about $\mathbf{f}_H(\mathbf{s})$: $\phi_A = \phi_H + \alpha$ and $\phi_B = \phi_H - \alpha$, where α is the angle COA or COB. If however we are dealing with a centrosymmetric projection, for which the structure factors of the corresponding centric zone are all real with phases 0 or π, then the two possibilities coincide (Fig. 8), and single isomorphous replacement provides

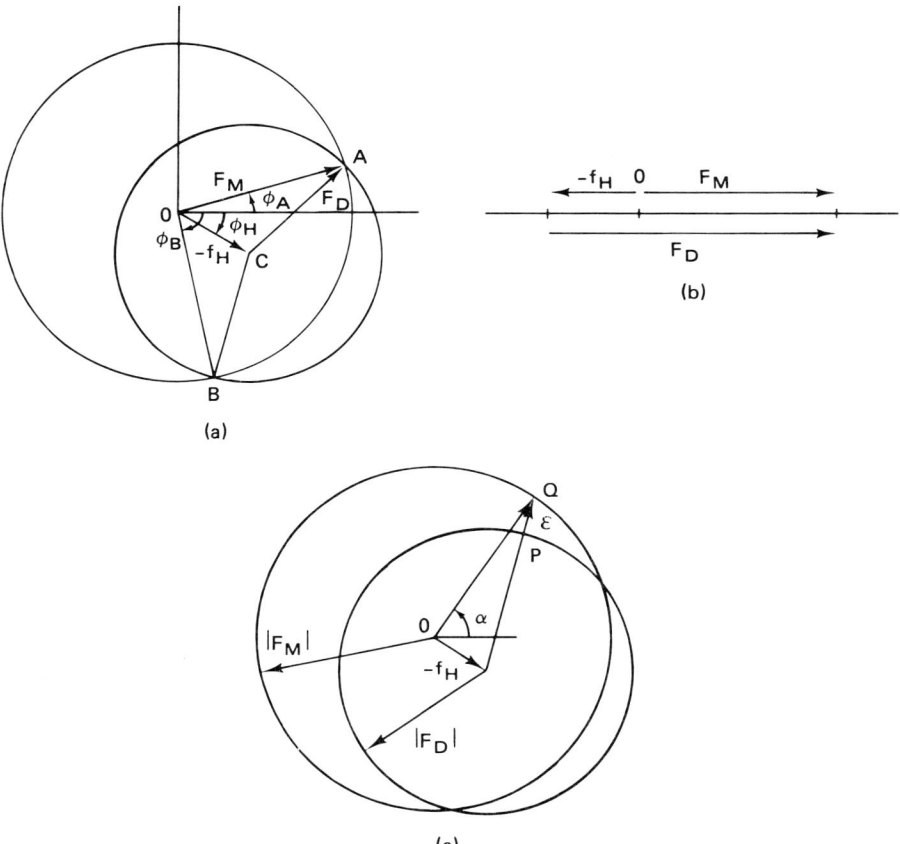

Fig. 8. (a) The Harker construction: a vector diagram that illustrates phasing by single isomorphous replacement for a reflection **s**. The circle with center O and radius $|\mathbf{F}_M(\mathbf{s})|$ represents the native data for this reflection, where $|\mathbf{F}_M(\mathbf{s})|^2$ is measured experimentally but the phase is unknown. Likewise, the circle with center C, where $OC = -\mathbf{f}_H(\mathbf{s})$, and radius $|\mathbf{F}_D(\mathbf{s})|$ represents the derivative data. The isomorphous replacement equation is satisfied by the vectors **OA** and **OB** since, at both A and B, $\mathbf{F}_M(\mathbf{s}) = (-\mathbf{f}_H(\mathbf{s})) + \mathbf{F}_D(\mathbf{s})$. The real axis is horizontal and the imaginary axis is vertical. (b) Single isomorphous replacement in a centrosymmetric projection, where $\mathbf{F}_M(\mathbf{s})$, $\mathbf{F}_D(\mathbf{s})$, and $\mathbf{f}_H(\mathbf{s})$ must be collinear. Here, $|\mathbf{F}_D| > |\mathbf{F}_M|$, and thus the sign of $\mathbf{F}_M$ is the same as that of $\mathbf{f}_H$. (c) Treatment of errors in single isomorphous replacement.

a unique sign determination. If $|F_D(s)| > |F_M(s)|$, then the sign of $F_M(s)$ is the same as that of $f_H(s)$; if $|F_D(s)| < |F_M(s)|$, then the signs are opposite. In both cases, $|f_H(s)| = |F_D(s) - F_M(s)|$. (If however, $|F_M(s)| < |f_H(s)|$, as is possible for a small number of very weak reflections, these relations may not hold; in such cases, $|f_H(s)| = |F_D(s) + F_M(s)|$. Thus in general, $|f_H(s)| = |F_D(s) \pm F_M(s)|$.)

The phase ambiguity which remains from single isomorphous replacement (SIR) may be resolved in two ways: addition of further heavy-atom derivatives to yield multiple isomorphous replacement (MIR), or by supplementing isomorphous-replacement data by anomalous-scattering data (Section IV.C). In MIR, we prepare N different heavy-atom derivatives, either by using different heavy-atom reagents or by altering the conditions so that a single reagent binds in different ways to the macromolecule. Then, Eq. (8) can be generalized to a family of N equations, for each reflection s:

$$F_{Di}(s) = F_M(s) + f_{Hi}(s),$$

where i runs from 1 to N. If $f_{Hi}(s)$ is known and $|F_{Di}(s)|$ is measured for all N derivatives, then there are $(N + 1)$ phase circles, one for the native and one for each derivative. If the experimental data are perfect, all phase circles will

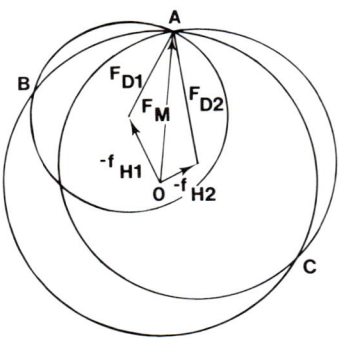

Fig. 9. Phase determination by multiple isomorphous replacement. In the absence of experimental error, the phase circles (three here, one for the native and one for each of two derivatives) intersect at a unique point A. Thus the vector **OA** gives the native-structure factor in both amplitude and phase. The vector **OB** (not drawn) is a second possible solution of the isomorphous replacement equation for the native data and for the first derivative; the vector **OC** is also a second possible solution for the native and second derivative. However, the vectors **OB** and **OC** do not satisfy the data for both derivatives simultaneously, unlike vector **OA**.

intersect at a single point for $N \geq 2$, which defines the vector $\mathbf{F}_M(\mathbf{s})$. The phase of $\mathbf{F}_M(\mathbf{s})$ has thus been uniquely determined (Fig. 9).

C. Anomalous Scattering

We originally assumed that, in the scattering of x rays by electrons, the electrons behave as if they were free. This assumption cannot be strictly valid since electrons are bound to atoms, and for the K and L shell electrons, this binding is quite tight. Bound electrons scatter x rays as if they possessed certain natural or resonance frequencies which correspond to the absorption frequencies of the electrons. The effect of resonance is that the atomic scattering factor becomes complex:

$$f = f_0 + \Delta f' + i\,\Delta f''$$
$$= f' + i\,\Delta f'',$$

where f_0 is the scattering factor which would obtain if the electrons were free; $\Delta f''$ the absorption component of anomalous scattering which is always $\pi/2$ ahead of f_0 in phase; and $\Delta f'$ the dispersion component, which is in phase with f_0 (Fig. 10). The magnitude of $\Delta f''$ is proportional to the x-ray absorption coefficient, and is therefore large only when the incident x-ray frequency (or energy) is at or just above an absorption edge, such as the K edge or one of the L edges. Both $\Delta f'$ and $\Delta f''$ depend strongly on the incident x-ray energy, as shown in Fig. 11. Their magnitudes are usually less than one-fifth that of f_0, but for incident energies very near the absorption edges of certain atoms, they can be much larger, approaching one-half f_0.

For the x-ray range with which we are concerned (energies of around 5–25 keV, or wavelengths of 0.5–2.5 Å), the values of $\Delta f'$ and $\Delta f''$ are very small for light atoms such as H, C, N, and O, which are most abundant in macro-

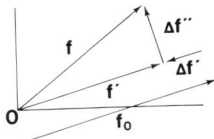

Fig. 10. The complex atomic scattering factor. The scattering factor f_0 found far from x-ray absorption edges for the atom in question is modified near x-ray absorption edges by both a dispersion component $\Delta f'$, which is collinear with but antiparallel to f_0, and an absorption component $\Delta f''$ which is $\pi/2$ ahead of f_0 in phase. The value of f_0 is equal to (at zero scattering angle) or less than the atomic number Z of the atom.

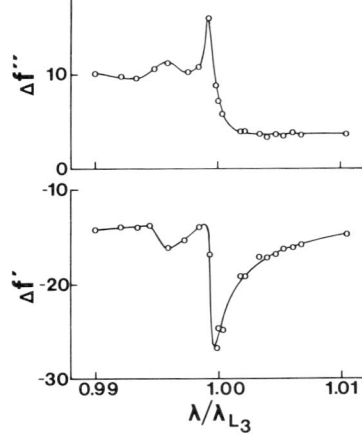

Fig. 11. Variation of $\Delta f'$ and $\Delta f''$ with wavelength for the L_3 absorption edge of Cs. Data taken from Phillips and Hodgson (1980). The units of the ordinate are electrons.

molecules, and anomalous scattering from these elements can safely be ignored. For S and P atoms, the values are somewhat larger but must be considered only under unusual circumstances (see, for example, Section VII.B); for atoms such as Fe and Cu found in metalloproteins, and for typical heavy atoms such as Hg, U, and Pt, the values are much larger, and anomalous scattering effects must be explicitly considered.

Consider a single isomorphous-replacement heavy-atom derivative prepared by the addition of Hg, which has a significant anomalous component, to a native crystal. For the native crystal, we measure $|\mathbf{F}_M(\mathbf{s})|$ and $|\mathbf{F}_M(-\mathbf{s})|$; if the native lacks intrinsic anomalous scatterers, Friedel's law holds and, therefore, $|\mathbf{F}_M(\mathbf{s})| = |\mathbf{F}_M(-\mathbf{s})|$. For the derivative, we measure $|\mathbf{F}_D(\mathbf{s})|$ and $|\mathbf{F}_D(-\mathbf{s})|$. Assume that $\mathbf{f}_H(\mathbf{s})$ and $\mathbf{f}_H(-\mathbf{s})$ are known, as in the isomorphous replacement case (Section IV.B), where now

$$\mathbf{f}_H(\mathbf{s}) = \mathbf{f}'_H(\mathbf{s}) + i\Delta \mathbf{f}''_H(\mathbf{s})$$

and

$$\mathbf{f}_H(-\mathbf{s}) = \mathbf{f}'_H(-\mathbf{s}) + i\Delta \mathbf{f}''_H(-\mathbf{s}).$$

For structure factors derived in the absence of an absorption coefficient, the phase of a reflection $-\mathbf{s}$ is the negative of the phase of a reflection $\mathbf{s}$ (Section III.C). Thus, if the structure factors for the reflection $-\mathbf{s}$ are reflected across the real axis, the reflected vector $\mathbf{F}_M(-\mathbf{s})$ will coincide with $\mathbf{F}_M(\mathbf{s})$, and the reflected vector $\mathbf{f}'_H(-\mathbf{s})$ with $\mathbf{f}'_H(\mathbf{s})$, as shown in Fig. 12. However, since the absorption component of anomalous scattering is always $\pi/2$ in advance of the dispersion and real components, the reflected vector $\Delta \mathbf{f}''_H(-\mathbf{s})$ will not

5. MACROMOLECULAR CRYSTALLOGRAPHY

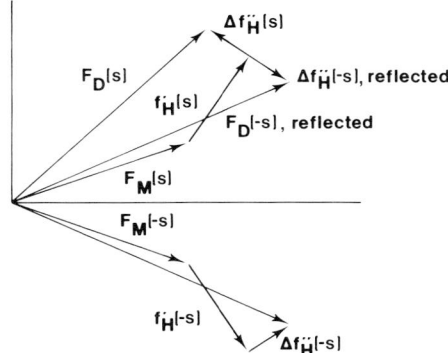

Fig. 12. The breakdown of Friedel's law when anomalous scatterers are present. The vectors $F_M(s)$ and $F_D(-s)$ coincide with $F_M(s)$ and $F_D(s)$ when reflected across the real (x) axis. However, since $\Delta f''$ is always $\pi/2$ in advance of $\Delta f'$ in phase, $\Delta f''_H(-s)$ when reflected does not coincide with $\Delta f''_H(s)$, but rather is antiparallel to it. Thus, $F_D(s) \neq F_D(-s)$ either in magnitude or in phase.

coincide with $\Delta f''_H(s)$. Indeed, these vectors are antiparallel and of equal magnitude, perpendicular to $f'_H(s)$. It follows (Fig. 12) that

$$|F_D(s)| \neq |F_D(-s)|.$$

Thus, in general, Friedel's law no longer holds for crystals which contain anomalous scatterers, and the diffraction pattern does not contain a center of symmetry. The intensity difference $I(s) - I(-s)$, which expresses the deviation from centrosymmetry, is known as the anomalous scattering difference or Bijvoet difference. Reflections which differ in intensity solely by virtue of the breakdown of Friedel's law are known as Bijvoet pairs. The Bijvoet difference will be a maximum when $f'_H(s)$ and $F_M(s)$ are perpendicular, and zero when they are collinear, as they may be by coincidence. However, they will always be collinear if the structure (or its projection) is centrosymmetric, since all phases are then real, 0, or π. Hence, all reflections in centric zones, even of noncentrosymmetric structures of macromolecules containing anomalous scatterers, obey Friedel's law.

The use of anomalous scattering in phase determination can now be shown graphically (Fig. 13), in a manner exactly analogous to isomorphous replacement (Fig. 8). The three phase circles, one for the native of radius $|F_M(s)|$ and two for the single derivative, $|F_D(s)|$ and $|F_D(-s)|$, intersect at a unique point C, and thus determine the phase of the native structure factor, $F_M(s)$.

In general, three or more phase circles are required to define a unique

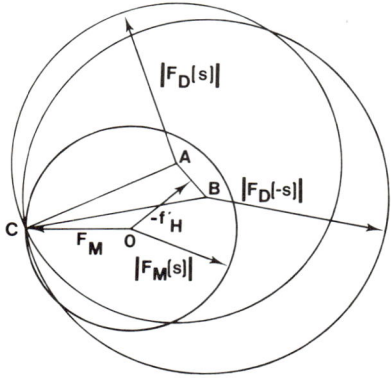

Fig. 13. Phase determination by single isomorphous replacement plus anomalous scattering. The three phase circles, of radii $|F_M(s)|$ (native), $|F_D(s)|$ and $|F_D(-s)|$ (derivative), intersect at the unique point C and thus define the native structure factor F_M.

intersection and the phase of the native. If only anomalous scattering is used, then the two possible phase solutions are symmetrically disposed about $\Delta f''_H(s)$. If only single isomorphous replacement is used, the two possible solutions are symmetrically disposed about $f'_H(s)$, as noted (Section IV.B). Now $\Delta f''_H(s)$ is perpendicular to $f'_H(s)$; hence, the contribution to phase determination of anomalous scattering is exactly complementary to that of isomorphous replacement (Fig. 13). This complementarity makes the combination of anomalous scattering with isomorphous replacement particularly powerful.

There are thus two strategies for phase determination based on the preparation of isomorphous derivatives: multiple isomorphous replacement (MIR) or single isomorphous replacement plus anomalous scattering (SIRAS). Both strategies obviously require that one (SIRAS) or more (MIR) isomorphous derivatives be prepared. Since it turns out to be difficult to ensure true isomorphism for several derivatives, as will be discussed, SIRAS has an advantage. However, the average Bijvoet intensity differences $\langle I_D(s) - I_D(-s) \rangle$ are small, roughly one-tenth of the average isomorphous intensity differences $\langle I_D(s) - I_M(s) \rangle$. (The scale of f'_H and $\Delta f''_H$ is exaggerated relative to F_M in Figs. 12 and 13). Thus, unusually accurate measurements of the structure amplitudes are essential if anomalous scattering is to be effective in phase determination. Anomalous scattering measurements are made by comparison of structure amplitudes from the same derivative crystal and are usually recorded close together in time, in a manner which minimizes errors due to, for example, unequal absorption of x-rays in the crystal or capillary in which the crystal is mounted. Thus, although anomalous scattering produces small intensity differences, these may often be measured nearly as accurately as the larger isomorphous replacement differences, which require comparison of intensities on different crystals, native and derivative, recorded at very different times, on crystals which may be of different shape and which may not be truly isomorphous. Most phase determination now depends on a combination of isomorphous replacement and anomalous scattering, usually involving several derivatives.

D. Treatment of Errors

In practice, intensities cannot be measured with complete accuracy; heavy atoms are never perfectly isomorphous, and their structure factors are never accurately known. Phase circles such as those depicted in Figs. 8A and 13 are therefore "fuzzy"; there is uncertainty in their centers due to lack of perfect knowledge of $f_H(s)$, and uncertainty in their radii due to experimental errors in the structure amplitudes $|F(s)|$. How can this imperfect information be combined to produce the "best" phase? Indeed, is there a precise definition of the "best" phase? These questions were first approached by Blow and Crick (1958), and their treatment, with later refinements, is still standard. They showed that it was valid to regard $|F_M(s)|$ and $f_H(s)$ as error free and therefore to assume that all errors could be lumped together in $|F_D(s)|$. Now consider two phase circles in SIR (Fig. 8). For a trial native phase α of $F_M(s)$, there is a "lack of closure" error $\epsilon(\alpha)$, where $\epsilon = |F_D(s)|_{obs} - |F_D(s)|_{calc}$, or PQ in Fig. 8. Assuming a Gaussian distribution of error in $|F_D(s)|$, then the probability $P(\alpha)$ that the native phase is α is

$$P(\alpha) = \exp[-\epsilon^2(\alpha)/2E^2], \qquad (9)$$

where E is the standard deviation of the error distribution. In centric zones for perfect data, $|f_H(s)| = |F_D(s) - F_M(s)|$ for most reflections, as noted (Section III.B). Hence E may be estimated from

$$\langle E^2 \rangle = \langle [|F_D(s) - F_M(s)| - |f_H(s)|]^2 \rangle.$$

When j derivative phase circles are involved from MIR, from SIR plus AS, or from MIR plus AS, then the j phase-probability distributions may be multiplied together, on the assumption that they are independent:

$$P(\alpha) = \prod_j P_j(\alpha) = \exp\left[-\sum_j \epsilon_j^2(\alpha)/2E_j^2\right].$$

The total probability distribution $P(\alpha)$ is a smoothly varying function of α, as shown by representative examples in Fig. 14. A possible choice of phase would then be the value of α for which $P(\alpha)$ is a maximum, the most probable phase. Although this might be appropriate for roughly unimodal

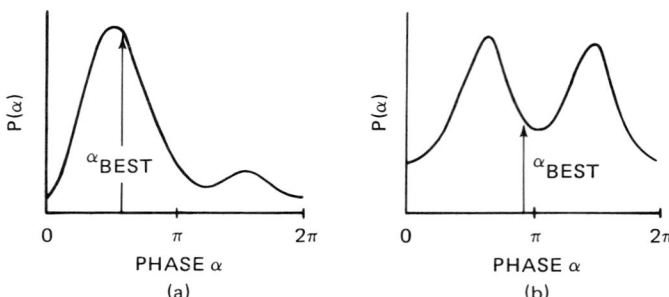

Fig. 14. Examples of two phase-probability distributions: (a) a nearly unimodal distribution which would yield a high figure of merit; (b) a bimodal distribution which would yield a low figure of merit.

probability distributions, many distributions are bimodal. Blow and Crick (1958) showed that the "best" phase was the centroid of the phase-probability distribution, given by

$$\mathbf{F}_M(\mathbf{s})_{best} = |\mathbf{F}_M(\mathbf{s})| \int_0^{2\pi} \exp[i\alpha] P(\alpha)\, d\alpha \Big/ \int_0^{2\pi} P(\alpha)\, d\alpha$$
$$= m(\mathbf{s})|\mathbf{F}_M(\mathbf{s})| \exp[i\alpha_{best}]. \tag{10}$$

The centroid is located at a radius $m|\mathbf{F}_M(\mathbf{s})|$ from the origin and phase angle α_{best}. The quantity m is known as the figure of merit.

If the probability distribution is sharp, reflecting little uncertainty in the phase angle, $m \simeq 1.0$; but if the distribution is broad, reflecting considerable uncertainty, $m < 1.0$. Thus, m functions as a weighting factor, to diminish the contribution of terms for which the phase is uncertain. Fourier syntheses using $\mathbf{F}_M(\mathbf{s})_{best}$ are "best" in the sense that use of these phases leads to a minimum least-mean-square error in electron density over the entire unit cell, $\langle \Delta\rho^2 \rangle$. Indeed,

$$\langle \Delta\rho^2 \rangle = V^{-2} \sum_{\mathbf{s}} |\mathbf{F}_M(\mathbf{s})|^2 [1 - m^2(\mathbf{s})],$$

which shows that the larger the values of m, the smaller the error in the electron density and vice versa (Dickerson et al., 1961).

Of course, m varies from reflection to reflection and is usually lower for weak reflections for which the magnitude of the amplitude error E may be comparable with the structure amplitude itself, $|\mathbf{F}_M(\mathbf{s})|$. Hence, the figure of merit is lowest at high resolution, where there is the largest proportion of weak reflections and where the effects of nonisomorphism are most apparent. Phasing becomes inaccurate when the error E is comparable with the mean structure amplitude of the heavy atoms alone, $\langle |\mathbf{f}_H(\mathbf{s})| \rangle$, and the "phasing power" is often expressed by the ratio $\langle |\mathbf{f}_H(\mathbf{s})| \rangle / E$, which should always be greater than 1.0.

It will always happen that certain reflections have small values of $|\mathbf{f}_H(\mathbf{s})|$ for that derivative, and such reflections cannot be phased by that derivative. MIR circumvents this problem, since it is unlikely that $|\mathbf{f}_H(\mathbf{s})|$ will be small for all derivatives.

In practice, it is not appropriate to use the same value of E for both isomorphous replacement and anomalous scattering data; the latter is usually associated with an E value roughly one-third that of the former, reflecting the greater precision of measurements made on the same crystal at nearly the same time (Section IV.C). Hence, a more complicated phase-probability expression is used (Matthews, 1966) in which the overall phase-probability distribution is given by the product of two terms analogous to Eq. (9), one each for the IR and AS contributions:

$$P(\alpha) = P_{IR}(\alpha) \cdot P_{AS}(\alpha).$$

Details are given by Matthews (1966, 1977).

E. Molecular Replacement and Noncrystallographic Symmetry

A single macromolecule may often crystallize in several different space groups, under the same or very slightly different solvent conditions; that is, the crystallization is polymorphic. Closely related molecules (for example, the same molecule from different species, with similar but not identical amino acid or nucleotide sequence and closely related tertiary structures) often crystallize in different space groups. If the structure in one space group is determined, can the structures in the others be inferred, without the necessity for laborious, independent phase determination? Alternatively, a macromolecule may crystallize with more than one copy of the structure in the crystallographic asymmetric unit. That is, the crystal possesses noncrystallographic symmetry, in which each copy is in a different crystallographic environment, subject to different packing constraints. Can the fact that the diffraction pattern arises from the superposition of different "views" of the structure be exploited to assist in phase and structure determination?

The answer to both these important questions is, fortunately, yes. As mentioned previously (Section II), crystal packing forces are weak, and thus chemically identical (or closely related) molecules retain basically the same tertiary structure, even when subjected to the varied forces present in different space groups or in different environments arising from noncrystallographic symmetry. To put it another way, intermolecular and solvent-dependent interactions are of lesser importance than the intramolecular interactions which stabilize the tertiary structure of the macromolecule. The only local aspects of structure which may not be retained are certain side-chain orientations near the intermolecular contacts, or side chains and local loops which are directly influenced by different solvent conditions, such as sulfate-binding sites near basic side chains. These represent only a small fraction of the structure, and make at most a minor contribution to its diffraction pattern. In general, tertiary structure is closely conserved. (The only important exception arises for macromolecules consisting of several flexibly connected domains such as immunoglobulins; although tertiary structure within the domains is found to be conserved, the interdomain relationships often depend on crystal packing and solvent effects).

If structure is conserved, then its Fourier transform is also conserved. Consider two related macromolecules which crystallize in different space groups, such as hemoglobin A and hemoglobin F (Frier and Perutz, 1977). The transform of the crystal is the transform of the contents of one unit cell sampled at the transform of the real lattice, the reciprocal lattice (Section III.C). Since the contents are related in structure, the transforms of the contents in the two space groups are also related but are sampled at different locations, arising from the different real lattices.

To establish the relationship between the transforms of the contents, it is necessary to relate the molecules in one unit cell in the first space group to those in the second space group, in both rotation and translation— molecular replacement. Rossmann and Blow (1962) pioneered these techniques in macromolecular crystallography and noted that the relative angular orientation of two molecules could be determined by examining the superposition of their Patterson functions, P_1 and P_2, via the rotation function R:

$$R = \int_U P_2(\mathbf{x}_2) \, P_1(\mathbf{x}_1) \, d\mathbf{x}_1.$$

The integral is taken over a volume U which is intended to include the intramolecular vectors but exclude the longer intermolecular vectors. The coordinates $\mathbf{x}_1$ and $\mathbf{x}_2$ within the Pattersons are related by a rotation matrix $[C]$: $\mathbf{x}_2 = [C] \, \mathbf{x}_1$. The rotation function R will be a maximum when peaks arising from intramolecular atomic vectors in the first Patterson P_1 coincide with peaks in P_2; that is, when the orientations of the molecules coincide. In the present example, the atomic structure of hemoglobin A was known, and hence its Patterson function P_A could be completely calculated. P_A could then be compared with the observed Patterson function P_F from the crystals of unknown structure, hemoglobin F, via the rotation function, to establish the orientation of the hemoglobin F molecules in their unit cell. Solving the translation problem (that is, locating the center of mass of the molecules in the unknown unit cell) is more difficult; for hemoglobin F it might be achieved by locating only the iron atoms through their anomalous scattering, or by packing constraints.

Finally, with the hemoglobin A molecules now correctly oriented and positioned within the hemoglobin F unit cell, the appropriate chemical substitutions to convert the primary structure of the β chains of hemoglobin A to the related structure of the γ chains of hemoglobin F would be made. The diffraction pattern would be calculated, $|\mathbf{F}_{calc}(\mathbf{s})| \exp[i\alpha_{calc}]$, and converted to $|\mathbf{F}_{obs}(\mathbf{s})| \exp[i\alpha_{calc}]$ by insertion of the observed structure amplitudes from the hemoglobin F crystal; a new electron-density map calculated; adjustments made to the model and new phases α'_{calc} calculated; and a process of cyclic refinement initiated. This process leads rapidly to a structure for hemoglobin F, without necessarily requiring any heavy-atom derivatives or explicit phase determination. In fact, SIR was used in combination with molecular replacement to facilitate the analysis (Frier and Perutz, 1977).

The rotation function suffers from a poor signal-to-noise ratio, as the correct relative orientation of the two Pattersons often produces a peak not much larger than those arising from other, spurious orientations. Also, the translation problem has lacked a satisfactory, general solution. Nevertheless, specific solutions are available, and the correct peak in the rotation function has usually been identifiable. Molecular replacement is the technique of choice when closely related structures are to be determined.

The similar problem which arises when the asymmetric unit contains more than one copy of a structure has been investigated in great detail by Bricogne (1974, 1976), based on earlier studies by Rossmann and Blow, Crowther and Jack. There are several constraints on electron density: it is everywhere real (in the absence of anomalous scattering), positive, arising from discrete, nearly equal atoms contained within molecular boundaries, and satisfying both space-group and (in certain cases) noncrystallographic symmetry. Bricogne considers the constraints on phases which arise from purely geometric sources: the existence of a molecular boundary, noncrystallographic symmetry, and polymorphism, and demonstrates with mathematical rigor (Bricogne, 1974) that the process of phase refinement is assured of success if there are three or more copies of the structure to be compared in the asymmetric unit or polymorphs.

For noncrystallographic symmetry, the strategy is as follows. Obtain a set of trial phases, for example by SIR; compute an electron density map using $|F_{obs}|$ and the trial phases; identify the symmetry operation which relates the various copies of the structure and a trial molecular boundary; average the electron density of the copies through this symmetry operation, and set the density outside the boundary (solvent) to a uniform level; reconstruct the original unit cell with the new, average electron density, and compute its transform; combine these new phases with the original, trial phases in an appropriately weighted manner due to Sim (1959); compute a second electron-density map; and reiterate until convergence is attained, usually in a few cycles. The same strategy may be applied to polymorphs, where a known electron-density distribution (such as hemoglobin A) may be averaged in real space with an unknown (such as a trial electron-density map for hemoglobin F).

Although the strategy is straightforward to describe, its successful implementation involves highly sophisticated computation, also described by Bricogne (1976). The power of the Bricogne strategy may be appreciated from the example of hemagglutinin, the antigenic protein from the surface of influenza virus, which crystallizes with a trimer, three copies of molecular weight 69,474 each, in the asymmetric unit (Wilson et al., 1981). Single isomorphous replacement data were collected to 3-Å resolution and 6 heavy-atom sites per molecule, 2 per monomer, were identified by difference Patterson techniques (Section V.B). SIR phases to 3-Å resolution had a mean figure of merit of only 0.34 and hence yielded an essentially uninterpretable electron-density map. However, 11 cycles of threefold symmetry averaging, interspersed with redetermination of the molecular boundary as the electron-density map became cleaner, led to a marked increase in the figure of merit and a fully interpretable map (Wilson et al., 1981). The truly remarkable improvement in the map is illustrated in Fig. 15.

A second example is provided by tomato bushy-stunt virus, where only four cycles of icosahedral averaging of an electron-density map at 2.9-Å resolution, obtained by MIR from two derivatives, led to an interpretable map (Harrison et al., 1978).

It has generally been believed that the larger the asymmetric unit and the larger the solvent content, the more difficult it is to solve a crystal structure. Although more data of course have to be collected, these recent results make it clear that, paradoxically, phase determination may be significantly easier when a large asymmetric unit contains both several copies of a structure and appreciable disordered solvent of truly uniform electron density.

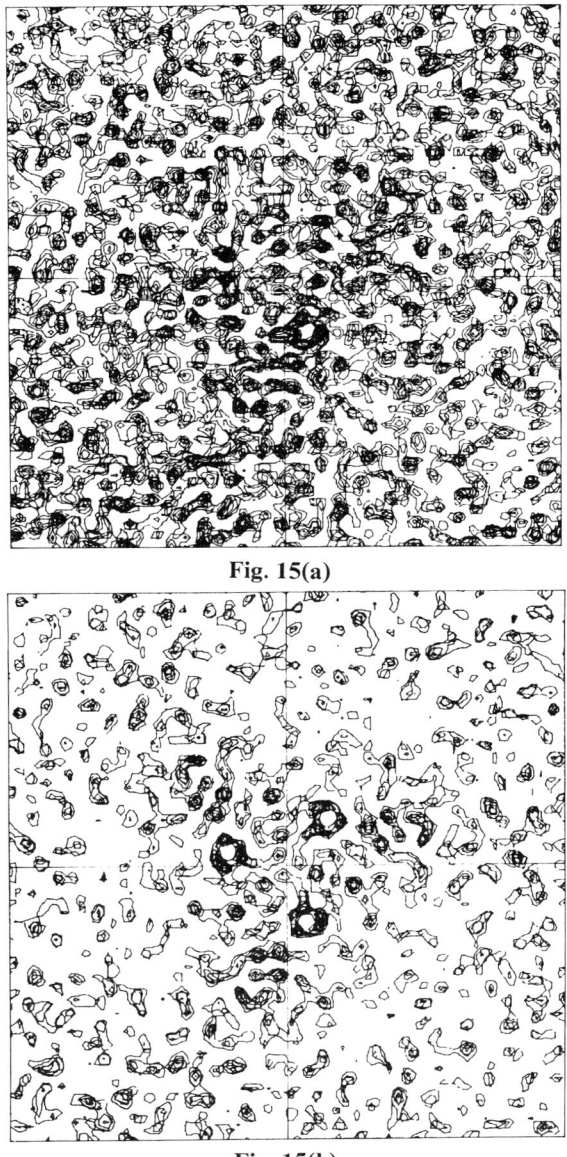

Fig. 15(a)

Fig. 15(b)

Fig. 15. Improvement in the quality of the electron-density map for hemagglutinin by noncrystallographic symmetry averaging and boundary determination. All panels show some superimposed sections of the electron-density map, viewed down the noncrystallographic threefold axis of symmetry. (a) The original electron-density map, phased by single isomorphous re-

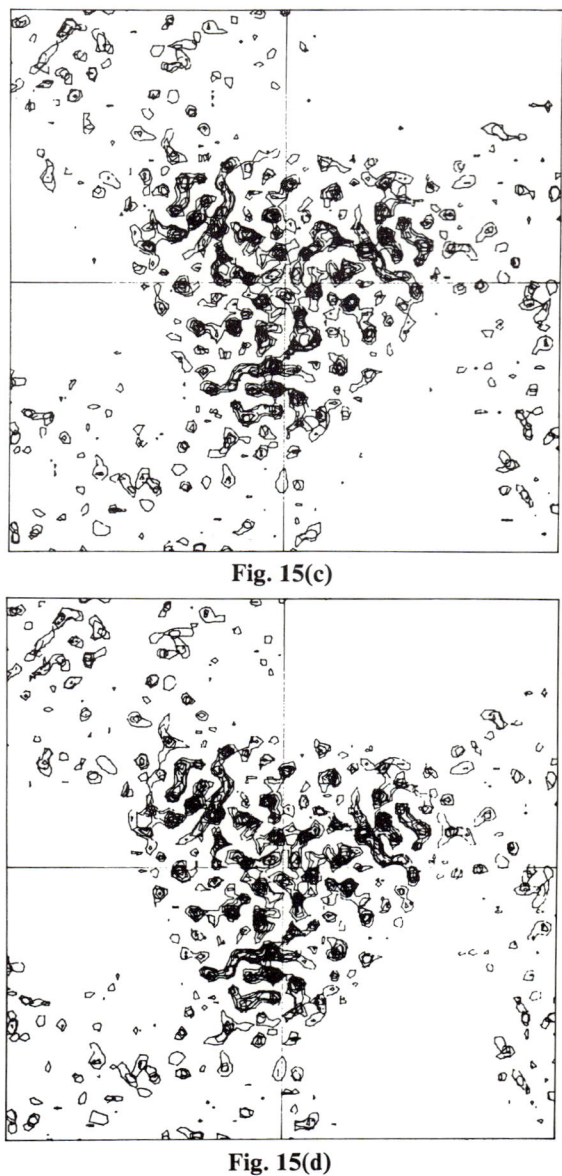

Fig. 15(c)

Fig. 15(d)

placement to 3-Å resolution, with a mean figure of merit of 0.34. The molecule is barely distinguishable from solvent, and, due to experimental error, the density does not exhibit threefold symmetry. (c) Threefold symmetry averaging of the original electron-density map shown in (a). The molecular boundary is becoming apparent. (c) Application of a molecular boundary to the map in (b). (d) Further threefold symmetry aver-

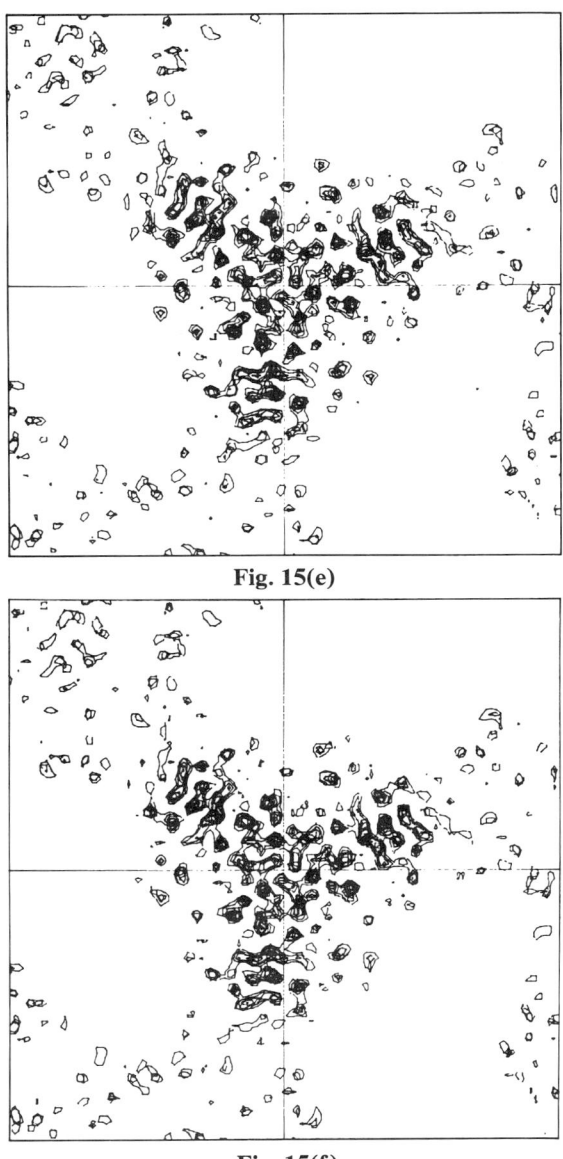

Fig. 15(e)

Fig. 15(f)

aging of the map in (c). (e) The electron-density map after seven cycles of symmetry averaging and boundary identification. (f) The final electron-density map after eleven cycles. This map is fully interpretable, allowing the polypeptide backbone to be clearly traced. Contrast with (a). See Wilson *et al.* (1981) for a full description of the structure determination of hemagglutinin. (Illustrations kindly provided by D. C. Wiley.)

V. Heavy-Atom Derivatives

A. Preparation

As we have seen, phase determination requires the preparation of isomorphous heavy-atom derivatives. Despite intense efforts, no single strategy or reagent has been developed which will be appropriate for all proteins or all nucleotides. Indeed, obtaining suitable derivatives has proved to be a stumbling block second only to initial crystallization.

There are several reasons for this difficulty. If derivatives are to be isomorphous, they must interact (covalently or noncovalently) with surface groups or in surface crevices. As the interior of macromolecules is largely close packed, introduction of an additional, bulky group is not usually possible without perturbation of the structure. However, reactive surface groups such as the ϵ-amino group of lysine are both numerous and mobile. Heavy-atom compounds designed to react with them are therefore likely to produce extensive substitution, and may not be well localized; both factors make location of the heavy atoms by difference Patterson techniques (Section V.B) difficult. Compounds designed to react with less numerous groups such as α-amino groups, the 3'-OH of nucleotides, or imidazole have often exhibited much less specificity than desired. The solvent from which the crystals are grown may interfere with the desired reaction with the macromolecule; for example, ammonium ions are a notorious offender in this regard. Finally, even if the chemistry is well understood, there is no means of predicting whether a particular compound will produce an isomorphous or a nonisomorphous derivative. Nonisomorphism and partial reaction at numerous sites are the most common reasons for rejection of a possible derivative.

Derivatives may be sought either be reacting in free solution, purifying the derivative and then attempting to crystallize it, or by soaking native crystals in solutions of the heavy-atom compound. The former has the advantage that the reaction can be carried out in any suitable solvent under controllable conditions, but nonisomorphism of the derivative or, worse yet, complete failure to crystallize are common outcomes. The latter strategy has proved more popular, although the reaction is obviously restricted to those solvents which maintain the integrity of the crystal lattice.

Roughly one hundred compounds have yielded covalent or noncovalent isomorphous derivatives of *some* macromolecule, and many more have been tried unsuccessfully. Lists of compounds examined and strategies are given by Holmes and Blow (1965), Eisenberg (1970), McPherson (1976a), and Blundell and Johnson (1976); the last is particularly comprehensive. One widely used approach is the shotgun: soak native crystals in numerous different compounds, at different concentrations (usually in the millimolar range) for a fixed period of time, say 24 hours. Diffusion of small molecules through the liquid intermolecular channels into the body of the crystal is generally complete in one hour or so, although chemical equilibrium may be attained more slowly. Several results are possible: no reaction, in which case the diffraction pattern is basically unaltered; extensive reaction, in which case the crystal may be shattered or the diffraction pattern grossly changed; or moderate reaction, producing changes in the intensities of the diffraction pattern clearly visible by eye. If extensive reaction occurs, repeat the soak at a series of lower concentrations until crystal cracking is clearly absent. If moderate reaction occurs, ensure that the changes in the diffraction pattern arise solely from changes in the structure factors and not from changes in the cell dimensions (which produce changes in intensities resulting from altered sampling of the molecular transforms). Small changes in cell dimensions are normal, but must not exceed about 0.3 Å if isomorphism is to be retained at high resolution.

A second strategy is to make use of some unusual aspect of the chemistry of the macromolecule. For example, several heavy-atom compounds react reasonably specifically with the sulphydryl group of cysteine residues (see McPherson, 1976a); indeed, the first heavy-atom derivative of any protein crystal was of this type (Green *et al.*, 1954). Successful derivatives have also been prepared by exploiting the functional properties of the macromolecule, as in the reaction with inhibitor or substrate analogs tagged with a heavy atom. Alternatively, many proteins bind ions such as Ca or Zn. In certain proteins it has proved possible to replace Ca ions by lanthanides with complete retention of isomorphism, for example, in thermolysin (Matthews and Weaver, 1974). Unfortunately, this is not generally applicable; all our attempts to prepare lanthanide derivatives of a small intestinal calcium-binding protein were hindered by nonisomorphism when extensive substitution for Ca was achieved (Szebenyi *et al.*, 1981). Similarly, the Zn at the active sites of carboxypeptidase A and carbonic anhydrase could be replaced by Hg (Hartsuck *et al.*, 1965) and in insulin by Pb (Adams *et al.*, 1969).

A severe drawback to approaches which place the heavy atom at the active site or ion-binding site is that local nonisomorphism around the heavy-atom sites is almost inevitable, and this particularly interesting region of the structure will as a consequence be somewhat uncertain in the final electron-density map. Such can be labeled "Trojan horse" derivatives.

B. Location: The Difference Patterson Function

As shown in Section IV.B, for centric zones the relationship

$$|f_H(s)| = |F_D(s) - F_M(s)|$$

holds for all but a few very weak reflections. Now

$$|f_H(s)| = ||F_D(s)| - |F_M(s)|| = \Delta F,$$

say, and hence

$$|f_H(s)|^2 = (\Delta F)^2.$$

Thus a difference Patterson function, defined as a Patterson function (Section III.D) with coefficients $(\Delta F)^2$, will possess peaks at locations which represent vectors between the heavy atoms only. The far more numerous peaks arising from vectors between the light atoms of the native macromolecule, or between the heavy atoms and the light atoms, have effectively been subtracted out. If there are a sufficiently small number of heavy atoms per asymmetric unit, it may be possible to deduce their locations by direct inspection of the difference Patterson function, in exactly the same way that certain structures of small molecules may be solved by inspection of their Patterson functions. It is this requirement for ready interpretability of the difference Patterson function that limits the number of sites of heavy-atom substitution which can be tolerated.

This derivation applies of course only to centric zones, the projections of noncentrosymmetric macromolecular structures. Some heavy atoms have been located solely by examination of the difference Patterson functions obtained from centric zones. However, it is more common to use the more abundant noncentric data. In noncentric zones, for which F_D and F_M can have any phase, the exact expression for $(\Delta F)^2$ is (Fig. 16):

$$(\Delta F)^2 = 4|F_M|^2 \sin^4 \alpha/2 + |f_H|^2 \cos^2 \gamma - 4|F_M||f_H| \sin^2(\alpha/2) \cos \gamma$$

Thus a difference Patterson for noncentric zones contains three terms, the first of which arises from macromolecule–macromolecule vectors, $(|F_M|^2)$; the second from heavy-atom–heavy-atom vectors, $(|f_H|^2)$; and the third from macromolecule–heavy-atom cross vectors, $(|F_M||f_H|)$. The second is the desired term, the signal, and the first and third contribute only noise.

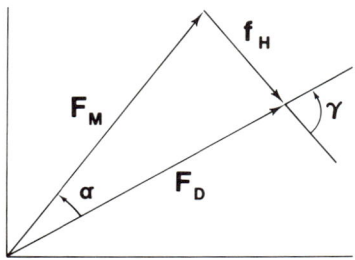

Fig. 16. The relationship between F_M, F_D, and f_H in a noncentric zone. The magnitude of f_H has been exaggerated relative to F_M and F_D.

Now if $|f_H| \ll |F_M|$ as is usually the case, then the angle α is small, as is $\sin^4 \alpha/2$; the first term is greatly weighted down. Furthermore, the angle γ will take all values between 0 and π with equal probability, and hence the third term will take both positive and negative values, averaging to zero in the complete difference Patterson summation over all reflections. Thus, the signal-to-noise ratio of the complete difference Patterson, employing both centric and noncentric data, may be favorable enough to allow identification of the desired heavy-atom–heavy-atom vectors, and hence deduction of the heavy-atom locations. Interpretation is often further aided by the presence of a simple pattern of peaks in the complete difference Patterson function on certain sections — the so-called Harker sections — which display the relation between atoms related by crystallographic symmetry.

Further useful difference Patterson functions can be calculated if anomalous scattering data is available in addition to the purely isomorphous replacement data assumed above. For example, an anomalous difference Patterson function with coefficients

$$(\Delta F'')^2 = (|\mathbf{F}_D(\mathbf{s})| - |\mathbf{F}_D(-\mathbf{s})|)^2$$

can be calculated. This function uses data from a single type of crystal (here, the derivative), and the question of isomorphism or nonisomorphism clearly does not arise. Such an anomalous difference Patterson contains identifiable anomalous-scatterer–anomalous-scatterer vectors — from which the location of the anomalous scatterers can be deduced. These can be the heavy atoms or atoms intrinsic to the native molecule, such as the iron atoms in hemoglobin.

Finally, the complementary nature of isomorphous replacement and anomalous scattering data can be exploited to produce a "combined" difference Patterson function:

$$|f_H|^2 = (|\mathbf{F}_D(\mathbf{s})| - |\mathbf{F}_M(\mathbf{s})|)^2 + k^2(|\mathbf{F}_D(\mathbf{s})| - |\mathbf{F}_D(-\mathbf{s})|)^2/4$$
$$= (\Delta F)^2 + k^2(\Delta F'')^2/4,$$

where k is the ratio of the real to the anomalous scattering of the heavy atoms. That is, $k = f'/\Delta f''$ (Section IV.C). A more accurate form of the combined difference Patterson function, known as the $|f_{HLE}|^2$ Patterson where HLE denotes "heavy-atom lower estimate" is given by Blundell and Johnson (1976; see pp. 338–340).

C. Refinement of Heavy-Atom Parameters

Once the approximate location(s) of heavy atoms have been established by examination of difference Patterson functions, the heavy-atom parameters which contribute to f_H should be determined as accurately as possible before employing these values of f_H in final phase determination. These parameters are the location, occupancy, and "shape" (or extent of order) of the heavy atom. Not all heavy-atom sites may be fully occupied; and the heavy atoms may be somewhat "smeared out" due to static or dynamic disorder, which may be isotropic or anisotropic in character.

Two techniques are in general use for parameter refinement: least-squares and phase refinement. Consider first least-squares refinement for a centric zone. The desired parameters are those which minimize the sum of squares of the residuals:

$$\Sigma = \sum_s W(|\mathbf{F}_D| - |\mathbf{F}_M \pm \mathbf{f}_H|)^2,$$

where W is a weighting factor, and the sum is taken over all reflections s. Minimization may be carried out either by conventional least-squares equations or by a "brute force" technique introduced by Hart (1961), in which the parameters are individually and systematically varied. This expression is not appropriate for noncentric zones. However, the availability of an estimate $|\mathbf{f}_H^{obs}| = |\mathbf{f}_{HLE}|$ allows a suitably weighted sum over $(|\mathbf{f}_{HLE}| - |\mathbf{f}_H^{calc}|)^2$ to be minimized, where $|\mathbf{f}_H^{calc}|$ is calculated from the current heavy-atom parameters. This approach has been widely and successfully used; it has the major advantage that the parameters of only one heavy-atom derivative are refined at a time, and hence errors arising from other derivatives have no influence.

By contrast, phase refinement utilizes data from all available derivatives and refines the parameters for each sequentially. For example, for the jth derivative, approximate native phases ϕ can be calculated from the current set of heavy-atom parameters, by the techniques described in Section IV.B and IV.C. One then seeks to minimize

$$\Sigma = \sum_j \sum_s W_j [\mathbf{F}_D^j - (\mathbf{F}_M + \mathbf{f}_H^j)]^2 = \sum_s W_j \epsilon_j^2,$$

where ϵ_j is the lack-of-closure error (PQ in Fig. 8c). After each derivative has been refined, usually through several cycles, native phase angles are recalculated from the new heavy-atom parameters and sequential refinement of each derivative repeated. Convergence is achieved when the phase angles settle to constant values. Problems arise when one derivative dominates the phase determination; the parameters of that derivative may not refine satisfactorily. If sufficient derivatives are available, the derivative currently being refined should be omitted from the phase calculation. If it is retained, its parameters obviously contribute to the phase determination and may greatly hinder convergence or seriously bias the final phase values.

The course of refinement is usually followed by calculating reliability factors, or R factors. Very commonly used for centric zones is

$$R_c = \sum_s ||\mathbf{F}_D \pm \mathbf{F}_M| - f^{\text{calc}}| / \sum_s |\mathbf{F}_D \pm \mathbf{F}_M|,$$

which, for good data on a well-refined derivative, will approach 0.40. For noncentric data, the R-factor originally given by Kraut is used:

$$R_K = \sum_s |\mathbf{F}_D - (\mathbf{F}_M \pm \mathbf{f}_H)| / \sum_s \mathbf{F}_D = \sum_s |\epsilon| / \sum_s \mathbf{F}_D.$$

The final value of R_K is not diagnostic of the quality of the refinement, as R_K is larger for more heavily substituted derivatives. However, its decrease during the course of refinement of a single derivative indicates the progress of the refinement.

D. Difference Fouriers

Once a refined set of heavy-atom parameters is available, trial native phases may be obtained, α_M. A difference Fourier synthesis can then be calculated with coefficients

$$m(|\mathbf{F}_D| - |\mathbf{F}_M|) \exp[i\alpha_M] = m\Delta F \exp[i\alpha_M]. \tag{11}$$

In a centric zone, a difference Fourier synthesis yields the difference in the (projected) electron density between the derivative and the native crystals. That is, it displays very directly the electron-density distribution of the heavy atoms, and is much easier to interpret than a difference Patterson function. In noncentric zones, the difference Fourier coefficient ΔF resembles the projection of $\mathbf{F}_D$ on $\mathbf{F}_M$ (Fig. 17) and yields again the difference in electron density between the native and the derivative, although weighted down and with increased noise (Henderson and Moffat, 1971). By inspection of a difference Fourier synthesis, hitherto unsuspected heavy-atom sites may be revealed, which can then in turn be refined, yielding new phases, a new difference Fourier, and so forth.

There is one major problem in the use of difference Fouriers for heavy-atom identification and refinement: their usefulness depends almost entirely on the quality of the phases, which is certainly low at an early stage of refinement. Qualitative errors introduced by poor phases (such as identification of totally spurious heavy-atom sites) can persist through many subsequent cycles of refinement. To guard against this possibility, it is prudent to calculate cross-difference Fouriers, in which the phases independently de-

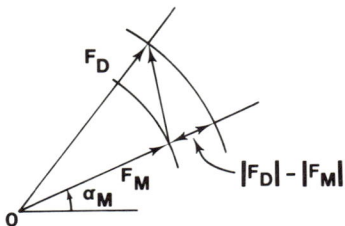

Fig. 17. The difference Fourier coefficient $|F_D| - |F_M|$. If the angle between F_D and F_M is small (and it is exaggerated here), then $|F_D| - |F_M|$ is approximately equal to the projection of the true difference structure factor, $F_D - F_M$, along F_M.

rived from other derivatives are combined with ΔF values from the single derivative in question. Furthermore, all heavy-atom parameters, whether initially derived from difference Patterson or difference Fourier analysis, must be shown to be consistent with the original, experimentally based difference Patterson functions.

VI. The Electron-Density Map

A. Interpretation

Once phases have been obtained, the three-dimensional electron-density distribution $\rho(x,y,z)$ can be calculated by Eq. (3), Section III.C, in the form of a series of two-dimensional sections, which are usually presented as contour maps for ease in interpretation. The nature of the electron-density map depends critically on the resolution to which the x-ray diffraction data have been obtained (Table II). The resolution is given by the minimum value of the interplanar spacing d in Bragg's law (Section III), d_{min}. Formally, groups may be resolved only if they are farther apart than $0.71\ d_{min}$. In practice, electron-density maps calculated at $d_{min} = 5.5-6$ Å, such as the initial maps of myoglobin and hemoglobin, reveal α helices as clearly identifiable continuous rods of electron density; other secondary structural features are completely indistinct. The course of the polypeptide backbone is not generally apparent until $d_{min} < 3.5$ Å, when bulky side chains such as Tyr, Phe, and Trp may be revealed as dense "branches." By $d_{min} = 2.0-2.5$ Å, the polypeptide backbone and side chains will generally be clearly visible, with protuberances on the backbone marking the carbonyl oxygens. Polynucleotide electron-density maps have proved somewhat easier to interpret at lower resolution, due to the electron-dense phosphate groups, 5–7 Å apart in the

Table II Interpretation of Electron Density Maps

d_{min} (Å)	$1/d_{min}^{3}$ [a]	$2\theta_{max}$ [b] (deg)	Notes
20	0.005	4.4	Typical resolution of negatively stained electron microscope samples. External form of molecule detectable.
7	0.13	12.6	Phosphate backbone of polynucleotides just detectable; α-helices visible as rods in well-phased, unstained electron microscope studies of bacteriorhodopsin.
5.5	0.26	16.1	α-Helices visible as rods in initial myoglobin and hemoglobin studies.
3.5	1.00	25.4	Lowest resolution at which polypeptide backbone may be traced; some discontinuities likely; bulky side chains evident. Polynucleotide backbone and base-pairing evident.
2.8	1.96	31.9	Polypeptide backbone continuous and readily traceable; almost all side chains discernible but not clearly formed.
2.0	5.36	45.3	Side chains well formed; peptide carbonyls detectable.
1.5	12.6	61.8	Aromatic side chains and bases "dimpled."
1.0	42.9	100.7	Almost all atoms visible, nearly resolved.

[a] Relative to d_{min} of 3.5 Å = 1.00.
[b] For $\lambda = 1.54$ Å.

polynucleotide backbone, and the characteristic bulky, planar bases. For example, an early electron-density map of yeast tRNA-Phe at 4 Å resolution clearly revealed the L-shaped outline of the molecule, and exhibited about 80 peaks 5–7 Å apart in an identifiable linear sequence, assigned to the backbone phosphate groups (Kim et al., 1973).

True atomic resolution has not been achieved for any macromolecular crystal, though it may ultimately be attainable for crambin (Hendrickson and Teeter, 1981) and avian pancreatic polypeptide (Blundell et al., 1981).

Since the amount of diffraction data varies as d_{min}^{3}, it has been usual to calculate an initial, low-resolution map at around 3.5 Å resolution before proceeding to process the very large amounts of extra data required at higher resolution (Table II). However, since the biochemical information at low resolution is limited, this step is now often bypassed unless the exigencies of poor diffraction quality, inadequately isomorphous derivatives (or even scientific grantsmanship!) require it.

Interpretation of a high-resolution electron-density map demands that a set of atoms be fitted to it, a process which depends heavily on *a priori* knowledge of backbone stereochemistry and the sequence of the monomer units, the primary structure, or the nucleotide sequence. In the absence of sequence information, experience shows that, in a well-phased map at $d_{min} \simeq 2.5$ Å, only 50–70% of the protein residues can be correctly identified. It is not surprising that this percentage is low; many amino acids have essentially the same electron-density distribution, such as acids and their corresponding amides, or Thr and Val, or Glu and Leu. Furthermore, side chains, particularly those on the outside of the molecule, may adopt many conformations and hence appear indistinct in the electron-density map. Thus, Lys or Arg may appear truncated, similar to Ala or even Gly. If the amino-acid sequences of a set of peptides are known, it may however be possible to deduce the order in which they are joined, by examining the distribution of bulky side chains or Pro residues which produce a characteristic high, flattened electron-density distribution in the backbone. The electron-density map of papain was consistent only with a rearrangement of the tryptic peptides and an alteration in the total number of residues; the map of intestinal calcium-binding protein (Szebenyi *et al.*, 1981) was inconsistent with what proved to be a totally false amino-acid sequence (Huang *et al.*, 1975) but confirmed a later sequence (Fullmer and Wasserman, 1981).

Interpretation is facilitated by superimposing an image of the electron-density map on a trial atomic model via a half-silvered mirror, in a device known as a Richards box (Richards, 1968) or, more informally, Fred's Folly. The usual strategy involves tracing the electron-density sections on transparent sheets, to a scale of 2 cm = 1 Å, which coincides with that of the Kendrew–Watson Cambridge wire models. The models are designed with fixed bond lengths and angles, derived from crystallographic studies of small peptides or nucleotides, and are manually adjusted by variation of the dihedral angles only. The experimenter seeks the best subjective fit between the atomic positions and the regions of high electron density, working outward along the sequence from regions of well-defined structure towards the chain termini (which are often indistinct because of chain motion). An advantage of wire models is that they are very open and permit visualization of the interior of the structure and ready measurement of atomic coordinates. However, a more realistic representation is provided by space-filling CPK models, in which the bulk and noninterpenetrating nature of the atoms is clearly displayed.

Computer graphics has become more widely used, both as a replacement for the Richards box and for the display in backbone ball-and-stick or space-filling form of the resulting structure. The atomic coordinates of a snake-venom toxin were deduced solely by computer manipulation of the electron-density map and a ball-and-stick atomic representation of trial structural segments (Tsernoglou *et al.*, 1977), without recourse to a physical wire model. It is not yet clear whether poorer quality electron-density maps, which could be fitted only with difficulty (if at all) in a Richards box, can be more readily interpreted by graphics techniques, especially if aided by sophisticated image-processing analysis (Greer, 1974).

B. Refinement

The initial deduction from an electron-density map is a set of atomic coordinates, which of course contains numerous errors. To improve these coordinates, they are subjected to the process of refinement in which the coordinates themselves, or some function directly calculable from them, are systematically varied in such a way as to improve agreement with certain experimental observations. Although refinement is routine in small-molecule crystallography, it is not so in macromolecular crystallography, where the number of parameters is very much larger and where the ratio of observations to parameters is not much greater than unity. A modest-sized protein of molecular weight 20,000 may be characterized by roughly 6000 parameters, yet may exhibit only 14,000 reflections to 2.0-Å resolution, for an observation-to-parameter ratio of 2.3. A more typical value in small-molecule crystallography would be 20. Refinement of macromolecular structures is thus made difficult both by the sheer volume of data and parameters, which tax powerful computers, and by the fact that convergence is slow as the problem is not strongly overdetermined.

Three modes of refinement may be distinguished: real-space, reciprocal-space, and energy-space refinement. Although until recently these were applied more or less independently in refining a crystal structure, each has disadvantages, and a strategy which combines them, stereochemically restrained least-squares refinement (Konnert, 1976; Hendrickson and Konnert, 1980), is finding great use.

In real-space refinement, the original electron-density map $\rho_{obs}(\mathbf{r})$ is regarded as the experimental observable. To each atom in a trial model structure can be ascribed an electron-density distribution. The extensive knowledge of relevant small-molecule structures of amino acids, dipeptides, and nucleotides may also be applied to establish likely values for bond lengths and bond angles. The model atoms (or groups of atoms) are then systematically moved to minimize the discrepancy between the observed electron-density distribution and that calculated from the current coordinates, $\rho_{calc}(\mathbf{r})$. Usually, this is conducted in a relatively small region of the sequence — the "molten zone" — which then progressively moves along the sequence, maintaining continuity with its sequence neighbors, until the whole sequence has been refined (Diamond, 1974). Problems arise in accommodating atoms which may be near the molten zone in space, though not in sequence. Furthermore, completion of refinement can lead to a set of coordinates whose Fourier transform (the calculated diffraction pattern) is manifestly a poor fit to the observed structure amplitudes $|F_{obs}(s)|$. That is, improvement of the fit in real space (electron-density space) may result in no improvement in reciprocal space (diffraction space).

In reciprocal-space refinement, the structure amplitudes $|F_{obs}(s)|$ are regarded as the experimental observables. The Fourier transform of a trial atomic model, calculated from Eq. (4), Section III.C, then yields $|F_{calc}(s)|$. The difference between $|F_{obs}|$ and $|F_{calc}|$ may then systematically be minimized by moving the atoms and hence altering the Fourier transform of the model. Here, a set of final coordinates may be obtained which yields an excellent fit to the diffraction pattern but which leads to physically unreasonable atomic positions: highly unusual bond lengths and angles or interpenetrating atoms.

In energy refinement, the energy is partitioned into a series of terms, each described by a potential function. For example, the energies associated with variation in bond length about a "standard" value a_o may be given by an expression of the form $k(a - a_0)^2$, where a is the trial bond length and k a constant derived from studies of small molecules. Other terms describe torsion angles, van der Waals interactions, electrostatic interactions, and the like. Any model atomic conformation thus has an energy associated with it which is readily calculated; this energy can be minimized (Levitt, 1982). As in real-space refinement, the final, refined coordinates may not yield good agreement with the diffraction pattern, and indeed may lead to atoms unrealistically positioned outside the electron density in the original map. The failures of energy refinement are often ascribed to inadequacies in the potential functions, to the lack of cross terms, or to the lack of a suitable model for the solvent. Indeed, the solvent is often ignored completely, so that the structure being refined is essentially the structure *in vacuo*.

If the macromolecular refinement problem were more overdetermined, with a larger ratio of observations to parameters, then it is likely that all three refinement strategies would converge quite rapidly to essentially the same structure. Since this happy state is manifestly not achieved, it is reasonable to ask if a blend of the three strategies might not produce better results. The diffraction amplitudes $|F_{obs}|$ are the real observables; electron-density maps contain the effects of phase errors; and model structures may not always be completely appropriate. The fluctuations in crystal structure mean that attempts to fit what is truly a superposition of a very large number of structures with a single set of atomic coordinates are prone to error at best, and are futile at worst.

Despite this, the stereochemically restrained least-squares refinement approach offers promise (Hendrickson and Konnert, 1980). Here, an expression of the form $\Phi = \phi_1 + \phi_2 + \phi_3 + \cdots$ is established, where the terms ϕ_i

arise from real-space, reciprocal-space, and (indirectly) energy-space variables, and where each contains an adjustable weighting factor. For example, ϕ_1 is a reciprocal-space expression of the form

$$\phi_1 = \sum_s (|\mathbf{F}_{obs}| - |\mathbf{F}_{calc}|)^2 / \sigma_F^2,$$

where σ_F is the standard deviation of $|\mathbf{F}_{obs}|$, or another suitable weight. Other terms ϕ_i relate to bonding distances, coplanarity, chiral centers, nonbonded contacts, torsion angles, temperature factors, and noncrystallographic symmetry. The quantity Φ is then minimized by standard least-squares techniques. By suitable adjustment of the weighting factors, the blend of real, reciprocal, and energy-space contributions can be adjusted as the refinement proceeds. The inclusion of real-space contributions assures the maintenance of respectable stereochemistry without allowing them to dominate, to the detriment of agreement with $|\mathbf{F}_{obs}|$.

Whatever the basic refinement strategy, the general procedure remains the same. Data to relatively low resolution, say, 2.8 Å, are included first (though the very lowest resolution data, $d_{min} > 10$ Å can be omitted as it is greatly influenced by poorly modeled solvent effects). At each stage, a residual or crystallographic R factor is calculated:

$$R = \sum_s ||\mathbf{F}_{obs}| - |\mathbf{F}_{calc}|| / \sum_s |\mathbf{F}_{obs}|,$$

which will decrease as refinement proceeds. It is usual to begin with only those atoms or groups which can be placed with confidence. As the R factor levels off, additional shells of higher-resolution data are introduced, and missing atoms are sought by inspection of difference Fourier maps with coefficients $(|\mathbf{F}_{obs}| - |\mathbf{F}_{calc}|)$, which yield the difference in electron density between the structure and the current model. Thus, omitted atoms should be revealed as positive peaks and misplaced ones as adjacent positive and negative peaks. Derivative maps are also often calculated; these have coefficients given by

$$|\mathbf{F}_{calc}| + 2(|\mathbf{F}_{obs}| - |\mathbf{F}_{calc}|) = 2|\mathbf{F}_{obs}| - |\mathbf{F}_{calc}|,$$

which represent the superposition of twice the difference and Fourier map on the model map. The factor of two compensates for the lower peak height in difference Fouriers (Henderson and Moffat, 1971). Atoms should not wander outside the electron density in the derivative map. Visual inspection of difference Fourier and derivative maps at regular intervals during the course of refinement, followed by manual intervention to correct gross errors in the positioning of atoms, is essential. Automated refinement procedures cannot be relied on to rescue all atoms or groups of atoms from local, false minima. For example, correct positioning, as evidenced by features on

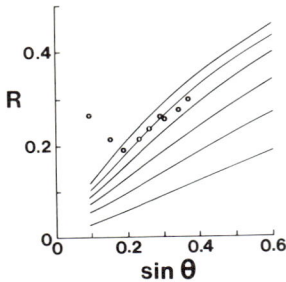

Fig. 18. The dependence of R factor on resolution, for the case in which the sole error is random positional error in the atomic coordinates. The curves (———) represent random positional errors of (from bottom to top) 0.10, 0.15, 0.20, 0.25, 0.30, and 0.35 Å. Data points (○) are those for refinement of bovine intestinal calcium-binding protein at 2.0-Å resolution (sin θ_{max} = 0.38) and are consistent with a positional error of 0.30 Å. (The R values at low resolution, sin θ > 0.15, are increased by inability to model the solvent correctly.) (From unpublished results of D. M. E. Szebenyi and K. Moffat.)

a difference Fourier map, may require flipping a peptide bond 180°. Computationally, this is hard to achieve; it requires traversing a large energetic or steric barrier, though the actual shift in atomic positions, apart from the carbonyl O and amide H, is small.

Once the atoms of the macromolecule have been located, it is generally found that there is an array of hitherto unaccounted for positive peaks on the final difference Fourier map, covering the surface of the molecule. These can be incorporated into the refinement, as they are identifiable as water molecules, hydrogen-bonded to groups on the macromolecule and to each other. Two shells can generally be identified: the first, relatively well ordered, is hydrogen-bonded to the macromolecule, and the second, less well ordered, is hydrogen-bonded to the first but not directly to the macromolecule. Other peaks can be identified by their higher electron density as, for example, sulfate, phosphate, sodium, or calcium ions. Often, hundreds of well-ordered water molecules can be located, which emphasizes that the structure determined is not just that of the macromolecule, but that of the macromolecule–solvent complex; the solvent is very important to the macromolecular structure and cannot be ignored.

The end product is a set of atomic coordinates. How accurate are they? An assessment of their random positional error may be obtained from a plot of the R factor versus resolution (Fig. 18). However, it cannot be stressed too highly that systematic errors, arising from an initial incorrect model which refinement has been unable to correct fully, can be much larger. This problem seems to have particularly afflicted the structures of prosthetic groups or metal environments in macromolecules. Chemically unprecedented structures derived from limited crystallographic refinement (or from an unrefined electron-density map) have been used as evidence for an "entatic state" at the active site or other important region, and major biochemical theories of mechanisms of action have been erected around these structures. This is dangerous. Other structural techniques such as EXAFS (*e*xtended *x*-ray *a*bsorption *f*ine *s*tructure) spectroscopy may subsequently provide more accurate information on the structural environment of metals (Shulman *et al.*, 1978), although the detailed interpretation of EXAFS spectra is also controversial. To give two examples, subsequent crystallographic refinement of the Fe—S cluster in rubredoxin and of the heme in deoxyhemoglobin showed that the initial crystallographic interpretation contained errors of detail. In the fully refined structures, the metal environments were by no means as chemically unusual as the initial interpretation had suggested.

The conservative approach is to insert the prosthetic group or metal into the macromolecular model in the chemically most plausible structure, and then to demonstrate that refinement inexorably drives this initial structure toward some more unusual structure. This unusual structure must be obtained, even if minor modifications are made in the initial structure. In the absence of this approach, claims of peculiar stereochemistry must be approached with caution.

C. Relation to Biochemical Problems

As noted in Section I.B, the results of a crystal-structure determination are essentially static, a time and space average. However, all biochemically interesting events require structural change. What bearing do crystallographic results have on structural changes, such as those occurring during catalysis or during the binding and release of oxygen by hemoglobin?

Two approaches to this important question have been taken. The first consists of studying additional, static structures which are believed to be related to the inaccessible, transient structures. For example, the structure of an enzyme can be examined, with products, inhibitors, virtual substrates, or transition-state analogs bound. These additional structures can often be simply determined by difference Fourier techniques, making use of the originally determined phases for the enzyme crystal alone, and the amplitudes for the enzyme and the complex. From this array of structures — snapshots or single frames — the continuous process of structural change — the movie — may be inferred. In the case of hemoglobin, the end points of the reaction, deoxy and oxyhemoglobin, are accessible, as are such intermediates as deoxyhemoglobin in the normal oxy quaternary structure and oxyhemoglobin in the normal deoxy quaternary structure. In addition, numerous mutant and chemically modified hemoglobins have been studied whose oxygen-binding or other functional properties are in some way altered. Alteration in function implies alteration in structure in some key region(s) of the molecule, which such studies may reveal. (Notice however that the converse is not true: alteration in structure need not produce alteration in function, since it may occur in a functionally irrelevant region of the structure).

In all such cases, exemplified by chymotrypsin (Steitz and Shulman, 1982), carboxypeptidase A (Lipscomb et al., 1970), and hemoglobin itself (Baldwin, 1975), the degree of confidence which attaches to the biochemical inferences is determined by the careful choice of the related structures, by the chemical plausibility of the reaction mechanism, and by the extent of agreement with independent structural data (derived from, for example, nuclear magnetic resonance or EXAFS data). This approach is necessarily indirect and is certainly challenging, but it has proved the most successful to date.

The second approach is to slow down the biochemical reactions and to speed up the crystallography (Moffat et al., 1984), so that direct crystallographic measurements can be made on structural intermediates. Cooling the crystal is one obvious way of slowing down reaction rates, but this has proved technically challenging. The discipline of cryoenzymology has developed to devise solvents which can be cooled without drastic alteration in the effective pH or dielectric constant (which would lead to profound structural changes and likely disruption of the crystal lattice), and which retain viscosity low enough to permit the diffusion of substrates in and products out. A few low-temperature mechanistic studies have been conducted on serine proteases (Alber et al., 1976), lysozyme (Artymiuk et al., 1979), myoglobin (Frauenfelder et al., 1979), and ribonuclease (G.A. Petsko, pers. comm.), and more can be anticipated.

The problems of initiation of the reaction in the crystal by diffusion can be minimized if the reaction is photo-activable, since a light pulse may then initiate turnover throughout the volume of the crystal. In a heroic effort, the structural changes in myoglobin crystals following photolysis of carboxymyoglobin, and recombination in the dark, are being followed using a synchrotron x-ray source (Bartunik et al., 1981). The problem here lies in the fact that only a few reflections can be measured at a time, and that numerous light pulses are necessary. It may also turn out that no intermediates are visible, only the immediate photoproduct (which can be more simply studied by continuous illumination) and the stable carboxymyoglobin.

It must be admitted that it has proved easier to study structure than to deduce structural mechanisms. As crystallographers become more familiar with enzymology, as enzymologists understand crystallographic results more thoroughly, and as cryoenzymology becomes more widely used, more soundly based mechanisms will emerge.

VII. New Directions

Macromolecular structure determination is constantly evolving. Three promising new directions will be considered: an experimental data collection technique, the use of synchrotrons as x-ray sources; a novel strategy for phase determination, partial structure resolved anomalous scattering; and the crystallization of a hitherto unstudied class of macromolecules, membrane proteins. A fourth new direction, not considered here because it deals with an application of the results of macromolecular crystallography rather than with an aspect of the technique itself, should not be ignored; protein dynamics. Here, crystallographic coordinates and potential-energy functions are used to calculate the force on each atom and, hence, the trajectory which it will follow on perturbation. This area was reviewed by Karplus and McCammon (1981).

A. Uses of Synchrotron Radiation

The electron or positron accelerators known as synchrotrons, and more particularly their associated storage rings, are unusual sources of x-rays. Since the particles are continuously accelerated by being constrained in (roughly) circular orbits, they are continuously emitting radiation, known as synchrotron radiation.

This radiation has several characteristic features which make it suitable for conventional macromolecular crystallography and which permit novel experiments (see, for example, Winick and Doniach, 1980; Greenhough and Helliwell, 1983). First, it is extremely intense, and thus permits the use of much smaller, more weakly diffracting crystals, or a great reduction in the exposure time necessary to acquire data on more normally sized crystals. Factors of between one and three orders of magnitude reduction in exposure time are possible, particularly if additional devices graphically called wigglers are inserted into the orbit. Second, the radiation is polychromatic, with great intensity over a wide range of wavelengths (energies). This is in striking contrast to the radiation emitted by conventional laboratory x-ray generators, which consists almost entirely of quasi-monochromatic x-rays, the characteristic x-rays of the target anode of the generator. Thus, by the insertion of suitable tunable x-ray monochromators at a synchrotron source, x-rays of any desired wavelength can be selected in exactly the same way as a grating monochromator selects visible radiation for optical spectroscopy. A typical synchrotron radiation spectrum, that emitted by CHESS, the *Cornell*

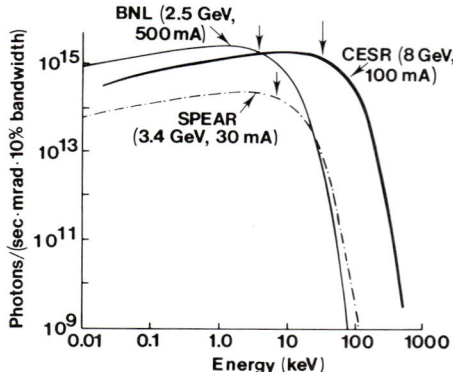

Fig. 19. The spectra of synchrotron radiation emitted by the Cornell storage ring CESR, the Stanford storage ring SPEAR, and that anticipated from the National Synchrotron Light Source at Brookhaven, BNL, at the indicated energies and currents. (From Batterman and Ashcroft, 1979. © 1979 by the American Association for the Advancement of Science.)

*H*igh *E*nergy *S*ynchrotron *S*ource, is shown in Fig. 19. Third, the radiation is largely polarized in the plane of the orbit; and last, it has a definite time structure, with very brief bursts emitted each time a pulse of electrons passes by. For example, at CHESS, bursts 150 nsec long are emitted every 2.56 μsec.

The first and second features have been the most useful to date. As noted in Section II, it is often very difficult to grow large, well-ordered crystals. Hence, any technique which permits the use of small crystals brings problems which were previously inaccessible into the realm of the feasible. For example, we can routinely grow crystals of the polypeptide hormone, human chorionic somatomammotropin, which have dimensions of roughly $350 \times 150 \times 75$ μm, with one molecule of molecular weight 22,000 per asymmetric unit (Hunt *et al.*, 1981). These crystals yield poor-quality diffraction patterns, requiring up to 40 hours of exposure on a rotating-anode x-ray generator, yet give patterns of excellent quality with exposures of 10–40 minutes at CHESS. Structure determination is routinely possible only when data can be collected at a synchrotron source. If such were not available, we would have to rely on the very small number of big crystals we have been able to grow.

The tunability permits the selection of x-ray wavelengths which will enhance anomalous scattering effects by maximizing $\Delta f'$ and $\Delta f''$ (Section IV.C; Fig. 11) and hence their contribution to phase determination in combination with isomorphous replacement. More important, it is theoretically possible to determine phases by means of anomalous scattering alone,

5. MACROMOLECULAR CRYSTALLOGRAPHY 425

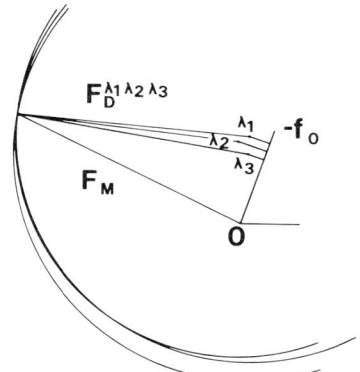

Fig. 20. Phasing at multiple wavelengths. Since both $\Delta f'$ and $\Delta f''$ vary with wavelength (Fig. 11), the atomic scattering factors for a heavy atom at three x-ray wavelengths are different, as shown by the points λ_1, λ_2, and λ_3. Measurement of the structure amplitude $|F_D|$ at these three wavelengths yields three phase circles centered at λ_1, λ_2, and λ_3, which (in the absence of error) will possess a common intersection. Thus, F_M is determined in both amplitude and phase.

without isomorphism being in any way involved. In this approach, a crystal containing an intrinsic heavy atom, such as Fe, Cu, or Zn, or an extrinsic, added heavy atom such as Hg or Pt, is used, and x-ray data are collected at several wavelengths which span an absorption edge of the heavy atom, either the K edge for the lighter atoms (say, $Z = 24-47$) or the L_{III} edge for heavier atoms (say, $Z = 53-92$). As noted (Section IV.C), the variation of $\Delta f'$ and $\Delta f''$ with wavelength around absorption edges is pronounced. Hence, the phase circles have different centers (Fig. 20), and a unique phase determination is possible. Practically speaking, difficulties arise due to absorption and to the fact that anomalous scattering effects, even when maximized, are of a magnitude comparable with experimental error. Nevertheless, direct phase determination through the use of multiple wavelength anomalous scattering data may become routine, as ingenious x-ray optics are devised to enhance it (Arndt et al., 1982; Moffat et al., 1984).

B. Partial Structure Resolved Anomalous Scattering

Crambin, a plant protein of unknown biological function, contains 46 amino-acid residues, including 6 cysteines forming 3 disulfide bonds, and forms exceptionally well-ordered crystals (Teeter and Hendrickson, 1979). Hendrickson and Teeter originally planned to solve the crystal structure of crambin in a conventional manner, by preparing isomorphous derivatives;

however, they were unable to obtain any. They turned this failure into spectacular success (Hendrickson and Teeter, 1981) by devising a strategy based on a successful technique in small-molecule crystallography, the heavy-atom method, and utilizing highly accurate data including anomalous scattering on the native crystals alone. Their new strategy, which they call partial structure resolved anomalous scattering, represents one of the major advances in macromolecular crystallography since the 1950s.

Briefly, the six sulfur atoms could be located by an anomalous difference Patterson synthesis alone (Section V.B). Great care had been taken to obtain accurate intensity differences, as the value of $\Delta f'''$ for sulfur is very low at Cu K_α wavelengths, only 0.557 electrons. The average sulfur anomalous scattering signal was only 1.7 electrons, 2.1% of the root-mean-square native structure amplitude.

Once the sulfurs had been located, their structure factors could be calculated, and parameters refined against the largest anomalous differences. Anomalous scattering at a single wavelength, like single isomorphous replacement, leads to an inherent phase ambiguity, with two equally possible solutions symmetrically disposed about $\Delta f''$. To resolve this ambiguity, use was made of the fact that the six sulfurs provide a significant fraction of the total structure amplitude of the protein, some 29%. Thus, they constitute a "heavy atom," and it might be assumed that the phase of the total structure amplitude will tend to be close to that of the heavy-atom structure, the phase of the sulfur structure amplitude. In practice, a probabilistic representation was used to combine the phase information from sulfur anomalous scattering with the (suitably weighted) information from the heavy atom. After a choice of enantiomer had been made, by chemical reasonableness, the resultant phases yielded a visually interpretable electron-density map, from which atomic coordinates could be extracted and refined in the usual manner.

It should be emphasized that this strategy is not unique to small, highly-ordered structures like crambin. Indeed, calculations based on the heavy-atom content of such proteins as high-potential iron protein (which contains a 4Fe-8S cluster), a snake neurotoxin (8S), a hexanucleotide (10 P), a single Hg derivative of myoglobin, or a U derivative of lysozyme, suggest that the identical strategy would apply successfully to them (see Table I of Hendrickson and Teeter, 1981). Indeed, prospects are further enhanced if the anomalous scattering data are collected at a synchrotron source.

C. Membrane Proteins

Proteins of two classes have proved remarkably difficult to crystallize: fibrous proteins and membrane proteins. Membrane proteins and, in particular, integral membrane proteins, have unusual surface properties which mean that they are not simply soluble in aqueous solutions. That region of the membrane protein surface which is normally in contact with the alkane side chains of the membrane lipids is believed to be extremely hydrophobic, but regions which are normally exposed to the solvent, or in contact with the polar lipid head groups, are hydrophilic like the surface of soluble, nonmembrane proteins. The existence in a single molecule of both extremely hydrophobic and hydrophilic surface regions means that detergents have to be used to extract, isolate, and purify membrane proteins, and to render them soluble in aqueous media by embedding them in detergent micelles. The major problem is then to identify detergents which will allow the retention of protein structure, without partial or complete denaturation, and which will be compatible with the three-dimensional order necessary for conventional crystallization.

Certain membrane proteins such as bacteriorhodopson occur naturally in two-dimensionally ordered arrays and hence can yield structural information via a sophisticated combination of x-ray and electron microscope techniques (Henderson and Unwin, 1975); others such as cytochrome oxidase may be induced to form two-dimensional arrays (Deatherage et al., 1982). Side-to-side interactions in these arrays are presumably mediated by hydrophobic interactions between the hydrophobic regions on the protein surface, embedded in the membrane lipid. Acquisition of full crystalline order in the third dimension, perpendicular to the plane of the membrane, is much harder to achieve. It would require simultaneously maximizing the hydrophobic interactions which convey two-dimensional order, and polar, hydrophilic interactions which would stabilize interactions within a stack of membranes. There is no convincing evidence for three-dimensional crystallization via the ordered stacking of two-dimensional arrays.

Two groups have had great success in crystallizing membrane proteins solubilized in particular small detergents such as β-D-octylglucopyranoside or N, N-dimethyldodecylamine-N-oxide. Crystals of high diffraction quality have been obtained of bacteriorhodopsin (Michel and Osterhelt, 1980) and especially of porin (Garavito and Rosenbusch, 1980) and of photosynthetic reaction centers from *Rhodopseudomonas viridis* (Michel, 1982). It is believed (Michel, 1983) that these small detergents bind to the protein in the same manner as larger detergents, but are much better able to fit into the crystal lattice. Precipitation or crystallization of the protein within these small detergent micelles is then accomplished in the usual way (Section II), using salt or polyethylene glycol. Crystal quality may be further improved by the addition of small amphiphilic molecules such as heptane-1,2,3-triol, cyclooctane-1,2,3,4-tetraol, and triethylammonium phosphate.

For porin and photosynthetic reaction centers, high-resolution crystallographic structures will soon be available and will certainly provide major insights into our understanding of a whole new class of macromolecules.

Despite the pessimistic remark attributed to Francis Crick: "If you can't study function, you'd better study structure!," crystallographic results have provided the basic underpinnings for much of modern biochemistry and molecular biology. With the structures of membrane proteins in prospect, and those of fibrous proteins involved in cell motility also conceivable, a similar underpinning for cell biology will be achieved.

Appendix A. The Basic Mathematics of Crystallography

1. Notation

A *vector* is a quantity possessing both a *magnitude* and a *direction*. Vectors are represented by lines whose length gives the magnitude and whose orientation relative to convenient axes gives the direction. Vector symbols are in boldface, thus: **r**. Magnitude is denoted by |**r**|.

The symbol **r** is reserved to denote vectors whose length is measured in angstroms (Å) (or another real space unit); thus $\rho(\mathbf{r})$ denotes the electron density at a point $\mathbf{r} = (x, y, z)$. Symbols like **k**, **k**$_0$, and **d*** denote vectors whose length is measured in Å^{-1}, reciprocal space. The *scalar product* of two vectors, say **r** and **k**, is written **r** · **k** and means: (magnitude of the vector **r**) × (magnitude of the vector **k**) × (cosine of the angle between them, ψ). That is, $\mathbf{r} \cdot \mathbf{k} = |\mathbf{r}||\mathbf{k}|\cos \psi$.

2. Waves

A *wave* (such as water wave on a pond) is characterized by four quantities: its *amplitude* (the crest-to-trough distance), its *wavelength* (the distance between crests), its *phase* (the relative time of arrival of a crest at the observer), and its *velocity*.

If we consider also the direction in which the wave is traveling, then a fifth quantity is applicable: the *wave vector,* parallel to the direction of travel and of magnitude $1/\lambda$, where λ is the wavelength.

Algebraically, a wave A may be depicted as a cosine function:

$$A = A_0 \cos(\omega t + \phi),$$

where A_0 is the amplitude, ω the frequency [where the frequency = 2π(velocity)/ (wavelength)], and ϕ the phase. Adding waves algebraically turns out to be inconvenient; however, waves may also be represented as vectors. The amplitude of the wave is given by the length of the vector (its magnitude), and its phase is given by the direction of the vector with respect to any convenient direction, such as the x axis. Plotting the vector on the complex plane, where the x axis denotes real values and the y axis denotes imaginary values, the wave **A** is given by $\mathbf{A} = A_0 \exp[i\phi]$, where A_0 is the amplitude and ϕ the phase. Therefore,

$$\mathbf{A} = A_0 \cos \phi + iA_0 \sin \phi = P + iQ \quad \text{(by de Moivre's theorem)},$$

say, where P is the real part, Q the imaginary part, and $i = \sqrt{-1}$.

3. Fourier Transforms

An expression of the form of Eq. (1), Section III.A,

$$\mathbf{F}(\mathbf{s}) = \int \rho(\mathbf{r}) \exp(2\pi i \mathbf{r} \cdot \mathbf{s}) \, d\mathbf{r},$$

states mathematically that the function $\mathbf{F}(\mathbf{s})$ is the Fourier transform of the function $\rho(\mathbf{r})$. In shorthand, we may write this as $\mathbf{F} = T(\rho)$. Four properties of Fourier transforms are useful here: inversion, addition, convolution, and the relation between sections and projections.

Inversion. Define an inverse Fourier transform T^* by the relation

$$\mathbf{G}(\mathbf{r}) = \int \lambda(\mathbf{s}) \exp(-2\pi i \mathbf{r} \cdot \mathbf{s}) \, d\mathbf{s} \quad \text{or} \quad \mathbf{G} = T^*(\lambda).$$

Note the minus sign in the exponent. Making use of the inversion property, we have $T^*(\mathbf{F}) = T^*[T(\rho)]$, since $\mathbf{F} = T(\rho)$. Therefore, $T^*(\mathbf{F}) = \rho$. Writing this expression in full, we have

$$\rho(\mathbf{r}) = \int \mathbf{F}(\mathbf{s}) \exp(-2\pi i \mathbf{r} \cdot \mathbf{s}) \, d\mathbf{s}.$$

Additivity. The Fourier transform of the sum of two functions is the sum of their individual Fourier transforms. Thus

$$T(\rho_1 + \rho_2) = T(\rho_1) + T(\rho_2).$$

Convolution. The operation of convolution of two functions may be described in words. Set down the origin of the first function at a point of the second; multiply the two functions; repeat for all points in the second; add the results. Mathematically, for two functions g_1 and g_2, the convolution of g_1 with g_2, denoted $g_1 * g_2$, is given by

$$g_1 * g_2(\mathbf{x}) = \int_{\text{all } \mathbf{y}} g_1(\mathbf{y}) \, g_2(\mathbf{x} - \mathbf{y}) \, d\mathbf{y}.$$

Convolution gives a particularly simple result if one of the functions is a *lattice,* namely, a function whose value is 1 at a set of points and 0 everywhere else (see Fig. 6).

There are two results. The Fourier transform of the product of two functions is the convolution of the transforms of the individual functions; and conversely, the Fourier transform of the convolution of two functions is the product of the transforms of the individual functions. Thus

$$T(g_1 g_2) = T(g_1) * T(g_2)$$

and

$$T(g_1 * g_2) = T(g_1) T(g_2).$$

Central Sections and Projections. A plane *projection* is defined as follows: Take a plane through a function and a set of parallel lines which intersect that plane. Add up the value of the function at all points along each line, and place the sum at the point of intersection of each line with the plane. The resulting array of sums on the plane is known as the *projection* of the function in the direction of the lines onto the plane.

A *central section* of a function is the value of that function on a plane passing through the origin. Then, if two functions are related by a Fourier transform, say, $\mathbf{F} = T(\rho)$, then the Fourier transform of a central section of the function $\mathbf{F}$ is the projection of ρ in the direction of a set of lines perpendicular to the plane of the section.

Appendix B. A Specific Strategy for Protein Crystallization

If only limited amounts of pure protein are available, the following strategy is suggested [adapted from Moffat (1980)]:

(1) Purify the protein to the greatest extent feasible, to homogeneity as judged by gel electrophoresis, if possible.

(2) Review the information already available on solubility in (for example) $(NH_4)_2SO_4$ or organic solvents, as a function of pH; review the conditions under which the protein is known to be stable; modify the subsequent four sections as appropriate.

(3) Set up vapor-diffusion experiments in multiwell tissue-culture plates, at a final protein concentration of 10–15 mg ml^{-1} in a total volume of 5 μl (50–75 μg per droplet). Try the following precipitants: $(NH_4)_2SO_4$, at several pH values; polyethylene glycol 400, 1000, 2000, and 6000, at several pH values; precipitation near the isoelectric point by pH variation, starting

from both the acid and alkaline side, and ending at several pH values; and organic solvents, if used in the purification scheme.

(4) Set up microdiffusion experiments with 5 μl of 10-15 mg ml^{-1} protein, against deionized water, and 1-2% NaCl.

(5) Set up controlled cooling experiments in various buffers and pH values.

(6) If sufficient protein is available, repeat part or all of (3)-(5), adding such reagents as metal ions, phosphate, citrate, or dithiothreitol; or try experiments at 4°C; or try crystallization of protein previously subjected to very limited proteolytic digestion, if there is reason to suspect extreme molecular flexibility.

Working on this scale, roughly 20 experiments can be conducted per milligram of protein, assuming none can be recycled. Since this assumption is certainly pessimistic, steps (3)-(5) can be conducted with 2-3 mg protein.

Appendix C. A Crystallographic Paper Primer

Like other scientific papers, those dealing with macromolecular crystallography should be read critically. Although the process of obtaining an initial electron-density map is reasonably objective, interpretation and refinement may involve key, subjective decisions by the crystallographer which affect the validity of the results.

Some questions to ask of an initial paper which presents an unrefined structure are the following. Is the resolution sufficiently low ($\lesssim 3.5$ Å) and the quality of the map sufficiently high to permit an unambiguous tracing of the polypeptide backbone? Or are there branches or gaps which confuse the issue? Is the sequence known, and if so, are almost all bulky side chains and small side chains (Gly, Ala, Ser) identifiable? Are specific, quantitative claims made ("N3 of the imidazole side chain is located 3.3 Å from the carbonyl oxygen of the substrate and can therefore protonate the leaving group . . . "), or are the claims more general ("An imidazole side chain is located near the presumed substrate binding site and may be involved in proton-transfer processes during catalysis. A detailed study of the enzyme-inhibitor complex is needed to establish its role."). The former is hard to justify; the latter, conservative approach is more likely to be correct. In the initial rush of enthusiasm for interpretation of a new structure and deduction of mechanistic information, crystallographers have almost always erred on the side of overinterpretation, as subsequent refinement showed.

For a more detailed paper which describes a crystallographically refined structure, different questions are appropriate: What is the resolution of the refinement, and what is the resolution of the experimental phases? (It is rare for experimental phases to extend beyond 2.0 Å; at higher resolution, phases must be calculated from the current model, and indeed may be calculated at lower resolution also, to supplement or supplant the experimental phases). If experimental phases do not extend beyond 2.3–2.5 Å, extension to much higher resolution has to be undertaken with caution. If unusual stereochemistry is present (in peptide bonds, or distorted metal environments), is this substantiated by refinement in which a normal stereochemistry was imposed at the sites in question and demonstrated to be erroneous? Sometimes, unusual stereochemistry is ignored. This is a danger signal, as it implies that the crystallographer does not attach sufficient weight to the coordinates to feel the unusual feature worthy of comment. Is the R factor "reasonable" given the resolution, the quality of the data, and the data-to-parameters ratio? It is not possible to give rigid guidelines; if in doubt, consult an unbiased crystallographic colleague. Errors here are usually flagrant, as when an energy refinement fails to diminish the R factor appreciably. What precautions have been taken to avoid false minima? A final difference Fourier map should be largely featureless, without a concentration of peaks and holes in the region of the molecule itself.

Appendix D. Suggestions for Further Reading

Elementary crystallography is covered in many *basic textbooks,* among them:

Glusker, J., and Trueblood, K. (1982). "Crystal Structure Analysis." Oxford University Press.
Ladd, M. F. C., and Palmer, R. A. (1977). "Structure Determination by X-ray Crystallography." Plenum, New York.
Stout, G. H., and Jensen, L. H. (1968). "X-ray Structure Determination." Macmillan, New York.
Woolfson, M. M. (1970). "An Introduction to X-ray Crystallography." Cambridge University Press.

There are two *textbooks* which specifically address the techniques of protein crystallography:

Blundell, T. L., and Johnson, L. M. (1976). "Protein Crystallography." Academic Press, New York.
McPherson, A., Jr. (1982). "Preparation and Analysis of Protein Crystals." Wiley (Interscience), New York.

The *basic data* for crystallography, space groups and symmetry, mathematics, and factors relating to the production and scattering of x rays, are tabulated in several volumes of

"International Tables for X-ray Crystallography" (K. Lonsdale, General ed.). Kynoch, Birmingham, England.

Two excellent small *books* deal more with the results of macromolecular crystallography and their impact on biochemistry:

Dickerson, R. E., and Geis, I. (1969). "The Structure and Action of Proteins." Harper and Row, New York. A revised and much expanded version is in press.

Schulz, G., and Schirmer, R. H. (1979). "Principles of Protein Structure." Springer-Verlag, Berlin. Highly recommended.

Review articles appear regularly which deal with aspects of macromolecular crystallography and its results. A classic is

Holmes, K. C., and Blow, D. M. (1965). *Methods Biochem. Anal.* **13,** 113.

A more recent, thorough review is provided in

Matthew, B. N. (1977). *In* "The Proteins" (H. Neurath, ed.), Vol. III, p. **403,** 3rd ed.

Others appear in series such as Annual Review of Biochemistry, Annual Review of Biophysics and Bioengineering, and Progress in Biophysics and Molecular Biology.

The major *journals* in crystallography (especially small-molecule crystallography) are *Acta Crystallographica* and the *Journal of Applied Crystallography*. Macromolecular crystallographic results appear frequently in *Nature,* with more detailed papers in the *Journal of Molecular Biology,* the *Journal of Biological Chemistry,* and *Biochemistry.*

Protein crystallographic *results,* in the form of computerized lists of atomic coordinates or x-ray structure factors, are available from the Protein Data Bank, Brookhaven National Laboratory in the United States and from related data banks in other countries.

Acknowledgments

I thank Professor D. C. Wiley for supplying Fig. 15, Excerpta Medica for permission to adapt sections of an earlier article (Moffat, 1980), and Professor B. W. Batterman and the American Association for the Advancement of Science for permission to reproduce Fig. 19. My colleague, Dr. Marian Szebenyi, provided useful comments on the entire manuscript, and James Wenban expertly prepared the figures. Research was supported by a grant from the National Institutes of Health, GM29044.

References

Adams, M. J., Blundell, T. L., Dodson, E. J., Dodson, G. G., Vijayan, M., Baker, E. N., Harding, M. M., Hodgkin, D. C., Rimmer, B., and Sheat, S. (1969). *Nature* **224,** 491.
Alber, T., Petsko, G. A., and Tsernoglou, D. (1976). *Nature* **263,** 297.
Amzel, M., and Poljak, R. (1979). *Annu. Rev. Biochem.* **48,** 961.
Arndt, U. W., Greenhough, T. J., Helliwell, J. R., Howard, J. A. K., Rule, S. A., and Thompson, A. W. (1982). *Nature* **298,** 835.
Artymiuk, P. J., Blake, C. C. F., Grace, D. E. P., Oatley, S. J., Phillips, D. C., and Steinberg, M. J. E. (1979). *Nature* **280,** 563.
Baldwin, J. M. (1975). *Prog. Biophys. Mol Biol.* **29,** 225.

Bartunik, H. D., Clout, P. N., and Robrahn, B. (1981). *J. Appl. Crystallogr.* **14**, 134.
Batterman, B. W., and Ashcroft, N. (1979). *Science* **206**, 157.
Bernal, J. D., and Crowfoot, D. (1934). *Nature* **133**, 794.
Blow, D. M., and Crick, F. H. C. (1958). *Acta Cryst.* **12**, 794.
Blundell, T. L., and Johnson, L. N. (1976). "Protein Crystallography." Academic Press, New York.
Blundell, T. L., Pitts, J. E., Tickle, I. J., Wood, S. P., and Wu, C. -W. (1981). *Proc. Natl. Acad. Sci. USA* **78**, 4175.
Bricogne, G (1974). *Acta Crystallogr.* **A30**, 395.
Bricogne, G. (1976). *Acta Crystallogr.* **A32**, 832.
Deatherage, J. F., Henderson, R., and Capaldi, R. A. (1982). *J. Mol. Biol.* **158**, 487.
Diamond, R. (1974). *J. Mol. Biol.* **82**, 371.
Dickerson, R. E., Kendrew, J. C., and Strandberg, B. E. (1961). *Acta Crystallogr.* **14**, 1188.
Eisenberg, D. (1970). *In* "The Enzymes" (P. D. Boyer, ed.), Vol I, 3rd ed. Academic Press, New York.
Feldman, R. J. (1976). *Annu. Rev. Biophys. Bioeng.* **5**, 477.
Frauenfelder, H., Petsko, G. A., and Tsernoglou, D. (1979). *Nature* **280**, 558.
Frier, J. A., and Perutz, M. F. (1977). *J. Mol. Biol.* **112**, 97.
Fullmer, C. S., and Wasserman, R. H. (1981). *J. Biol. Chem.* **256**, 5669.
Garavito, R. M., and Rosenbusch, J. P. (1980). *J. Cell Biol.* **86**, 327.
Green, D. W., Ingram, V. M., and Perutz, M. F. (1954). *Proc. Soc. London* **A225**, 287.
Greenhough, T. J., and Helliwell, J. R. (1983). *Prog. Biophys. Mol. Biol.* **41**, 67.
Greer, J. (1974). *J. Mol. Biol.* **82**, 279.
Harrison, S. C., Olson, A. J., Schutt, C. E., Winkler, F. K., and Bricogne, G. (1978). *Nature* **276**, 368.
Hart, R. G. (1961). *Acta Crystallogr.* **14**, 1194.
Hartsuck, J., Ludwig, M. L., Muirhead, H., Steitz, T. A., and Lipscomb, W. N. (1965). *Proc. Natl. Acad. Sci. USA* **53**, 396.
Henderson, R., and Moffat, K. (1971). *Acta Crystallogr.* **B27**, 1414.
Henderson, R., and Unwin, P. N. T. (1975). *Nature* **257**, 28.
Hendrickson, W. A., and Konnert, J. H. (1980). *In* "Computing in Crystallography" (R. Diamond, S. Ramaseshan, and K. Venkatesan, eds.), pp. 13.01–13.23. Indian Institute of Science, Bangalore.
Hendrickson, W. A., and Teeter, M. (1981). *Nature* **290**, 107.
Holmes, K. C., and Blow, D. M. (1965). *Methods Biochem Anal.* **13**, 113.
Huang, W. Y., Cohn, D. V., and Hamilton, J. W. (1975). *J. Biol. Chem.* **250**, 7647.
Hunt, R. E., Moffat, K., and Golde, D. W. (1981). *J. Biol. Chem.* **256**, 7042.
Karplus, M., and McCammon, J. A. (1981). *CRC Crit. Rev. Biochem.* **9**, 293.
Kim, S. H., Quigley, G. J., Suddath, F. L., McPherson, A., Sneden, D., Kim, J. J., Weinzierl, J., and Rich, A. (1973). *Science* **179**, 285.
Konnert, J. J. (1976). *Acta Crystallogr.* **A32**, 614.
Lagerkvist, V., Rymo, L., Linquist, O., and Anderson, E. (1972). *J. Biol. Chem.* **247**, 3897.
Langridge, R., Ferrin, T. E., Kuntz, I. D., and Connolly, M. L. (1981). *Science* **211**, 661.
Levitt, M. (1982). *Annu. Rev. Biophys. Bioeng.* **11**, 251.
Lipscomb, W. N., Reeke, G. N., Jr., Hartsuck, J. A., Quiocho, F. A., and Bethge, P. H. (1970). *Philos. Trans. R. Soc. London* **B257**, 177.
McPherson, A. (1976a). *Methods Biochem. Anal.* **23**, 249.
McPherson, A. (1976b). *J. Biol. Chem.* **251**, 6300.
McPherson, A. (1982). "Preparation and Analysis of Protein Crystals." Wiley (Interscience), New York.

Matthews, B. W. (1966). *Acta Crystallogr.* **20,** 82.
Matthews, B. W. (1975). *Biochem. Biophys. Acta* **405,** 442.
Matthews, B. W. (1976). *Annu. Rev. Phys. Chem.* **27,** 493.
Matthews, B. W. (1977). *In* "The Proteins" (H. Neurath, ed.), Vol. III, 3rd ed., p. 403. Academic Press, New York.
Matthews, B. W., and Bernhard, S. A. (1973). *Annu. Rev. Biophys. Bioeng.* **2,** 257.
Matthews, B. W., and Weaver, L. H. (1974). *Biochemistry* **13,** 1719.
Michel, H. (1982). *J. Mol. Biol.* **158,** 567.
Michel, H. (1983). *Trends Biochem. Sci.* **8,** 56.
Michel, H., and Osterhelt, D. (1980). *Proc. Natl. Acad. Sci. USA* **77,** 1283.
Moffat, K. (1980). *In* "Growth Hormone and other Biologically Active Peptides" (A. Pecile and E. E. Müller, eds.), p. 19. Excerpta Medica, Amsterdam.
Moffat, K., Bilderback, D. H., and Szebenyi, D. M. E. (1984). *Science* **223,** 1423.
Pabo, C. O., Sauer, R. T., Sturtevant, J. M., and Ptashne, M. (1979). *Proc. Natl. Acad. Sci. USA* **76,** 1608.
Phillips, J. C., and Hodgson, K. O. (1980). *In* "Synchrotron Radiation Research" (H. Winick and S. Doniach, eds.), p. 565. Plenum, New York.
Phizackerley, R. P., Cork, C. W., Hamlin, R. C., Nielsen, C. P., Vernon, W., Xuong, Ng. H., and Perez-Mendez, V. (1980). *Nucl. Instrum. Methods* **172,** 393.
Richards, F. M. (1968). *J. Mol. Biol.* **37,** 225.
Rossmann, M. G., and Blow, D. M. (1962). *Acta Crystallogr.* **15,** 24.
Sasaki, K., Dockerill, S., Adamiak, D. A., Tickle, I. J., and Blundell, T. L. (1975). *Nature* **257,** 751.
Schulz, G. E., and Schirmer, R. H. (1979). "Principles of Protein Structure", Ch. 6. Springer-Verlag, New York and Heidelberg.
Schulz, G. E., Barry, C. D., Friedman, J., Chou, P. Y., Fasman, G. D., Finkelstein, A. V., Lim, V. I., Ptitsyn, O. B., Kabat, E. A., Wu, T. T., Levitt, M., Robson, B., and Nagano, K. (1974). *Nature* **250,** 140.
Shulman, R. G., Eisenberger, P., and Kincaid, B. M. (1978). *Annu. Rev. Biophys. Bioeng.* **7,** 559.
Sim, G. A. (1959). *Acta Crystallogr.* **12,** 813.
Steitz, T. A., and Shulman, R. G. (1982). *Annu. Rev. Biophys. Bioeng.* **11,** 419.
Szebenyi, D. M. E., Obendorf, S. K., and Moffat, K. (1981). *Nature* **294,** 327.
Takano, T., and Dickerson, R. E. (1981). *J. Mol. Biol.* **153,** 95.
Teeter, M., and Hendrickson, W. A. (1979). *J. Mol. Biol.* **127,** 219.
Tsernoglou, D., Petsko, G. A., McQueen, J. E., Jr., and Hermans, J. (1977). *Science* **197,** 1378.
Weber, B. H., and Goodkin, P. E. (1968). *Arch. Biochem. Biophys.* **141,** 489.
Wiley, D. C., and Lipscomb, W. L. (1968). *Nature* **218,** 1119.
Wilson, I. A., Skehel, J. J., and Wiley, D. C. (1981). *Nature* **289,** 366.
Winick, H., and Doniach, S. (1980). "Synchrotron Radiation Research." Plenum, New York.
Zeppezauer, M., Eklund, H., and Zeppezauer, E. S. (1968). *Arch. Biochem. Biophys.* **126,** 564.

6

Small-Angle X-Ray Scattering and Diffraction

J. STAMATOFF

Celanese Research Co.
Summit, New Jersey

I.	Introduction	437
II.	X-Ray Scattering and Diffraction	438
	A. Scattering	439
	B. Diffraction	445
III.	X-Ray Technology	447
	A. X-Ray Sources	447
	B. X-Ray Detectors	449
IV.	Biological Applications	451
V.	Summary	467
	References	468

I. Introduction

X-ray diffraction and scattering are two primary tools available to research scientists for direct structure determination. Methods developed in the first half of this century have found wide application in material-science problems. Many investigations have focused on biological areas. An excellent example of the use of small-angle x-ray scattering is given by Sardet *et al.* (1976). These authors studied rhodopsin, the major protein found in retinal disk membranes which are the site of the primary photoreaction in vision. The protein was extracted as a detergent complex from the other membrane components and used as an x-ray scattering specimen. Using general analysis (e.g., Guinier's law), the study provided the size, shape, and regions of the protein which bound detergent molecules. This information suggested that the protein spans the disk membrane.

Diffraction analysis of lipid bilayers is now a common procedure. Torbet and Wilkins (1976) studied changes in the small-angle x-ray diffraction pattern of a pure lipid bilayer of dipalmitoyl phosphatidyl choline molecules. Their analysis provided a high spatial resolution electron-density profile across the lipid bilayer. This profile not only confirms the formation of a bilayer but also gives the projected structure of the lipid head-group region, the angle of hydrocarbon-chain tilt, and many other features. Using similar analytical methods, Pachence *et al.* (1979) studied small-angle x-ray diffraction from membranes composed of bacterial photoreaction centers reconstituted in a phosphatidyl choline lipid bilayer. The electron-density profiles which they determined for a variety of reaction-center–lipid ratios provided a structural profile for both lipid and protein components across the width of the membranes. This structural work provides a key model of the way major membrane components fit together.

These three examples show the very fundamental impact that small-angle x-ray methods have had on a wide range of biological research. Tremendous numbers of detailed and excellent studies have been published. The progress that has been made serves to establish small-angle x-ray methods as a practical and routine tool. Development of computer methods and refined experimental techniques have added greater accuracy and precision to the measurements. This chapter is not intended to review this important type of progress in small-angle methods or to provide a bibliography of the many investigations of merit; reviews are available [see, e.g., Glatter and Kratky (1982)].

Significant advances in experimental technology have opened a new vista of small-angle scattering studies. For instance, dynamic structural studies can now be performed with nanosecond time resolution. Structural determination of the location of trace elements at a concentration of one atom in ten thousand is also practical. Of the possibilities, only a few have been performed. Based on these positive results, it is clear that these new methods will find wide application in the coming decade.

The scope of this chapter is to introduce the technology and methods made available since the early 1970s. A very brief review of elementary theory will first be given, to provide a necessary background, followed by a description of the technology. Application of the new experimental techniques will then be presented to illustrate the methods which have been spawned.

II. X-Ray Scattering and Diffraction

The distinction between scattering and diffraction is not well defined (particularly in biological systems). Both areas are really one, from a theoretical point of view. If a sample is placed in a collimated beam of x rays, a pattern will be formed on an x-ray-sensitive film placed at some distance behind the specimen. The central beam is prevented from striking the film and obliterating the pattern by a beam stop. If the pattern consists of sharp spots or rings, then the phenomena is known as diffraction. If, on the other hand, the pattern consists of very diffuse features, then the term scattering applies. *Diffraction* occurs if the atomic structure is arranged in a periodic fashion with long-range order. *Scattering* occurs if the structure lacks such order (e.g., a gas or a liquid). Biological samples display a range of structural order extending from highly crystalline (e.g., arrays of lipid bilayers) to nearly total disorder (e.g., a protein solution).

X-ray scattering and diffraction are techniques used to determine the structure of a material. If the material is crystalline, so that the diffraction case occurs, the detailed molecular structure of the repeating unit which forms the crystal can be obtained. In addition, the degree of order in the crystalline material can be deduced (e.g., the degree of orientation of crystallites along a given direction, the crystallite size, or the degree of strain or disorder within crystallites). If the material is not crystalline, so that the scattering case occurs, useful information can still be derived. For example, the radius of protein molecules in solution, the radially averaged structure of protein molecules in solution, or the degree of association of several protein molecules in a solution can be determined.

The angle of x-ray scattering determines the degree of spatial resolution. Wide-angle x-ray scattering or diffraction produces higher-resolution results, whereas small-angle phenomena are restricted to lower resolution. As a consequence, small-angle studies apply to large distances (typically greater than 20 Å). Because many biological molecules are large, small-angle X-ray diffraction and scattering is particularly useful in biology. Furthermore, natural assemblies of biological molecules (e.g., biological membranes) typically do not contain the perfection of crystalline order necessary to produce extensive wide-angle diffraction. (One notable exception is those proteins which truly crystallize and are the basis for protein crystallography.)

A. Scattering

X-ray scattering is the subject of a number of detailed texts [see, e.g., Vainshtein (1966), Ramachandran and Srinivasan (1970), and Alexander (1979)]. For the development of detailed theory, reference should be made to one of these works. The following is a simplified review to provide a mathematical framework for further discussion.

In Fig. 1, two electron-dense particles are placed in an x-ray beam. The particles contain different numbers of electrons given by $N(\mathbf{r}_1)$ and $N(\mathbf{r}_2)$. The primary beam is described here as a set of planar electromagnetic waves. Scattering from each particle can then be described by the emission

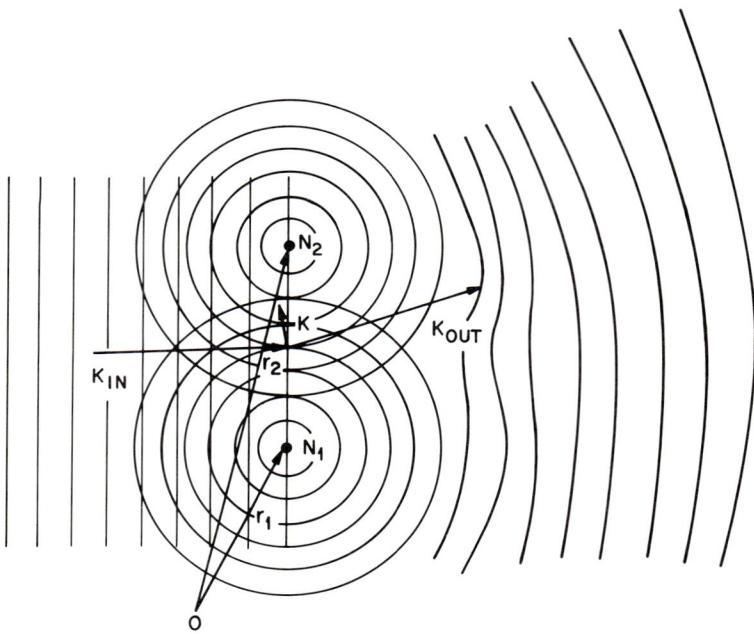

Fig. 1. The scattering of x-rays from two electron-dense particles. Each particle contains a different number of electrons (N_1 or N_2) at a given distance (r_1 or r_2) from the origin O. Incident X-rays are represented by planar wavefronts associated with the wave vector $\mathbf{K}_{in}$. Each particle scatters spherical waves which interfere with each other to produce the diffracted wave fronts associated with the wave vector $\mathbf{K}_{out}$; $\mathbf{K}$ is defined as $\mathbf{K}_{out} - \mathbf{K}_{in}$.

of spherical wavefronts from each particle. A simple formula gives the final observed waveform as the sum of the two spherical waves:

$$F(\mathbf{K}, t) = N(\mathbf{r}_1) \exp[i(\mathbf{K}\cdot\mathbf{r}_1 + \omega t)] + N(\mathbf{r}_2) \exp[i(\mathbf{K}\cdot\mathbf{r}_2 + \omega t)]. \quad (1)$$

In Eq. (1), $F(\mathbf{K}, t)$ is a complex function which describes the final scattered wave in both amplitude and phase, and $\mathbf{K}$ is the wave vector whose direction is described relative to the particles which scatter and whose length is given by $|\mathbf{K}| = (4\pi \sin\theta)/\lambda$, where λ is the wavelength of the incident and scattered radiation and 2θ the scattering angle. The terms t and ω are time and frequency, which are usually neglected.

The salient points to grasp from Eq. (1) is that the amplitude of each spherical wave is determined by the number of electrons which scatter, $N(\mathbf{r}_1)$. The phase of one spherical wave relative to the other is determined by the position of the particles $\mathbf{r}_1$ and $\mathbf{r}_2$.

The intensity is always observed by "square-law" detectors. It is given by

$$I(\mathbf{K}) = F(\mathbf{K},t)^* F(\mathbf{K},t). \quad (2)$$

From Eq. (1) it is seen that the time and frequency dependence may be factored out as the term $\exp(i\omega t)$. The product given in Eq. (2), which involves the complex conjugate of $F(\mathbf{K},t)$, eliminates the term $\exp(i\omega t)$. For elastic scattering of x rays, the time dependence of $F(\mathbf{K},t)$ can therefore be ignored†. $I(\mathbf{K})$ does depend on $\mathbf{K}$ and, implicitly, on both the number of electrons within each particle and the spatial relationships of the two particles.

Equation (1) can easily be generalized to a continuum of density. If $\rho(\mathbf{r})$ is the density of electrons at $\mathbf{r}$, then $\int \rho(\mathbf{r}) \, d\mathbf{r} = N(\mathbf{r})$ (i.e., the number of electrons). Thus, for any electron-density distribution we have

$$F(K) = \int_{\text{all space}} \rho(\mathbf{r}) \exp(i\mathbf{K}\cdot\mathbf{r}) \, d\mathbf{r}. \quad (3)$$

Recognizing that electrons are concentrated in atoms, another common formulation is

$$F(\mathbf{K}) = \sum_j f_j \exp(i\mathbf{K}\cdot\mathbf{r}_j), \quad (4)$$

where f_j is the atomic scattering factor of the Jth atom, which is (at small angles) approximately equal to the number of electrons in the atom. At larger angles, f_j is diminished due to the finite size of the atom. For the general case, the observable intensity is still given by

$$I(\mathbf{K}) = F(\mathbf{K})^* F(\mathbf{K}). \quad (5)$$

The wave vector $\mathbf{K}$ can be used to define a space, typically called reciprocal space due to the reciprocal-length relationship it maintains with regard to physical space. For any distribution of matter placed in an x-ray beam, the physical space is defined by the vector $\mathbf{r}$ relative to some three-dimensional coordinate system. The electron-density distribution is defined by the function $\rho(\mathbf{r})$. Associated with a particular distribution $\rho(\mathbf{r})$ is a unique function $F(\mathbf{K})$, given by Eq. (3). $F(\mathbf{K})$ exists in reciprocal space, which is defined by the vector $\mathbf{K}$. Figure 2 shows a pair of physical and reciprocal spaces. Angular orientation is preserved between both spaces. However, a point $\rho(\mathbf{r})$ is not uniquely mapped, one to one, into a point $F(\mathbf{K})$. Rather,

† The time dependence which is removed is due to the propagation of an x-ray electromagnetic wave of frequency ω. $F(\mathbf{K})$ and $I(\mathbf{K})$ may adopt another time dependence if the structure changes with time. In this case, the relative position of the two electron-dense particles could oscillate with time. Such changes are of current interest in biology, and will be discussed in this chapter as dynamic x-ray scattering or diffraction studies.

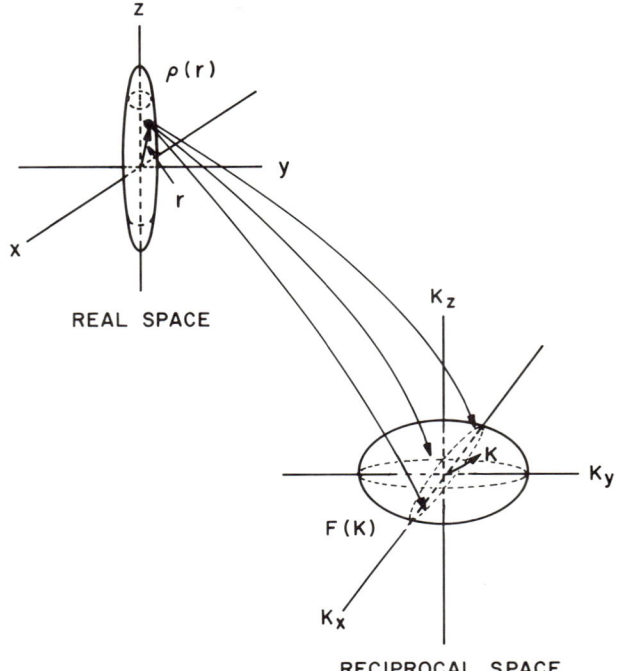

Fig. 2. A pair of physical and reciprocal spaces. Physical space is defined by the vector **r** (coordinates x, y, and z). Reciprocal space is defined by the vector **K** (coordinates K_x, K_y, and K_z). The ellipsoidal object in each space is related to the other by Fourier transformation. A single point in real space contributes to all reciprocal space. Although angular orientation is retained between each space, Fourier transformation results in a reciprocal length scale. Thus, the elongated ellipsoid depicted in real space becomes a flattened disk in reciprocal space.

owing to the Fourier transformation relationship of Eq. (3), a point $\rho(\mathbf{r})$ contributes to the entire reciprocal space.

The term **K** has been introduced rather arbitrarily. Through the use of the concept of reciprocal space, its physical meaning can be clarified. The incident x-ray beam (planar wave fronts in Fig. 1) can be represented in reciprocal space by a vector $\mathbf{K}_{in}$ whose direction is given by the propagation direction of the x-ray beam and whose magnitude is given by

$$|\mathbf{K}_{in}| \equiv 4\pi/\lambda, \qquad (6)$$

where λ is the wavelength of the incident radiation. The scattered x-ray beam can also be represented by a vector, $\mathbf{K}_{out}$. Because scattering may occur in many directions, it is described by many $\mathbf{K}_{out}$ vectors with different angular orientations. However, the scattering which will be considered in this chapter is elastic scattering, so that

$$|\mathbf{K}_{out}| = 4\pi/\lambda. \tag{7}$$

The surface which $\mathbf{K}_{out}$ describes, over the entire 4π steradian angular range, is a sphere known as the Ewald sphere. For a given $\mathbf{K}_{in}$, the sphere defines the surface in reciprocal space, which can be observed for various $\mathbf{K}_{out}$ vectors.

The vector $\mathbf{K}$ can now be defined in terms of $\mathbf{K}_{in}$ and $\mathbf{K}_{out}$ as

$$\mathbf{K} = \mathbf{K}_{out} - \mathbf{K}_{in}. \tag{8}$$

From Eq. (8) it immediately follows that

$$|\mathbf{K}| = (4\pi \sin \theta)/\lambda, \tag{9}$$

where 2θ is the angle between $\mathbf{K}_{in}$ and a particular $\mathbf{K}_{out}$.

Figure 3 depicts the Ewald sphere and the relationship of the three vectors $\mathbf{K}_{in}$, $\mathbf{K}_{out}$, and $\mathbf{K}$. It is clear that the ends of $\mathbf{K}$ must lie on the sphere. For small-angle x-ray scattering, the angle 2θ is very small, so that usually the sphere may be regarded as a flat surface and

$$|\mathbf{K}| \simeq 4\pi\theta/\lambda. \tag{10}$$

Having introduced the basic concepts of elastic scattering theory, we can examine various small-angle equations. The radius of gyration, for example, is a quantity which is typically measured by small-angle scattering. From Eqs. (4) and (5), we have

$$I(\mathbf{K}) = \sum_h \sum_j f_h f_j \exp(i\mathbf{K} \cdot \mathbf{r}_{hj}), \tag{11}$$

where $\mathbf{r}_{hj} = \mathbf{r}_h - \mathbf{r}_j$.

For a solution of protein molecules, the sum in Eq. (11) extends over the atoms in one molecule. Assuming that other molecules are structurally identical but randomly oriented, the observed intensity for the solution is computed by performing angular averaging of Eq. (11), producing the following general Debye scattering formula:

$$I(K) = \sum_h \sum_j f_h f_j \frac{\sin(Kr_{hj})}{Kr_{hj}}, \tag{12}$$

where K is now a scalar because, due to angular disorientation, the intensity is spherically symmetric. Expanding $\sin(Kr_{hj})/Kr_{hj}$ and noting that for sufficiently small angles Kr_{hj} is always small, we obtain the following approximation, known as Guiner's law:

$$I(K) \simeq MN^2 \exp(-K^2 R^2/3), \tag{13}$$

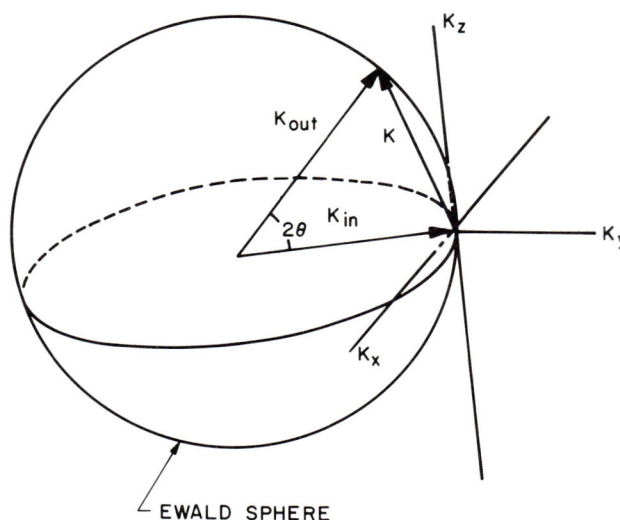

Fig. 3. The Ewald sphere construction. Reciprocal space is represented by the coordinates K_x, K_y, K_z. The incident x-ray beam is represented by the vector $\mathbf{K}_{in}$. Because angular orientation is preserved between physical and reciprocal spaces, $\mathbf{K}_{in}$ represents the actual direction of the incident x-ray beam. In this construction, the end of $\mathbf{K}_{in}$ is at the origin of reciprocal space. Diffracted x-rays are represented by $\mathbf{K}_{out}$. Because the scattering is elastic, the length of $\mathbf{K}_{out}$ equals that of $\mathbf{K}_{in}$ and, thus, describes a spherical surface for all angles 2θ between the incident and diffracted beams. The sphere itself can be moved in reciprocal space by altering the direction of $\mathbf{K}_{in}$. For a given $\mathbf{K}_{in}$, diffraction can be observed only on the surface of the Ewald sphere.

where M is the number of molecules in the beam, N the number of electrons in one molecule, and R the radius of gyration defined by

$$R^2 = \sum f_j r_j^2 \Big/ \sum f_j,$$

where r_j is defined relative to the electronic center of mass. The electronic center of mass is a point which, if taken as the origin of physical space, has the property that $\Sigma_j\ f_j r_j^2 = 0$ where the sum extends over all vectors located within a single molecule.

Enormous numbers of x-ray scattering studies have been performed to determine many structural parameters (particularly the radius of gyration). A few examples will be discussed in detail in Section IV.

B. Diffraction

Phenomenologically, diffraction is distinguished from scattering by the appearance of sharp spots or rings in an observed pattern. Mathematically, diffraction is distinguished from scattering by the coherent addition of scattered x-rays, which occurs if the structure is periodic. To describe diffraction, three mathematical relationships must be introduced. First, a Dirac delta function may be defined in the usual way by

$$\delta(\mathbf{r}) = 0 \quad \text{for} \quad \mathbf{r} \neq 0,$$

$$\int_{\text{all space}} \delta(\mathbf{r}) = 1 \tag{14}$$

Second, the operation known as convolution is described by the symbol $*$ and is defined by

$$f(\mathbf{x}) * g(\mathbf{x}) = \int f(\mathbf{x} - \mathbf{y}) g(\mathbf{y}) \, d\mathbf{y}. \tag{15}$$

Third, it is noted that the Fourier transform of the convolution of two functions is the product of their individual transforms. Thus

$$\int_{\text{all space}} f(\mathbf{r}) * g(\mathbf{r}) \exp(i\mathbf{K} \cdot \mathbf{r}) \, d\mathbf{r} = \left[\int f(\mathbf{r}) \exp(i\mathbf{K} \cdot \mathbf{r}) \, d\mathbf{r} \right]$$
$$\times \left[\int g(\mathbf{r}) \exp(i\mathbf{K} \cdot \mathbf{r}) \, d\mathbf{r} \right]. \tag{16}$$

A three-dimensional lattice may be described by

$$L(\mathbf{r}) = \sum_{lmn} \delta(\mathbf{r} - l\mathbf{a} - m\mathbf{b} - n\mathbf{c}), \tag{17}$$

where $\mathbf{a}$, $\mathbf{b}$, and $\mathbf{c}$ are translation vectors which define a unit-cell dimension. A structure can now be given by $\rho_u(\mathbf{r})$ (which must fit within the unit cell). Placement of this structure into a three-dimensionally periodic lattice is accomplished by convolution so that the final periodic structure is given by

$$\rho(\mathbf{r}) = \rho_u(\mathbf{r}) * L(\mathbf{r}). \tag{18}$$

The Fourier transform is found using Eq. (16):

$$F(\mathbf{K}) = F_u(\mathbf{K}) \cdot F_l(\mathbf{K}), \tag{19}$$

where the subscripts u and l define the transform of ρ_u and L, respectively. Performing the Fourier integral for L explicitly gives

$$F_l(\mathbf{K}) = \sum_{hkl} \delta(\mathbf{K} - h\mathbf{a}^* - k\mathbf{b}^* - l\mathbf{c}^*). \tag{20}$$

Equation (20) is simply the sum of a set of Dirac delta functions. Thus, Eq. (20) describes a function which is zero everywhere except at a discrete set of points which are separated by the translation vectors $\mathbf{a}^*$, $\mathbf{b}^*$, and $\mathbf{c}^*$ (i.e., a lattice). The reciprocal unit-cell vectors $\mathbf{a}^*$, $\mathbf{b}^*$, and $\mathbf{c}^*$ are directly related to the real-space unit-cell vectors $\mathbf{a}$, $\mathbf{b}$, and $\mathbf{c}$. From Eq. (19), $F(\mathbf{K})$ is, like the lattice, zero except at the reciprocal lattice points, where it assumes the value of $F_u(\mathbf{K})$. The intensity function, $I(\mathbf{K}) = F(\mathbf{K}) * F(\mathbf{K})$, is also zero except at the reciprocal lattice points. Thus, x-ray diffraction patterns show spots of variable intensity. The position of the spots is determined by the lattice $L(\mathbf{r})$, whereas the intensity of the spots is determined by the detailed structure $\rho_u(\mathbf{r})$, which is periodically arranged on the lattice. The mathematical description given shows that the phenomenological observation of diffraction (i.e., the appearance of sharp spots or rings in the recorded pattern) is a direct consequence of long-range translational periodicity within the sample.

In reality, there is not a sharp division between scattering and diffraction. Especially, biological systems display a range of order from ideal solutions to nearly perfect crystals. Membrane lamella stacked to form an ordered array usually show considerable disorder, so that the system may not be considered ideally crystalline. In many systems, crystalline order permits high-resolution structural determination. In other systems, the type of disorder is itself of great interest.

Putting together Eqs. (19), (20), and (5), $I(\mathbf{K})$ or $F(\mathbf{K})$ is simply described as a set of numbers $\{I_{hkl}\}$ or $\{F_{hkl}\}$ where h, k, and l given in Eq. (20) are integers which locate the reflections in reciprocal space; (h,k,l are known as Miller indices). As is the case for $F(K)$, the set of numbers $\{F_{hkl}\}$ are, in general, complex. Because F_{hkl} is a complex number, it may be expressed as

$$F_{hkl} = |F_{hkl}| \exp(i\Phi_{hkl}), \tag{21}$$

where $|F_{hkl}|$ is the amplitude and Φ_{hkl} the phase angle of F_{hkl}. From Eq. (5), it follows that

$$|F_{hkl}| = (I_{hkl})^{1/2}. \tag{22}$$

Thus, only the amplitude of F_{hkl} is obtained from the intensity function.

If $\{F_{hkl}\}$ is known, then, by inverting the Fourier transformation given in Eq. (3), we have

$$\rho(\mathbf{r}) = \int_{\text{all space}} F(\mathbf{K}) \exp(-i\mathbf{K}\cdot\mathbf{r}) \, d\mathbf{K}, \tag{23}$$

or

$$\rho(\mathbf{r}) = \sum_{hkl} F_{hkl} \exp(-i[ha*x + kb*y + lc*z]). \tag{24}$$

Therefore, if $\{F_{hkl}\}$ can be found, then the electron density function $\rho(\mathbf{r})$ (i.e., the structure) can be directly computed.

There are a variety of methods for finding phase angles $\{\Phi_{hkl}\}$ for x-ray reflections and thus, in combination with $\{I_{hkl}\}$, for completely determining $\{F_{hkl}\}$. The methods used typically depend on the nature of the structure to be solved.

Examples of the case of small-angle x-ray diffraction are also large in number. A few selected applications, including methods to determine x-ray phase angles, will be described in detail in Section IV.

III. X-Ray Technology

Although x-ray theory has changed very little since the early 1970s, x-ray experimental technology has rapidly advanced. The following section describes the most significant changes. The section is divided into two parts (x-ray sources and x-ray detectors). Other areas of technological change have also had considerable impact. For example, the availability of precision orientational stages and accurate thermal controllers have made possible many biological studies.

A. X-Ray Sources

From the discovery of x-rays by Roentgen in 1895 to around 1970, x-rays were generally produced by the same method. Namely, current is passed through a metallic filament which boils free electrons from the surface. Under vacuum, these electrons are accelerated at high potential to a metal target. The deceleration of the electrons on striking the target produces radiation over a wide range of energies up to the accelerating potential applied. This radiation is known collectively as *Bremsstrahlung*. In addition, radiation is concentrated into narrow spikes at discrete energies characteristic of the metal target. This characteristic radiation is produced by exciting, via the incident electron beam, a tightly bound inner-shell electron of the metal target to the continuum. Subsequent relaxation occurs within the metal atom, producing x-rays at a discrete wavelength. For example, a copper target (or anode) produces CuK_α radiation by a relaxation transition from the L shell to the K vacancy. The energy difference between the copper L and K levels is 8.048 keV. The wavelength is conveniently related to energy by

$$\lambda(CuK_\alpha) = hc/E = 12.3981/E \quad \text{keV} = 1.5405 \text{ Å}, \tag{25}$$

where h is Planck's constant, c the speed of light, and E the energy (in keV).

This equation holds, of course, for all light. It can be used together with Eq. (9) to redefine the magnitude of the wave vector **K** as

$$|\mathbf{K}| = 4\pi(12.3981)(\sin\theta)E. \tag{26}$$

Many modern x-ray generators are of the fixed target (or anode) design. These generators are of basically the same design as the original x-ray sources. Water cooling is provided for the anode to permit increased power (higher accelerating voltage and current for the electron beam which strikes the anode). Since the 1950s, more powerful generators have been designed, known as rotating-anode x-ray generators. In this design the metallic anode is in the form of a wheel which rotates at speeds up to 6000 rpm. Water cooling is still provided on the inner metallic surface. The outer surface is exposed to the electron beam under high vacuum (2×10^{-6} Torr). A bearing and lubrication system is necessary. This superior design results in the ability to produce higher x-ray intensity without melting the metallic anode. Because x-ray sources are specifically designed for each application, it is difficult to directly compare the improvement achieved by the rotating-anode design. Roughly, the rotating anode produces enhancements of 10–100 times the intensity of a fixed-tube target of identical source size. The rotating-anode x-ray generator is currently the most advanced generator for small-laboratory applications. It has found wide application.

In this past decade, a unique new source of x-rays has become available. Developed for high-energy physics, this source was designed not to produce x-rays but to store highly energetic electron and positron beams for the purpose of studying the nature of matter. These devices, known as storage rings, are injected with bunches of highly energetic electrons. Each electron bunch follows an approximately circular path, consisting of a series of straight lines with bends at intervals accomplished by strong bending magnets. For the high-energy physics experiment, a positron beam travels the same path in an opposite direction.

The energy of the stored-electron beam is very large. For example, at the Stanford Positron Electron Accelerating Ring (SPEAR), typical energies of larger than 3 GeV are achieved for beam currents of up to 100 mA. For this high energy, the speed of electrons approaches the speed of light. It is well known that acceleration of charge produces radiation. Larmor first described the radiation pattern produced by electrons traveling in a circular orbit in a magnetic field. At high speed, relativistic effects become dominant, and the radiation produced at each bending magnet within the storage ring has unusual properties (Jackson, 1962; Davis, 1981). This type of radiation is known as synchrotron radiation. [See Davis (1981) for a review].

There are several properties of synchrotron radiation which make it attractive for x-ray diffraction or scattering studies. First, the radiation is very intense. Again, it is difficult to directly compare specifically designed x-ray sources. However, synchrotron radiation produces enhancements of 100–1000 times the intensity of a rotating-anode generator using characteristic radiation. Much greater enhancements exist if synchrotron radiation is compared with the intensity of noncharacteristic radiation from rotating-anode generators. Substantially larger enhancements are available from radiation devices inserted into the storage ring, known as wigglers or undulators [see Davis (1981)]. The tremendous intensity of synchrotron radiation permits dynamical or time-resolved biological x-ray diffraction or scattering studies.

A second unique feature of synchrotron radiation is that it is tunable. Unlike conventional sources, synchrotron radiation is intense over a wide range of energies. A particular energy or wavelength can be chosen using a crystal monochromator. This availability of wavelengths makes possible resonance (or anomalous) x-ray diffraction and scattering studies. A third unique feature is that the radiation is very bright and parallel, which permits performing very small-angle scattering studies with exceptionally low background. Other properties include polarization of the x-ray beam and a unique time structure, resulting from the bunch structure of the stored electrons, which produces a burst of x-rays as each electron bunch passes a bending magnet (e.g., single-bunch running at Stanford produces an ~ 1 nsec burst of x rays every ~ 2 μsec).

Another unusual source of x-rays was developed for laser fusion research. This source produces ClK_α x-rays by striking a hexochloroethane target with a very high-intensity, focused, pulsed laser beam. The surface is stripped of its electrons, creating a plasma. Upon relaxation, a short, intense burst of x-rays is produced. Dynamical small-angle diffraction patterns of membranes have been produced in one nanosecond, using the Rochester laser plasma x-ray source.

B. X-Ray Detectors

Historically, scattered or diffracted x-rays have been detected by photographic methods. The x-ray film is then scanned by a densitometer to provide quantitative data. These techniques are still in use and, in many instances, are the methods of choice.

X-ray photon-counting methods began prior to 1950 with the widespread use of Geiger–Müller (GM) detectors. These detectors are known as point detectors because they simply count x-rays with no position sensitivity. Introduced after the GM detectors, proportional gas detectors, scintillation detectors, and solid-state detectors have found wide application. Each detector is a point detector but varies considerably in counting rate versus energy resolution (higher counting-rate capability results in poorer energy resolution). For example, NaI scintillation detectors can count 8 keV x-rays at approximately 10^5 cps (counts per second) with small dead-time losses. The energy resolution of this detector is approximately 2 keV. On the other extreme, a Si(Li) solid-state detector has an energy resolution of approximately 0.15 keV but can count 8 keV x-rays at only 10^4 cps with small dead-time losses.

Point detectors make possible diffractometry studies. As previously shown, for a given $\mathbf{K}_{in}$ a range of $\mathbf{K}_{out}$ diffracted beams defines the Ewald sphere [see Eq. (8) and Fig. 3]. To obtain diffracted intensity along a line in reciprocal space (as is desired in many crystal studies), it is necessary to rotate both the sample (by some angle θ) and the detector (by some angle 2θ) with respect to the incident beam. The resulting $\theta - 2\theta$ scan is linear in reciprocal space. Other scans are also possible with a diffractometer. In addition, incident-beam monochromators combined with analyzer-beam monochromators provide exceptionally low background and high angular resolution.

As briefly mentioned, for small-angle diffraction or scattering studies, the small section of the Ewald sphere which is observed may be regarded in some cases as flat. Particularly for specimens which possess angular disorder, the curvature of the Ewald sphere in the small-angle region can be neglected. In this case, detectors which have position sensitivity are useful.

Several types of position-sensitive detectors have been developed since 1970. Only a few will be mentioned here. (For a review, see Bordas, 1982.) The first class are linear detectors which are position sensitive in one dimension. Photon-counting detectors encode position information either by charge-division, time-delay, or rise-time methods. All of these detectors are filled with a proportional counter gas (e.g., 90% Ar, 10% CH_4). An anode wire is suspended in the gas perpendicular to the scattered x-ray radiation. Upon absorption of an x-ray by the counter gas, electrons travel to the anode under high voltage. In one design, the anode is resistive, and the electron current divides on striking the anode. Preamplifiers at either end of the detector determine the amount of charge which arrives at either end. The charge divides resistively, and the amount received at each end is directly related to the position along the wire. Typical resolutions are 100–500 μm along a 10-cm length. Other encoding schemes or detector designs produce similar resolutions. Compared with point detectors, these devices collect scattered x rays at many angles simultaneously. Thus with 100-μm resolution over a 10-cm length, a single linear detector reduces the time required to collect a diffraction pattern by 10^3. Photographic film also detects all angles simultaneously; however, the detection efficiency is only ~5%, whereas it is nearly 100% for a proportional linear detector. Thus the same pattern may be collected 20 times faster with such a device. In addition, photographic film has a limited dynamic range. For diffraction patterns which contain very intense peaks and very weak ones, multiple exposures are required to avoid film saturation by the intense peak. Further, the intrinsic background or fog in photographic film seriously limits the lower level of detection. Linear detectors have lower counting-rate capability than point detectors. Typical rates of 2×10^4 cps may be increased to 5×10^4, or 10^5 cps for newer designs (Boie *et al.*, 1982).

Two-dimensional detectors have also been developed. Counting detectors have been developed but counting-rate limitations pose a serious difficulty for these large-area devices. Two-dimensional silicon-intensified target (SIT) detectors (Reynolds *et al.*, 1978) do not have any counting-rate limitations. For the SIT detector, charge which is proportional to x-ray intensity is accumulated on a silicon grid. Resolution exceeding 100 μm is easily achieved. However, electronic noise dominates the low-level detection limit and saturation limits the dynamic range. Neither of these problems is a serious limitation, and good progress has been made to avoid both difficulties in experiments. The SIT detector has found wide use in dynamical biological diffraction studies based on synchrotron or plasma sources.

In summary, a tremendous range of detectors has been developed which have a wide variety of characteristics. Detectors suitable to the new x-ray sources, for almost any type of experiment, now exist. Care must be taken to choose the detector compatible with experimental characteristics.

IV. Biological Applications

Having introduced new sources and detectors, it should not be surprising that the capabilities of small-angle x-ray diffraction and scattering have been remarkably enhanced. Impact of the technology can be divided into two areas: the improvement of standard measurements and the development of nonstandard measurements.

Measurement of the radius of gyration of a soluble protein is a standard technique. With a standard fixed-anode x-ray tube design and photographic methods, determining the R_G of cytochrome c at 0.5 mM concentration would require ~ 24 hours of exposure. Following the exposure, the film would be developed, dried, and scanned by a microdensitometer. The densitometer trace would then be digitized to provide $I(K)$. Finally, a plot of ln $I(K)$ versus $(K)^2$ would be constructed, fit with a line, and R_G would be computed from the slope of this line. [See Eq. (13).] In practice, the entire procedure would require approximately two days.

Study of the radius of gyration R_G of cytochrome c as a function of pH, or as a function of weakly binding substrates such as $Ru(CN)_6$, can reveal information about the structure of this protein in solution. Using the two-day process described here, such a study would be impossible. In a recent study, Borso and Stamatoff (1980) examined cytochrome c in solution. The measurements utilized a linear position-sensitive x-ray detector, a rotating-anode generator, and a unique doubly curved point-focusing monochromator (Berreman et al., 1977). As a result, each measurement required only one half hour to provide R_G via a computer data-acquisition system. Using Eq. (13), the observed scattered intensity (after removal of instrumental background scattering) was plotted according to Guinier's law. Figure 4 shows plots of the logarithm of the scattered intensity for cytochrome c and cytochrome c in the presence of 50 mM $K_4Ru(CN)_6$. Changes in the observed radius (computed from the slope of these plots) were found to be small and dominated by the experimental error determined by least-square analysis. Borso applied the method of Simon (1971) and examined a plot of the $\ln(I_1/I_2)$ versus K^2, where I_1 and I_2 are the scattered intensities for the two experiments. By considering the ratio of intensities, the stochastic error (due to counting statistics) contributed in the same manner as if the two radii were computed separately and compared, in Fig. 4. However, the use of ratios of intensities considerably reduced systematic errors associated with finite-slit effects, nonlinear detector response, and multiple-particle (diffraction) effects. As a result, a linear fit of $\ln(I_1/I_2)$ versus K^2, which is shown in Fig. 5, produces a far smaller total error, even though the error associated with individual data points (due to stochastic error) is, as expected, significantly larger than the error of an individual data point, as given in Fig. 4. Therefore, small changes, ΔR_G, could be measured to an accuracy of approximately ± 0.1 Å using those methods. Long exposure using photographic methods and requiring densitometer scanning of the film would preclude this type of analysis.

Figure 6 gives the apparent radius of gyration of cytochrome c as a function of $Ru(CN)_6$ concentrations. The change in R_G was used by Borso and Stamatoff (1980) to model the counterion cloud surrounding the protein. No other known technique could provide this information. Of course, R_G measurements can be used to establish the formation of dimers, trimers, or higher oligimers of protein subunits. Using techniques similar to those described here, the equilibrium constants for dimer formation as a function of temperature or pH could be easily, quickly, and routinely determined.

Protein crystallography gives a high-resolution spatial view of the three-dimensional structure for an ever-increasing number of extraordinarily complex protein molecules. These structures require a significant effort to solve, and sometimes key structures cannot be crystallized. This is the case

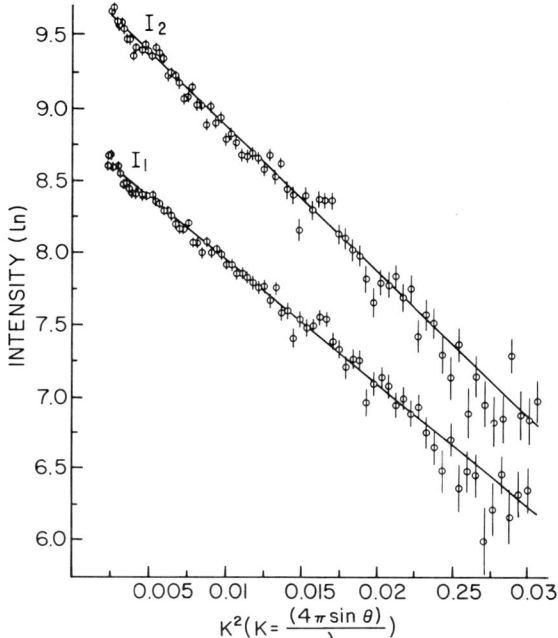

Fig. 4. I_1: Guinier plot of logarithmic intensity versus K^2 for 1 mM cytochrome c in 2.5 mM Tris-Tes buffer with 50 mM K_4 Ru(CN)$_6$ at pH 7.6. I_2: Guinier plot for 1 mM cytochrome c in 2.5 mM tris buffer at pH 7.6 without K_4 Ru(CN)$_6$. (From Borso and Stamatoff, 1980.)

for yeast hexokinase. Two different forms of the same protein (A and B isozymes) were studied by crystallography. Both forms showed a cleft in the structure. The A form showed the cleft closed about the bound glucose (Bennett and Steitz, 1978), whereas the B form, which is not bound with glucose, showed an open cleft (Anderson *et al.*, 1978). The difference between the two conformations could be due to the isozyme used, or could be the direct consequence of the binding of glucose.

McDonald *et al.* (1979) performed extensive small-angle x-ray scattering experiments to decide the question. Calculating the radius of gyration R_G from the crystallographic data, they found a decrease of approximately 1 Å between the open and the closed structures. In solution, the B isozyme radius of gyration could be determined both with and without glucose. The use of a modern position sensitive x-ray detector made possible achieving the necessary accuracy so that a change of approximately 1 Å (with accuracy of 0.2 Å) was found. The results demonstrate that, in solution, a conforma-

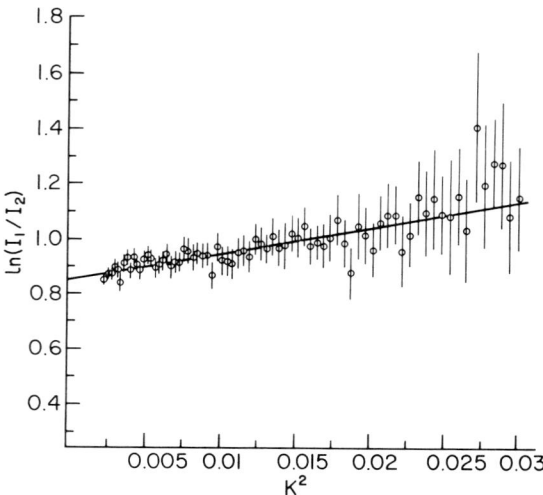

Fig. 5. Ratio plot of $\ln(I_1/I_2)$ versus K^2, where I_1 and I_2 are from Fig. 4. (From Borso and Stamatoff, 1980.)

tional change does occur, and suggest that the binding of glucose to hexokinase does cause the cleft in its structure to close.

Small-angle x-ray scattering is very sensitive to molecular aggregation. Scattering at small angles in general increases dramatically with aggregation consistent with a larger radius of gyration for the molecular aggregate. Mandelkow *et al.* (1980) utilized this phenomena along with modern x-ray methods to answer a fundamental question concerning microtubule forma-

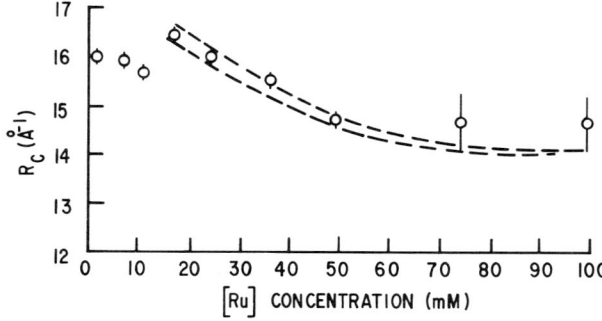

Fig. 6. The radius of gyration R_G of cytochrome c as a function of the Ru ion concentration at pH 7.4–7.6. Each point represents the results of an individual diffraction pattern. (From Borso and Stamatoff, 1980.)

tion during spindle development in mitotic cell division. Microtubules are composed of individual globular proteins (tubulin and other associated proteins). The proteins are arranged in rings, stacked to form the microtubule. The question is whether proteins form rings which are then stacked or whether the microtubule is formed continuously from smaller protein aggregates. At the European Molecular Biology Laboratory station at the DESY storage ring in Hamburg, Mandelkow et al. (1980) recorded the small-angle x-ray scattering during microtubule formation in 15-sec intervals. Formation of microtubules was initiated by rapidly increasing the temperature of the protein solution from 4° to 36°C and observing the time dependence of the scattering curve. The very small-angle forward scattering is proportional to the degree of polymerization of tubulin.

The forward scatter was found to decrease slightly and then dramatically increase after the thermal jump. This data suggests that ring structures which exist at 4°C disassociate to smaller protein aggregates and then reassociate to form microtubules. Differences in the x-ray scattering curve before and immediately after the thermal jump can be modeled in terms of a ring structure of tubulin (which disassociates with increasing temperature). Much more detailed analysis of these data is possible. However, even this straightforward interpretation of x-ray scattering from a very complex association of proteins has proven invaluable. The key feature, the ability to perform the experiment rapidly, relies on the use of a storage ring. Once application of this very significant technological advance is made, then even this simplified and direct analysis becomes a powerful analytical tool.

The examples chosen for studies of small-angle x-ray scattering show the progress which new technology has elicited. Sardet et al. (1976) (whose work is described at the beginning of this chapter) applied sophisticated analysis of intensity data. By altering the electron density of the solvent, these authors established important biological facts about rhodopsin. Borso and Stamatoff (1980) used modern detectors and a unique crystal monochromator which made possible applying a ratio method of analysis for intensity data to achieve highly accurate comparison of R_G, to establish key information about protein–ion interactions.

McDonald et al. (1979) used similar modern detectors to achieve high accuracy, which they utilized in combination with protein crystallography data to answer a fundamental biological question. Finally, Mandelkow et al. (1980) used a new source and new detector design to develop the technique of time-resolved x-ray scattering. This technique was used to answer a fundamental biological question and opens many possibilities for further study.

Storage rings are powerful sources of x-rays. As previously mentioned, intensities exceed conventional sources by $\sim 10^3$. However, synchrotron radiation is also polychromatic (i.e., the radiation is intense over a broad band of wavelengths). Conventional small-angle scattering measurements for the purpose of determining R_G are conducted with monochromatic radiation. Stuhrmann (1978) and Stamatoff (1979) showed that polychromatic radiation can be used to determine R_G by a nonstandard method. The basic approach to the use of polychromatic radiation can be illustrated by expanding Eq. (13) so that

$$I(K) \simeq MN_2(1 - \tfrac{1}{3}K^2R^2 + \cdots). \tag{27}$$

For small angles of scattering, Eq. (26) becomes $K = \alpha\theta E$, where α is a constant. Thus

$$I \simeq MN^2(1 - \tfrac{1}{3}\alpha^2\theta^2 E^2 R^2). \tag{28}$$

For a range of x-ray energies E, the observed intensity becomes

$$\begin{aligned}I &= \int If(E)\,dE \\ &= MN^2\left(\int f(E)\,dE - \tfrac{1}{3}\alpha^2\theta^2 R^2 \int E^2 f(E)\,dE\right),\end{aligned} \tag{29}$$

where $f(E)$ is the spectral distribution of incident energies. For a known spectral distribution (which can be calibrated), R can be determined from a plot of I versus θ^2. Future experiments using storage rings and SIT detectors will provide R_G with time resolution of less than 1 msec. Many dynamic experiments involving solution structure and function kinetics are open for investigation. Thus, new sources and detectors have contributed by improving standard methods of solution scattering as well as by making possible entirely new techniques.

Examples of the use of small-angle x-ray diffraction in biology are also large in number. Pure lipids have been extensively studied to determine structural principles relevant for biological membranes. For example, dipalmitoylphosphatidyl choline (DPPC) has been examined in great detail. In water, this lipid spontaneously forms lipid bilayers which are periodically stacked to form a lipid multilayer. The periodicity of approximately 64 Å makes the system amenable to small-angle x-ray diffraction. As a function of temperature and water content, a phase diagram can be determined. For DPPC three lamellar phases exist, and each phase produces unique diffraction phenomena. The low-temperature phase (known as $L_{\beta'}$) consists of relatively immobile lipid molecules packed in a bilayer to minimize exposure of the hydrophobic aliphatic chains to water. The chains are tilted with respect to a normal to the bilayer surface. The high-temperature phase, L_α, appears to be a smectic A liquid-crystalline phase. In the L_α phase, the hydrocarbon chains are melted and fluid. The lipid molecules remain packed in a bilayer, but molecular mobility within the plane of the bilayer is significantly increased. At intermediate temperatures and higher water content, a unique phase P_β is found to exist.

Several diffraction studies of the $L_{\beta'}$ phase of DPPC have been performed. One excellent example is the study by Torbet and Wilkins (1976). Lipid bilayers can be oriented, and the observed small-angle periodicity corresponding to the distance between bilayers is found to change with humidity (from 64 Å in excess water to 56 Å under vacuum). Torbet and Wilkins (1976) determined the electron-density variation across the lipid bilayer at various states of hydration by analyzing intensity changes in up to 12 orders of diffraction. To do this, $\rho(\mathbf{r})$ was computed using Eq. (24); $|F_{00\ell}|$ was determined using Eq. (22). Because the lipid bilayer is a symmetric structure along the multilayering axis, $\rho(\mathbf{r}) = \rho(-\mathbf{r})$. From Eq. (3) it follows that $F(K)$ must be a real function so that $\Phi_{00\ell}$, given in Eq. (21), is either 0 or π. Therefore, only the algebraic signs of $F_{00\ell}$ need to be determined. By plotting the observed $F_{00\ell}$ for various spacings associated with different water contents, and by requiring that the resulting curve $F(\mathbf{K})$ be smooth, the authors determined most of the phase angles. Further analytical requirements lead the authors to obtain the remaining phase angles. Figure 7 shows the calculated electron-density profiles across the lipid bilayer. The lipid head-group region is significantly more electron dense than the hydrocarbon-chain region. In the center of the bilayer is a sharp dip in electron density corresponding to the terminal methyl group at the end of the aliphatic chain. The profile remains approximately constant as a function of water content. At low water content, the head-group region of a neighboring lipid layer is seen to approach the central bilayer. Important small changes in the profiles do occur with increasing water content. In particular, the distance between head groups on either side of the structure is found to decrease with increasing water content. Because the aliphatic chains are rigid in this phase, the observed decrease in bilayer thickness supports the idea of increasing hydrocarbon-chain tilt angle with increasing water content. This small-angle x-ray diffraction result is consistent with a large number of tilt-angle studies made by other techniques.

Janiak et al. (1976) studied small-angle x-ray diffraction from DPPC as a function of temperature and water content. When the intermediate-temperature, higher-water-content phase was entered, these authors observed an additional set of x-ray reflections with spacings exceeding 150 Å. By indexing the strong $I_{00\ell}$ reflections as well as the new weaker-intensity reflections, Janiak et al. (1976) concluded that the lipid bilayers had become rippled in the plane of the bilayer. The ripple wavelength was found to be approximately three times the bilayer thickness.

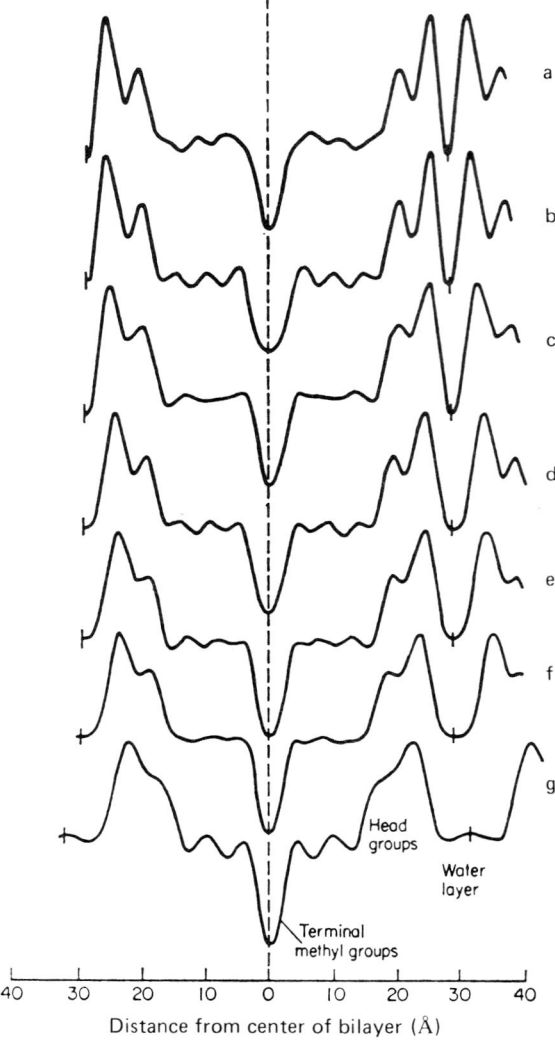

Fig. 7. Electron density profiles of the DPPC bilayer at various humidities: curve a, vacuum, 12 orders, $d = 56.6$ Å; curve b, 0% RH, 11 orders, $d = 57.0$ Å; curve c, 23% RH, 10 orders, $d = 57.5$ Å; curve d, 58% RH, 12 orders, $d = 57.7$ Å; curve e, 66% RH, 11 orders, $d = 57.7$ Å; curve f, 100% RH, 11 orders, $d = 58.8$ Å; curve g, vesicles in excess water, 10 orders, $d = 64.0$ Å. Note that as the humidity increases, the width of the water layer increases and that of the bilayer decreases. (From Torbet and Wilkins, 1976.)

Thermal transitions (endotherms) are observed by heating DPPC in water from the $L_{\beta'}$ to P_β and finally to the L_α phases. A broad pretransitional peak at 34°C occurs between the $L_{\beta'}$ and P_β phases for DPPC in excess water. Another lipid, dipalmitoylphosphatidyl ethanolamine (DPPE), exhibits no pretransitional peak. This lipid is very similar to DPPC, except that the head group is much smaller and less bulky.

McIntosh (1980) undertook a structural comparison of the two lipids. By comparing electron-density profiles across the bilayer, McIntosh (1980) concluded that DPPE had no hydrocarbon-chain tilt, whereas DPPC did. This finding was confirmed by larger-angle x-ray diffraction analysis. The author explained the observed difference in terms of the presence of a bulky choline head group for DPPC. In this case, hydrocarbon chains tilt to match the surface area presented by the lipid head group.

These three examples demonstrate the central role that small-angle x-ray diffraction continues to play in the study of lipids. These studies have progressed to the point where reasonable structural models based on the detailed chemical structure of a particular lipid can be used to predict the existence of thermal transitions.

Similar structural diffraction studies have been used on naturally occurring membranes. For example, phase transition studies have been performed for retinal disk membranes (Gruner *et al.*, 1982). To determine profile structures for natural membranes, a method of computing phase angles is required. Several workers have used various methods to regain $\Phi(\mathbf{K})$. In one method, use is made of the fact that the periodicity of the lamella is a function of water content, as it is for the lipid case. Stamatoff and Krimm (1976) developed a general alogrithm for the determination of $\rho(\mathbf{r})$ using this swelling method. However, the method requires that the structure remain constant as water is added between the lamella.

From the study by Torbet and Wilkins (1976) on DPPC, this assumption is found to be appropriate, provided the changes in spacing due to water content are considered only over a small spacing interval. Pachence *et al.* (1979) used this alogrithm to obtain electron-density profiles from reconstituted bacterial reaction-center membranes, as described in the introduction.

The profile structure of several other membranes has also been determined by small-angle x-ray diffraction. For example, structural determination of nerve myelin has been considered using swelling experiments (Worthington, 1972). In other systems, the stacking of membranes is not perfect, so that the lattice $L(\mathbf{r})$ [Eq. (17)] becomes disordered. In this case, disorder analysis was successfully applied by Schwartz *et al.* (1975).

New x-ray technology, which has so significantly impacted small-angle x-ray scattering, has also made possible new types of small-angle diffraction experiments. For example, Herbette *et al.* (1981) have determined the electron-density profile of sarcoplasmic reticulum membranes, which they succeeded in stacking in a semicrystalline array. Combined with neutron diffraction studies, these results provide a profile structure of the lipid protein sheet. Blasie *et al.* (1982a) used intense synchrotron radiation at the Stanford Synchrotron Radiation Laboratory and a SIT detector system to record the same diffraction information in less than one second. Using a unique caged ATP compound (McCray *et al.*, 1980), these scientists succeeded in stimulating the membrane by releasing ATP from the compound via a laser flash. By uniformly stimulating the membrane stack, Herbette and Blasie recorded diffraction patterns during the calcium-ion transport cycle of the membrane. Figure 8 shows the changes in the diffraction pattern observed after a laser flash released the ATP. The results suggest that observable membrane structural changes do occur during the pumping cycle. Future work should permit determining the detailed movements involved.

As another example of the use of advanced x-ray technology, consider the purple membrane of H. Halobium, which consists of hexagonally arranged trimers of bacteriorhodopsin protein within a lipid framework. The protein bacteriorhodopsin is a photon-activated proton pump. Upon stimulation, major changes in the optical absorption spectrum occur. A key question is what structural changes are associated with the pumping process.

Small-angle x-ray diffraction patterns were first recorded using photographic methods by Blaurock and Stoeckenius (1971). This pattern demonstrates the hexagonal arrangement of protein. Further work by Unwin and Henderson (1975) using low-dose electron-microscopy methods showed that the hexagonal unit cell consists of a trimer of bacteriorhodopsin molecules, each of which consists of seven helical segments transversing the lipid membrane. Figure 9 shows a diffraction pattern obtained using a rotating anode generator and linear position-sensitive detector. The pattern required a total exposure of 3.5 hours. However, the unusual characteristic is that the diffraction pattern was collected at $-40°C$ from a specimen with an optical density at 570 nm of ~ 4 trapped in a spectroscopic intermediate state. For a specimen of optimal thickness for diffraction, the optical density at 570 nm would exceed 100. A sample of this thickness would be impossible to trap uniformly in a spectroscopic intermediate state. Even with a thick specimen, photographic methods would require exposures of several hours. Using more advanced x-ray technology permits examining thin samples, which may be uniformly trapped in various intermediate states, and permits obtaining diffraction patterns in reasonably short time intervals. Thus, direct comparisons of membrane structural changes can be correlated with functionally meaningful spectroscopic states.

Using even more advanced x-ray technology, Frankel and Forsythe (1982)

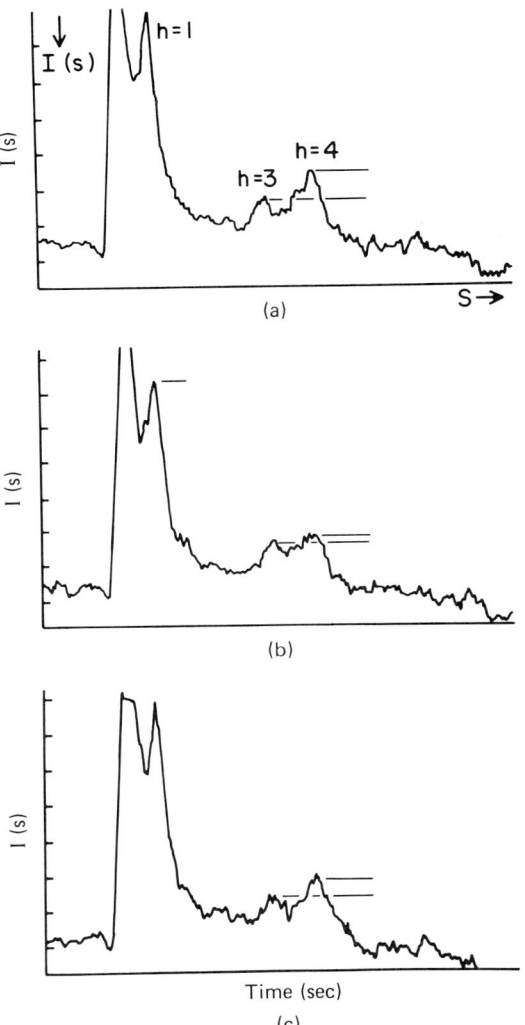

Fig. 8. The lamellar x-ray diffraction obtained from partially dehydrated, oriented multilayers of isolated sarcoplasmic reticulum membranes using a 200-msec x-ray exposure. The patterns were recorded at the Stanford Synchrotron Radiation Laboratory: (a) immediately prior to flash photolysis of caged ATP; (b) immediately following flash photolysis of caged ATP; (c) 1 min after flash photolysis of caged ATP. (From Blasie *et al.,* 1982a.)

462 J. STAMATOFF

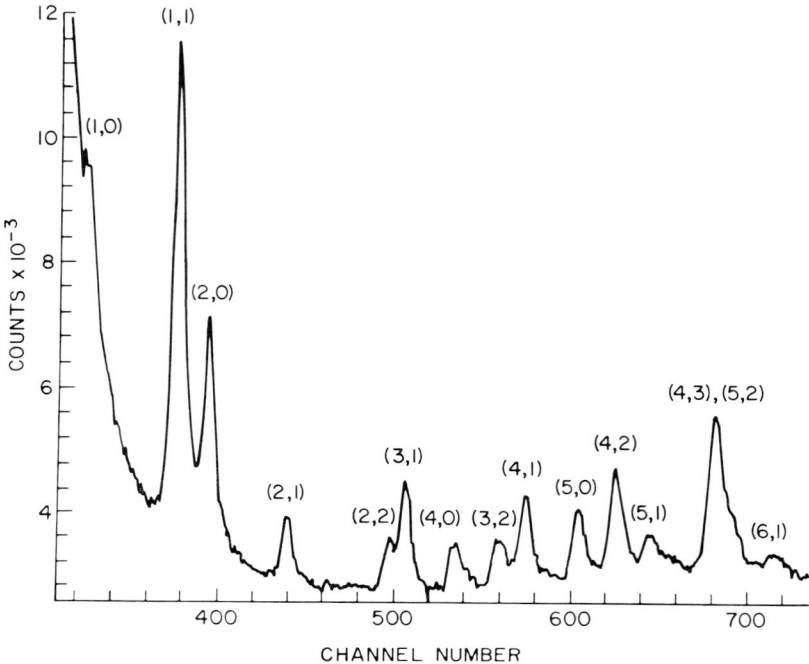

Fig. 9. A small-angle x-ray diffraction obtained from an oriented multilayer of purple membranes from H. Halobium. The thin layer of membranes had an optical density at 570 nm of ~4 (the optimal thickness for x-ray diffraction has an optical density which exceeds 100). The pattern was recorded with a 3.5-hr exposure at $-40°C$.

examined changes in this membrane in real time without resorting to low-temperature trapping methods. Using the unique laser-plasma source at the University of Rochester, Frankel and Forsythe (1982) obtained patterns on the nanosecond time scale. Figure 10 shows a diffraction pattern obtained using this unique source of x-rays and a SIT detector from purple membranes. The source produces chlorine K_α x-rays, which significantly increases the scattering cross section of the membranes. Both the low-temperature trapping experiment and the room-temperature nanosecond time-scale experiment showed diffracted intensity changes associated with the formation of membrane intermediates. In summary, new sources and detectors of x-rays have substantially improved standard small-angle x-ray diffraction measurements and have permitted dynamic experiments which were previously not possible.

As a final example of new methods for biological x-ray diffraction and scattering studies, anomalous or resonance x-ray studies will be considered. These types of experiments are entirely new and have no analog with standard small-angle methods. Typically, small-angle studies are performed at only one wavelength. Synchrotron radiation makes possible selecting a narrow (2 eV) band of radiation over a broad range (3–20 keV). From Eq. (21) it follows that the elastic scattering angle 2θ changes as a function of the incident energy. However, if the intensity is expressed as a function of $\mathbf{K}$, then no change would be observed in $I(\mathbf{K})$ at different energies E (provided appropriate absorption corrections are made). An important exception to this generality occurs if the incident x-ray energy is close to an atomic absorption-edge energy within the specimen. In this case, the atomic scattering factor f, given in Eq. (4), becomes a complex energy-dependent function given by

$$f = f_0 + f'(E) + if''(E), \tag{30}$$

where f' is the change in the real part of the atomic scattering factor and f'' the change in the imaginary part. This effect changes the observed x-ray

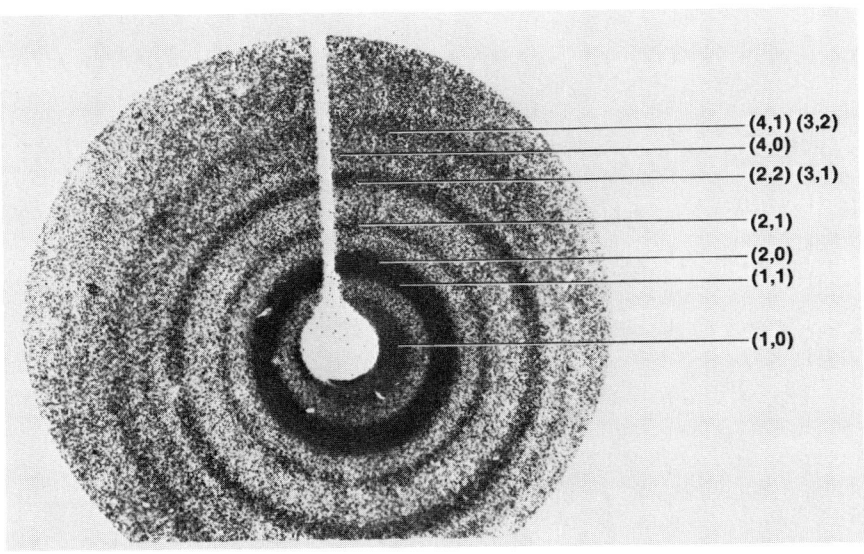

Fig. 10. Small-angle x-ray diffraction pattern of purple membranes from H. Halobium recorded at the Laboratory for Laser Energetics, Rochester, New York. The pattern was recorded using a flash of x-rays from a laser-plasma x-ray source: laser pulse energy, 213 J; laser pulse width, 700 psec; x-ray source, Cl^{15+} laser-heated plasma; x-ray wavelength, 4.45 Å. (From Frankel and Forsyth, 1982.)

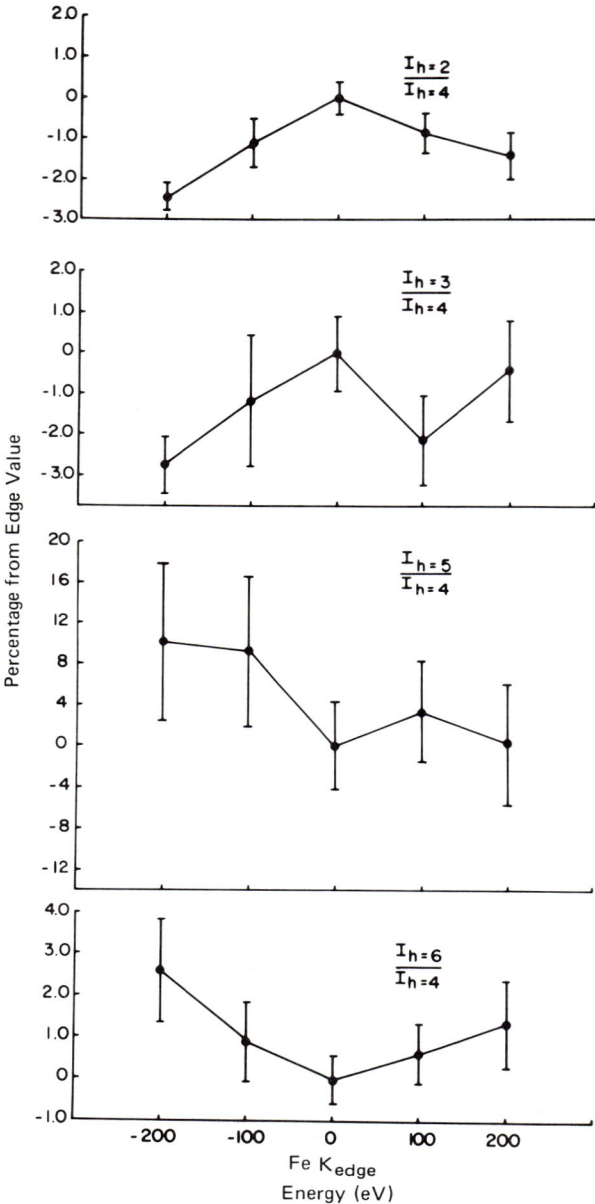

Fig. 11. The ratios of the integrated intensities of the lamellar reflections, $R_E(h/h')$, for a reconstituted cytochrome oxidase membrane multilayer. (a) $R_E(h/h')$ is shown for $h' = 4$ and $h = 2, 3, 5,$ and 6 at incident x-ray energies about the iron K-absorption edge. The error bars indicated are based on the errors in the estimated background scattering. These plots show resonance diffraction effects at the iron K-absorption edge. (b) The ratios

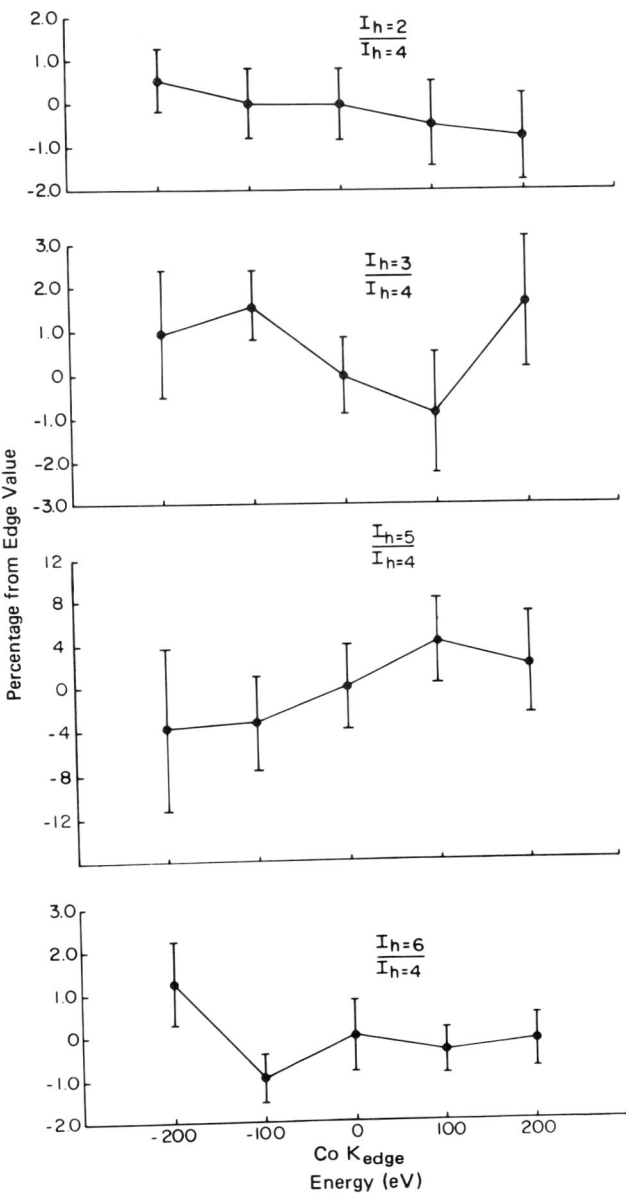

$R_E(h/h')$ for $h' = 4$ and $h = 2, 3, 5,$ and 6 for a reconstituted cytochrome oxidase membrane multilayer at incident x-ray energies about the cobalt K-absorption edge. These plots do not show any resonance diffraction effects, which is consistent with the absence of cobalt in this membrane protein. (From Stamatoff *et al.*, 1982.)

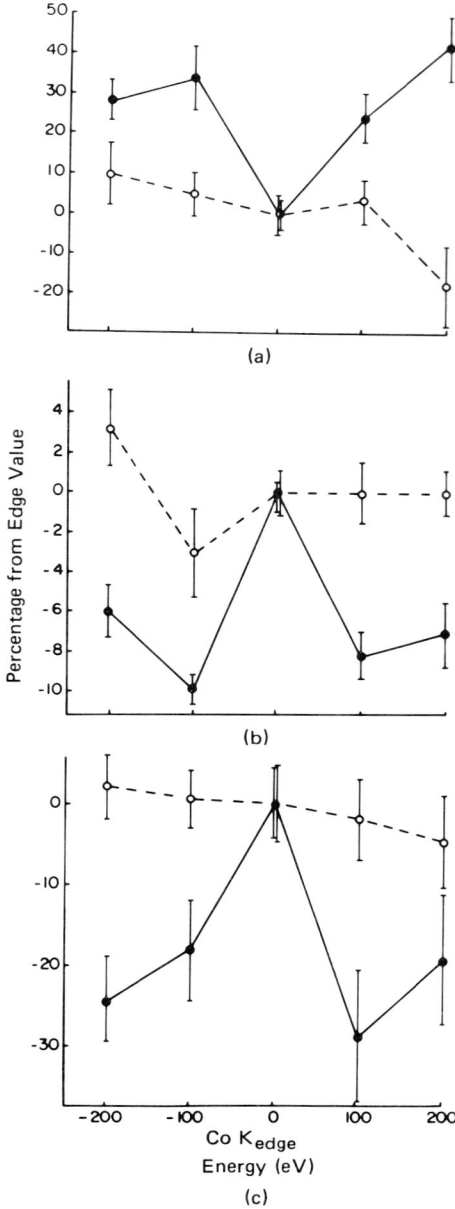

Fig. 12. The ratios $R_E(h/h')$ for $h' = 2$ and (a) $h = 3$, (b) $h = 4$, and (c) $h = 6$ for a reconstituted reaction-center cytochrome c membrane multilayer at incident x-ray energies about the cobalt K-absorption edge. The points connected by a solid line (———) are for a sample in which the iron atom of the

scattering via Eqs. (4) and (5). The changes are known collectively as anomalous diffraction and have been used in crystallography for many years.

Sturhmann combined the advantages of synchrotron radiation with the resonance effect to perform small-angle x-ray scattering studies on the proteins ferritin (Sturhmann, 1980) and hemoglobin (Sturhmann and Notkohm, 1981). By conducting experiments near and far from the iron-absorption edge, he was able to model the iron-containing portion of the proteins. Near the absorption edge $f' \simeq -8e$, whereas $f_0 = 26e$, and thus iron scatters as a much lighter element.

Stamatoff et al. (1982) and Blasie et al. (1982b) developed the method of resonance small-angle x-ray diffraction from biological membranes. Using synchrotron radiation, the method showed extreme sensitivity, which is required for the study of dilute metal atoms in biological systems. Specifically, the iron atoms in cytochrome oxidase containing lipid membranes and bacterial photoreaction-center membranes were observed. For cytochrome oxidase membranes, a change of $\sim 2\%$ in diffracted intensity was observed in the ratio of the strong second and fourth orders of lamellar diffraction as a function of energy. Figures 11 and 12 show the results for the two membranes. The technique permits locating specific metal atoms within biological structures without altering the structure.

V. Summary

Small-angle x-ray scattering and diffraction has been significantly enlarged in its scope and depth of application in biology through a series of remarkable technical developments. New x-ray sources and detectors have improved standard measurements so that detailed studies requiring many parameter changes can be conducted in a reasonable time. Standard experiments may also be conducted with greater precision and accuracy, providing detailed comparisons. Most important, new methods such as dynamic x-ray scattering and diffraction, as well as resonance x-ray scattering and diffraction, have been developed. Two new fields of structural biology are thus open for further and potentially very exciting investigation.

---

cytochrome c heme group has been replaced by cobalt. In this case, resonance diffraction effects are observed. The points connected by a dashed line (- - -) are for a sample in which the iron atom of the cytochrome c heme group was not substituted. For this case, no resonance diffraction effects were observed. (From Stamatoff et al., 1982.)

References

Anderson, C. M., Stenkamp, R. E., and Steitz, T. A. (1978). *J. Mol. Biol.* **123**, 15–33.
Alexander, L. E. (1979). "X-ray Diffraction Methods in Polymer Science." R. E. Krieger Publishing Company, New York.
Bennett, W. S., and Steitz, T. A. (1978). *Proc. Nat. Acad. Sci.* **75**, 4848–4852.
Berreman, D. W., Stamatoff, J., and Kennedy, S. J. (1977). *Appl. Opt.* **16**, 2081–2085.
Blasie, J. K., et al. (1982a). *Biochim. Biophys. Acta.* **679**, 188–197.
Blasie, J. K., Herbette, L., Pierce, D., Pascolini, D., Scarpa, A., and Fleischer, S. (1982b). *Ann. N. Y. Acad. Sci.* **402**, 478–484.
Blaurock, A. E., and Stoeckenius, W. (1971). *Nature New Biol.* **233**, 152–155.
Boie, R. A., et al. (1982). "High Resolution X-Ray Gas Proportional Detectors with Delay Line Position Sensing for High Counting Rates." Brookhaven National Laboratories Report No. 30707, Upton, New York.
Bordas, J. (1982). *Nucl. Instrum. Method Phys. Res.* **201**, 209–220.
Borso, C. S., and Stamatoff, J. B. (1980). *Biopolymers* **19**, 1887–1897.
Davis, H. L., ed. (1981). *Physics Today* **34** (5), 28–71.
Frankel, R. D., and Forsyth, J. M. (1982a). *Biophys. J.* **37**, 232a.
Frankel, R. D., and Forsyth, J. M., (1982b). "Methods in Enzymology," vol. 88 (L. Packer, ed.), pp. 276–281. Academic Press, New York.
Glatter, O., and Kratky, O. (1982). "Small Angle X-Ray Scattering." Academic Press, New York.
Gruner, S. M., Rothschild, K. J., and Clark, N. A. (1982). *Biophys. J.* **39**, 241–251.
Herbette, L., Scarpa, A., Blasie, J. K., Wang, C. T., Saito, A., and Fleischer, S. (1981). *Biophys. J.* **36**, 47–72.
Jackson, J. D. (1962). In "Classical Electrodynamics," pp. 468–472. Wiley, New York.
Janiak, M. J., Small, B. M., and Shipley, G. G. (1976). *Biochemistry* **15**, 4575–4580.
Mandelkow, E. M., Harmsen, A., Mandelkow, E., and Bordas, J. (1980). *Nature* **287**, 595–599.
McCray, J. A., Herbette, L., Kihara, T., and Trentham, D. R. (1980). *Proc. Nat. Acad. Sci.* **77**, 7237–7241.
McDonald, R. C., Steitz, T. A., and Engelman, D. M. (1979). *Biochemistry* **18**, 338–342.
McIntosh, T. J. (1980). *Biophys. J.* **29**, 237–246.
Pachence, J. M., Dutton, P. L., and Blasie, J. K. (1979). *Biochim. Biophys. Acta.* **548**, 348–373.
Ramachandran, G. N., and Srinivasan, R. (1970). "Fourier Methods in Crystallography." Wiley, New York.
Reynolds, G. T., Milch, J. R., and Gruner, S. M. (1978). *Rev. Sci. Instrum.* **49**, 1241–1249.
Sardet, C., Tardieu, A., and Luzzati, V. (1976). *J. Mol. Biol.* **105**, 383–407.
Schwartz, S., Cain, J. E., and Dratz, E. A. (1975). *Biophys. J.* **15**, 1201–1233.
Simon, I. (1971). *J. Appl. Crystal.* **4**, 317–318.
Stamatoff, J. (1979). *Biophys. J.* **26**, 325–328.
Stamatoff, J. B., and Krimm, S. (1976). *Biophys. J.* **16**, 503–516.
Stamatoff, J., et al. (1982). *Biochim. Biophys. Acta.* **679**, 177–187.
Sturhmann, H. B. (1978). *Q. Rev. Biophys.* **11**, 71–98.
Sturhmann, H. B. (1980). *Acta Cryst. A* **36**, 996–1001.
Sturhmann, H. B., and Notkohm, H. (1981). *Proc. Nat. Acad. Sci.* **78**, 6216–6220.
Torbet, J., and Wilkins, M. H. F. (1976). *J. Theor. Biol.* **69**, 447–458.
Unwin, P. N. T., and Henderson, R. (1975). *J. Mol. Biol.* **94**, 425–440.
Vainshtein, B. K. (1966). "Diffraction of X-rays by Chain Molecules." Elsevier, New York.
Worthington, C. R. (1972). *Ann. N. Y. Acad. Sci.* **195**, 293–308.

Index

A

Absorption coefficient
  in EXAFS, 300–302
  in transmission x-ray absorption spectroscopy, 318, 328
  in x-ray crystallography, 396
Absorption edges, 295, 297–299, 306–307
  bound-state transitions, 299, 307
  K edge, 298, 395
  L edge, 298, 395
Adenylate kinase, studied by NMR, 60
Adrenocortitropic hormone (ACTH), crystallization, 371
Aggregation, effect on x-ray scattering, 454
Alkaline phosphatase, studied by NMR, 60
Ammonium sulfate fractionation, 370
Angular amplitude of rotation, measured by ESR, 164
Anomalous scattering and diffraction, 395–399, 448, 463–467
  difference, 397
  partial structure resolved, 425–426
Aspartate transcarbamylase
  studied by EXAFS, 358
  studied by x-ray crystallography, 388–389
Asymmetric unit, 377
Atomic falloff, 336–337
Atomic scattering factor, 395, 441, 463
ATP (adenosine triphosphate)
  studied by NMR, 43–46, 80
    ATPase reaction, 70–71
    configurations, 44–45
    effect of catecholamines, 46
    effect of divalent cations, 44
    effect of dopamine, 46
  studied by x-ray diffraction, caged compound, 460
Axial splitting parameter, 114–116, 148–149
Azobacter vinelandii nitrogenase MoFe protein, 192–194
  studied by ESR, 192–194
  studied by EXAFS, 356–357
  studied by Mössbauer spectroscopy, 275
Azurin
  studied by EXAFS, 353
  studied by NMR, 50

B

Backscattering, in EXAFS, 300
Bacteriorhodopsin
  studied by x-ray crystallography, 427
  studied by x-ray diffraction, 460
Batch method of crystallization, 371–372
B_{12} coenzyme, 139–142
Beat pattern in EXAFS, 349
Bijvoet difference, *see* Anomalous scattering and diffraction, difference
Bloch equations, 10–15
Bohr magneton β, 91
Boltzmann distribution, 9, 22
Bone samples, studied by Mössbauer spectroscopy, 284
Bovine superoxide dismutase
  studied by ESR, 178–179, 207–209
  studied by EXAFS, 356
BPTI, studied by NMR, 52, 56, 63
Bragg reflection, 312, 333, 382
  harmonics, 315
Bravais lattice, 377
Bremsstrahlung, 447
Bricogne strategy, 403–404
Bulk magnetization, 9–10

C

Calcium, EXAFS, 357–358
Characteristic radiation, 447
Chemical exchange, 29–30
  "fast," 30–31, 51
  "slow," 30–31

Chemical shift
  in Mössbauer spectroscopy, 251–253
  of iron–sulfur proteins, 267–268
  in NMR, 15–18
    anisotropy (CSA), 24–25
    due to pH, 68–70
    of benzene, 16–17
    of toluene, 16–17
    "upfield" and "downfield," 18
Chloroperoxidase, studied by EXAFS, 351
Chlorophyll, studied by NMR, 74
Chymotrypsin
  studied by NMR, 79
  studied by x-ray crystallography, 422
Chymotrypsinogen, studied by ESR, 157
Cobalamine, studied by Mössbauer
  spectroscopy, 274
Cochlea, studied by Mössbauer spectroscopy, 289–290
Coherent scattering, 374
Collision frequency, in ESR, 174
Computer graphics, in x-ray crystallography, 391, 416–417
Conalbumin
  studied by ESR, 210–212
  studied by Mössbauer spectroscopy, 271
Contact shifts, 21–23
"Continuum resonance," 307
Convolution, 382–383, 429–430, 444
Coordination, determined by ESR, 111–119
Crambin, studied by x-ray crystallography, 425–426
Creatine phosphokinase, studied by NMR, 72
Cross-difference Fourier, 413–414
Cryoenzymology, 422–423
Crystal growth, 367–374, 430–431
Cu(II) proteins, studied by ESR, 112–114
Curie's law, 9–10, 22
Cyanobacteria, studied by Mössbauer
  spectroscopy, 277–279, 285
Cytochrome c
  studied by anomalous x-ray diffraction, 466–467
  studied by NMR, 22, 60, 63
  studied by x-ray scattering, 451–452
Cytochrome c oxidase
  studied by anomalous diffraction, 464–467
  studied by ESR, 219
  studied by EXAFS, 335, 337, 351–352
  studied by x-ray crystallography, 427

Cytochrome P-450
  studied by ESR, 173, 213–216, 220–221
  studied by EXAFS, 351
  studied by Mössbauer spectroscopy, 278–280

D

Data analysis
  of crystallographic data, 390–391
  of EXAFS spectra, 341–344
  of Mössbauer spectra, 263–265
Data reduction, of x-ray absorption spectra, 333–341
  background subtraction, 334–337
Debye scattering formula, 443
Debye–Waller factor, 303, 305–306, 341–342, 344, 348, 353
Densitometer, 390
Derivative maps, 419–420
Deuterium substitution, in electron spin-echo technique, 216–217
Dialysis technique in crystallization, 373
Difference Fourier synthesis, 413–414
Diffractometry studies, 450
Dipolar coupling, 20–21, 23–24, 151
Dipolar relaxation, 81–84
Dispersion spectra, 223
Dispersive x-ray absorption spectroscopy, 316–317
DNA, studied by NMR, 52, 63
  B-DNA, 52
  Z-DNA, 52
Double labeling, in ESR, 169–174
  biradicals, 169
Dynamic diffraction, 449
Dynamic studies, see Molecular motion

E

Effective field, 11, 13
Electric-field gradient, 253–254, 256
Electric-monopole interaction, 251–252
Electric-quadrupole interaction, 251–253
Electron density, 374–376
  distribution, 441
  function, 447
  maps, 414–423
Electron–electron exchange, 170
Electron-exchange term, 151

Electron–nuclear double resonance (ENDOR), 109–111, 121, 184–199
Electron-spin echo, 199–222
  envelope modulation, 201–203
  time evolution of amplitude, 214
Electron-spin relaxation, 256
Electron Stark effect, 217–218
  frequency shift, 218
  "half-fall" value, 218
Enterochelin, studied by Mössbauer spectroscopy, 287–288
Escherichia coli
  studied by Mössbauer spectroscopy, 287–288
  studied by NMR, ATPase reaction, 70–71
Ewald sphere, 380, 443–444
Exchange frequency, 173
Extended x-ray absorption fine structure (EXAFS), 295, 421–422
  theory, 297–307

F

Fermi contact term, 256
Fe(III) porphyrins
  studied by ESR, 114–117
  studied by EXAFS, 349
Ferredoxin
  studied by EXAFS, 349
  studied by Mössbauer spectroscopy, 247, 267–270, 278–280
Ferritin
  studied by anomalous diffraction, 467
  studied by EXAFS, 352
  studied by Mössbauer spectroscopy, 272–273, 289
Field-dependent transitions, 103
Figure of merit, 400–401
Filter fluorescence, 324
First coordination sphere, 304
Fluorescence
  data-acquisition technique in x-ray absorption spectroscopy, 320–326, 345
    "barrel monochromator," *or* crystal analyzer, 322–324
    scatter-rejection scheme, 323–324
    scintillation counter, 322, 324
    solid-state detector, 322–323
  yield, 320
Focusing effect, 305, 353

Formica pratensis (ant), studied by Mössbauer spectroscopy, 290
Four-circle diffractometer, 390
Fourier filtering, 339–341
Fourier summation, 383, 385
Fourier transform
  in crystallography, 375–376, 429–430
  in EXAFS, 339
  of free induction decay signal, 37, 81
  in NMR, 80–81
  spectrometer, 14–15
  of three-pulse spectrum, in ESR, 204
  two-dimensional, 52–53
  in x-ray diffraction, 445–446
Free induction decay (FID), 15, 36–37
Free-radical intermediate, probed by ESR, 138–142
Friedel's law, 384–385
  breakdown, 397

G

Galactose oxidase, studied by ESR, 117–119, 205–207, 219–221
Glucose metabolism, studied by NMR, 73–76
Gold, EXAFS, 358
Guinier's law, 437, 443–444, 451
g value, 91–97
Gyromagnetic ratio, 7, 11

H

Halobacterium, studied by Mössbauer spectroscopy, 288
Harker construction, 392–394
  sections, 411
Harmonic discrimination, 315
Harmonic radiation, 330–331
Heavy-atom derivative, 391–399, 408–414
  preparation, 408–409
Hemagglutinin, studied by x-ray crystallography, 404–407
Hemerythrin, studied by EXAFS, 352
Hemocyanin, studied by EXAFS, 332, 353–355
Hemoglobin
  crystallization, 368, 373
  deoxy- and oxy-Co(II), studied by ESR, 195–199
  studied by anomalous diffraction, 467

studied by EXAFS, 350
studied by Mössbauer spectroscopy, 265–266
studied by saturation-transfer ESR, 225–228
studied by x-ray crystallography, 391, 402–404, 411, 421–422
Hepatocytes, studied by NMR, 74–76
Hexokinase (yeast), studied by x-ray scattering and diffraction, 453–454
High-potential iron proteins, studied by Mössbauer spectroscopy, 267
Human chorionic somatomammotropin, studied by x-ray crystallography, 424
Hydrogenase, studied by Mössbauer spectroscopy, 279–280
Hyperfine coupling constant, 98
Hyperfine interactions, 250
Hyperfine splitting, 97–101
constant, 99–100

I

Imidazole
group-fitting technique, in EXAFS, 354
studied by ESR, 203–216, 220–221
Immunoglobins, crystallization, 371
Intensity ratio (IR), 149
"Internal" calibration, 320
Iron dioxigenases, studied by EXAFS, 352
Iron–sulfur proteins, studied by Mössbauer spectroscopy, 266–270, 279, 281–282
Isoelectric precipitation, 370
Isomer shift, see Chemical shift
Isomorphous replacement, 390–395
Isotropic rotation, 26, 54

J

J coupling, see Scalar coupling

K

Kinetic data, using NMR, 57–67, 70–73
isotopic exchange, 61–67
linewidth analysis, 58
magnetization transfer, 58–61

L

Lactoferrin, studied by Mössbauer spectroscopy, 271

Lambda repressor, crystallization, 371
Larmor frequency, 8–10
Laue equation, 379
Leakage radiation, 326–328
Least-squares fitting
in Mössbauer spectroscopy, 264
refinement, in crystallography, 390–391, 417–421
Ligand
coordination, 111–119
exchange, 119–123
"Linear" electric field effect (LEFE), 218–222
Lipid bilayers, see Membrane bilayers
Local field, 15–16
Longitudinal relaxation time, see Spin–lattice relaxation time
"Loose-gap" resonator, 104
Lorentzian lines, 81, 263
Lung tissue, studied by Mössbauer spectroscopy, 282–283
Lysozyme
crystallization, 368
studied by NMR, 6, 51–52, 63
studied by x-ray crystallography, 422, 426

M

Magnetic-dipole interaction, 254–255
Magnetic hyperfine field, 255–256
Magnetic splitting, 254–257
in hemoglobin, 265–266
in iron-storage proteins, 272
in iron–sulfur proteins, 267
Magnetite particles, 272
Magnetotactic bacteria, 272
Membrane bilayers
studied by anomalous x-ray diffraction, 467
studied by ESR, 164–169
studied by NMR, 57
studied by x-ray scattering and diffraction, 437–438, 456–459
dipalmitoyl phosphatidyl choline, 437–438, 456–459
dipalmitoyl phosphatidyl ethanolamine, 459
Membrane proteins, studied by x-ray crystallography, 426–428
Metabolic studies, by NMR, 67–79

Metalloproteins
  studied by ESR, 109–142, 205–222
  studied by EXAFS, 344–358
  studied by x-ray absorption, 208
Metal metabolism, studied by ESR, 132–134
Microtubule formation, studied by x-ray scattering, 454–455
Miller indices, 445
Modulation of field, 104–106
Molecular motion
  studied by NMR, 26, 29, 53–56
  studied by x-ray scattering, 438, 456
Molecular replacement, 402
Molybdenum enzymes, studied by EXAFS, 356–357
Monochromator, x-ray, 312–316
  crystal, 312–314
  "fixed-in, fixed-out," 314
  focusing, 314–315
  "glitches," 319, 326–332
  Johansson geometry, 314
  resolution, 314–316
  rocking curve, 314–315
Mössbauer effect, 246, 249–250
Multiple scattering, in EXAFS, 304, 352
Multiplet structure, 19–20
Muscle, metabolic studies by NMR, 67–70; see also Creatine phosphokinase
  intracellular Mg^{2+} concentration, 69–70
  intracellular pH, 68–69
Myoglobin
  studied by ESR, 180–183, 194–199, 221
  studied by Mössbauer spectroscopy, 289
  studied by x-ray crystallography, 422, 426
Myosin, studied by ESR, 153

N

Néel temperature, 272
Nerve myelin, studied by x-ray diffraction, 459
Nitrogenase, studied by Mössbauer spectroscopy, 274–276
Nitroxide spin labels, 152–169
  as pH indicator, 163
  in saturation-transfer ESR, 222–225
Noncrystallographic symmetry, 401–402, 404–407
Nuclear frequency, 188–191

Nuclear magnetic resonance (NMR)
  ^{13}C studies, 73–76
  ^{1}H studies, 76–79
  probe, 33–35
  surface coil, 35, 73
  ^{31}P studies, 67–73
  theory, 6–9
Nuclear Overhauser effect (NOE), 27–29, 51, 81–84
  enhancement factor (NOEF), 27
  heteronuclear, 83–84
Nuclear-quadrupole coupling tensor, 190–191

O

Optimal signal-to-noise for resonance, 39
Order parameter, in ESR, 164–169
Organic solvent precipitation, 370
Oxygenase, studied by Mössbauer spectroscopy, 280–282

P

Papain, studied by x-ray crystallography, 416
Paramagnetic metal ions, in NMR, 26, 49–51, 80
Patterson function, 386–388
  anomalous difference, 411
  difference, 409–411
Pepsinogen, studied by ESR, 158
Peptide conformation, studied by NMR, 47–51
  backbone conformation, 51
Phase angle, 446
Phase determination, 391–407
Phase probability distribution, 399–401
Phase problem, 376
Phase shift, in EXAFS, 303, 305, 341–342
Photoelectric effect, 298–300
Photosynthesis, 276–279
Phycomyces blakesleanii (fungus), studied by Mössbauer spectroscopy, 285
Plastocyanin
  studied by EXAFS, 353
  studied by NMR, 50
Platelets, nucleotide and amine storage in, 72
Point symmetry (point group), 378
Poisson statistics, 260
Polarization measurement, 345
Polychromatic nature of synchrotron radiation, 456

Poly-L-lysine, studied by NMR, 55
Polynomial spline, 336
Position-sensitive detector, 316, 450
Potentiometric titration, 128
Powder spectrum, in ESR, 96–97, 146, 159–161
Precession photograph, 381–382
Primitive lattice, 377
Proportional counter, in Mössbauer spectroscopy, 258
Prostaglandin synthase, 138
Proteolysis, effect on crystallization, 371
Proton imaging, 79
Pseudocontact shifts, 22–23
Purple membrane of *H. Halobium*, studied by x-ray diffraction, 460–463; see also *Halobacterium*

Q

Quadrupole-split doublet, 253
Quadrupole splitting, 251, 253–254
  effect on magnetic splitting, 256
  in hemoglobin, 265–266
  in iron–sulfur proteins, 269–270
Quadrupolar nuclei, 56–57

R

Radial distribution function, 306, 339, 344
Radiation damage, 389
Radius of gyration, 444, 451
Reaction centers
  studied by anomalous diffraction, 466–467
  studied by Mössbauer spectroscopy, 276–279
  studied by x-ray crystallography, 427
  studied by x-ray scattering, 438, 459
Reciprocal lattice, 379–382
Reciprocal space, 441–442
Recoil-free fraction, 248, 260–261
Red blood cells
  studied by Mössbauer spectroscopy, 283–284
    reticulocytes, 283–284
  studied by NMR, 70, 76–79
Redox titrations and kinetics, 122–134
  continuous-flow method, 124
  stopped-flow method, 124–125

Refinement
  of electron-density map, 417–421
    energy, 418
    real-space, 417–418
    reciprocal-space, 418
    stereochemically restrained least-squares, 418–419
  of heavy-atom parameters, 411–413
    least-squares, 412
    phase, 412
Reliability *(R)* factor, 413, 419–420
Resolution, of x-ray monochromator, 314–316
Resonance lines
  area, 38
  equation for, 81
  frequency, 38
  homogeneously broadened, 183
  inhomogeneously broadened, 183–184
  linewidth, 38
Resonance x-ray diffraction and scattering, *see* Anomalous scattering
Retinal disk membranes, studied by x-ray diffraction, 459
Rhodopsin, studied by x-ray scattering, 437, 455
Rhus vernicifera laccase, studied by EXAFS, 355
Ribonuclease
  studied by NMR, 63
  studied by x-ray crystallography, 422
Ribonucleotide reductase, 134–138
Richards box, 416–417
Ring-current shifts, 18, 43
"Rotating frame," 11
Rotational correlation time, 26, 154, 156
  anisotropic motion, 157
  fast-tumbling regime, 156
  intermediate regime, 159
  isotropic motion, 156
  slow-motion regime, 157, 159
Rotation function, 402–403
Rubredoxin
  studied by EXAFS, 346–348
  studied by Mössbauer spectroscopy, 267, 269, 289
  studied by x-ray crystallography, 421

S

Sample holders, for Mössbauer spectroscopy, 262

Sarcoplasmic reticulum membrane, studied by x-ray diffraction, 460–461
Saturation phenomenon
  in ESR, 103, 106–108
  in NMR, 27–28
Saturation-recovery technique, 65
Saturation transfer
  in ESR, 222–229
    turning points, 223
  in NMR
    inversion transfer, 60–61
    magnetization transfer, 58–60
    used to study enzyme kinetics, 70–73
Scalar coupling, 19–20, 47–48
Scattering amplitude, in EXAFS, 341
Scattering factor, 395
Scattering function, in EXAFS, 303
Scattering vector, 375
Scintillation detector
  in Mössbauer spectroscopy, 258
  for x-rays, 318, 449–450
Secular line broadening, 108
Selection rules, 307
Selective excitation methods, 63
Serine proteases
  studied by NMR, 63
  studied by x-ray crystallography, 422
"Shielding constant," 15–16
Siderochromes, studied by Mössbauer spectroscopy, 271
Single-crystal experiments, in ESR, 177–183
Single-point calibration, 319
Single-scattering theory of EXAFS, 296
Solid-state detectors, 258, 449–450
Soller slits, 324
Solvent exchange, 63–67
Space group, 377–378
Spectral distribution of incident energy, 456
Spectral doublet, 140–141
Spectrometers
  Mössbauer, 257–259
  NMR, 31–36
Spin echo, 41–42
Spin labels, 142–177
  metals, 142–152
    Co(II), 143–144
    Cu(II), 143–144
    Mn(II), 142, 147–152
    VO^{2+}, 144–147
Spin–lattice relaxation time (T_1), 10, 23–26, 107–108, 256

dependence on molecular correlation times, 24–25
  measurement
    inversion recovery, 40–41
    saturation recovery, 40
Spin-orbit coupling, 94–95
Spin relaxation, 21
  exchange limits, 30
  parameters (T_1 and T_2), 3–6; see also Spin–lattice relaxation time; Spin–spin relaxation time
Spin–spin interactions, 19–23
Spin–spin relaxation time (T_2), 10, 23–26, 107–108, 256
  dependence on molecular correlation times, 24–25
  measurement, 41–42
Spin trapping, 175–177
Spin vector s, 90–91
Splitting diagram, 92–93
"Square-law" detectors, 441
Steady-state magnetization, 39
Stellacyanin
  studied by ESR, 205–207, 219–221
  studied by EXAFS, 353
Stereographic diagram, 182–183
Stimulated-emission sequence, see Three-pulse sequence
Stokes–Einstein relation, 146, 160–161
  in NMR, 26
  in saturation-transfer ESR, 223–225
Structure factor, 375, 383–386
  equation, 385–386
Sulfite oxidase, studied by EXAFS, 356–357
Superhyperfine splitting, 101–102
  ligand, 117–119
Superoxide dismutase, studied by NMR, 63
Supersaturation, 367

T

T_1, see Spin–lattice relaxation time
T_2, see Spin–spin relaxation time
Thermal vibration, 252
"Thickness effect" in x-ray absorption spectroscopy, 326–328, 330
Three-pulse sequence, in ESR, 203–205
Threshold energy, 338
Thyroxine, studied by Mössbauer spectroscopy, 273–274

INDEX

Tissue studies
  by Mössbauer spectroscopy, 248
  by NMR, 57, 67–79
  whole body, 72–73, 78–79
Tobacco mosaic virus (TMV), studied by Mössbauer spectroscopy, 274
Tomato bushy-stunt virus, studied by x-ray crystallography, 404
Transferrin
  studied by ESR, 210–212
  studied by Mössbauer spectroscopy, 271
Transition probability, 9
Transverse relaxation time, *see* Spin–spin relaxation time
t-RNA
  studied by NMR, 51, 63–65
  studied by x-ray crystallography, 415

U

Unit cell, 376–377
  vector, real-space, 446
  vector, reciprocal, 446

V

Vanadium, edge and EXAFS, 358
Vapor diffusion, 372–373
Vicinal coupling, 48–49
Victoreen equation, 334, 337
Vitamin B_{12}, studied by Mössbauer spectroscopy, 274

W

Wave vector
  in crystallography, 374, 428–429
  in EXAFS, 300, 338
  in x-ray scattering, 441
Weighting factor, in x-ray absorption spectroscopy, 325–326

X

Xanthine dehydrogenase, studied by EXAFS, 356–357

Xanthine oxidase
  studied by ESR, 125–132
  studied by EXAFS, 356–357
X-ray absorption
  coefficient, 395
  near-edge structure (XANES), 306
X-ray detection, 317–318, 449–451
  digital photon counter, 390, 449
  electronic area detector, 390, 450
    silicon-intensified target (SIT), 451
  film, 389–390, 450–451
    precession, 389
    rotation–oscillation, 389–390
  gas-ionization chamber, 317, 320, 324, 327–330
  Geiger–Müller, 449–450
  nonlinearity of response, 326
  position-sensitive, 450
  scintillation counters, 318, 450
  solid-state Si(Li), 450
  transmission, 317
  x-ray fluorescence, 317–318
X-ray diffraction, 379–386, 445–447
X-ray mirror, 316
  focusing mirror, 316
X-ray monochromator, *see* Monochromator, x-ray
X-ray scattering, 439–444
X-ray sources, 296–297, 308–312, 447–449
  copper anode, 447–448
  fixed target, 448
  Rochester laser plasma, 449
  rotating anode, 308, 389, 448
  sealed tube, 389
  storage ring, 297, 309–312, 423–425, 448, 455
  synchrotron radiation, 297, 309–312, 423–425, 448–449, 455–456
  undulator, 449
  wiggler, 312, 449

Z

Zeeman energy, 6–9, 91, 97–98
  nuclear Zeeman term, 97, 185, 189
Zero-field splitting, 114, 147–149, 169
  parameter, 115
Zinc, EXAFS, 358

DATE DUE

APR 2 1992		

DEMCO 38-297